**Springer Proceedings in Mathematics & Statistics**

Volume 88

**Springer Proceedings in Mathematics & Statistics**

This book series features volumes composed of selected contributions from workshops and conferences in all areas of current research in mathematics and statistics, including OR and optimization. In addition to an overall evaluation of the interest, scientific quality, and timeliness of each proposal at the hands of the publisher, individual contributions are all refereed to the high quality standards of leading journals in the field. Thus, this series provides the research community with well-edited, authoritative reports on developments in the most exciting areas of mathematical and statistical research today.

More information about this series at http://www.springer.com/series/10533

De-Jun Feng · Ka-Sing Lau
Editors

# Geometry and Analysis of Fractals

## Hong Kong, December 2012

Springer

*Editors*
De-Jun Feng
Ka-Sing Lau
Department of Mathematics
The Chinese University of Hong Kong
Hong Kong

ISSN 2194-1009 ISSN 2194-1017 (electronic)
ISBN 978-3-662-43919-7 ISBN 978-3-662-43920-3 (eBook)
DOI 10.1007/978-3-662-43920-3

Library of Congress Control Number: 2014944719

Mathematics Subject Classification: 28-XX, 31-XX, 37-XX, 60-XX

Springer Heidelberg New York Dordrecht London



Printed on acid-free paper

Springer is part of Springer Science+Business Media (www.springer.com)

# Foreword

It was about half a century ago when Mandelbrot first formulated a revolutionary idea in the notion of fractals, and used it to describe irregular geometric objects and physical phenomena in nature. The importance of this concept was immediately recognized in many disciplines such as physics, geosciences, and biological sciences. In mathematics, it emerged as a new area named fractal geometry. Over the years, rather complete theories on fractal dimensions, multifractals, and random fractals have been developed; their connections with dynamical systems, ergodic theory, and probability theory are apparent. In recent developments, the close interplay of fractals with harmonic analysis, complex analysis, geometric measure theory, graph theory, and algebra also enriches the area greatly.

In China, we share with the world the great enthusiasm in the development of the area. In 2011–2012, we were grateful to have support from the Focused Investment Scheme of the Chinese University of Hong Kong on a two-year special program, *Advances on Fractals*. The program was intended to bring together interested researchers from China and abroad to discuss and carry out fractal research. There were seminars and lectures throughout, which culminated in the *International Conference on Advances of Fractals and Related Topics* on December 10–14, 2012. The conference was attended by 150 participants from all over the world and 90 lectures were presented.

This monograph is the fruit of the conference. It contains 13 polished versions of lectures from the conference, covering a wide range of current research directions:

- dimension theory—multifractal analysis—geometric measure theory—harmonic analysis and spectral analysis—analysis on metric spaces and heat kernels.

They are in the form of a survey with sufficient introduction to the topics and with selected proofs. We regret that we are not able to include many of the interesting lectures because of the limited size of the monograph.

Dimension theory is classical in fractal geometry. We have four articles on this topic. The Marstrand's projection theorem in the setup of Heisenberg groups is being surveyed by Mattila. Another consideration is due to Rams and Simon on fractal percolation and random self-similar sets. For self-similar sets of overlapping IFS, Hochman provides a heuristic argument to illustrate his in-depth result on the comparison of Hausdorff dimension and self-similar dimension. Also, Shmerkin outlines his joint work with Feng on the self-affinity dimension and its continuity on self-affine sets.

On multifractal analysis, Falconer extends a potential theoretic method and uses it to estimate the $L^p$-dimensions of self-affine measures, images of measures under Brownian-type processes, and moments of random cascade measures. The random cascade, introduced by Mandelbrot, is intended to give a simple tractable model to turbulence and is an origin of multifractals. In his survey, Barral gives an exposition of this theory as well as the recent development on branching random walks and directed polymers on disordered trees. Fan, on the other hand, provides a study of the multifractals from the point of view of ergodic average, especially on his recent work on multiple ergodic averages.

On the geometric measure theory, Zähle reviews the fractal extension of the curvature measures in classical convex geometry and differential geometry, and investigates the stability problem on different approximations.

Self-similarity plays a special role in harmonic analysis. Dutkay and Jorgensen give an operator theoretic approach to consider the affine function systems, and build a Cantor-wavelet analysis as an application. In another direction, Dai, He, and Lai review some of the recent development concerning the spectral problem of self-similar measures, namely the existence of exponential orthonormal bases and Fourier frames.

Analysis on metric spaces generalizes the analysis on manifolds, graphs, as well as fractals. In this exciting topic, Kigami brings in the quasisymmetric classification of metrics on self-similar sets, and applies this to estimate the quasiconformal dimension of the Sierpinski carpet. Grigor'yan, Hu, and Lau give an expository survey on the heat kernel estimates on metric measure spaces, and the new development on nonlocal Dirichlet forms. Furthermore, Grigor'yan and Huang give a detailed discussion of a stochastic completeness criterion of a nonlocal Dirichlet form based on volume growth.

We are grateful to Profs. Chi-Wai Leung, Sze-Man Ngai, Yan-Hui Qu, and Ms. Annie Wong for their help in preparing the conference, and to the Department of Mathematics of CUHK for providing the facilities and secretarial support. Finally, we would like to thank the following organizations for their generous financial support to the conference: CRS 701 of the German Research Council, Hong Kong Pei Hua Education Foundation, K. C. Wong Education Foundation, and the Hong Kong Mathematical Society.

De-Jun Feng
Ka-Sing Lau

# Contents

# Mandelbrot Cascades and Related Topics

**Julien Barral**

**Abstract** This article is an extended version of the talk given by the author at the conference "Advances in fractals and related topics", in December 2012 at the Chinese Hong-Kong University. It gathers recent advances in Mandelbrot cascades theory and related topics, namely branching random walks, directed polymers on disordered trees, multifractal analysis, and dynamical systems.

## 1 Introduction

In the late sixties, motivated by Kolmogorov's work [Kol62] on turbulence in which the "lognormal hypothesis" appeared, Mandelbrot introduced lognormal multiplicative processes (see [Man72]) to build random measures obtained as limit of martingales, and describing the distribution of the energy dissipation in intermittent turbulence. As his model turned out to be too difficult to found and study in complete rigor, Mandelbrot defined (not necessarily log-normal or conservative) multiplicative cascades on homogeneous trees [Man74a, Man74b], now called Mandelbrot cascades, to provide a mathematically easier to define and, a priori, tractable model of turbulence. Some of his conjectures on the model behavior were then proved by Kahane and Peyrière in [Kah74, Pey74, KP76], and some questions remained open for a long time. In the eighties, Kahane developed multiplicative chaos theory [Kah85, Kah87a] to give a completely rigorous framework to Mandelbrot original lognormal multiplicative processes, and go beyond. This theory (which covers the case of Mandelbrot cascades) and its applications have been particularly enriched by the regular introduction of new models of multiplicative cascades [BF05, BJM10, BJM10a, BM02, BM03, BM04, BM04a, BM09, CRV13, Fan89, Fan97, LRV00, Pey77, Pey79, RV10, WW69], and during the last five years by the rigorous

The author thanks Dr Xiong Jin for his kind help in the figures elaboration.

J. Barral (✉)
LAGA (UMR 7539), Département de Mathématiques, Institut Galilée, Université Paris 13, Paris-Sorbonne Cité, 99 avenue Jean-Baptiste Clément, 93430 Villetaneuse, France
e-mail: barral@math.univ-paris13.fr

 
D.-J. Feng and K.-S. Lau (eds.), *Geometry and Analysis of Fractals*, Springer Proceedings in Mathematics & Statistics 88,
DOI 10.1007/978-3-662-43920-3_1

connexion between lognormal multiplicative chaos and KPZ relations in quantum gravity [BJRV13, BS09, DS11, KPZ98, RV08] and the connexion between this chaos and SLE curves [AJKS11, She00], as well as by fine renormalization results for degenerate multiplicative chaos [AS14, BRV12, DRSV00a, DRSV00, JW11, Mad00, Web11] (the interested reader should consult the recent survey [RV13] on lognormal multiplicative chaos for more details).

This paper focuses on Mandelbrot cascades, and aims at gathering recent advances in its theory and related topics, namely branching random walks, directed polymers on disordered trees, multifractal analysis, and dynamical systems (other surveys dealing with Mandelbrot cascades, multiplicative chaos and applications are [BFP10, Fan04, Kah91, Pey00]).

## 2 Mandelbrot Cascades

Consider the set $\mathscr{A} = \{0, \ldots, b-1\}$, where $b \geq 2$. Set $\mathscr{A}^* = \bigcup_{n\geq 0} \mathscr{A}^n$, where, by convention, $\mathscr{A}^0$ is the singleton $\{\epsilon\}$ whose the only element is the empty word $\epsilon$. If $w \in \mathscr{A}^*$, we denote by $|w|$ the integer such that $w \in \mathscr{A}^{|w|}$. If $n \geq 1$ and $w = w_1 \cdots w_n \in \mathscr{A}^n$ then for $1 \leq k \leq n$ the word $w_1 \cdots w_k$ is denoted by $w|k$. By convention, $w|0 = \epsilon$.

Given $v$ and $w$ in $\mathscr{A}^n$, $v \wedge w$ is defined to be the longest prefix common to both $v$ and $w$, i.e., $v|n_0$, where $n_0 = \sup\{0 \leq k \leq n \,:\, v|k = w|k\}$.

Let $\mathscr{A}^\omega$ stand for the set of infinite sequences $w = w_1 w_2 \cdots$ of elements of $\mathscr{A}$. Also, for $x \in \mathscr{A}^\omega$ and $n \geq 0$, let $x|n$ stand for the projection of $x$ on $\mathscr{A}^n$.

If $w \in \mathscr{A}^*$, we consider the cylinder $[w]$ consisting of infinite words in $\mathscr{A}^\omega$ whose $w$ is a prefix.

We index the closed $b$-adic subintervals of $[0, 1]$ by $\mathscr{A}^*$: for $w \in \mathscr{A}^*$, we set

$$I_w = \left[\sum_{1\leq k\leq |w|} w_k b^{-k}, \sum_{1\leq k\leq |w|} w_k b^{-k} + b^{-|w|}\right].$$

Let $W$ be a non negative random variable such that $\mathbb{E}(W) = 1$ and $\big(W(w)\big)_{w\in\mathscr{A}^*}$ a family of independent random variables, identically distributed with $W$. Denote by $(\Omega, \mathcal{A}, \mathbb{P})$ the probability space over which these random variables are defined.

Then the non negative martingale

$$Y_n = b^{-n} \sum_{w\in\mathscr{A}^n} W(w|1)W(w|2)\cdots W(w|n) \tag{2.1}$$

converges to a non negative random variable $Y$, and the sequence of Borel positive measures $(\mu_n)_{n\geq 1}$ defined on $[0, 1]$ by

$$\mu_n(\mathrm{d}x) = W(w|1)W(w|2)\cdots W(w|n)\,\mathrm{d}x \quad \text{if } x \in I_w,\ w \in \mathscr{A}^n, \tag{2.2}$$

weakly converges to a non negative measure $\mu = \mu_W$ with total mass $\|\mu\| = Y$ almost surely. In the literature, the terminology *Mandelbrot cascade*, or *Mandelbrot martingale* denotes the martingale $(Y_n)_{n\geq 1}$ or the measure-valued sequence $(\mu_n)_{n\geq 1}$, while $\mu$ is called a *Mandelbrot measure*.

For each $0 \leq k \leq b-1$, substituting $\big(W(kw)\big)_{w\in\mathscr{A}^*}$ to $\big(W(w)\big)_{w\in\mathscr{A}^*}$ yields a copy $Y(k)$ of $Y$, so that the $Y(k)$ are independent, and independent of $(W(0), \ldots, W(b-1))$. The statistical self-similarity of the construction is summarized in the fundamental almost sure equation

$$Y = b^{-1}\sum_{k=0}^{b-1} W(k)Y(k). \tag{2.3}$$

Moreover, defining more generally

$$Y_n(w) = b^{-n}\sum_{v\in\mathscr{A}^n} W(w\cdot v|1)W(w\cdot v|2)\cdots W(w\cdot v|n), \quad (w \in \mathscr{A}^*),$$

and $Y(w) = \lim_{n\to\infty} Y_n(w)$, $Y(w)$ is a copy of $\|\mu\|$ and one gets the following multiplicative structure for the the mass of $b$-adic intervals:

$$\mu_{|w|+n}(I_w) = b^{-|w|}Y_n(w)\prod_{1\leq j\leq |w|} W(w|j),$$

which leads

$$\mu(I_w) = b^{-|w|}Y(w)\prod_{1\leq j\leq |w|} W(w|j), \tag{2.4}$$

since $b$-adic numbers cannot be atoms of $\mu$ (indeed, fixing $x_0 \in [0,1]$, and, for $\epsilon > 0$, $f_\epsilon$ a non negative function bounded by 1, with compact support in $[x_0-\epsilon, x_0+\epsilon]$ and taking the value 1 over $[x_0-\epsilon/2, x_0+\epsilon/2]$, we have $\mathbb{E}(\mu(\{x_0\})) \leq \mathbb{E}(\mu(f_\epsilon)) \leq \liminf_{n\to\infty}\mathbb{E}(\mu_n(f_\epsilon)) \leq \mathrm{Leb}(f_\epsilon) \leq 2\epsilon$, where we used the martingale property of $(\mu_n)_{n\geq 1}$).

Finally, $\mu$ possesses the statistical self-similar structure:

$$\mu_{|I_w} = \Big(b^{-|w|}\prod_{1\leq j\leq |w|} W(w|j)\Big)\mu^{(w)}\circ f_w^{-1} \quad (w \in \mathscr{A}^*), \tag{2.5}$$

where $f_w$ is the direct similitude mapping $[0,1]$ onto $I_w$ and $\mu^{(w)}$ is the copy of $\mu$ built from $(W(w\cdot v))_{v\in\mathscr{A}^*}$.

It turns out that $\mu$ may vanish almost surely. This had been observed by Mandelbrot, who considered the function

$$\varphi(q) = \varphi_W(q) = q - 1 - \log_b \mathbb{E}(W^q) \quad (q \in \mathbb{R}_+), \tag{2.6}$$

and conjectured that $\mathbb{P}(\mu \neq 0) > 0$ if and only if $\varphi'(1^-) > 0$. It was Kahane who proved this result in [Kah74, KP76] (this was also proved independently and a bit later by Biggins [Big77] in the more general context of the branching random walk, with a different approach; another approach is presented in [WW69]).

**Theorem 1** *One has* $\mathbb{P}(\mu \neq 0) > 0$ *if and only if* $\mathbb{E}(W \log W) < \log(b)$, *i.e.* $\varphi_W'(1^-) > 0$. *Moreover, in this case,* $(Y_n)_{n\geq 1}$ *is uniformly integrable*: $\mathbb{E}(Y) = 1$.

The necessity of $\varphi'(1^-) \geq 0$ is easy to see, for otherwise since $\varphi(1) = 0$ one has $\varphi(h) > 0$ for $h$ close to $1^-$, so that

$$\mathbb{E}(Y_n^h) \leq b^{-nh}\mathbb{E} \sum_{w \in \mathscr{A}^n} W(w|1)^h W(w|2)^h \cdots W(w|n)^h = b^{-n\varphi(h)}$$

tends to 0, and $Y = 0$ almost surely by applying Fatou's lemma. The sufficiency of $\varphi'(1^-) > 0$ is also quite direct if one assumes $\varphi(h) > -\infty$ for some $h > 1$. Indeed, in this case choosing $h$ close enough to $1^+$ yields $\varphi(h) < 0$. Then, following Kahane in [KP76], if we write $Y_n = b^{-1}\sum_{k=0}^{b-1} W(k)Y_{n-1}(k)$ and take $h \in (1, 2]$, then $Y_n^h = ((b^{-1}\sum_{k=0}^{b-1} W(k)Y_{n-1}(k))^{h/2})^2 \leq b^{-h}(\sum_{k=0}^{b-1} W(k)^{h/2}Y_{n-1}(k)^{h/2})^2$, so $\mathbb{E}(Y_n^h) \leq b^{-h}b\mathbb{E}(W^h)\mathbb{E}(Y_{n-1}^h) + b^{-h}b(b-1)\mathbb{E}(W^{h/2})^2\mathbb{E}(Y_{n-1}^{h/2})^2 \leq b^{\varphi(h)}\mathbb{E}(Y_{n-1}^h) + b^{-h}b(b-1)$ since $\mathbb{E}(W) = \mathbb{E}(Y_{n-1}) = 1$, and finally $\mathbb{E}(Y_n^h) \leq (1 - b^{\varphi(h)})^{-1}b^{-h}b(b-1)$ after noting that $\mathbb{E}(Y_{n-1}^h) \leq \mathbb{E}(Y_n^h)$ by Jensen's inequality. Thus $(Y_n)_{n\geq 1}$ is bounded in $L^h$, hence uniformly integrable. The sharp result is much more delicate.

When $\mathbb{E}(W \log W) \geq \log(b)$, i.e. $\varphi'(1^-) \leq 0$, Mandelbrot naturally asked in [Man74a] for the existence of a normalizing positive sequence $(A_n)_{n\geq 1}$ so that $(Y_n/A_n)_{n\geq 1}$ converges, at least in law, to some non degenerate random variable. He observed that the convergence in law of $Y_n/A_n$ to some random variable $Z$ imposes that $A_{n+1}/A_n$ converges to a positive constant $c$, so that $Z$ must satisfy the equation

$$Z \stackrel{\text{dist}}{=} \sum_{k=0}^{b-1} \frac{W(k)}{bc} Z(k), \tag{2.7}$$

where $\stackrel{\text{dist}}{=}$ means equality in distribution, the $Z(k)$ are independent, identically distributed with $Z$, and independent of $(W(0), \ldots, W(b-1))$.

The non trivial solutions of this equation and its generalization to the branching random walk context, also called fixed points of the smoothing transformation, have been studied intensively, starting with the fundamental paper [DL83] by Durrett and Liggett (their motivation came from interacting particle systems) and followed by regular notable advances. We will see (Sect. 3.1) that its general solutions are natural combinations of stable laws and the laws of random variables of the same nature as $\|\mu\|$, or combinations of stable laws and the laws of the total mass of *critical* Mandelbrot measures. These solutions have counterparts in terms of statistically self-similar measures, namely derivatives of stable Lévy subordinators in Mandelbrot time, or critical Mandelbrot time. These statistically self-similar measures, as well as Mandelbrot and critical Mandelbrot measures provide fundamental illustrations of the multifractal formalism, a notion that was pointed out by Frisch and Parisi to provide a fine geometric description of the energy dissipation in intermittent turbulence, with Mandelbrot measures as main illustration [FP88]. This will be explained in Sect. 5, where we present recent progress in the multifractal analysis of these objects, as well as in the control of the modulus of continuity of Mandelbrot and critical Mandelbrot measures. These moduli of continuity controls are partly based on the remarkable advances achieved in the solution to the original normalization question (see Sect. 3.2). It turns out that combining these renormalization results with fixed points of the smoothing transformation theory yields, in the case of a second order phase transition, a precise description of the asymptotic behavior of the partition functions and the limiting laws of Gibbs measures associated with polymers on disordered trees expressed in terms of the statistically self-similar measures mentioned above (Sect. 4).

In fact, in his notes on multiplicative cascades [Man74a], Mandelbrot starts by raising the following general problem: assume only that $W$ is a random variable taking values in $\mathbb{R}$, and consider the sequence of functions

$$F_n(x) = F_{W,n}(x) = \int_0^x Q_n(u)\,\mathrm{d}u, \tag{2.8}$$

where $Q_n(x) = W(w|1)W(w|2)\cdots W(w|n)\,\mathrm{d}x$ if $x \in I_w,\ w \in \mathscr{A}^n$; under which condition does there exist a normalizing sequence $(A_n)_{n\geq 1}$ such that $F_n(x)/A_n$ converges in law, or in a stronger sense?

We will give some results in this direction in Sect. 6. In Sect. 7 we explain how Mandelbrot cascades define a natural dynamical system on fixed points of the smoothing transformation with finite expectation, to which is associated a functional CTL whose limit process is obtained as the limit of an *additive cascade* on $\mathscr{A}^*$.

Finally, let us mention that a lot of results presented in this paper are not specific to multiplicative cascades on regular trees, and have extensions in the context of branching random walks on Galton-Watson trees. Also, while we consider measures on the interval $[0, 1]$, the results of Sect. 5 can be extended to Mandelbrot measures on $[0, 1]^d$ ($d \geq 2$). The only difference in that one must use a different way to define the Lévy-Mandelbrot measures defined in Sect. 3.1.2; the procedure is explained in [BJRV13, BRV12].

In Sects. 3–5, to simplify the discussion we assume that $W > 0$ and the probability distribution of $\log(W)$ is non lattice.

# 3 Fixed Points of the Smoothing Transformation and Associated Statistically Self-similar Measures; Renormalization of Mandelbrot Cascades

This section gathers old and new information related to the solutions of the functional equation pointed out by Mandelbrot in connexion with the renormalization of the martingale $(Y_n)_{n\geq 1}$ when $\varphi'(1) \leq 0$.

## 3.1 *Fixed Points of the Smoothing Transformation*

If $Z$ is a non negative random variable, denote by $\phi_Z$ the Laplace transform of its probability distribution.

Consider the positive random variable $W$ and do not assume a priori that $\mathbb{E}(W) = 1$. Then consider the equation

$$Z \overset{\text{dist}}{=} b^{-1} \sum_{k=0}^{b-1} W(k)Z(k), \tag{3.1}$$

where the random variables $W(k), 0 \leq k \leq b-1$, are independent copies of $W$, the random variables $Z(k)$ are independent copies of the non negative random variable $Z$, and $(W(0), \ldots, W(b-1))$ and $(Z(0), \ldots, Z(b-1))$ are independent. If $Z$ is not identically equal to 0, then it is easily deduced from (3.1) that in fact $Z$ is positive almost surely; moreover any positive multiple of $Z$ satisfies (3.1).

In terms of Laplace transform, (3.1) means

$$\phi_Z = \left(\mathbb{E}(\phi_Z(b^{-1}tW))\right)^b, \tag{3.2}$$

so that $\phi_Z$ is a fixed point of the *smoothing transformation* $T_W$ defined on the space of Laplace transforms of probability distributions on $\mathbb{R}_+$ as

$$T_W(\phi) = (\mathbb{E}(\phi(b^{-1}tW)))^b.$$

As indicated in the previous section, non trivial solutions of Eq. (3.1) have been studied intensively [DL83, Gui90, BK97, Liu98, Liu00, Liu01, BK04, BK05, AM12, ABM12] (see also [AM13] for the study of non necessarily positive solutions, still with $W > 0$).

Let $\varphi$ be defined as in Sect. 5.1.3.

**Theorem 2** *Equation* (3.1) (*or equivalently* (3.2)) *has a non trivial solution if and only if there exists* $\alpha \in (0, 1]$ *such that* $\varphi_W(\alpha) = 0$ *and* $\varphi'_W(\alpha^-) \geq 0$. *Moreover, if* $Z_1$ *and* $Z_1$ *are two such solutions, there exists* $c > 0$ *such that* $Z_1 \overset{\text{dist}}{=} cZ_2$.

This was proved by Durrett and Liggett under the assumption $\varphi(1+\epsilon) > -\infty$ for some $\epsilon > 0$. This assumption has been removed thanks to recent advances due to Alsmeyer, Biggins, Kyprianou and Meiners works in [ABM12, AM12, BK04, BK05]. We do not try to outline the extremely involved proof of this result; we just mention that renewal theory and fluctuation theory of random walks on $\mathbb{R}$ play a central role here.

It turns out that assuming that $\varphi''(\alpha^-) > -\infty$ whenever $\varphi'(\alpha^-) = 0$ simplifies the description of the solutions. This is what we will do in the next subsections.

### 3.1.1 Special Solutions from Mandelbrot and Critical Mandelbrot Cascades When $\alpha = 1$

For $q \in \mathbb{R}_+$ and $n \geq 1$ set

$$Y_{q,n} = b^{n(\varphi(q)-q)} \sum_{w \in \mathscr{A}^n} W(w|1)^q W(w|2) \cdots W(w|n)^q,$$

and

$$\widetilde{Y}_n = -\frac{\mathrm{d}Y_{q,n}}{\mathrm{d}q}\Big|_{q=1}.$$

Notice that if $\varphi(1) = 0$, i.e. $\mathbb{E}(W) = 1$, then $Y_{1,n} = Y_n$ (see (2.1)). If, moreover, $\varphi'(1) = 0$, we are in the critical case of degeneracy of Mandelbrot cascades, and $\widetilde{Y}_n$ takes the form

$$\widetilde{Y}_n = -\sum_{w \in \mathscr{A}^n} (b^{-n} W(w|1)W(w|2)\cdots W(w|n)) \log(b^{-n} W(w|1)W(w|2)\cdots W(w|n));$$

in this case it is a martingale with respect to $(\sigma(W(w) : w \in \mathscr{A}^k,\ 1 \leq k \leq n))_{n\geq 1}$ and it is called *derivative martingale*.

The following result gathers information about the non trivial solutions of (3.1) when $\varphi(1) = 0$ and $\varphi'(1) \leq 0$.

**Theorem 3** *Assume that $\varphi(1) = 0$.*

(1) *Suppose that $\varphi'(1^-) > 0$. Let $Y$ be the almost sure limit of $(Y_n)_{n\geq 1}$ given by Theorem 1.*

  (a) *For $q > 1$ one has $\mathbb{E}(Y^q) < \infty$ if and only if $\varphi(q) > 0$;*
  (b) *If $\varphi(q_0) = 0$ and $\varphi'(q_0) > -\infty$ for some $q_0 > 1$ (such a $q_0$ is necessarily unique due to the concavity of $\varphi$), one has $\mathbb{P}(Y > x) \sim_\infty Ax^{-q_0}$ for some $A > 0$. (c) For $q < 0$, $\varphi(q/b) > -\infty$ implies $E(Y^{q'}) < \infty$ for all $q' \in (q, 0)$, and $E(Y^q) < \infty$ implies $\varphi(q/b) > -\infty$.*

*Moreover, any non trivial solution $Z$ of (3.1) satisfies $Z \stackrel{dist}{=} cY$ for some $c > 0$.*

(2) *Suppose that* $\varphi'(1^-) = 0$ *and* $\varphi''(1^-) > -\infty$. *Then* $(\widetilde{Y}_n)_{n\geq 1}$ *is a martingale which converges almost surely to a positive random variable* $\widetilde{Y}$ *satisfying the following properties*:

   (a) *for* $q > 0$ *one has* $\mathbb{E}(Y)^q < \infty$ *if and only if* $q < 1$;
   (b) $\phi_{\widetilde{Y}}(t) \sim_{0^+} 1 - ct\log(1/t)$ *for some* $c > 0$;
   (c) *if* $\varphi(1+\epsilon) > -\infty$ *for some* $\epsilon > 0$, $\mathbb{P}(\widetilde{Y} > x) \sim_\infty Ax^{-1}$ *for some* $A > 0$.
   (d) *For* $q < 0$, $\varphi(q/b) > -\infty$ *implies* $E(Y^{q'}) < \infty$ *for all* $q' \in (q, 0)$, *and* $E(Y^q) < \infty$ *implies* $\varphi(q/b) > -\infty$.

*Moreover, any non trivial solution* $Z$ *of* (3.1) *satisfies* $Z \stackrel{\text{dist}}{=} c\widetilde{Y}$ *for some* $c > 0$.

Under the assumption that $\varphi(1+\epsilon) > -\infty$ for some $\epsilon > 0$, Durrett and Liggett showed that when $\varphi(1) = 0$ and $\varphi'(1^-) = 0$, Eq. (3.2) possesses non trivial solutions satisfying $\phi(t) \sim_{0^+} 1 - ct\log(1/t)$ for some $c > 0$, and unique up to a scaling factor in their argument. Then Liu [Liu98] used this behavior of $\phi$ at $0^+$ to identify the solutions as the Laplace transform of $\widetilde{Y}$ by proving that $\widetilde{Y}_n$ converges to a non trivial limit (the point is that for any fixed $t > 0$, $\prod_{w\in\mathscr{A}^n} \phi(tW(w|1)W(w|2)\cdots W(w|n))$ is a bounded non negative martingale with expectation in (0, 1), which is asymptotically equivalent to $e^{-tc\widetilde{Y}_n}$). The weaker condition used in Theorem 3 is established in [BK04, BK05].

The necessary and sufficient condition for the finitness of moments of order greater than 1 when $\varphi'(1) > 0$ (conjectured in [Man74a]) was established by Kahane in [Kah74, KP76] by generalizing the argument presented just after Theorem 1. The right tail behaviors of solutions when $\varphi'(1) > 0$ and $\varphi'(1) = 0$ are due Guivarc'h [Gui90] (conjectured in [Man74a]) and Buraczewski [Bur09] respectively. The proofs strongly rely on random difference equations and renewal theories.

The result on moments of negative orders is first established in [Kah91] (see also [Liu01, Mol96]). Its proof consists in estimating from (3.2) the asymptotic behavior of $\phi_Y$ or $\phi_{\widetilde{Y}}$ at $\infty$.

**Associated statistically self-similar measures** Under the assumptions of Theorem 3(1), we already now the Mandelbrot measure $\mu$.

Under the assumptions of Theorem 3(2), defining

$$\widetilde{Y}_n(w) = -\sum_{v\in\mathscr{A}^n} \frac{W(w\cdot v|1)\cdots W(w\cdot v|n)}{b^n}\log\frac{W(w\cdot v|1)\cdots W(w\cdot v|n)}{b^n} \quad (w\in\mathscr{A}^*)$$

and $\widetilde{Y}(w) = \lim_{n\to\infty}\widetilde{Y}_n(w)$, $\widetilde{Y}(w)$ is a copy of $\|\widetilde{\mu}\|$ and

$$\widetilde{\nu}([w]) = b^{-|w|}\widetilde{Y}(w)\prod_{1\leq j\leq |w|} W(w|j), \tag{3.3}$$

defines almost surely a measure on $\mathscr{A}^\omega$. One can show (see [Bar00]) $\widetilde{\nu}$-amost every $t\in\mathscr{A}^\omega$ is normal, so that $\widetilde{\nu}$ has no atom in the set of infinite branches of $\mathscr{A}^\omega$

encoding $b$-adic numbers of $[0, 1]$. Thus $\widetilde{\nu}$ naturally projects on $[0, 1]$ to a measure $\widetilde{\mu} = \widetilde{\mu}_W$, called *critical Mandelbrot measure*, such that for any $w \in \mathscr{A}^*$,

$$\widetilde{\mu}_{|I_w} = \Big(b^{-|w|} \prod_{1 \le j \le |w|} W(w|j)\Big)\widetilde{\mu}^{(w)} \circ f_w^{-1} \quad (w \in \mathscr{A}^*), \tag{3.4}$$

where $\widetilde{\mu}^{(w)}$ is the copy of $\widetilde{\mu}$ built from $(W(w \cdot v))_{v \in \mathscr{A}^*}$.

### 3.1.2 General Form of the Solutions

**Theorem 4** *Assume that $\varphi(\alpha) = 0$ and $\varphi'(\alpha^-) \ge 0$ for some $\alpha \in (0, 1]$ (this $\alpha$ is then unique by concavity of $\varphi$). Let $L_\alpha$ be a stable Lévy subordinator of index $\alpha$ if $\alpha \in (0, 1)$ and the identity map of $\mathbb{R}_+$ if $\alpha = 1$. Assume that $L_\alpha$ is independent of $(W(w))_{w \in \mathscr{A}^*}$. Let $W_\alpha = W^\alpha/\mathbb{E}(W^\alpha)$.*

(1) *Suppose that $\varphi'(\alpha^-) > 0$. Then $W_\alpha$ satisfies the assumptions of Theorem* 3(1). *Let $Y_\alpha$ be the limit of the associated Mandelbrot cascade built from $(W_\alpha(w))_{w \in \mathscr{A}^*}$. Any non trivial solution $Z$ of* (3.1) *satisfies $Z \overset{dist}{=} cL_\alpha(Y_\alpha)$ for some $c > 0$. Moreover, $\phi_Z(t) \sim_{0^+} 1 - c't^\alpha$ for some $c' > 0$.*
(2) *Suppose that $\varphi'(\alpha^-) = 0$ and $\varphi''(\alpha^-) > -\infty$. Then $W_\alpha$ satisfies the assumptions of Theorem* 3(2). *Let $\widetilde{Y}_\alpha$ be the limit of the associated derivative martingale built from $(W_\alpha(w))_{w \in \mathscr{A}^*}$. Any non trivial solution $Z$ of* (3.1) *satisfies $Z \overset{dist}{=} cL_\alpha(\widetilde{Y}_\alpha)$ for some $c > 0$. Moreover, $\phi_Z(t) \sim_{0^+} 1 - c't^\alpha \log(1/t)$ for some $c' > 0$.*

Durrett and Liggett obtained the same sets of solutions under the stronger assumption mentioned above, and the sharp result, without the assumption $\varphi''(\alpha^-) > -\infty$ when $\varphi'(\alpha^-) = 0$, is obtained in [ABM12]. Notice that checking that $cL_\alpha(Y_\alpha)$ is solution of (3.1) in case (1) and $cL_\alpha(\widetilde{Y}_\alpha)$ is solution of (3.1) in case (2) is a simple exercise using that $\phi_{L_\alpha}(t) = e^{-\gamma t^\alpha}$ for some $\gamma > 0$.

**Associated statistically self-similar measures** Denote by $\mu$ and $\widetilde{\mu}$ the Mandelbrot measure and critical Mandelbrot measure considered at the end of Sect. 3.1.1. Let $\alpha \in (0, 1)$ and $L_\alpha$ an $\alpha$-stable Lévy subordinator independent of $(W(w))_{w \in \mathscr{A}^*}$. Then let $\mu^\alpha$ (resp. $\widetilde{\mu}^\alpha$) be the positive measure obtained as the derivative of $L_\alpha(\mu([0, \cdot]))$ (resp. $L_\alpha(\widetilde{\mu}([0, \cdot]))$). Let us also call $\mu^\alpha$ a *Lévy-Mandelbrot measure* and $\widetilde{\mu}^\alpha$ a *critical Lévy-Mandelbrot measure*. These measures are statistically self-similar in the sense that we have for any $w \in \mathscr{A}^*$

$$(\mu^\alpha)_{|I_w} \overset{\text{dist}}{=} \Big(b^{-|w|} \prod_{1 \le j \le |w|} W(w|j)\Big)^{1/\alpha} \mu^{\alpha,(w)} \circ f_w^{-1} \quad (w \in \mathscr{A}^*), \tag{3.5}$$

where $\mu^{\alpha,(w)}$ is a copy of $\mu^\alpha$ independent of $\prod_{1 \le j \le |w|} W(w|j)$, and the same holds for $\widetilde{\mu}_\alpha$.

In the context of Theorem 4, (3.5) reads

$$(\mu_{W_\alpha}^\alpha)_{|I_w} \stackrel{\text{dist}}{=} \Big(b^{-|w|} \prod_{1\le j\le |w|} W(w|j)\Big)\mu_{W_\alpha}^{\alpha,(w)} \circ f_w^{-1} \quad (w \in \mathscr{A}^*),$$

which is similar to (2.5) and (3.4).

## 3.2 *Renormalization of Mandelbrot Cascades*

We now suppose that $\varphi(1) = 0$ and $\varphi'(1^-) \le 0$ and come to the existence of a sequence $(A_n)_{n\ge 1}$ such that $Y_n/A_n$ converges in law to a non trivial limit. When $\log W$ is Gaussian, the solution to this question has been conjectured by Derrida and Spohn [DS88] in the context of directed polymers on disordered trees and rigorously established recently by Webb [Web11], while in the general result presented below, the first part has been obtained by Aidekon and Shi [AS14], and the second one combines a convergence in law result obtained by Madaule [Mad00] with an identification of the limiting law using Theorem 4.

**Theorem 5**

(1) *Suppose that* $\varphi(1) = 0$, $\varphi'(1^-) = 0$ *and* $\varphi''(1^-) > -\infty$. *Then* $(\sqrt{n}\, Y_n)_{n\ge 1}$ *converges in probability to* $\sqrt{\dfrac{-2}{\pi\varphi''(0)}}\,\widetilde{Y}$.

(2) *Suppose that* $\varphi(1) = 0$ *and* $\varphi'(1^-) < 0$. *Let* $\alpha$ *be the unique solution of* $\varphi'(\alpha) = \varphi(\alpha)/\alpha$ *in* $(0, 1)$. *The random variable* $W_\alpha = W^\alpha/\mathbb{E}(W^\alpha)$ *satisfies the assumptions of Theorem* 4(2). *Let* $\widetilde{Y}_\alpha$ *be the associated limit of the derivative martingale built from* $(W_\alpha(w))_{w\in\mathscr{A}^*}$.

*The sequence* $((n^{3/2}b^{n(\varphi(\alpha)-\alpha)})^{1/\alpha}\, Y_n)_{n\ge 1}$ *converges in law to* $L_\alpha(\widetilde{Y}_\alpha)$, *where* $L_\alpha$ *is a stable Lévy subordinator of index* $\alpha$ *independent of* $\widetilde{Y}_\alpha$.

Theorem 5(2) is a special case of a more general renormalization results for $Y_n$ when one assumes only that $W > 0$ and there exists $\alpha \in (0, 1)$ such that $\varphi'(\alpha) = \varphi(\alpha)/\alpha$ (see Sect. 4.2).

The limiting law of $\mu_n/A_n$ is described in the next section.

# 4 Directed Polymers on $\mathscr{A}^*$: Partition Functions, Free Energies and Gibbs Measures

Here we consider the random variables $(W(w))_{w\in\mathscr{A}^*}$ of the previous sections and we only assume that $W > 0$. If we define the potential $V = \log(b) - \log W$ and set $V(w) = \log(b) - \log W(w)$ $(w \in \mathscr{A}^*)$, in the setting of [DS88] the branching random walk $H(w) = V(w|1) + \cdots + V(w|n)$, $w \in \mathscr{A}^n$, defines for each $n \geq 1$ a polymer on the tree $\mathscr{A}^n$ in the random medium $(V(w))_{w \in \bigcup_{1 \leq k \leq n} \mathscr{A}^n}$ by associating the energy $H(w)$ to each path of length $n$; moreover, this model possesses *logarithmic correlations*.

Then, of first importance is the asymptotic behavior, as $n$ tends to $\infty$, of thermodynamical objects such as the partition function

$$Z_n(\beta) = \sum_{w \in \mathscr{A}^n} e^{-\beta H(w)} \quad (\beta \geq 0),$$

the free energy

$$P_n(\beta) = \frac{\log Z_n(\beta)}{n \log(b)},$$

and the Gibbs measures defined on [0, 1] by

$$\nu_{\beta,n}(\mathrm{d}x) = b^{-n} \frac{e^{-\beta H(w)}}{Z_n(\beta)} \mathrm{d}x \quad \text{if } x \in I_w \text{ and } w \in \mathscr{A}^n,$$

where $\beta$ stands for the inverse of the temperature.

Writing $e^{-\beta H(w)} = (b^{-n} W(w|1) \cdots W(w|n))^{\beta}$ shows the direct connexion with Mandelbrot cascades.

The understanding of these asymptotic behaviors has made enormous progress in the recent years. We still define $\varphi(\beta)$ as in Sect. 5.1.3.

Four different situations can occur; they are described in the Sects. 4.1–4.4 and depend on the behavior of $\varphi$ at

$$\beta_c = \sup\{\beta > 0 : \varphi'(\beta^-)\beta - \varphi(\beta) > 0\},$$

with the convention $\sup(\emptyset) = 0$.

For $\beta \geq 0$ and $n \geq 1$, we denote by $\mu_{\beta,n}$ the measure defined by (2.2) when the weight $W$ is taken equal to

$$W_\beta = \frac{W^\beta}{\mathbb{E}(W^\beta)}. \tag{4.1}$$

Notice that by construction we have

$$\|\mu_{\beta,n}\| = \frac{Z_n(\beta)}{\mathbb{E}(Z_n(\beta))} \quad \text{and} \quad \varphi'_{W_\beta}(1^-) = \varphi'(\beta^-)\beta - \varphi(\beta).$$

Section 4.5 presents a unified result for the free energy behavior.

### 4.1 No Phase Transition: $\beta_c = +\infty$

**Theorem 6** *With probability 1, for all $\beta \in [0, \beta_c)$, $(\mu_{\beta,n})_{n\geq 1}$ weakly converges to a non degenerate Mandelbrot measure $\mu_\beta$.*

*Consequently, $\left(\frac{Z_n(\beta)}{\mathbb{E}(Z_n(\beta))}\right)_{n\geq 1}$ converges to $\|\mu_\beta\|$, $P_n(\cdot)$ converges to the analytic function $-\varphi(\cdot)$ uniformly on compact subintervals of $[0, \beta_c)$, and $(\nu_{\beta,n})_{n\geq 1}$ weakly converges to $\frac{\mu_\beta}{\|\mu_\beta\|}$, as $n \to \infty$.*

That the result holds for each fixed $\beta$ almost surely essentially follows from Theorem 1 applied when $W$ is taken equal to $W_\beta$. The uniform version of this result essentially follows from [Big92] and [Bar00]. Biggins considers in [Big92] the analytic extensions of the mappings $\mathscr{Y}_n : \beta \mapsto \frac{Z_n(\beta)}{\mathbb{E}(Z_n(\beta))}$ in a complex neighborhood of $\{\beta : \varphi'(\beta^-)\beta - \varphi(\beta) > 0\}$. He proves the almost sure uniform convergence of these extensions on a common domain $U$ of $\mathbb{C}$ by combining Cauchy's formula with the fact that, for each compact subset of $U$, $\mathscr{Y}_n(z) - \mathscr{Y}_{n-1}(z)$ converges uniformly exponentially to 0 in $L^p$ for some $p > 1$. It remains to prove that almost surely, for all $\beta$ such that $\varphi'(\beta^-)\beta - \varphi(\beta) > 0$, we have $\lim_{n\to\infty} \mathscr{Y}_n(\beta) > 0$; this is done in [Bar00] (Fig. 1).

### 4.2 Second Order Phase Transition: $\beta_c \in (0, \infty)$ and $\varphi'_{W_{\beta_c}}(1^-) = \varphi'(\beta_c^-)\beta_c - \varphi(\beta_c) = 0$

This situation is illustrated by the case where the potential $V$ is Gaussian (Fig. 2).

**Theorem 7**

(1) *The same conclusions as in Theorem 6 hold over $[0, \beta_c)$.*

(2) *Suppose that $\mathbb{E}(W^{\beta_c}|\log(W)|^2) < \infty$, i.e. $\varphi''_{W_{\beta_c}}(1^-) > -\infty$. Let $\widetilde{\mu}_{\beta_c}$ be the critical Mandelbrot measure built from $(W_{\beta_c}(w))_{w\in\mathscr{A}^*}$ in Sect. 3.1.1. Then $\left(\sqrt{n}\frac{Z_n(\beta_c)}{\mathbb{E}(Z_n(\beta_c))}\right)_{n\geq 1}$ converges in probability to $c\|\widetilde{\mu}_{\beta_c}\|$ for some explicit $c > 0$, and $(\nu_{\beta_c,n})_{n\geq 1}$ weakly converges in probability to $\frac{\widetilde{\mu}_{\beta_c}}{\|\widetilde{\mu}_{\beta_c}\|}$ as $n \to \infty$.*

(3) *Suppose that $\mathbb{E}(W^{\beta_c}|\log(W)|^3) < \infty$. For $\beta > \beta_c$, let $\widetilde{\mu}_{\beta_c}^{\frac{\beta_c}{\beta}}$ be the critical Lévy-Mandelbrot measure built from $\widetilde{\mu}_{\beta_c}$ and a stable Lévy subordinator of index $\beta_c/\beta$ in Sect. 3.1.2.*

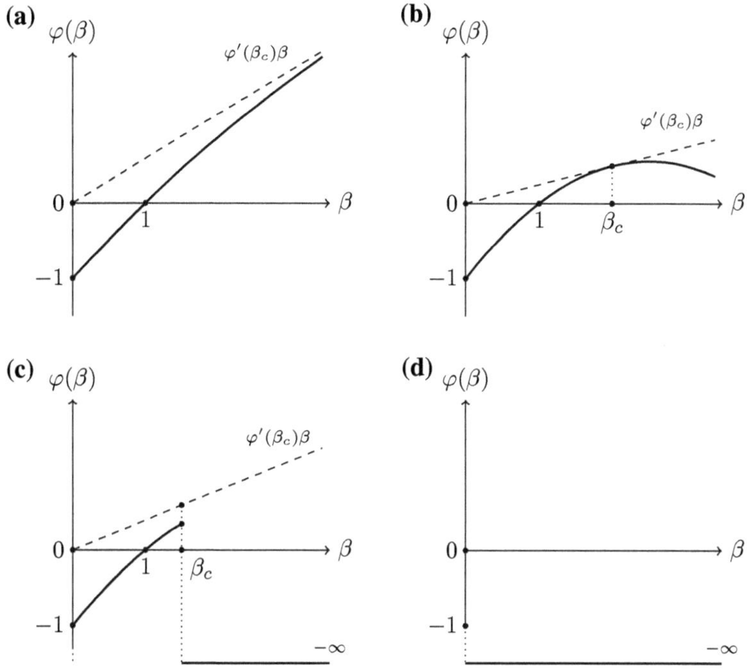


**Fig. 1** The four possible behaviors of $\varphi$ (by design, when $\beta_c > 0$, we make the pictures with $\varphi(1) = 0$ and $\varphi'(1) > 0$ to link these behaviors with multifractal analysis of Mandelbrot measures in Sect. 5). **a** No phase transition. **b** Second order phase transition. **c** First order phase transition. **d** The degenerate case

*Then, for all* $\beta > \beta_c$, $\left(n^{\frac{3}{2}\frac{\beta}{\beta_c}} \frac{Z_n(\beta)}{(\mathbb{E}(Z_n(\beta_c)))^{\frac{\beta}{\beta_c}}}\right)_{n\geq 1}$ *weakly converges in distribution to* $c\|\widetilde{\mu}_{\beta_c}^{\frac{\beta_c}{\beta}}\|$ *for some* $c > 0$, *and* $(\nu_{\beta,n})_{n\geq 1}$ *weakly converges in distribution to* $\widetilde{\mu}_{\beta_c}^{\frac{\beta_c}{\beta}}/\|\widetilde{\mu}_{\beta_c}^{\frac{\beta_c}{\beta}}\|$ *as* $n \to \infty$.

(4) $P_n(\cdot)$ *converges almost surely, uniformly over the compact subsets of* $\mathbb{R}_+$ *to*

$$P(\beta) = \begin{cases} -\varphi(\beta) & \text{if } 0 \leq \beta \leq \beta_c, \\ \dfrac{-\varphi(\beta_c)}{\beta_c}\beta = -\varphi'(\beta_c^-)\beta & \text{if } \beta > \beta_c \end{cases}.$$

*Thus* $P$ *is analytic except at* $\beta_c$ *where it is differentiable with a discontinuity of its second derivative, hence it presents a second order phase transition at* $\beta_c$.

Part (1) is obtained in a similar way as in the previous case.

For parts (2) and (3), the convergences of the renormalized partition functions are established in [Web11] for the log-normal case (with convergence in law for the

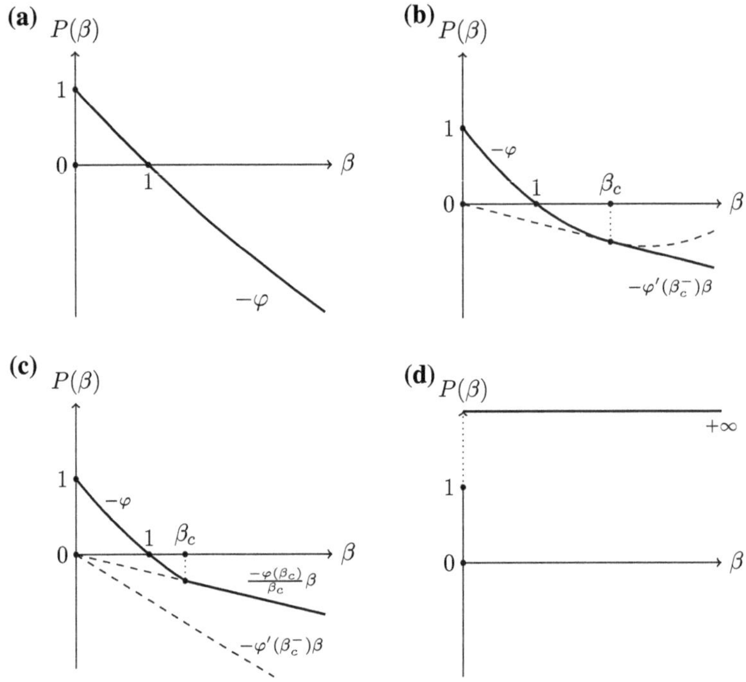


**Fig. 2** The corresponding pressure function. **a** No phase transition. **b** Second order phase transition. **c** First order phase transition. **d** The degenerate case

part (2)), and [AS14] and [Mad00] for part (2) and (3) respectively. The identification of the limit in case (3) follows easily from (2.7) and Theorem 4.

In part (3), the exponent $\frac{3}{2}$ is reminiscent from the asymptotic behavior of $M_n = \min\{\beta_c H(w) - n\varphi(\beta_c) : w \in \mathscr{A}^n\}$, for which one has, due to [Web11] for the log-normal case, and [Aid13] for the general case.

**Theorem 8** *Suppose that* $\mathbb{E}(W^{\beta_c}|\log(W)|^2) < \infty$. *Then, there exists* $c > 0$ *such that for all* $x \in \mathbb{R}$,

$$\lim_{n\to\infty} \mathbb{P}(M_n \geq \frac{3}{2}n + x) = \mathbb{E}(e^{-ce^x\|\widetilde{\mu}_{\beta_c}\|}).$$

The asymptotic behavior of the Gibbs measures reflects the fact that when the temperature $1/\beta$ becomes lower than $1/\beta_c$, the systems is in its glassy phase: asymptotically with $n$, the energy concentrates around a few local minima of $H(x|n)$, a phenomenon amplified as $\beta$ tends to $\infty$, and also called freezing transition (see [CD01]).

It is out of our scope to describe the techniques developed in [Aid13, AS14, Mad00, Web11]. The convergence of the Gibbs measures in case (2) and (3) requires additional arguments provided in [JW11] for the case $\beta = \beta_c$, and in [BRV12] for $\beta \geq \beta_c$.

Regarding point (4), this result or its analogue for Mandelbrot measures, has been obtained by several authors [CK92, Fra95, HW92, Mol96, OW00, WW13]. Due to (1), only the case $\beta \geq \beta_c$ must be considered. Due to the convexity of the functions $P_n$, it is quite direct that $\liminf_{n\to\infty} P_n(\beta) \geq -\varphi'(\beta_c^-)\beta$ for $\beta \geq \beta_c$. Moreover, for $\beta \geq \beta_c$, if $0 < \beta' < \beta_c$, we have $Z_n(\beta) = \sum_{w\in\mathscr{A}^n} e^{-\beta H(w)} = \sum_{w\in\mathscr{A}^n} (e^{-\beta' H(w)})^{\beta/\beta'} \leq Z_n(\beta')^{\beta/\beta'}$. Consequently, since $P_n(\beta')$ converges to $-\varphi(\beta')$, we get $\limsup_{n\to\infty} P_n(\beta) \leq \frac{\beta}{\beta'}(-\varphi(\beta'))$. Letting $\beta'$ tend to $\beta_c$ yields the result.

### 4.3 First Order Phase Transition: $\beta_c \in (0, \infty)$ and $\varphi'_{W_{\beta_c}}(1^-) = \varphi'(\beta_c^-)\beta_c - \varphi(\beta_c) > 0$

Notice that in this case one necessarily has $\varphi(\beta) = -\infty$ for $\beta > \beta_c$. Also, the measures $\mu_{\beta_c,n}$ weakly converge almost surely to the non degenerate Mandelbrot measure $\mu_{\beta_c}$, for which Theorem 1 is optimal in the sense that $\varphi_{W_{\beta_c}}(\beta) = -\infty$ for $\beta > 1$, hence in this case $(\|\mu_{\beta_c,n}\|)_{n\geq 1}$ is not bounded in any $L^q$, $q > 1$.

We have the following result.

**Theorem 9**

(1) *The same conclusions as in Theorem* 6 *hold over* $[0, \beta_c]$.
(2) *With probability* 1, $P_n(\cdot)$ *converges almost surely, uniformly over the compact subsets of* $\mathbb{R}_+$ *to*

$$P(\beta) = \begin{cases} -\varphi(\beta) & \text{if } 0 \leq \beta \leq \beta_c, \\ \dfrac{-\varphi(\beta_c)}{\beta_c}\beta > -\varphi'(\beta_c^-)\beta & \text{if } \beta > \beta_c \end{cases}.$$

*Thus $P$ is analytic except at $\beta_c$ where it is continuous and not differentiable, hence it presents a first order phase transition at $\beta_c$.*

*Remark 1* The expression of $P$ shows that the concave Legendre-Fenchel transform of $-P$, i.e. the mapping $\alpha \in \mathbb{R} \mapsto \inf\{\alpha\beta + P(\beta) : \beta \geq 0\}$ is non negative at $\alpha = \varphi(\beta_c)/\beta_c$. This proves that $|\beta_c H(w) - n\varphi(\beta_c)|$ behaves sub-linearly at some $w \in \mathscr{A}^n$, like in the case of the previous section.

Consequently, this raises the following questions: does $\min\{\beta_c H(w) - n\varphi(\beta_c) - s_n : w \in \mathscr{A}^n\}$ converge in law for some sub-linear sequence $(s_n)_{n\geq 1}$? If so, is there a freezing phenomenon like in the previous case?

Then, one can wondering if for $\beta > \beta_c$, $\left(e^{\frac{\beta}{\beta_c}s_n} \dfrac{Z_n(\beta)}{(\mathbb{E}(Z_n(\beta_c)))^{\frac{\beta}{\beta_c}}}\right)_{n\geq 1}$ weakly converges in distribution to $c\|\mu_{\beta_c}^{\frac{\beta_c}{\beta}}\|$ for some $c > 0$, and $(\nu_{\beta,n})_{n\geq 1}$ weakly converges in distribution to $\mu_{\beta_c}^{\frac{\beta_c}{\beta}}/\|\mu_{\beta_c}^{\frac{\beta_c}{\beta}}\|$ as $n \to \infty$, where $\mu_{\beta_c}^{\frac{\beta_c}{\beta}}$ is the Lévy-Mandelbrot measure built from $\mu_{\beta_c}$ and a stable Lévy subordinator of index $\beta_c/\beta$ in Sect. 3.1.2.

The upper bound for $\limsup_{n\to\infty} P_n(\beta)$ over $(\beta_c, \infty)$ is obtained like in the previous case. The lower bound for $\liminf_{n\to\infty} P_n(\beta)$ is more involved. It was obtained by Molchan in [Mol96] only along a deterministic subsequence. However, we believe that Molchan's approach can give the convergence along the whole sequence. Anyway, in [AB14], an alternative approach gives the result in connection with the multifractal analysis of $H(w)$ (see Sect. 5.1.3).

### 4.4 The Degenerate Case: $\beta_c = 0$, i.e. $\varphi = -\infty$ Over $\mathbb{R}_+^*$

In this case, we have:

**Theorem 10** *With probability* 1, $P_n$ *converges pointwise to* $-\varphi$ *as* $n \to \infty$.

This convergence result is shown in [Mol96] along a deterministic subsequence, and in this form in [AB14].

*Remark 2* One can wonder if there is a precise superexponential speed of divergence of $Z_n(\beta)$ to $\infty$ when $\beta > 0$.

### 4.5 A Uniform Point of View for the Free Energy

We can deduce from the previous results the following synthetic presentation:

**Theorem 11** *With probability* 1, $\lim_{n\to\infty} P_n(\beta) = \beta \inf\{-\varphi(\beta')/\beta' : \beta' \in (0, \beta]\} = \inf\{-\varphi(\theta\beta)/\theta : \theta \in (0, 1]\}$ *for all* $\beta > 0$.

## 5 Fine Geometric Properties of Statistically Self-similar Measures

This section presents recent results about the modulus of continuity of Mandelbrot and critical Mandelbrot measures, as well as recent progress in their multifractal analysis, with consequences for the multifractal analysis of their discrete companions, i.e. Lévy-Mandelbrot and critical Lévy-Mandelbrot measures. We also give results about the dimension of these measures. Finally, we say a word about KPZ formula.

## *5.1 Dimension, Modulus of Continuity, and Multifractal Analysis of Mandelbrot and Critical Mandelbrot Measures*

Here we suppose that $\mathbb{E}(W) = 1$.

### 5.1.1 Dimension

Recall that a positive and finite Borel measure $\nu$ on $[0, 1]$ is said to be exact dimensional with dimension $D$ if $\lim_{r\to 0^+} \frac{\log(\mu(B(x,r)))}{\log(r)} = D$, $\nu$-almost everywhere, or equivalently $\lim_{n\to\infty} \frac{\log(\mu(I_n(x)))}{-n\log(b)} = D$, $\nu$-almost everywhere, where $I_n(x)$ stands for the closure of the semi-open to the right $b$-adic interval of generation$x$ which contains $x$ when $x \in [0, 1)$ and $[1 - b^{-n}, 1]$ if $x = 1$.

**Theorem 12**

(1) *Suppose that $\varphi'(1^-) > 0$. Then the Mandelbrot measure $\mu$ is exact dimentional with dimension $\varphi'(1^-)$; in particular it is continuous.*
*If, moreover, $\mathbb{E}(W|\log W|^3) < \infty$ and $-\varphi''(1^-) > 0$ (i.e. $W \overset{dist}{\neq} 1$), then for $\mu$-almost every $x$ we have*

$$\liminf_{n\to\infty} (\mathit{resp.}\ \limsup_{n\to\infty}) \frac{\log(\mu(I_n(x))) + n\log(b)\varphi'(1^-)}{\sqrt{2\log(b)(-\varphi''(1^-))n\log\log(n)}} = -1 (\mathit{resp.}\ -1).$$

(2) *Suppose that $\varphi'(1^-) = 0$ and $\varphi''(1^-) > -\infty$. Then the critical Mandelbrot measure $\widetilde{\mu}$ is exact dimensional with dimension 0. If, moreover, $\varphi(1+\epsilon) > -\infty$ for some $\epsilon > 0$, then $\widetilde{\mu}$ is continuous.*
*Also, if $\log(W)$ is Gaussian, for $\widetilde{\mu}$-almost every $x$, for all $\alpha > 1/3$ and $k \in \mathbb{N}$, for $n$ large enough, we have*

$$\exp\left(-\sqrt{6\log(b)}\sqrt{n\,(\log n + \alpha \log\log n)}\right) \le \widetilde{\mu}(I_n(x)) \le n^{-k}. \tag{5.1}$$

For part (1), the fact that $\mu$ is exact dimensional, conjectured in [Man74b], was proved by Peyrière in [KP76, Pey74] under the assumption $\mathbb{E}(Y\log^+(Y)) < \infty$, which is shown to be equivalent to $\varphi''(1) > -\infty$ in [Big79]. In [Kah87], Kahane used a powerful percolation argument combined with Theorem 1 to eliminate the assumption $\mathbb{E}(Y\log^+(Y)) < \infty$.

The law of the iterated logarithm is stated in [Liu00] assuming $\mathbb{E}(W|\log W|^2) < \infty$. To get such a law, consider the now called “Peyrière measure” $\mathbb{Q}$ introduced in [Pey74] and defined on $(\Omega \times [0, 1], \mathcal{A} \otimes \mathcal{B}([0, 1]))$ by

$$\mathbb{Q}(A) = \mathbb{E}\left(\int_{[0,1]} \mathbf{1}_A(\omega, x)\, \mu(\mathrm{d}x)\right),$$

so that "$\mathbb{Q}$-almost everywhere" means "almost surely, $\mu$-almost everywhere". Its proof first uses the standard law of the iterated logarithm applied to the sequence of centered i.i.d. random variables $\log(W(x|k)) - \log(b) + \log(b)\varphi'(1^-)$ with variance $-\log(b)\varphi'(1^-)$ with respect to $\mathbb{Q}$, to control $\frac{\sum_{k=1}^n \log(W(x|k)) - \log(b) + \log(b)\varphi''(1^-)}{\sqrt{2\log(b)(-\varphi''(1^-))n\log\log(n)}}$. However the control of $Y(x|n)$ as $o(\sqrt{n\log\log(n)})$ has a gap in [Liu00]. Assuming $\mathbb{E}(W|\log W|^3) < \infty$, we know from [Big79] that $\mathbb{E}(Y\log(Y)^2) < \infty$. Then, for $\epsilon > 0$, a calculation yields

$$\begin{aligned}
\mathbb{Q}\left(\{|\log Y(x|n)| \geq \sqrt{n}\epsilon\}\right) &= \mathbb{Q}\left(\{Y(x|n) \geq e^{\sqrt{n}\epsilon}\}\right) + \mathbb{Q}\left(\{Y(x|n) < e^{-\sqrt{n}\epsilon}\}\right) \\
&= \mathbb{E}\left(Y \cdot \mathbf{1}_{\{Y \geq e^{\sqrt{n}\epsilon}\}}\right) + \mathbb{E}\left(Y \cdot \mathbf{1}_{\{Y < e^{-\sqrt{n}\epsilon}\}}\right) \\
&\leq \mathbb{E}\left(Y \cdot \mathbf{1}_{\{Y \geq e^{\sqrt{n}\epsilon}\}}\right) + e^{-\sqrt{n}\epsilon}.
\end{aligned}$$

Applying the elementary inequality $\sum_{n\geq 1} \mathbf{1}_{\{X \geq \sqrt{n}\}} \leq X^2$ we get

$$\begin{aligned}
\sum_{n\geq 1} \mathbb{Q}\left(\{|\log Y(x|n)| \geq \sqrt{n}\epsilon\}\right) &\leq \sum_{n\geq 1} \mathbb{E}\left(Z \cdot \mathbf{1}_{\{Z \geq e^{\sqrt{n}\epsilon}\}}\right) + \sum_{n\geq 1} e^{-\sqrt{n}\epsilon} \\
&= \mathbb{E}\left(Z \cdot \sum_{n\geq 1} \mathbf{1}_{\left\{\frac{\log Z}{\epsilon} \geq \sqrt{n}\right\}}\right) + \sum_{n\geq 1} e^{-\sqrt{n}\epsilon} \\
&\leq \epsilon^{-2}\mathbb{E}(Z(\log Z)^2) + \sum_{n\geq 1} e^{-\sqrt{n}\epsilon} < \infty.
\end{aligned}$$

Then the Borel-Cantelli lemma yields $|\log Y(x|n)| = o(\sqrt{n})$, $\mu$-almost everywhere.

For part (2), the fact that $\widetilde{\mu}$ is exact dimensional with dimension 0 is established in [Bar00] using large deviations estimates to prove that $\sum_{n\geq 1} \widetilde{\mu}(\{x : \widetilde{\mu}(I_n(x)) \leq b^{-n\epsilon}\}) < \infty$ for all $\epsilon > 0$. This is refined in [BKNSW14a], using Theorem 3(2, 3), to get the lower bound in (5.1) (this bound is easy to extend to general distributions for $W$). The fact that $\widetilde{\mu}$ is atomless under the assumption $\varphi(1+\epsilon) > -\infty$ for some $\epsilon > 0$ is also established in [BKNSW14a]; this exploits Theorems 3(2, 3) and 5(1) to prove that $n^\gamma \max_{w\in\mathscr{A}^n} \mu(I_w)$ converge to 0 in probability as $n \to \infty$ for all $\gamma \in [0, 1/2)$. The upper bound in (5.1) is more involved than the lower one; we refer to [BKNSW14a] for the details of (5.1).

*Remark 3* The bounds (5.1) are not completely satisfactory since we do not know whether they are sharp or not; in fact, we believe that at least the second one is not sharp, and also that the order of magnitude of the sharp upper bound should differ from that of the lower bound and reflect the fluctuations of random walks conditioned to stay positive.

### 5.1.2 Modulus of Continuity

At first we notice that if $\mu$ is a non degenerate Mandelbrot measure then $\beta_c \geq 1$ and $\beta_c > 1$ if $\varphi'_{W_{\beta_c}}(1^-) = 0$, and if $\widetilde{\mu}$ is a critical Mandelbrot measure, then $\beta_c = 1$, where $\beta_c$ is defined in Sect. 4.

Also, it is important to have in mind that for either of these measures, the smallest pointwise Hölder exponent is $\varphi(\beta_c)/\beta_c$, which equals $\varphi'(\beta_c^-)$ if $\varphi'_{W_{\beta_c}}(1^-) = 0$ (see Sect. 5.1.3). It is natural to complete this information by estimating the modulus of continuity of these measures.

**Theorem 13**

(1) *Assume that* $\varphi'(1) > 0$, $\varphi'_{W_{\beta_c}}(1^-) = 0$, *and there exists* $\beta > \beta_c$ *such that* $\varphi(\beta) > -\infty$.
*With probability* 1, *for all* $\gamma \in (0, 1/2)$, *there exists* $C > 0$ *such that for all subintervals* $I \subset [0, 1]$, *the Mandelbrot measure* $\mu$ *satisfies*

$$\mu(I) \leq C|I|^{\varphi(\beta_c)/\beta_c}\left(\log\left(1+\frac{1}{|I|}\right)\right)^{-\gamma/\beta_c}. \tag{5.2}$$

(2) *Assume that* $\varphi(-\beta) > -\infty$ *and* $\varphi(1+\beta) > -\infty$ *for some* $\beta > 0$, *as well as* $\varphi'(1) = 0$.
*With probability* 1, *for all* $\gamma \in (0, 1/2)$, *there exists* $C > 0$ *such that for all subintervals* $I \subset [0, 1]$, *the critical Mandelbrot measure* $\widetilde{\mu}$ *satisfies*

$$\widetilde{\mu}(I) \leq C\left(\log\left(1+\frac{1}{|I|}\right)\right)^{-\gamma}. \tag{5.3}$$

*Moreover, one cannot take* $\gamma > 1/2$ *in the above statement.*

This result is proved in [BKNSW14a].

For (1), one needs that $\mathbb{E}(Y^{\beta}) < \infty$ for some $\beta > \beta_c$ to use the fact that $h(x) = \mathbb{P}(Y \geq x) \leq cx^{-\beta}$ for such a $\beta$; this is the case by Theorem 3(1) since $\varphi(\beta_c) > 0$ and our assumption imply $\varphi(\beta) > 0$ near $\beta_c^+$. This is combined in a non obvious way with the upper bound for $\mathbb{E}\Big(\Big(n^{\frac{3}{2}\frac{\beta}{\beta_c}}\frac{Z_n(\beta)}{(\mathbb{E}(Z_n(\beta_c)))^{\frac{\beta}{\beta_c}}}\Big)^{\theta}\Big)$ provided for $\theta \in (0, 1)$ in [Mad00] (and used here with $\theta = \beta_c/\beta$) to prove that

$$\begin{aligned}\mathbb{P}\big(\max_{w\in\mathscr{A}^n}\{\mu(I_w)\} \geq 2^{-n\frac{\varphi(\beta_c)}{\beta_c}}n^{-\frac{\gamma}{\beta_c}}\big) &= \mathbb{E}\Big(\prod_{w\in\mathscr{A}^n}(1-h(e^{\beta_c H(w)-n\varphi(\beta_c)}n^{-\gamma}))\Big)\\ &\leq C_\epsilon n^{(1-\epsilon)(\gamma-3/2)}\end{aligned}$$

for all $\epsilon > 0$ and $\gamma \in (0, 3/2)$. This yields (5.2).

Similarly, the upper bound for $\mathbb{E}\big((n^{\frac{1}{2}} Z_n(1))^\theta\big)$ provided for $\theta \in (0, 1)$ in [HS09] is combined with the information $\mathbb{P}(\widetilde{Y} \geq x) \leq cx^{-1}$ provided by Theorem 3(2, 3) to prove that $\mathbb{P}(\max_{w\in\mathscr{A}^n} \mu(I_w) \geq n^{-\gamma}) \leq C_\epsilon n^{(1-\epsilon)(\gamma-1/2)}$ for all $\epsilon > 0$ and $\gamma \in (0, 1/2)$. This yields (5.3). The fact that one cannot take $\gamma \geq 1/2$ in (5.3) is proved by using Theorems 5(1) and 8, as well as the information $\mathbb{P}(\widetilde{Y} \geq x) \geq c'x^{-1}$ $(x \geq 1)$ provided by Theorem 3(2, 3) to obtain that $\mathbb{P}(\max_{w\in\mathscr{A}^n} \mu(I_w) < n^{-\gamma})$ tends to 0 for all $\gamma > 1/2$.

*Remark 4*

(1) It is not known if the choice $\gamma < 1/2$ in (5.2) can be improved.
(2) Property (5.3) is extended to critical lognormal multiplicative chaos measures in [BKNSW14+].
(3) Theorem 13 only deals with the second order phase transition case (in the frame of Sect. 4). To get similar results in the first order phase transition case is desirable. This is related to Remark 1.

### 5.1.3 Multifractal Analysis

Recall that given a positive Borel measure $\nu$ supported on [0, 1], its multifractal analysis consists in computing the Hausdorff dimension, denoted dim, of the level sets of the pointwise Hölder exponent of $\nu$, namely the sets

$$E_\nu(\gamma) = \Big\{x \in [0, 1] : \gamma(\nu, x) = \gamma\Big\} \quad (\gamma \geq 0),$$

where

$$\gamma(\nu, x) = \liminf_{r\to 0^+} \frac{\log \nu(B(x, r))}{\log(r)}.$$

The $L^q$-spectrum of $\nu$ is defined as

$$\tau_\nu : q \in \mathbb{R} \mapsto \liminf_{r\to 0^+} \frac{\log \sup\Big\{\sum_i \mu(B(x_i, r))^q\Big\}}{\log(r)}, \tag{5.4}$$

where the supremum is taken over all the centered packing of [0, 1] by closed balls of radius $r$. Define also

$$\widetilde{\tau}_\nu : q \in \mathbb{R} \mapsto \liminf_{n\to\infty} \frac{-1}{n\log(b)} \log \sum_{w\in\mathscr{A}^n} \mu(I_w))^q, \tag{5.5}$$

Throughout, we adopt the convention that a set has a negative dimension if and only if it is empty.

One always has

$$\dim E_\nu(\gamma) \le \tau_\nu^*(\gamma),$$

and one says that $\nu$ obeys the multifractal formalism at $\gamma$ if $\dim E_\nu(\gamma) = \tau_\nu^*(\gamma)$ (see [Ols95]).

Let us naturally extend $\varphi$ to the real line as

$$\varphi(q) = \varphi_W(q) = q - 1 - \log_b \mathbb{E}(W^q) \quad (q \in \mathbb{R}),$$

Given a function $\psi : \mathbb{R} \to \mathbb{R} \cup \{-\infty\}$, define its concave Legendre-Fenchel conjugate as

$$\psi^* : \gamma \in \mathbb{R} \mapsto \inf\{\gamma q - \psi(q) : q \in \mathbb{R}\}.$$

Let

$$f : \gamma \mapsto \begin{cases} \varphi^*(\gamma) & \text{if } \varphi^*(\gamma) \ge 0 \\ -\infty & \text{otherwise} \end{cases}.$$

For non degenerate Mandelbrot measures and critical Mandelbrot measures, one has the following result (for point (3) we refer the reader to appendix for the definition of Hausdorff measures).

**Theorem 14** *Let $\nu$ be the non degenerate Mandelbrot measure $\mu$ if $\varphi'(1^-) > 0$, or the critical Mandelbrot measure $\widetilde{\mu}$ if $\varphi'(1^-) = 0$ and $\varphi''(1^-) > -\infty$.*

*Suppose that $\varphi(-\epsilon) > -\infty$ for some $\epsilon > 0$. Then, with probability 1,*

(1) $\tau_\nu(0) = -1$ *and for all* $q \in \mathbb{R}^*$, $\tau_\nu(q) = \sup\{\varphi(\theta q)/\theta : \theta \in (0, 1]\}$. *Moreover,* $\tau_\nu = f^*$, $f = \tau_\nu^*$, *and in* (5.4), $\liminf$ *can be replaced by* $\lim$.
(2) *For all* $\gamma \ge 0$, $\dim E_\nu(\gamma) = \tau_\nu^*(\gamma) = f(\gamma)$.
(3) $E_\nu(\tau_\nu'(0))$ *is of full Lebesgue measure;*
(0–$\infty$ *law) for all* $\gamma \ge 0$ *such that* $\dim 0 < \tau_\nu^*(\gamma) < 1$, *for all gauge functions* $g$, *we have* $\mathcal{H}^g(E_\nu(\gamma)) = \infty$ *if* $\limsup_{t\to 0^+} \log(g(t))/\log(t) \le \tau_\nu^*(\gamma)$ *and* $\mathcal{H}^g(E_\nu(\gamma)) = 0$ *otherwise.*
(4) *For* $n \ge 1$, $\gamma \in \mathbb{R}_+$ *and* $\epsilon > 0$ *let*

$$f(n, \gamma, \epsilon) = n^{-1} \log_b \# \Big\{ w \in \mathscr{A}^n : \frac{\log \mu(I_w)}{-n \log(b)} \in [\gamma - \epsilon, \gamma + \epsilon] \Big\}.$$

*We have*

$$\text{for all } \gamma \ge 0, \quad \lim_{\epsilon\to 0^+} \liminf_{n\to\infty} f(n, \gamma, \epsilon) = \lim_{\epsilon\to 0^+} \limsup_{n\to\infty} f(n, \gamma, \epsilon) = f(\gamma).$$

Let us sketch the ideas of the proof of this result, and start with a remark.

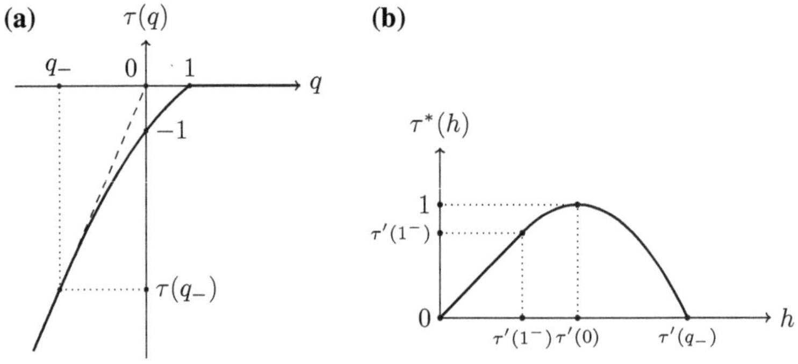


**Fig. 3** Multifractal nature of a Mandelbrot measure with a second order phase transition at $q_-$ and a first order phase transition at $q_+ = 1$, i.e. in the case where Theorem 1 is sharp. **a** The $L^q$ spectrum of $\mu$. **b** Its Legendre transform

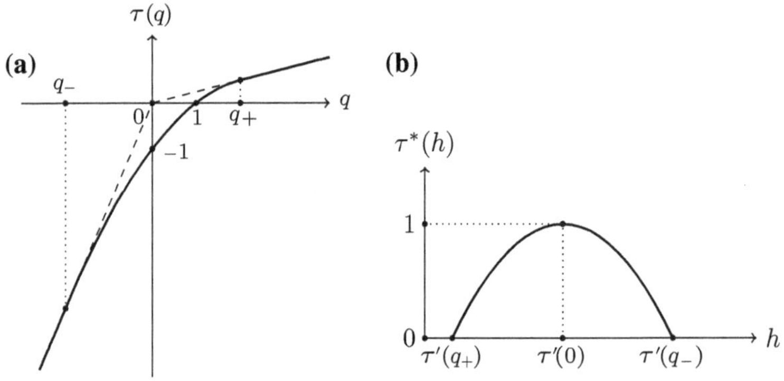


**Fig. 4** Multifractal nature of a Mandelbrot measure with a second phase transition at both $q_-$ and $q_+$. This situation occurs when $W$ is lognormal. **a** The $L^q$ spectrum of $\mu$. **b** Its Legendre transform

*Remark 5*

(1) It turns out that $\widetilde{\tau}_\nu = \tau_\nu$, so according to Part(1), $\tau_\nu(q)$ is on $\mathbb{R}_+$ the opposite of the free energy $P(q)$ of the directed polymer associated with $H(w)$ defined in Sect. 4, and on $\mathbb{R}_-$, $\tau_\nu(q)$ is the opposite of $P(-q)$, where $P$ is the free energy of the directed polymer associated with $-H(w)$. Thus, there may be two phase transitions for $\tau_\nu$: one on $\mathbb{R}_+$ according to $q_+ = \sup\{q > 0 : \varphi'(q)q - \varphi(q) > 0\}$ is finite or not, and one on $\mathbb{R}_-$ according to $q_- = \inf\{q < 0 : \varphi'(q)q - \varphi(q) > 0\}$ is finite or not. This yields nine possible situations under our assumptions in the case of Mandelbrot measures, and three in the case of critical Mandelbrot measures, since on $\mathbb{R}_+$ there is automatically a second order phase transition at 1. Some of these possibilities are illustrated in Figs. 3, 4, 5 and 6.

(2) Due to (2.4) and (3.3), and Parts (2) and (4) of Theorem 14 are geometric and statistical counterparts in $\mathscr{A}^\omega$ and $\mathscr{A}^*$ of Cramer's theorem and its extension by Bahadur and Zabell (see [DZ98]), which ensures that for all $\gamma \in \mathbb{R}$, over any fixed

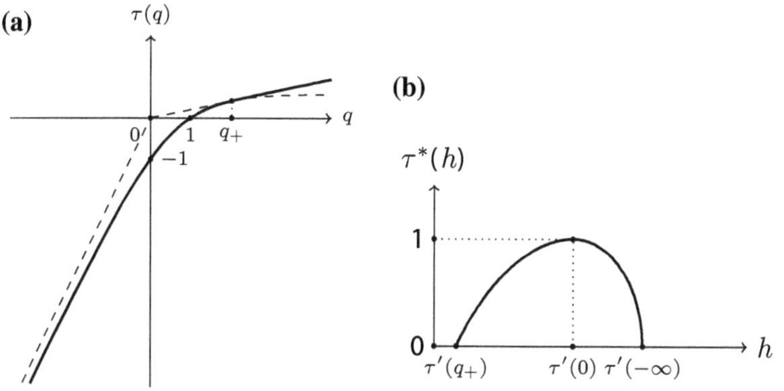


**Fig. 5** Multifractal nature of a Mandelbrot measure with $q_- = -\infty$ and a second order phase transition at $q_+$. **a** The $L^q$ spectrum of $\mu$. **b** Its Legendre transform

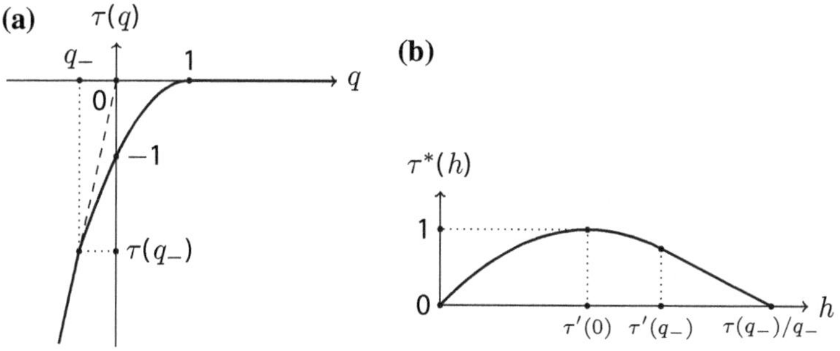


**Fig. 6** Multifractal nature of a critical Mandelbrot measure (one always has $q_+ = 1$, where a second order phase transition occurs), with a first order phase transition at $q_-$. **a** The $L^q$ spectrum of $\widetilde{\mu}$. **b** Its Legendre transform

infinite branch $w_1 \cdots w_n \cdots$ of $\mathscr{A}^\omega$, after setting $L(q) = \log_b \mathbb{E}((bW^{-1})^q)$, one has

$$\begin{aligned} r(\gamma) &:= \inf_{q\in\mathbb{R}} L(q) - \gamma q \\ &= \lim_{\epsilon\to 0} \liminf_{n\to\infty} n^{-1} \log_b \mathbb{P}(|\gamma - (n\log b)^{-1} \sum_{k=1}^n \log(b) - \log(W(w_1\cdots w_n)| \le \epsilon) \\ &= \lim_{\epsilon\to 0} \limsup_{n\to\infty} n^{-1} \log_b \mathbb{P}(|\gamma - (n\log b)^{-1} \sum_{k=1}^n \log(b) - \log(W(w_1\cdots w_n)| \le \epsilon). \end{aligned}$$

Indeed, heuristically, due to the rate of growth of the trees $\{\mathscr{A}\}^n$, $n \ge 1$, almost surely, for all $\gamma \in \mathbb{R}$ such that $1 + r(\gamma) \ge 0$, one should have the "logarithmic frequency"

$$\begin{aligned}\lim_{\epsilon\to 0^+}\liminf_{n\to\infty} f(n,\gamma,\epsilon) &= \lim_{\epsilon\to 0^+}\limsup_{n\to\infty} f(n,\gamma,\epsilon)\\ &= \lim_{\epsilon\to 0}\liminf_{n\to\infty} n^{-1}\log_b\Big(\#(\mathscr{A}^n)\cdot\mathbb{P}(|\gamma-(n\log b)^{-1}\sum_{k=1}^{n}\log(b)\\ &\qquad -\log(W(w_1\cdots w_n)|\le\epsilon)\Big)\\ &= 1+r(\gamma)=\varphi^*(\gamma).\end{aligned}$$

This observation about point (4) was made by Mandelbrot in [Man74a].

We notice that the formula given by Theorem 14 for $\tau_\nu$ simplifies to $\tau_\nu=\varphi$ over $(q_-,q_+)$, and also $\tau_\nu^*(\gamma)=\varphi^*(\gamma)$ for $\gamma=\varphi'(q)$, $q\in(q_-,q_+)$.

The lower bound $\tau_\nu(q)\ge T(q)=\sup\{\varphi(\theta q)/\theta:\theta\in(0,1]\}$ is quite easy. At first the problem reduces to packings by $b$-adic intervals, since $\tau_\nu=\widetilde{\tau}_\nu$. Then, notice that for $q\in(q_-,q_+)$, after Theorem 3 we have $\mathbb{E}(\|\nu\|^q)<\infty$, so (using (2.5) or (3.4))

$$\sum_{n\ge 1}\mathbb{E}\Big(\sum_{w\in\mathscr{A}^n}\nu(I_w)^q b^{n(\varphi(q)-\epsilon)}\Big)=\mathbb{E}(\|\nu\|^q)\sum_{n\ge 1} b^{-n\epsilon}<\infty$$

for all $\epsilon>0$. This yields $\widetilde{\tau}_\nu(q)\ge\varphi(q)$, but $\varphi(q)=\sup\{\varphi(\theta q)/\theta:\theta\in(0,1]\}$ if $q\in(q_-,q_+)$. For the other values of $q$, the argument giving the lower bound for $\widetilde{\tau}_\nu$ is similar to that used to obtain the upper bound for $\limsup_{n\to\infty}P_n$ in Sect. 4.2 when $\beta\ge\beta_c$. Moreover, one can show that $T^*=f$ and $f=T^*$ (see [AB14] for the details).

Thus, for all $\gamma\ge 0$, $\dim E_\nu(\gamma)\le\tau_\nu^*(\gamma)\le f(\gamma)\le\varphi^*(\gamma)$.

Let $\gamma\ge 0$ such that $\varphi^*(\gamma)\ge 0$. When $\nu=\mu$, the sharp lower bound $\varphi^*(\gamma)$ for $\dim E_\nu(\gamma)$ is determined in several papers (under different kind of assumptions, and sometimes for the level sets associated with $\widetilde{\gamma}(\nu,x)$ rather than $\gamma(\nu,x)$) in the case that $\gamma=\varphi'(q)$ with $q\in(q_-,q_+)$ [Bar00, BBP03, BHJ11, BJ10, Fal94, HW92, Kah91, Mol96, Ols94] or $q\in\{q_-,q_+\}$ [Bar00, BJ10]. This fully describes the range of $\gamma$ for which $E_\nu(\gamma)\neq\emptyset$ in the cases for which $q_-$ and $q_+$ correspond to no phase transition or a second order phase transition: one can prove that, with probability 1, for all $q\in(q_-,q_+)$, the Mandelbrot measure $\mu_q$ associated with the weights $W_q(w)$ (see (4.1)) is exact dimensional with a Hausdorff dimension equal to $\varphi^*(\varphi'(q))=\varphi'(q)q-\varphi(q)$ and is carried by $E_\nu(\alpha)$ [Bar00] (in [BHJ11, Fal94, HW92, Kah91, Mol96, Ols94], one finds the weaker version: for each fixed $q\in(q_-,q_+)$, $\dim E_\nu(\varphi'(q))\ge\varphi^*(\varphi'(q))$ almost surely). For $\gamma=\varphi'(q)$ with $q\in\{q_-,q_+\}$, if $q$ is finite then one shows that the critical Mandelbrot measure $\widetilde{\mu}_q$ is carried by $E_\nu(\varphi'(q))$ [Bar00], otherwise one builds an inhomogeneous Mandelbrot measure carried by $E_\nu(\varphi'(q))$ and whose Hausdorff dimension is $\varphi^*(\varphi'(q))$ [BJ10].

The same approach is used in [Bar00] to achieve the multifractal analysis of $\widetilde{\mu}$ in absence of first order phase transition on $\mathbb{R}_-$.

However, in case of first order phase transition at $q\in\{q_-,q_+\}$, the previous method does not make it possible to study $E_\nu(\gamma)$ for the exponents $\gamma$ in

$[\varphi(q)/q, \varphi'(q^-))$ if $q = q_+$ or in $(\varphi'(q^+), \varphi(q)/q]$ if $q = q_-$. Moreover, in any situation, they cannot provide the 0–$\infty$ law obtained in point (3).

In [AB14], a new approach is developed to treat all the possible types of phase transition in the multifractal analysis of branching random walks. Moreover, following the approach developed in [BBP03] or [BJ10], one can reduce the study of the level sets of $\gamma(\nu, x)$ to that of the level sets of $\liminf_{n\to\infty} \frac{\log \nu(I_n(x))}{-n\log(b)}$. Also, the assumption $\varphi(-\epsilon) > -\infty$ for some $\epsilon > 0$ assures that $\mathbb{E}(\|\nu\|^{-\epsilon}) < \infty$, and $\mathbb{E}(\|\nu\|^{\epsilon}) < \infty$ for $\epsilon \in (0,1)$, so that the study of the level sets of $\liminf_{n\to\infty} \frac{\log \nu(I_n(x))}{-n\log(b)}$ reduces to the level sets of $\liminf_{n\to\infty} \frac{H(x|n)}{n\log(b)}$, by using (2.4).

One can also assume that $W \overset{\text{dist}}{\neq} 1$ without loss of generality. Then (we still use the notations of Sect. 4), there exists $A_0 > 0$ such that for $A \geq A_0$, $\mathbb{P}(|V| \leq A) > 0$. One considers the random weights

$$W_{q,A}(w) = \frac{\mathbf{1}_{\{|V(w)|\leq A\}} W^q(w)}{\mathbb{E}(\mathbf{1}_{\{|V(w)|\leq A\}} W^q(w))}$$

and the functions

$$\varphi_A(q) = -1 + q - \log_b \mathbb{E}(\mathbf{1}_{\{|V(w)|\leq A\}} W^q).$$

One has $\varphi_A \searrow \varphi$ pointwise as $A \to \infty$, and it can be shown that this implies that $\varphi_A^* \nearrow \varphi^*$ pointwise on the interior of $\{\gamma : \varphi^*(\gamma) > -\infty\}$ as $A \to \infty$. It follows that one can find an increasing sequence $(A_k)_{k\geq 1}$ converging to $\infty$ and for each $k \geq 1$ a finite set $D_k \subset \{q : \varphi^*_{A_k}(\varphi'_{A_k}(q)) > 0\}$ such that for each $\gamma$ such that $\varphi^*(\gamma) \geq 0$, there exists $(q_k)_{k\geq 1} \in \prod_{k\geq 1} D_k$ such that $\lim_{k\to\infty} \varphi'_{A_k}(q_k) = \gamma$ and $\lim_{k\to\infty} \varphi^*_{A_k}(\varphi'_{A_k}(q_k)) = \varphi^*(\gamma)$.

Instead of considering Mandelbrot measures like the measures $\mu_q$, $q \in (q_-, q_+)$, one considers the family of inhomogeneous Mandelbrot measures obtained as follows.

Fix an increasing sequence of integers $(N_k)_{k\geq 0}$ with $N_0 = 0$. Set $M_k = \sum_{i=1}^k N_i$ and for $n \geq 1$ define $k_n$ so that

$$M_{k_n} + 1 = 1 + \sum_{k=1}^{k_n} N_k \leq n \leq \sum_{k=1}^{k_n+1} N_k.$$

For every sequence $B = (q_k)_{k\geq 1} \in \prod_{k=1}^{\infty} D_k$, consider the inhomogeneous branching random walk

$$H_B(w) = n\log(b) - \sum_{k=1}^{k_n} \sum_{i=1}^{N_k} \log W_{q_k,A_k}(w|M_{k-1}+i) - \sum_{i=1}^{n-M_{k_n}} \log W_{q_k,A_k}(w|M_{k_n}+i)$$

defined on $E_{B,n} = \{w \in \mathscr{A}^n : |V(w|M_k + i)| \le A_k, \forall k, i \text{ such that } M_k + i \le n\}$. Then define the random measures

$$\mu_{B,n}(\mathrm{d}x) = \mathbf{1}_{E_{B,n}}(w) e^{-H_B(w)} b^n \mathrm{d}x, \text{ if } x \in I_w.$$

They form a martingale which converges almost surely to a random measure $\mu_B$ supported on $E_B = \bigcap_{n\ge 1} \bigcup_{w\in E_{B,n}} I_w$. Moreover, it is possible to choose $(N_k)_{k\ge 1}$ suitably so that all the measures $\mu_B$ are simultaneously defined and non degenerate conditionally on $E_B \neq \emptyset$. The sequence $(N_k)_{k\ge 1}$ can also be fixed so that for all $\gamma$ such that $\varphi^*(\gamma) \ge 0$, each time $B = (q_k)_{k\ge 1}$ is such that $\lim_{k\to\infty} \varphi'_{A_k}(q_k) = \gamma$ and $\lim_{k\to\infty} \varphi^*_{A_k}(\varphi'_{A_k}(q_k)) = \varphi^*(\gamma)$, then, conditionally on $\{E_B \neq \emptyset\}$, $\mu_B$ is exact dimensional with dimension $\varphi^*(\gamma)$ and carried both by the set $E_H(\gamma) = \{x : \lim_{n\to\infty} \frac{H(x|n)}{n\log(b)} = \gamma\}$ and the set $\widetilde{E}_\nu(\gamma) = \left\{x \in [0,1] : \lim_{n\to\infty} \frac{\log \nu(I_n(x))}{-n\log(b)} = \gamma\right\} \cap \left\{x \in [0,1] : \lim_{r\to 0^+} \frac{\log \nu(B(x,r))}{\log(r)} = \gamma\right\}$. Moreover, there are uncountably many such measures for a given $\gamma$, and two such measures $\mu_B$ and $\mu_{B'}$ are mututally singular if $B$ and $B'$ are not ultimately equal. Since, moreover, $\mathbb{P}(\{E_B \neq \emptyset\})$ tends to 1 as the first terms $A_1$ of $(A_k)_{k\ge 1}$ tends to $\infty$, the measures $\mu_B$ are the main tool to get parts (2) of the theorem. Part (3) requires additional work, and uses in an essential way the reach family exhibited above of inhomogeneous measures of dimension $\varphi^*(\gamma)$ supported by $E_\nu(\gamma)$.

To get what remains of part (1), it is first not hard to deduce point (4) from the previous estimations for the Hausdorff dimensions: we have

$$\text{a.s., for all } \gamma \ge 0, \quad \lim_{\epsilon\to 0^+} \liminf_{n\to\infty} f(n,\gamma,\epsilon) = \lim_{\epsilon\to 0^+} \limsup_{n\to\infty} f(n,\gamma,\epsilon) = f(\gamma).$$

Then the conclusion follows from Varadhan's integral Lemma [DZ98, Theorem 4.3.1].

*Remark 6*

(1) The measures of type $\mu_B$ can be used to control the Hausdorff and packing dimensions of the wider family of sets

$$\mathbb{E}_\nu(\gamma;\gamma') = \left\{x \in [0,1] : \liminf_{n\to\infty} \frac{\log \nu(I_n(x))}{-n\log(b)} = \gamma,\ \limsup_{n\to\infty} \frac{\log \nu(I_n(x))}{-n\log(b)} = \gamma'\right\},$$

$0 \le \gamma \le \gamma'$, (see [AB14] for the details).

(2) Consider the branching random walk $H(w)$, $w \in \mathscr{A}^*$, for itself, and do not assume any integrability properties for $H$ (or $W$). Set

$$\widetilde{f}(n,\gamma,\epsilon) = n^{-1} \log_b \#\left\{w \in \mathscr{A}^n : \frac{H_n(w)}{n\log(b)} \in [\gamma-\epsilon, \gamma+\epsilon]\right\}.$$

The same approach as above yields that with probability 1,

$$\text{for all } \gamma \in \mathbb{R}, \quad \dim E_H(\gamma) = \lim_{\epsilon \to 0^+} \liminf_{n\to\infty} \widetilde{f}(n, \gamma, \epsilon)$$
$$= \lim_{\epsilon \to 0^+} \limsup_{n\to\infty} \widetilde{f}(n, \gamma, \epsilon) = f(\gamma).$$

(see [AB14]).
This includes in particular the both sided degenerate case $\varphi(q) = -\infty$ for all $q \neq 0$ (see Sect. 4.4 for the definiton of the degenerate case), and in this case one has $\dim E_H(\gamma) = 1$ for all $\gamma \in \mathbb{R}$.

## 5.2 *Multifractal Analysis of Lévy-Mandelbrot and Critical Lévy-Mandelbrot Measures*

We now describe the multifractal nature of the discrete statistically self-similar measures defined in Sect. 3.1.2 (notice that by construction these measures are exact dimensional with dimension 0) (Fig. 7).

**Theorem 15** *Let $\alpha \in (0, 1)$ and $\nu$ the non degenerate Mandelbrot measure $\mu$ if $\varphi'(1^-) > 0$, or the critical Mandelbrot measure $\widetilde{\mu}$ if $\varphi'(1^-) = 0$ and $\varphi''(1^-) > -\infty$. Let $\nu^\alpha$ be the associated Lévy-Mandelbrot or critical Lévy-Mandelbrot measure in Sect.* 3.1.2.

*Suppose that $\varphi(-\epsilon) > -\infty$ for some $\epsilon > 0$. With probability* 1, *we have*

(1) *for all* $q \in \mathbb{R}$, $\tau_{\nu^\alpha}(q) = \min(\tau_\nu(q/\alpha), 0)$;
(2) *for all* $\gamma \in \mathbb{R}_+$, $\dim E_{\nu^\alpha}(\gamma) = \tau^*_{\nu^\alpha}(\gamma)$.

*Remark 7* There is a phase transition at $q_+ = \alpha$; it is of first order when $\nu = \mu$ and of second order when $\nu = \widetilde{\mu}$. On $\mathbb{R}_-$, we have the same three possibilities as for $\nu$.

We are going to see that when $\nu = \mu$, the first order phase transition at $q = \alpha$ also corresponds to a transition in the geometric properties responsible for the Hausdorff

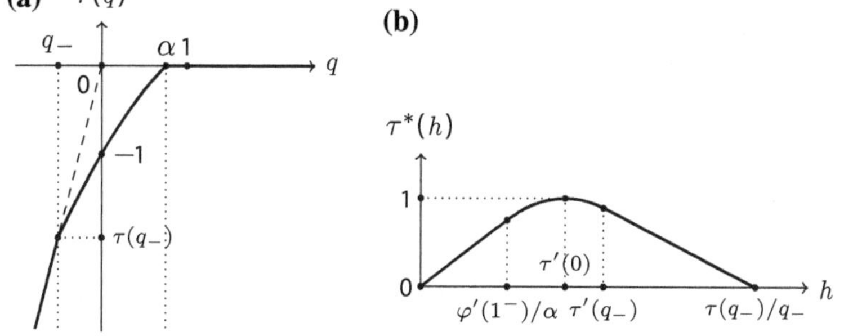


**Fig. 7** Multifractal nature of a Lévy-Mandelbrot measure, with the necessary first order phase transition at $\alpha = q_+$, and another first order phase transition at $q_-$. **a** The $L^q$ spectrum of $\mu^\alpha$. **b** Its Legendre transform

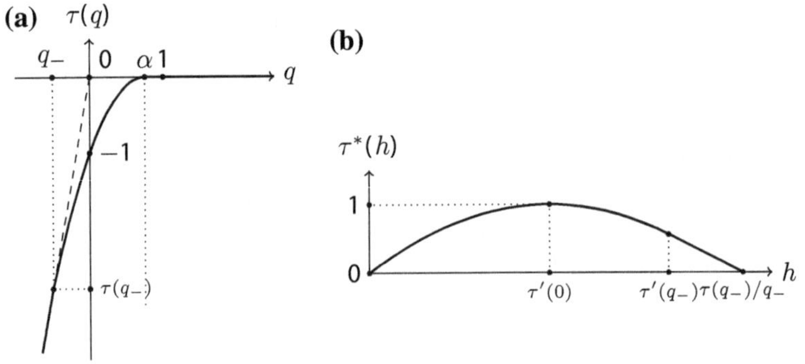


**Fig. 8** Multifractal nature of a critical Lévy-Mandelbrot measure, with the necessary second order phase transition at $\alpha = q_+$, and a first order phase transition at $q_-$. **a** The $L^q$ spectrum of $\widetilde{\mu}^\alpha$. **b** Its Legendre transform

dimensions of the sets $E_{\nu^\alpha}(\gamma)$ for $\gamma \in [0, \varphi'(1^-)/\alpha]$: for these exponents the rate of approximations of the elements of $E_{\nu^\alpha}(\gamma)$ by the atoms of $\nu^\alpha$ come into play, while for the other exponents, and the same holds when $\nu = \widetilde{\mu}$, the Hausdorff dimension of $E_{\nu^\alpha}(\gamma)$ can be captured essentially in the same way as for the sets $E_\nu(\alpha\gamma)$ (Fig. 8).

We can reduce the study of the $L^q$-spectrum to packings by $b$-adic intervals. For $\alpha q_- < q < \alpha$ we have, using (3.5)

$$\sum_{n\geq 1} \mathbb{E}\Big( \sum_{w\in\mathscr{A}^n} \nu^\alpha(I_w)^q b^{n(\varphi(q/\alpha)-\epsilon)}\Big) = \mathbb{E}(\|\nu^\alpha\|^q) \sum_{n\geq 1} b^{-n\epsilon} < \infty$$

for all $\epsilon > 0$. This yields $\tau_\nu(q) \geq \varphi(q/\alpha)$, and with manipulations similar to those use in the sketch of proof of Theorem 14, we get the lower bound $\tau_{\nu^\alpha}(q) \geq \min(\tau_\nu(q/\alpha), 0)$ almost surely. In particular, $\tau_{\nu^\alpha}(\alpha) \geq 0$, so that since $\tau_{\nu^\alpha}(1) = 0$ and $\tau_{\nu^\alpha}$ is non decreasing and concave, we must have $\tau_{\nu^\alpha} = 0$ over $[\alpha, \infty)$. The upper bound for $\tau_{\nu^\alpha}$ is a consequence of the lower bound for the Hausdorff dimension in part (2) of the theorem, as well as the inverse Legendre transform for $\varphi^*$.

Part (2) is proven in [Jaff99] when $\nu$ is the Lebesgue measure restricted to $[0, 1]$ ($W = 1$ almost surely), i.e. $\nu^\alpha$ is just the derivative of a Lévy subordinator, in the following form: $\dim E_{\nu^\alpha}(\gamma) = \alpha\gamma$ if $\gamma \in [0, 1/\alpha]$, and $\dim E_{\nu^\alpha}(\gamma) = -\infty$ otherwise.

Fix $T > 0$ and let $\mathcal{P} = \{(y_n, r_n) : n \geq 1\}$ be a Poisson point process with intensity $\mathrm{d}y \otimes \frac{\mathrm{d}r}{r^2}$ in $[0, T] \times \mathbb{R}_+^*$, so that the sequence $(r_n)_{n\geq 1}$ tends to 0 as $n \to \infty$. Then take for $L_\alpha$ the $\alpha$-stable Lévy subordinator $L_\alpha(y) = \sum_{n: y_n \leq y} r_n^{1/\alpha}$, so that $\nu^\alpha = \sum_{n\geq 1} r_n^{1/\alpha} \delta_{y_n} := \rho$ when $\nu$ is the Lebesgue measure. The multifractal analysis in this case uses the following facts. At first, it results from quite direct estimates that for any $y \in \mathbb{R}_+ \setminus \{y_n : n \geq 1\}$, the pointwise exponent $\gamma(\rho, y)$ equals $1/(\alpha s(y))$, where $s(y)$ is the rate of approximation of $y$ by the family $\{(y_n, r_n) : n \geq 1\}$, defined as

$$s(y) = \limsup_{n\to\infty} \frac{\log(|y - y_n|)}{\log(r_n)}.$$

By a theorem of Shepp [She72], one has $s(y) \geq 1$ for all $y \in [0, T] \setminus \{y_n : n \geq 1\}$, hence $\gamma(\rho, y) \leq 1/\alpha$, so $E_\rho(\gamma) = \emptyset$ if $\gamma > 1/\alpha$. Moreover, as a consequence of a theorem on the "ubiquitous systems" established independently in [Jaf00a] and [DMPV95] and applied to the family $\{(y_n, r_n) : n \geq 1\}$, one has $\dim\{y : s(y) : s\} = 1/s$ for all $s \geq 1$, hence the Hausdorff dimension of $E_\rho(\gamma) = \{y \in [0, T] : s(y) : 1/(\alpha\gamma)\} = \alpha\gamma$.

The general case is treated in [BS07] under stronger assumptions on $\varphi$ that only allow no phase transition or a second order phase transition on $\mathbb{R}_-$. Conditionally on $\nu$, the Lévy subirdinator $L_\alpha$ is considered over $[0, T]$, with $T = \|\nu\|$. The composition by the indefinite integral of $\nu$ when $W \overset{\text{dist}}{\neq} 1$ induces distorsions with respect to the situation explained above. For $\gamma \geq \varphi'(1^-)/\alpha$ such that $\varphi^*(\alpha\gamma) \geq 0$, one takes a measure $\mu_B$ of Hausdorff dimension $\varphi^*(\alpha\gamma) \geq 0$ carried by $E_\nu(\alpha\gamma)$ as in the proof of Theorem 14 and proves that for $\mu_B$-almost very $x$, noting $F(x) = \nu([0, x])$, we have $s(F(x)) = 1$, which combined with $\lim_{r\to 0^+} \frac{\log(\nu(B(x,r)))}{\log(r)} = \alpha\gamma$ and $\gamma(\varrho, F(x)) = 1/(\alpha s(F(x)))$ yields $\gamma(\nu^\alpha, x) = \gamma$. Hence $\dim E_{\nu^\alpha}(\gamma) \geq \varphi^*(\alpha\gamma) = (\min(\tau_\nu(q/\alpha), 0))^*(\gamma)$.

When $\nu = \widetilde{\mu}$, since $\varphi'(1^-) = 0$, this yields the whole spectrum for $\gamma > 0$. Since $E_{\nu^\alpha}(0)$ contains the atoms of $\nu^\alpha$, i.e. $\{F^{-1}(\{y_n : n \geq 1\})$, we also have $\dim E_{\nu^\alpha}(0) \geq 0 = (\min(\tau_\nu(q/\alpha), 0))^*(0)$, and this is also valid when $\nu = \mu$.

The most delicate sets are the $E_{\nu^\alpha}(\gamma)$ for $\gamma \in (0, \varphi'(1^-)/\alpha)$ when $\nu = \mu$, for which we must prove $\dim E_{\nu^\alpha}(\gamma) \geq \alpha\gamma$. This requires a non trivial extension of classical "ubiquitous systems" (which deserve to be called homogeneous) to "heterogeneous ubiquitous systems". This is achieved in [BS07], and applied to the present situation in [BS07]. The main tool provided by these papers is, for each $s > 1$, a Borel probability measure $\rho_s$ on $[0, 1]$ such that for $\rho_s$-almost every $x$, $\liminf_{r\to 0^+} \frac{\log(\rho_s(B(x,r)))}{\log(r)} \geq \frac{\varphi'(1^-)}{s}$ (which implies that for any Borel set $E$, one has $\rho_s(E) = 0$ if $\dim E < \frac{\varphi'(1^-)}{s}$) and there exists a decreasing sequence $(\epsilon_j)_{j\geq 1}$ and a subsequence $(n_j)_{j\geq 1}$ such that for each $j \geq 1$ one has $F^{-1}(y_{n_j}) \in B(x, r_{n_j}^{s/\varphi'(1^-)-\epsilon_j})$. This second property implies that $\nu^\alpha(B(x, r_{n_j}^{s/\varphi'(1^-)-\epsilon_j})) \geq r_{n_j}^{1/\alpha}$, hence $\gamma(\nu^\alpha, x) \leq \varphi'(1^-)/\alpha s$. In particular for $\gamma = \varphi'(1^-)/(\alpha s)$, $\rho_s(F_{\nu^\alpha}(\gamma)) = 1$, where $F_{\nu^\alpha}(\gamma) = \{x \in [0, 1] : \gamma(\nu^\alpha, x) \leq \gamma\}$.

Now, we notice that for $\gamma \in (0, \varphi'(1^-)/\alpha)$, we have $(\min(\tau_\nu(q/\alpha), 0))^*(\gamma) = \alpha\gamma$. Moreover, it also comes from the multifractal formalism [Ols95] that $\dim F_{\nu^\alpha}(\gamma') \leq \tau_{\nu^\alpha}^*(\gamma')$ for $\gamma' \in [0, \tau_{\nu^\alpha}'(0^-)]$. Since $\tau_{\nu^\alpha}(q) \geq \min(\tau_\nu(q/\alpha), 0)$ almost surely for all $q$, we get $\dim F_{\nu^\alpha}(\gamma') \leq \tau_{\nu^\alpha}^*(\gamma') \leq \alpha\gamma'$ for all $0 \leq \gamma' < \gamma < \varphi'(1^-)/\alpha)$. Now, setting $\gamma = \varphi'(1^-)/\alpha s$, since $\liminf_{r\to 0^+} \frac{\log(\rho_s(B(x,r)))}{\log(r)} \geq \frac{\varphi'(1^-)}{s} = \alpha\gamma$, we get $\rho_s(F_{\nu^\alpha}(\gamma')) = 0$ for all $0 \leq \gamma' < \gamma$, hence $\rho_s\Big(\bigcup_{0\leq\gamma'<\gamma} F_{\nu^\alpha}(\gamma')\Big) = 0$

noting that the sets $F_{\nu^\alpha}(\gamma')$ are non decreasing. Finally, writing $E_{\nu^\alpha}(\gamma) = F_{\nu^\alpha}(\gamma) \setminus \bigcup_{0\le\gamma'<\gamma} F_{\nu^\alpha}(\gamma')$, we obtain $\rho_s(E_{\nu^\alpha}(\gamma)) > 0$ hence $\dim E_{\nu^\alpha}(\gamma)) \ge \alpha\gamma$.

The construction of the measures $\rho_s$ is rather involved, and uses a combination of Shepp's theorem and the statistical self-similarity and exact dimensionality properties of $\mu$.

## *5.3 KPZ Formula*

Here we give, in the Mandelbrot cascade context, results related to KPZ formula [KPZ98] of two dimensional quantum gravity. The KPZ formula was reformulated and proved by Duplantier and Sheffield [DS09, DS11] as a relation between the box counting dimensions of sets in the Euclidean geometry and the expecting box counting dimensions of sets computed using a random measure given by exponential of the Gaussian Free Field (a fundamental example of log-Gaussian multiplicative chaos), and reformulated by Benjamini and Schramm [BS09] and Rhodes and Vargas [RV08] as a relation between Hausdorff dimensions of sets computed using the Lebesgue measure to measure the size of balls, and the Hausdorff dimensions of sets when the Lebesgue measure is replaced by a non degenerate Mandelbrot measure or the limit of a non degenerate log-infinitely divisible cascade. In dimension 1, this can be directly interpreted in terms of a change of metric. Then, a rigorous proof of the "dual" KPZ formula was given in [BJRV13], using discrete random measures which are the analogue of the measure $\mu^\alpha$ of Sect. 3.1.2 in the log-normal multiplicative chaos framework (see [RV13] for a review of this). This has been extended to the cascade case in [BKNSW14a], with similar formulas for critical Mandelbrot measures $\widetilde{\mu}$ and the associated discrete measures $\widetilde{\mu}^\alpha$.

Since we work in dimension 1, it is convenient to present KPZ formulas as a relation between the Hausdorff dimension of a set and the Hausdorff dimension of its image by the indefinite integral of a given statistically self-similar measure:

**Theorem 16** *Let $\nu$ be the non degenerate Mandelbrot measure $\mu$ if $\varphi'(1^-) > 0$, or the critical Mandelbrot measure $\widetilde{\mu}$ if $\varphi'(1^-) = 0$ and $\varphi''(1^-) > -\infty$. Denote also $\nu$ by $\nu^1$. Suppose that $\varphi(-q) < -\infty$ for all $q \in (0, 1/b)$.*

*Let $\alpha \in (0, 1]$ and define $F_\alpha(t) = \nu^\alpha([0, t])$ for $t \in [0, 1]$ (i.e. $F_\alpha$ is the $\alpha$-stable Lévy process $L_\alpha$ in independent multifractal time $F : t \mapsto \nu([0, t])$). If $E$ is a Borel subset of $[0, 1]$ of Hausdorff dimension $\xi_0$, then, with probability 1, the Hausdorff dimension of $F_\alpha(E)$ is the unique solution $\xi$ of the equation*

$$1 + \varphi(\xi/\alpha) = \xi_0.$$

Notice that when $\alpha = 1$, $F_1 = F$. In this case, when $\nu = \mu$, the result is proved in [BS09], since the authors prove that the dimension $\xi$ of $E$ under the metric $d(x, y) = \mu([x, y])$ is given by the above formula. The formula is quite easy to guess using a natural covering argument and the fact that for all $x, x' \in [0, 1]$,

$\mathbb{E}(d(x,y)^s) \le 8|x-y|^{1+\varphi(s)}$ for all $s \in [0,1]$ to find an upper bound of $\dim E$ under the metric $d$. For the lower bound, one first reduces the situation to $E$ being compact. If $\xi_0 = 0$ there is nothing to prove. Otherwise, for each $t \in [0,\xi_0)$ one fixes a Frostman Borel measure $\rho$ such that the energy $\int\int \frac{\rho(\mathrm{d}x)\rho(\mathrm{d}y)}{|y-x|^t}$ is finite, sets $s$ the solution of $1+\varphi(s) = t$, and then shows that the measures $\rho_n(\mathrm{d}x) = W_s(w|1)W_s(w|2)\cdots W_s(w|n)\,\rho(\mathrm{d}x)$ if $x \in I_w$, $w \in \mathscr{A}^n$ weakly converge to a non degenerate measure $\rho_s$ supported on $E$ almost surely and such that $\int\int \frac{\rho_s(\mathrm{d}x)\rho_s(\mathrm{d}y)}{d(x,y)^s} < \infty$ almost surely, which implies that the Hausdorff dimension of $E$ under $d$ is at least $s$. Since $s$ tends to $\xi$ as $t$ tends to $\xi_0$, this is enough to conclude.

This approach can be adapted to get the result when $\alpha = 1$ and $\nu = \widetilde{\mu}$ (see [BKNSW14a]). The general case then follows from the fact that a.s. $L_\alpha(A) = \alpha \dim A$ for all subsets $A$ of $[0,1]$ [Ber96, III.5].

## 6 On Signed and Complex Multiplicative Cascades

We give some convergence and renormalization results proved in [BM09, BJM10, BJM10a] for the continuous function-valued sequence $(F_n = F_{W,n})_{n\ge 1}$ defined in (2.8). The asymptotic behavior of $(F_n)_{n\ge 1}$ is far from being completely understood in general, and deserves to be further explored. An interesting related and earlier work is [DES93] about the mean field theory of directed polymers with random complex weights whose modulus is independent of the argument; under this kind of assumptions, renormalization results in the space of distributions have been obtained very recently in [LRV00] for complex Gaussian multiplicative chaos on $\mathbb{R}^d$ when the modulus and argument which are independent.

When $(F_n)_{n\ge 1}$ has a non degenerate limit $F$, the multifractal analysis of $F$ is similar to that of Mandelbrot measures, but new multifractal phenomena can emerge when working with conservative complex cascades (see [BJ10]), a situation excluded in the present setting. This classical multifractal analysis can be completed by the natural notion of multifractal analysis of the graph roughness, a notion explored for non degenerate limits of $(F_n)_{n\ge 1}$ in [Jin11].

For simplicity we suppose that $|W| > 0$ almost surely. Also, we assume that $\mathbb{P}(W \in \mathbb{C}\setminus\mathbb{R}_+) > 0$ and the normalization $E(W) = 1$ holds, so that $(F_n)_{n\ge 1}$ is a martingale. Then we set

$$\varphi(q) = \varphi_W(q) = q - 1 - \log_b \mathbb{E}(|W|^q).$$

## 6.1 Some Convergence Theorems

### 6.1.1 Strong Convergence of Complex Cascades

There are conditions sufficient to imply the convergence of $F_n = F_{W,n}$, which in view of the positive case (Theorems 1 and 3(b)) seem optimal. The following result is proved in [BJM10a] (see [BJM10] for an extended version to general complex multiplicative chaos).

**Theorem 17** *Assume that there exists $q > 1$ such that $\varphi_W(q) > 0$. Also suppose $q \in (1, 2]$ or $\varphi_W(2) > 0$.*

(1) *$(F_n)_{n\geq 1}$ uniformly converges, with probability 1, and in $L^q$ norm, when n goes to $\infty$, towards a Hölder function $F = F_W$.*

(2) $F = \sum_{i=0}^{b-1} \mathbf{1}_{[i/b,(i+1)/b]}\Big(F(i/b) + W(i)\, F_i \circ f_i^{-1}\Big)$, *where $(W(0), \dots, W(b-1))$, $F_0, \dots, F_{b-1}$ are independent, $F_i$ is equidistributed with $F$, and the equality holds with probability* 1.

With respect to the convergence of $(\mu_n)_{n\geq 1}$ when $W > 0$, the proof necessitates an additional compactness argument.

*Remark 8*

(1) The statistical self-affinity expressed by Theorem 17(2) implies, setting $Z = F(1) - F(0)$ and $Z(i) = F_i(1) - F_i(0)$:

$$Z = b^{-1} \sum_{k=0}^{b-1} W(k)\, Z(k),$$

which could be considered as an extension of (2.3) to the case of complex weights $W(k)$.

(2) When the weight $W$ is real and such that $|W| = b^{1-H}$, Theorem 17 yields a limit function $F$ which is a monofractal object obtained by a multiplicative cascade, which shares lots of properties with the fractional brownian motion of index $H$, but its construction is more straightforward. This remarkable fact was one of Mandelbrot's first motivations to consider the signed cascades. Nevertheless there is a constraint on the exponent $H$: it should lie within the interval $(1/2, 1]$ (see [BM09] for a specific study of the monofractal case).

For a pair $B_H = (F_1, F_2)$ of two independent copies of such a monofractal process, Jin has proved in [Jin14] an analogue of Kaufman's theorem about the Hausdorff dimension of the image of Borel subsets of $\mathbb{R}_+$, namely $\mathbb{P}(\forall\, E \in \mathcal{B}([0, 1]), \dim B_H(E) = H^{-1} \dim E) = 1$.

(3) In [BJM10a], when $|W|$ is not constant, the natural question of deciding whether the limit $F$ can be decomposed as a monofractal process $B$ composed with

the indefinite integral of a Mandelbrot measure $\mu$ is discussed. Under stronger assumptions than that of Theorem 17, denoting by $\beta$ the unique solution of $\varphi(q) = 0$ in $[1, 2]$, and setting $W_\beta = |W|^\beta/\mathbb{E}(|W|^\beta)$, such a decomposition exists, $B$ being of index $H = 1/\beta$, and $\mu = \mu_{W_\beta}$.

### 6.1.2 Weak Convergence Towards a Brownian Motion in Multifractal Time When $W$ Is Real Valued

When one cannot use Theorem 17, it is natural to seek for a suitable normalisation of the process $F_{W,n}$ in order it be convergent in distribution. We present one result in this direction.

Assume

(1) $W \in \mathbb{R}$ almost surely.
(2) $\varphi_W > -\infty$ on $\mathbb{R}_+$.
(3) $\varphi_W(2) \le 0$ and $\varphi_W$ is non-decreasing.

It follows from these hypotheses that $|W| \le b$ a.s., and the hypotheses of Theorem 17 are not fulfilled. Also, a direct computation shows that the martingale $F_n(1)$ is not bounded in $L^2$. More precisely

$$\sigma_n^2 = \mathbb{E}(|F_n(1)|^2) \sim \begin{cases} \sigma^2 b^{-n\varphi_W(2)} \text{ with } \sigma^2 = \dfrac{\mathbb{E}(|\sum_{i=0}^{b-1} W(i)|^2) - b^2}{\mathbb{E}(\sum_{i=0}^{b-1} |W(i)|^2) - b^2} & \text{if } \varphi_W(2) < 0 \\ \sigma^2 n \text{ with } \sigma^2 = b^{-2} \sum_{i \neq j} \mathbb{E}(W(i)W(j)) & \text{if } \varphi_W(2) = 0 \end{cases}.$$

Moreover, (3) implies $\varphi_{W_2}(q) > 0$ for all $q > 1$ (where $W_2 = W^2/\mathbb{E}(W^2)$. So, the non decreasing function $F_{W_2}$, which is nothing but the indefinite integral of the Mandelbrot measure $\mu_{W_2}$, is non degenerate, and $F_{W_2}(1)$ is bounded in $L^q$ for all $q > 1$ after Theorem 3(1).

**Theorem 18** *Under the above assumptions, the random continuous function $F_n/\sigma_n$ converges in distribution towards $B \circ F_{W^{(2)}}$, where $B$ is a standard brownian motion independent of $F_{W^{(2)}}$.*

This is established in [BJM10a]. The assumption (3) is used to identify the law of the limit process thanks to the moments method and the equation $F_n(1) = \sum_{k=0}^{b-1} b^{-1} W(k) F_{n-1,k}$, after proving the tightness of the normalized sequence. However, it is desirable to relax this assumption.

*Remark 9* A few additional information about the asymptotic behavior of $F_n$ when the assumptions of Theorems 17 and 18 fail can be found in [BJM10a]. When $W/|W|$ and $|W|$ are independent, precise information about the convergence in probability of the free energy $n^{-1} \log |F_n(1)|$ can be found in [DES93], with three main possible phases. The recent related work achieved in [LRV00] leads to believe that in general, possible limit of $F_n/A_n$ when the conditions of Theorem 17 fail are Brownian motion in critical Mandelbrot or critical Lévy-Mandelbrot time when $\varphi$ presents a second order phase transition at $\beta_c$.

## 6.2 Multifractal Analysis of Roughness in the Graph of F

The multifractal analysis of the limit function $F$ in Theorem 17 is a refinement of that of the Mandelbrot measure $\mu$: one considers the Hölder exponent associated with the oscillations of order 1 of $F$,

$$h_F(t) = \liminf_{r\to 0^+} \frac{\log \mathrm{Osc}_F([t-r, r+r])}{\log r}$$

and the corresponding level sets

$$E_F(h) = \{t \in [0,1] : h_F(t) = h\} \quad (h \geq 0).$$

The exponent $h_F(\cdot)$ should be considered as a measure of the aspect more or less rough of the graph of $F$ at each point: the smaller $h_F(t)$, the larger roughness at $t$ is.

Let

$$\tau_F(q) = \liminf_{r\to 0} \frac{\log \sup\left\{\sum_i \mathrm{Osc}_F(B_i)^q\right\}}{\log r},$$

where the supremum is taken over all families of pairwise disjoint closed intervals $B_i$, of diameter $2r$, and centered in supp$(F')$, and $F'$ is the derivative of $F$ in the distribution sense. The following statement is similar to Theorem 14. It is originally proved in [BJ10] under much stronger assumptions (also, in [BJ10] a finer multifractal analysis is achieved; also see [BFP10, BS05, Jaf98, Jaf00, Jaf04] to have a substantial overview of the multifractal formalism for functions), but the techniques introduced in [AB14] makes it possible to relax them, and now cover cases presenting first order phase transition.

**Theorem 19** *Under the assumptions of Theorem* 17, *if* $\varphi(-\epsilon) > -\infty$ *for some* $\epsilon > 0$, *then with probability* 1, *all the conclusions of Theorem* 14 *hold with* $\nu$ *replaced by* $F$.

The sets $E_F(h)$ are only indirectly linked to the apparent roughness of the graph because they lie in the support of $F$. This leads to consider the *roughness spectrum* which gives the Hausdorff dimension of the sets obtained by lifting on the graph of $F$ the sets $E_F(h)$. In other terms, this spectrum is the Hausdorff dimensions of the sets

$$G_F(h) = \{(t, F(t)) : t \in E_F(h)\} \quad (h \geq 0).$$

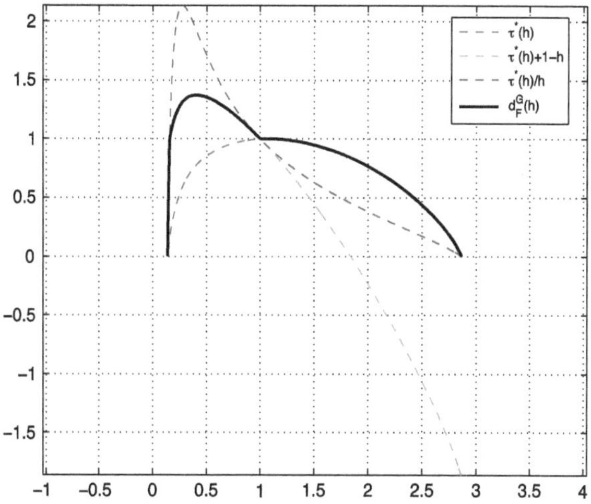


**Fig. 9** Example of roughness spectrum $d_F^G : h \mapsto \dim G_F(h)$

The following result [Jin11] is the first of this kind for multifractal stochastic processes. We present here a stronger version using the assumptions of Theorem 19.

**Theorem 20** *Under the assumptions of Theorem* 19*, if W is real valued, with probability* 1*, one has*

$$\dim \mathrm{Graph}(F) = 1 - \tau(1),$$

*and, for all* $h \geq 0$ *such that* $E_F(h) \neq \emptyset$*, one has*

$$\dim E_F(h) = \tau_F^*(h)$$

*and*

$$\dim G_F(h) = \Big(\frac{\dim E_F(h)}{h} \wedge \big(\dim E_F(h) + 1 - h\big)\Big) \vee \dim E_F(h).$$

The upper bounds for Hausdorff dimensions follows from quite standard arguments and is true in general. For the lower bound, the proof uses the auxiliary measures introduced to find a lower bound for $\dim E_F(h)$. These measures are lifted to the graph of $F$, and the lifted measures do have the desired lower Hausdorff dimension. This dimension is estimated by showing that they have a finite energy with respect to suitable Riesz kernels. However, while for the calculation of the dimension of the graph this method works quite straightforward, for the sets $G_F(h)$ in general the study is very delicate (see [Jin11] for details) (Fig. 9).

# 7 Mandelbrot Cascades as a Dynamical System

We present the results obtained in [BPW09], showing that Mandelbrot cascades define a dynamical system on a subset of the fixed points of the smoothing transformation. The asymptotic behavior of this dynamical systems exhibits a functional central limit theorem whose Gaussian limit process is, unexpectedly, the limit of an additive cascade on a tree. Fine properties of this process are also detailed.

## 7.1 A Dynamical System

Let $\mathcal{P}$ the set of Borel probability measures on $\mathbb{R}_+$. If $\mu \in \mathcal{P}$ and $p > 0$, we denote by $\mathbf{m}_p(\mu)$ the moment of order $p$ of $\mu$, i.e.,

$$\mathbf{m}_p(\mu) = \int_{\mathbb{R}_+} x^p \, \mu(\mathrm{d}x).$$

Then let $\mathcal{P}_1$ be the set of elements of $\mathcal{P}$ whose first moment equals 1:

$$\mathcal{P}_1 = \{\rho \in \mathcal{P} \, : \, \mathbf{m}_1(\rho) = 1\}.$$

Using the characterizations of the elements of $\mathcal{P}$ by their Laplace transform, the smoothing transformation $\mathsf{S}_\rho$ associated with $\rho \in \mathcal{P}$ considered in Sect. 3.1 is nothing but the mapping from $\mathcal{P}$ to itself so defined: If $\nu \in \mathcal{P}$, one considers $2b$ independent random variables, $Z(0),\ Z(1), \dots,\ Z(b-1)$, whose common probability distribution is $\nu$, and $W(0),\ W(1), \dots,\ W(b-1)$ whose common probability distribution is $\rho$; then $\mathsf{S}_\rho \nu$ is the probability distribution of the random variable $b^{-1} \sum_{0 \le j < b} W(j)\, Z(j)$.

Since the measure $\rho$ is in $\mathcal{P}_1$, $\mathsf{S}_\rho$ maps $\mathcal{P}_1$ into itself. We have seen (Theorem 1) that the condition $\int x \log(x)\, \rho(\mathrm{d}x) < \log b$ is necessary and sufficient for the weak convergence of the sequence $\mathsf{S}_\rho^n \delta_1$ (where $\delta_1$ stands for the Dirac mass at point 1) towards a probability measure $\nu$, which therefore is a fixed point of $\mathsf{S}_\rho$. Indeed, under this assumption $\mathsf{S}_\rho^n \delta_1$ is the probability distribution of the uniformly integrable non-negative martingale $(Y_n)_{n \ge 1}$ whose limit $Y$ has probability distribution $\nu$ belonging to $\mathcal{P}_1$ and satisfies $\mathsf{S}_\rho \nu = \nu$ due to (2.3). In this case, we denote the measure $\nu$ by $\mathsf{T}\rho$. It is natural to try and iterate $\mathsf{T}$. But, in general this is not possible because $\nu = \mathsf{T}\rho$ may not inherit the property $\int x \log(x)\, \nu(\mathrm{d}x) < \log b$. So, we have to find a domain stable under the action of $\mathsf{T}$. This is done by imposing conditions on moments.

Indeed, it is easily seen that the sequence $(Y_n)_{n \ge 1}$ defined by (2.1) remains bounded in $L^2$ norm if and only if $\mathbb{E}(W^2) = \mathbf{m}_2(\rho) < b$, and that in this case Formula (2.3) yields

$$\mathbb{E}Y^2 = \frac{b-1}{b - \mathbb{E}W^2} \tag{7.1}$$

(since the random variables $W(j)$ and $Y(j)$ are independent and of expectation 1, squaring both sides of Formula (2.3) yields $b^2\mathbb{E}Y^2 = b\mathbb{E}W^2\mathbb{E}Y^2 + b(b-1)$). It follows that if $b \geq 3$ and $1 \leq \mathbb{E}W^2 < b-1$, we have $\mathbb{E}Y^2 \leq \mathbb{E}W^2$ (the equality holding only if $W = 1$). Therefore, since the condition $\mathbb{E}W^2 < b$ is stronger than $\mathbb{E}(W \log W) < \log b$ when $\mathbb{E}W = 1$ (since the function $t \mapsto \log \mathbb{E}W^t$ is convex), $\mathsf{T}$ is a transformation on the subset of $\mathcal{P}_1$ defined by

$$\mathscr{P}_b = \left\{\rho \in \mathcal{P}_1 \ : \ 1 < \mathbf{m}_2(\rho) < b-1\right\}.$$

If $\rho \in \mathscr{P}_b$, due to (2.3), we can associate with each $n \geq 0$ a random variable $W_{n+1}$ as well as $2b$ independent random variables $W_n(0), \ldots, W_n(b-1)$ and $W_{n+1}(0), \ldots, W_{n+1}(b-1)$ such that

$$W_{n+1} = \frac{1}{b}\sum_{k=0}^{b-1} W_n(k) W_{n+1}(k), \tag{7.2}$$

$\mathsf{T}^n\rho$ is the probability distribution of $W_n(k)$ for every $k$ such that $0 \leq k \leq b-1$, and $\mathsf{T}^{n+1}\rho$ is the probability distribution of $W_{n+1}$ and $W_{n+1}(k)$ for every $0 \leq k \leq b-1$. We advise the reader that if one writes Formula (7.2) with $n-1$ instead of $n$, the variables $W_n(k)$ which then appear are different from the previous ones.

We have seen in Sect. 2 that the random variable $Y$ represents the increment between 0 and 1 of the non-decreasing continuous function $F$ on $[0, 1]$ obtained as the almost sure uniform limit of the sequence of non-decreasing continuous functions $F_n$ defined in (2.8), and $F$ is nothing but the indefinite integral of the Mandelbrot measure $\mu$. Thus, (2.4) rewrites, for $w \in \mathscr{A}^*$,

$$\Delta(F, I_w) = b^{-|w|} Y(w) \prod_{1 \leq j \leq |w|} W(w|j), \tag{7.3}$$

where for any bounded $f : [0, 1] \mapsto \mathbb{R}$, for every sub-interval $I = [\alpha, \beta]$ of $[0, 1]$, we denote by $\Delta(f, I)$ the increment $f(\beta) - f(\alpha)$ of $f$ over the interval $I$.

Let us denote by $\Phi(\rho)$ the probability distribution of the limit $F$, considered as a random continuous function.

We can describe the asymptotic behavior of the dynamical system $(\mathscr{P}_b, \mathsf{T})$, as well as the asymptotic behavior of $\left(\mathsf{T}^n\rho, \Phi(\mathsf{T}^{n-1}\rho)\right)_{n\geq 1}$ as $n$ goes to $\infty$.

We need some more definitions. For $b \geq 3$, set

$$w_2(b) = \min\left(b-1, \ b\frac{b^4 - 4b^2 + 12b - 8}{b^4 + 8b^2 - 12b + 4}\right)$$

and, for $t$ such that $1 < t < w_2(b)$,

$$w_3(b,t) = \frac{b^2}{2} + \frac{1}{2}\sqrt{\frac{b(b^4 - 4b^2 + 12b - 8) - t\big(b^4 + 8b^2 - 12b + 4\big)}{b-t}}.$$

One always has $w_3(b,t) < b^2 - 1$.

Also set

$$\mathscr{D}_b = \Big\{\rho \in \mathcal{P} \,:\, \mathbf{m}_1(\rho) = 1,\ 1 < \mathbf{m}_2(\rho) < w_2(b), \ \text{ and } \mathbf{m}_3(\rho) < w_3\big(b, \mathbf{m}_2(\rho)\big)\Big\}.$$

**Theorem 21** *Suppose* $b \geq 3$. *Let* $\rho \in \mathscr{P}_b$, *and, for* $n \geq 0$, *define* $\sigma_n = \left(\int (x-1)^2\, \mathsf{T}^n\mu(dx)\right)^{1/2}$. *Then*

(1) *The limit of* $(b-1)^{n/2}\sigma_n$ *exists and is positive; so* $\lim_{n\to\infty} \mathsf{T}^n\mu = \delta_1$. *More generally, the probability distributions* $\Phi(\mathsf{T}^n\rho)$ *weakly converges towards* $\delta_{Id}$.

(2) *Suppose that* $\rho$ *lies in the domain* $\mathscr{D}_b \subset \mathscr{P}_b$. *Then, if* $W_n$ *is a variable whose distribution is* $\mathsf{T}^n\mu$, $\dfrac{W_n - 1}{\sigma_n}$ *converges in distribution towards* $\mathcal{N}(0,1)$. *More generally, if* $h_n$ *is a random function distributed according to* $\Phi(\mathsf{T}^{n-1}\rho)$, *the distribution of* $\dfrac{h_n - Id}{\sigma_n}$ *weakly converges towards the distribution of the unique continuous Gaussian process* $(X_t)_{t\in[0,1]}$, *such that* $X(0) = 0$ *and, for all* $j \geq 1$, *the covariance matrix* $\mathsf{M}_j$ *of the vector* $\big(\Delta(X, I_w)\big)_{w\in\mathscr{A}^j}$ *is given by*

$$\mathsf{M}_j(w,w') = \begin{cases} b^{-2j}\big(1 + (b-1)|w|\big) & \text{if } w = w', \\ b^{-2j}(b-1)|w \wedge w'| & \text{otherwise.}\end{cases}$$

Part (1) of the theorem follows from an easy calculation. For part (2), setting $Z_n = \frac{W_n - 1}{\sigma_n}$ one first exploits (7.2) to prove that $(|W_n|)_{n\geq 1}$, and then $(|Z_n|)_{n\geq 1}$, is bounded in $L^3$. This uses a recursion in which the domain of attraction $\mathscr{D}_b$ is introduced. Then, using (7.2) again one gets, after setting,

$$R_n = \frac{1}{b}\sum_{j=0}^{b-1} Z_n(j) Z_{n-1}(j)\sigma_{n-1} + \frac{1}{b}\left(\frac{\sigma_{n-1}}{\sigma_n} - \sqrt{b-1}\right)\sum_{j=0}^{b-1} Z_{n-1}(j),$$

$$Z_n = R_n + \frac{\sqrt{b-1}}{b}\sum_{k=0}^{b-1} Z_{n-1}(k) + \frac{1}{b}\sum_{k=0}^{b-1} Z_n(k). \tag{7.4}$$

Due to the equivalence $\sigma_n \sim c(b-1)^{-n/2}$, the situation is essentially reducible to the relation $Z_n = \frac{\sqrt{b-1}}{b}\sum_{k=0}^{b-1} Z_{n-1}(k) + \frac{1}{b}\sum_{k=0}^{b-1} Z_n(k)$. Using the relation (7.4) recursively $n$ times yields a relation of the form $Z_n \overset{\text{dist}}{=} o(1) + \sum_{k=1}^{N_n} a_{n,k} Z_{n,k}$, where the $Z_{n,k}$ are independent centered variables, each of which is distributed like one of the $Z_k$, $1 \leq k \leq n$, $\lim_{n\to\infty} \sup\{a_{n,k} = 1 \leq k \leq N_n\} = 0$ and $\sum_{k=1}^{N_n} a_{n,k}^2 = 1$. An application of Lindeberg theorem (using the boundedness of $(|Z_k|)_{k\geq 1}$ in $L^3$) yields the convergence in law of $(Z_n)_{n\geq 1}$ to $\mathcal{N}(0,1)$.

For the functional central limit theorem, set $\mathcal{Z}_n = \dfrac{h_n - \mathrm{Id}}{\sigma_n}$. From (7.3) one deduces

$$\begin{aligned}\Delta(\mathcal{Z}_n, I_w) &= b^{-j}\sigma_n^{-1}\left[W_n(w)\prod_{k=1}^{j} W_{n-1}(w|k) - 1\right]\\ &= b^{-j} Z_n(w)\prod_{k=1}^{j} W_{n-1}(w|k)\\ &+ b^{-j}\sum_{l=1}^{j}\frac{\sigma_{n-1}}{\sigma_n} Z_{n-1}(w|l)\prod_{k=1}^{l-1} W_{n-1}(w|k). \qquad (7.5)\end{aligned}$$

From this relation one can both show the tightness of the distributions of the processes $\mathcal{Z}_n$, $n \geq 1$, and the convergence in distribution of the increments by simply passing to the limit and using the independences between the $W_{n-1}(w|k)$ and $W_n(w)$: there exist $\big(\mathcal{N}(v)\big)_{v\in\bigcup_{k=1}^{j}\mathscr{A}^k}$ and $\big(\widetilde{\mathcal{N}}(w)\big)_{w\in\mathscr{A}^j}$ two families of $\mathcal{N}(0,1)$ random variables so that all the random variables involved in these families are independent, and

$$\lim_{n\to\infty}\big(\Delta(\mathcal{Z}_n, I_w)\big)_{w\in\mathscr{A}^j} \overset{\text{dist}}{=} b^{-j}\left(\widetilde{\mathcal{N}}(w) + \sqrt{b-1}\sum_{k=1}^{j}\mathcal{N}(w|k)\right)_{w\in\mathscr{A}^j}. \qquad (7.6)$$

This is enough to derive the convergence in law of the $\mathcal{Z}_n$ to the Gaussian process $X$ of the statement.

*Remark 10* It is natural to seek for a larger domain of attraction than $\mathscr{D}_b$. This requires to be able to keep controls similar to the previous ones in $L^{2+\epsilon}$ (if not in $L^2$) rather than in $L^3$.

## 7.2 The Limit Process X as the Limit of an Additive Cascade

Recall that, if $v \in \mathscr{A}^*$, $[v]$ stands for the cylinder in $\mathscr{A}^\omega$ consisting of sequences beginning by $v$.

Let $(\xi(w))_{w\in\mathscr{A}^*}$ be a sequence of independent $\mathcal{N}(0,1)$ random variables. For each $w \in \mathscr{A}^*$,

$$\zeta_n(w) = \sqrt{b-1}\sum_{j=1}^{n} b^{-j}\sum_{v\in\mathscr{A}^j}\xi(wv),$$

is a martingale bounded in $L^2$ norm, whose limit $\zeta(w)$ is a $\mathcal{N}(0,1)$ random variable, independent of the $\xi(v)$, $|v|\le|w|$. Moreover, all the $\zeta(w)$, $w\in\mathscr{A}^n$ are independent.

One can then check that

$$M([w]) = b^{-|w|}\left(\zeta(w) + \sqrt{b-1}\sum_{1\le k\le |w|}\xi(w|k)\right) \tag{7.7}$$

defines a finitely additive Gaussian random measure defined on the cylinders of $\mathscr{A}^\omega$ Then, the limit process $X$ of the previous sections can be seen as the primitive of the projection of $M$ on $[0,1]$, and the structure of the increment $\Delta(X, I_w)$ given by (7.7) is an additive counterpart to the multiplicative structure of the increment of $F$ given by (7.3).

Of course (7.7) makes sense even for $b=2$.

## 7.3 Fine Properties of X

Due to the structure of the increments of $X$, it is natural to consider for all $\alpha\in\mathbb{R}$ the sets

$$\overline{E}_\alpha = \left\{t\in[0,1) : \limsup_{n\to\infty}\frac{\Delta(X, I_n(t))}{nb^{-n}} = \alpha\sqrt{b-1}\right\},$$

$$\underline{E}_\alpha = \left\{t\in[0,1) : \liminf_{n\to\infty}\frac{\Delta(X, I_n(t))}{nb^{-n}} = \alpha\sqrt{b-1}\right\},$$

and

$$E_\alpha = \underline{E}_\alpha \bigcap \overline{E}_\alpha,$$

where $I_n(t)$ stands for the semi-open to the right $b$-adic interval of generation $n$ containing $t$. One has

**Theorem 22** *With probability 1,*

(1) *the modulus of continuity of $X$ is a $O(\delta\log(1/\delta))$,*
(2) *$X$ does not belong to the Zygmund class,*
(3) *the set $E_0$ contains a set of full Lebesgue measure at each point of which $X$ is not differentiable,*
(4) $\dim E_\alpha = \dim \underline{E}_\alpha = \dim \overline{E}_\alpha = 1 - \frac{\alpha^2}{2\log b}$ *if* $|\alpha|\le\sqrt{2\log b}$*, and* $E_\alpha=\emptyset$ *if* $|\alpha|>\sqrt{2\log b}$.

The multifractal analysis part (4) is directly deduced from the multifractal analysis of the branching random walk $S(w) = \sum_{1\le j\le |w|}\xi(w|j)$ which follows from the approach explained in Sect. 5.1.3. Part (3) is a consequence of the law of the iterated logarithm with respect to $\mathbb{P} \otimes \text{Leb}$. Points (1) and (2) follow from quite direct calculations.

It would be good to decide whether or not $X$ is nowhere differentiable.

## Appendix: Hausdorff Measures and Dimension

Given $g : \mathbb{R}_+ \to \mathbb{R}_+$ a continuous non-decreasing function near 0 and such that $g(0) = 0$, and $E$ a subset of $[0, 1]$, the Hausdorff measure of $E$ with respect to the gauge function $g$ is defined as

$$\mathcal{H}^g(E) = \lim_{\delta \to 0^+} \inf \Big\{ \sum_{i \in \mathbb{N}} g(\text{diam}(U_i)) \Big\},$$

the infimum being taken over all the countable coverings $(U_i)_{i \in \mathbb{N}}$ of $E$ by subsets of $K$ of diameters less than or equal to $\delta$.

If $s \in \mathbb{R}_+^*$ and $g(u) = u^s$, then $\mathcal{H}^g(E)$ is also denoted $\mathcal{H}^s(E)$ and called the $s$-dimensional Hausdorff measure of $E$. Then, the Hausdorff dimension of $E$ is defined as

$$\dim E = \sup\{s > 0 : \mathcal{H}^s(E) = \infty\} = \inf\{s > 0 : \mathcal{H}^s(E) = 0\},$$

with the convention $\sup \emptyset = 0$ and $\inf \emptyset = \infty$.

For more information the reader is referred to [Fal03, Mat95].

## References


[Aid13] Aidekon, E.: Convergence in law of the minimum of a branching random walk. Ann. Probab. **41**, 1362–1426 (2013)

[AS14] Aidekon, E., Shi, Z.: The Seneta-Heyde scaling for the branching random walk. Ann. Probab. **42**, 959–993 (2014)

[ABM12] Alsmeyer, G., Biggins, J.-D., Meiners, M.: The functional equation of the smoothing transformation. Ann. Probab. **140**, 2069–2105 (2012)

[AM12] Alsmeyer, G., Meiners, M.: Fixed points of inhomogeneous smoothing transforms. J. Differ. Equ. Appl. **18**, 1287–1304 (2012)

[AM13] Alsmeyer, G., Meiners, M.: Fixed points of the smoothing transformation: two-sided case. Probab. Theory Relat. Fields **155**, 165–199 (2013)

[AJKS11] Astala, K., Jones, P., Kupiainen, A., Saksman, E.: Random conformal weldings. Acta Math. **207**(2), 203–254 (2011)

[AB14] Attia, N., Barral, J.: Hausdorff and packing spectra, large deviations, and free energy for branching random walks. in $\mathbb{R}^d$. To appear in Comm. Math. Phys (2014). arXiv:1305.2034

[BM03] Bacry, E., Muzy, J.F.: Log-infinitely divisible multifractal processes. Commun. Math. Phys. **236**, 449–475 (2003)

[Bar00] Barral, J.: Continuity of the multifractal spectrum of a random statistically self-similar measure. J. Theor. Probab. **13**, 1027–1060 (2000)

[BBP03] Barral, J., Ben Nasr, F., Peyrière, J.: Comparing multifractal formalisms: the neighboring boxes conditions. Asian J. Math. **7**, 149–166 (2003)

[BF05] Barral, J., Fan, A.H.: Covering numbers of different points in the Dvoretzky covering. Bull. Sci. Math. **129**, 275–317 (2005)

[BFP10] Barral, J., Fan, A.-H., Peyrière, J.: Mesures engendrées par multiplications. In: Quelques Interactions Entre Analyse, Probabilités et Fractals, Panoramas et Synthéses, vol. 32, Soc. Math., France (2010)

[BJ10] Barral, J., Jin, X.: Multifractal analysis of complex random cascades. Commun. Math. Phys. **219**, 129–168 (2010)

[BJM10] Barral, J., Jin, X., Mandelbrot, B.B.: Uniform convergence for complex [0, 1]-martingales. Ann. Appl. Probab. **20**, 1205–1218 (2010)

[BJM10a] Barral, J., Jin, X., Mandelbrot, B.B.: Convergence of signed multiplicative cascades. Ann. Appl. Probab. **20**, 1219–1252 (2010)

[BJRV13] Barral, J., Jin, X., Rhodes, R., Vargas, V.: Gaussian multiplicative chaos and KPZ duality. Commun. Math. Phys. **323**, 451–485 (2013)

[BKNSW14a] Barral, J., Kupiainen, A., Nikula, M., Saksman, E., Webb, C.: Critical Mandelbrot cascades. Commun. Math. Phys. **325**, 685–711 (2014)

[BKNSW14+] Barral, J., Kupiainen, A., Nikula, M., Saksman, E., Webb, C.: Basic properties of critical lognormal multiplicative chaos. To appear in Ann. Probab. arXiv:1303.4548

[BM02] Barral, J., Mandelbrot, B.: Multifractal products of cylindrical pulses. Probab. Theory Relat. Fields **124**, 409–430 (2002)

[BM04] Barral, J., Mandelbrot, B.: Introduction to infinite products of random independent functions (Random multiplicative multifractal measures, Part I). In: Lapidus, M.L., van Frankenhuysen, M. (eds.) Fractal Geometry and Applications: A Jubilee of Benot Mandelbrot. Proceedings of Symposia in Pure Mathematics, vol. 72, Part 2, pp. 3–16. AMS, Providence, RI (2004)

[BM04a] Barral, J., Mandelbrot, B.: Non-degeneracy, moments, dimension, and multifractal analysis for random multiplicative measures (Random multiplicative multifractal measures, Part II), pp. 17–52. In: Proceedings of Symposia in Pure Mathematics, vol. 72, Part 2. AMS, Providence, RI (2004)

[BM09] Barral, J., Mandelbrot, B.B.: Fractional multiplicative processes. Ann. Inst. Henry Poincaré Probab. Stat. **45**, 1116–1129 (2009)

[BPW09] Barral, J., Peyrière, J., Wen, Z.: Dynamics of Mandelbrot cascades. Probab. Theory Relat. Fields **144**, 615–631 (2009)

[BRV12] Barral, J., Rhodes, R., Vargas, V.: Limiting laws of supercritical branching random walks, C. R. Acad. Sci. Paris Ser. I **350**, 535–538 (2012)

[BS05] Barral, J., Seuret, S.: From multifractal measures to multifractal wavelet series. J. Fourier Anal. Appl. **11**, 589–614 (2005)

[BS07] Barral, J., Seuret, S.: Heterogeneous ubiquitous systems in $\mathbb{R}^d$ and Hausdorff dimension. Bull. Braz. Math. Soc. **38**, 467–515 (2007)

[BS07] Barral, J., Seuret, S.: The singularity spectrum of Lévy processes in multifractal time. Adv. Math. **214**, 437–468 (2007)

[BS09] Benjamini, I., Schramm, O.: KPZ in one dimensional random geometry of multiplicative cascades. Commun. Math. Phys. **289**, 653–662 (2009)

[Ber96] Bertoin, J.: Lévy Processes. Cambridge University Press, Cambridge (1996)

[Big77] Biggins, J.D.: Martingale convergence in the branching random walk. J. Appl. Probab. **14**, 25–37 (1977)

[Big79] Biggins, J.D.: Growth rates in the branching random walk. Z. Wahrsch. Verw. Geb. **48**(48), 17–34 (1979)

[Big92] Biggins, J.D.: Uniform convergence of martingales in the branching random walk. Ann. Probab. **20**, 137–151 (1992)

[BK97] Biggins, J.D., Kyprianou, A.: Seneta-Heyde norming in the branching random walk. Ann. Probab. **25**, 337–360 (1997)

[BK04] Biggins, J.D., Kyprianou, A.: Measure change in multitype branching. Adv. Appl. Probab. **36**, 544–581 (2004)

[BK05] Biggins, J.D., Kyprianou, A.: The smoothing transform: the boundary case. Electron. J. Probab. **10**, 609–631 (2005)

[BHJ11] Biggins, J.D., Hambly, B.M., Jones, O.D.: Multifractal spectra for random self-similar measures via branching processes. Adv. Appl. Probab. **43**, 1–39 (2011)

[Bov06] Bovier, A.: Statistical mechanics of disordered system. A mathematical perspective. Cambridge Series in Statistical and Probabilistic Mathematics, vol. 18. Cambridge University Press, Cambridge (2006)

[Bra83] Bramson, M.: Convergence of solutions of the Kolmogorov equation to travelling waves. Mem. Am. Math. Soc. **44**(285), iv+190 pp. (1983)

[Bur09] Buraczewski, D.: On tails of fixed points of the smoothing transform in the boundary case. Stoch. Process. Appl. **119**(11), 3955–3961 (2009)

[CD01] Carpentier, D., Le Doussal, P.: Glass transition of a particle in a random potential, front selection in nonlinear RG and entropic phenomena in Liouville and Sinh-Gordon models. Phys. Rev. E **63**, 026110 (2001)

[CRV13] Chevillard, L., Rhodes, R., Vargas, V.: Gaussian multiplicative chaos for symmetric isotropic matrices. J. Stats. Phys. **150**, 678–703 (2013)

[CK92] Collet, P., Koukiou, F.: Large deviations for multiplicative chaos. Commun. Math. Phys. **147**, 329–342 (1992)

[DZ98] Dembo, A., Zeitouni, O.: Large deviations techniques and applications. In: Applications of Mathematics, vol. 38, 2nd edn. Springer, New York (1998)

[DS88] Derrida, B., Spohn, H.: Polymers on disordered trees, spin glasses and traveling waves. J. Stat. Phys. **51**, 817 (1988)

[DES93] Derrida, B., Evans, M.R., Speer, E.R.: Mean field theory of directed polymers with random complex weights. Commun. Math. Phys. **156**(2), 221–244 (1993)

[DMPV95] Dodson, M.M., Melián, M.V., Pestana, D., Vélani, S.L.: Patterson measure and ubiquity. Ann. Acad. Sci. Fenn. Ser. A I Math. **20** 37–60 (1995)

[DS09] Duplantier, B., Sheffield, S.: Duality and KPZ in Liouville quantum gravity. Phys. Rev. Lett. **102**, 150603 (2009)

[DS11] Duplantier, B., Sheffield, S.: Liouville quantum gravity and KPZ. Invent. Math. **185**(2), 333–393 (2011)

[DRSV00a] Duplantier, B., Rhodes, R., Sheffield, S., Vargas, V.: Critical Gaussian multiplicative chaos: convergence of the derivative martingale. arXiv:1206.1671

[DRSV00] Duplantier, B., Rhodes, R., Sheffield, S., Vargas, V.: Renormalization of critical Gaussian multiplicative chaos and KPZ formula. arXiv:1212.0529

[DL83] Durrett, R., Liggett, T.: Fixed points of the smoothing transformation. Z. Wahrsch. Verw. Geb. **64**, 275–301 (1983)

[Fal94] Falconer, K.J.: The multifractal spectrum of statistically self-similar measures. J. Theor. Probab. **7**(3), 681–702 (1994)

[Fal03] Falconer, K.J.: Fractal Geometry. Mathematical Foundations and Applications, 2nd edn. Wiley, New York (2003)

[Fan89] Fan, A.H.: Chaos additifs et chaos multiplicatifs de Lévy. C. R. Acad. Sci. Paris Sér. I **308**, 151–154 (1989)

[Fan97] Fan, A.H.: Sur le chaos de Lévy d'indice $0 < \alpha < 1$. Ann. Sci. Math. Que. **21**, 53–66 (1997)

[Fan04] Fan, A.-H.: Some topics in the theory of multiplicative chaos. In: Bandt, Ch., Zaehle, M., Mosco, U. (eds.) Progress in Probability, vol. 57, pp. 119–134. Birkhäuser, Basel (2004)

[Fra95] Franchi, J.: Chaos multiplicatif: un traitement simple et complet de la fonction de partition. Séminaire Probab. Strasbourg **29**, 194–201 (1995)

[FP88] Frisch, U., Parisi, G.: Fully developed turbulence and intermittency in turbulence. Proceedings of the International School of Physics "Enrico Fermi". In: Ghil, M. (ed.) Turbulence and Predictability in Geophysical Fluid Dynamics and Climate Dynamics, course 88, pp. 84–88. North Holland, Amsterdam (1985)

[Gui90] Guivarch, Y.: Sur une extension de la notion de loi semi-stable. Ann. Inst. Henry Poincaré Probab. Stat. **26**, 261–285 (1990)

[HW92] Holley, R., Waymire, E.C.: Multifractal dimensions and scaling exponents for strongly bounded random fractals. Ann. Appl. Probab. **2**, 819–845 (1992)

[HS09] Hu, Y., Shi, Z.: Minimal position and critical martingale convergence in branching random walks, and directed polymers on disordered trees. Ann. Probab. **37**(2), 742–789 (2009)

[Jaf98] Jaffard, S.: Oscillations spaces: properties and applications to fractal and multifractal functions. J. Math. Phys. **39**(8), 4129–4141 (1998)

[Jaff99] Jaffard, S.: The multifractal nature of Lévy processes. Probab. Theory Relat. Fields **114**, 207–227 (1999)

[Jaf00a] Jaffard, S.: On lacunary wavelet series. Ann. Appl. Probab. **10**, 313–329 (2000)

[Jaf00] Jaffard, S.: On the Frisch-Parisi conjecture. J. Math. Pures Appl. **79**(6), 525–552 (2000)

[Jaf04] Jaffard, S.: Wavelets techniques in multifractal analysis. In: Lapidus, M., Frankenhuijsen, M.V. (eds.) Fractal Geometry and Applications: A Jubilee of Benoît Mandelbrot. Proceedings of Symposia in Pure Mathematics, vol. 72(2), pp. 91–151 (2004)

[Jin11] Jin, X.: The graph and range singularity spectra of $b$-adic independent cascade functions. Adv. Math. **226**, 4987–5017 (2011)

[Jin12] Jin, X.: Dimension result and KPZ formula for two-dimensional multiplicative cascade processes. Ann. Probab. **40**, 1–18 (2012)

[Jin14] Jin, X.: A uniform result for two dimensional fractional multiplicative processes. Ann. Inst. H. Poincaré Probab. Stat. **5**0(2), 512–523 (2014)

[JW11] Johnson, T., Waymire, E.: Tree polymers in the infinite volume limit at critical strong disorder. J. Appl. Probab. **48**, 885–891 (2011)

[Kah74] Kahane, J.-P.: Sur le modele de turbulence de Benoit Mandelbrot. C. R. Acad. Sci. Paris, **278**, 567–569 (1974)

[Kah85] Kahane, J.-P.: Sur le chaos multiplicatif. Ann. Sci. Math. Que. **9**, 105–150 (1985)

[Kah87a] Kahane, J.-P.: Positive martingales and random measures. Chi. Ann. Math. **8B1**, 1–12 (1987)

[KP76] Kahane, J.-P., Peyrière, J.: Sur certaines martingales de B. Mandelbrot Adv. Math. **22**, 131–145 (1976)

[Kah87] Kahane, J.-P.: Multiplications aléatoires et dimensions de Hausdorff. Ann. Inst. Henri Poincaré Probab. Stat. **23**, 289–296 (1987)

[Kah91] Kahane, J.-P.: Produits de poids aléatoires indépendants et applications. In: Fractal Geometry and Analysis (Montreal, PQ, 1989), pp. 277–324. NATO Adv. Sci. Inst. Ser. C Math. Phys. Sci., vol. 346. Kluwer, Dordrecht (1991)

[KPZ98] Knizhnik, V.G., Polyakov, A.M., Zamolodchikov, A.B.: Fractal structure of 2D-quantum gravity. Mod. Phys. Lett A **3**(8), 819–826 (1988)

[Kol62] Kolmogorov, A.N.: Précisions sur la structure locale de la turbulence dans un fluide visqueux aux nombres de Reynolds élevés, Mécanique de la turbulence, Colloq. Intern. CNRS, Marseille (1961) [Editions CNRS, pp. 447–451 (1962)]

[LRV00] Lacoin, H., Rhodes, R., Vargas, V.: Complex Gaussian multiplicative chaos. arXiv:1307.6117v1

[Liu98] Liu, Q.: Fixed points of a generalized smoothing transformation and applications to the branching random walk. Adv. Appl. Probab. **30**, 85–112 (1998)

[Liu00] Liu, Q.: On generalized multiplicative cascades. Stoch. Process. Appl. **86**(2), 263–286 (2000)

[Liu01] Liu, Q.: Asymptotic properties and absolute continuity of laws stable by random weightedmean. Stoch. Proc. Appl. **95**, 83–107 (2001)

[Mad00] Madaule, T.: Convergence in law for the branching random walk seen from its tip. arXiv:1107.2543

[Man72] Mandelbrot, B.B.: Possible refinement of the lognormal hypothesis concerning the distribution of energy in intermittent turbulence. Statistical models and turbulence. In: Rosenblatt, M., Atta, C.V. (eds.) Lectures Notes in Physics, vol. 12, 333–351. Springer, New York (1972)

[Man74a] Mandelbrot, B.B.: Multiplications aléatoires itérées et distributions invariantes par moyennes pondérées. C. R. Acad. Sci. Paris **278**, 289–292 and 355–358 (1974)

[Man74b] Mandelbrot, B.B.: Intermittent turbulence in self-similar cascades, divergence of high moments and dimension of the carrier. J. Fluid Mech. **62**, 331–358 (1974)

[Mat95] Mattila, P.: Geometry of sets and measures in Euclidean spaces, fractals and rectifiability. In: Cambridges Studies in Advanced Mathematics, vol. 44. Cambridge University Press, Cambridge (1995)

[Mol96] Molchan, G.M.: Scaling exponents and multifractal dimensions for independent random cascades. Commun. Math. Phys. **179**(3), 681–702 (1996)

[Ols94] Olsen, L.: Random geometrically graph directed self-similar multifractals, Pitman Res. Notes Math. Ser. **307** (1994)

[Ols95] Olsen, L.: A multifractal formalism. Adv. Math. **116**, 92–195 (1995)

[OW00] Ossiander M., Waymire, E.C.: Statistical estimation for multiplicative cascades. Ann. Stat. **28**, 1–29 (2000)

[Pey74] Peyrière, J.: Turbulence et dimension de Hausdorff. C. R. Acad. Sci. Paris **278**, 567–569 (1974)

[Pey77] Peyrière, J.: Calculs de dimensions de Hausdorff. Duke Math. J. **44**, 591–601 (1977)

[Pey79] Peyrière, J.: A singular random measure generated by spliting [0, 1]. Z. Wahrsch. Verw. Geb. **47**, 289–297 (1979)

[Pey00] Peyrière, J.: Recent results on Mandelbrot multiplicative cascades. In: Bandt, Ch., Graf, S., Zähle, M. (eds.) Fractal Geometry and Stochastics II, pp. 147–159. Progress in Probability, vol. 46. Birkhäuser, Basel (2000)

[RV10] Robert, R., Vargas, V.: Gaussian multiplicative chaos revisited. Ann. Probab. **38**(2), 605–631 (2010)

[RV08] Rhodes, R., Vargas, V.: KPZ formula for log-infinitely divisible multifractal random measures. arXiv:0807.1036

[RV13] Rhodes, R., Vargas, V.: Gaussian multiplicative chaos and applications: a review. arXiv:1305.6221

[She00] Sheffield, S.: Conformal weldings of random surfaces: SLE and the quantum gravity zipper. arXiv:1012.4797

[She72] Shepp, L.: Covering the line with random arcs. Z. Wahrsch. Verw. Geb. **23**, 163–170 (1972)

[WW69] Waymire, E.C., Williams, S.C.: A cascade decomposition theory with applications to Markov and exchangeable cascades. Trans. Am. Math. Soc. **348**, 585–632 (1996)

[WW13] Waymire, E.C., Williams, S.C.: $T$-martingales, size-biasing and tree polymer cascades. In: Barral, J., Seuret, S. (eds.) Further Developments in Fractals and Related Fields. Birkhäuser, Basel (2013)

[Web11] Webb, C.: Exact asymptotics of the freezing transitions of a logarithmically correlated random energy model. J. Stat. Phys. **145**, 1595–1619 (2011)

# Law of Pure Types and Some Exotic Spectra of Fractal Spectral Measures

**Xin-Rong Dai, Xing-Gang He and Chun-Kit Lai**

**Abstract** Let $\mu$ be a Borel probability measure with compact support in $\mathbb{R}^d$ and let $E(\Lambda) = \{e^{-2\pi\lambda\cdot x} : \lambda \in \Lambda\}$. We make a review on some recent progress about spectral measures. We first show that the law of pure types holds for spectral measures, i.e. if $E(\Lambda)$ is a frame for $L^2(\mu)$, then $\mu$ is discrete or absolutely continuous or singular continuous with respect to Lebesgue measure (see [HLL13]). And we discuss the spectral properties of Cantor measures (see [DaHL13]), where we focus on some exotic properties of the spectra of some Cantor measures.

**Keywords** Cantor measures fourier frames · Law of pure types · Spectral measures

**2010 Mathematics subject classification** 28A80 · 42C05

The research is partially supported by the RGC grant of Hong Kong and the Focused Investment Scheme of CUHK; The first two authors are also supported by the National Natural Science Foundation of 10871180 and 11271148, and Guangdong Province Key Laboratory of Computational Science at the Sun Yat-sen University.

X.-R. Dai
School of Mathematics and Computational Science, Sun Yat-Sen University,
Guangzhou 510275, P. R. China
e-mail: daixr@mail.sysu.edu.cn

X.-G. He (✉)
School of Mathematics and Statistics, Central China Normal University,
Wuhan 430079, P. R. China
e-mail: xingganghe@sina.com

C.-K. Lai
Department of Mathematics and Statistics, McMaster University, Hamilton,
ON L8S 4K1, Canada
e-mail: cklai@math.mcmaster.ca



D.-J. Feng and K.-S. Lau (eds.), *Geometry and Analysis of Fractals*,
Springer Proceedings in Mathematics & Statistics 88,
DOI 10.1007/978-3-662-43920-3_2

# 1 Introduction to the General Spectral Measures

Let $\mu$ be a Borel probability measure on $\mathbb{R}^d$ with compact support. We call a family $E(\Lambda) = \{e_\lambda := e^{-2\pi i\lambda\cdot x} : \lambda \in \Lambda\}$ ($\Lambda$ is a countable set) a *Fourier frame* for the Hilbert space $L^2(\mu)$ if there exist $A, B > 0$ such that

$$A\|f\|^2 \le \sum_{\lambda\in\Lambda} |\langle f, e_\lambda\rangle_\mu|^2 \le B\|f\|^2, \quad \forall\, f \in L^2(\mu). \tag{1.1}$$

Here the inner product is defined as usual, $x \cdot y = \sum_{i=1}^d x_i y_i$ for $x, y \in \mathbb{R}^d$ and

$$\langle f, e_\lambda\rangle_\mu = \int_{\mathbb{R}^d} f(x)e_{-\lambda}d\mu(x).$$

$E(\Lambda)$ is called an *(exponential) Riesz basis* if it is both a basis and a frame for $L^2(\mu)$. Fourier frames and exponential Riesz bases are natural generalizations of exponential orthonromal bases in $L^2(\mu)$. They have fundamental importance in non-harmonic Fourier analysis and wavelet. When (1.1) is satisfied, $f \in L^2(\mu)$ can be expressed as $f(x) = \sum_{\lambda\in\Lambda} c_\lambda e^{2\pi i\lambda x}$, and the expression is unique if it is a Riesz basis.

When $E(\Lambda)$ is an orthonormal basis (Riesz basis, or frame) for $L^2(\mu)$, we say that $\mu$ is a *spectral measure* (*R-spectral measure, or F-spectral measure respectively*) and $\Lambda$ is called a *spectrum* (*R-spectrum, or F-spectrum respectively*) of $\mu$. We will also use the term *orthonormal spectrum* instead of spectrum when we need to emphasize the orthonormal property. If $E(\Lambda)$ only satisfies the upper bound condition in (1.1), then it is called a *Bessel sequence*; for convenience, we also call $\Lambda$ a Bessel sequence of $L^2(\mu)$.

Since Fuglede proposed the spectral set conjecture [Fug74] and Jorgensen and Pedersen [JP98] discovered the first singular fractal spectral measure, there has been a lot of interest in understanding which kind of measures are spectral and its delicate connection with translational tiling. In this short note, we aim at giving a systematic survey on the recent progress in this line of research and some more detailed explanations about our discovery in [HLL13, DaHL13] will be given.

The first fundamental result is about the law of pure type. It was proved by He et al. [HLL13], which generalized the early investigation of spectral measures by Łaba and Wang [LaW06]. Recall that a $\sigma$-finite Borel measure $\mu$ on $\mathbb{R}^d$ can be decomposed uniquely as *discrete, singularly continuous* and *absolutely continuous* measures, i.e., $\mu = \mu_d + \mu_s + \mu_a$. The measure $\mu$ is said to be of *pure types* if $\mu$ equals only one of the three components.

**Theorem 1.1** *Let $\mu$ be an F-spectral measure on $\mathbb{R}^d$. Then it must be one of the three pure types: discrete (and finite), singularly continuous or absolutely continuous.*

By the law of pure types, we can study spectral measures according to its type. When $\mu$ is a discrete counting measure with finite support, it is an R-spectral measure

[HLL13]; When $\mu$ is absolutely continuous, Lai proved that $\mu$ is an F-spectral measure if and only if its density function is bounded above and bounded away from 0 almost everywhere on its support [Lai11]. Furthermore, Dutkay and Lai proved that if $\mu$ is a spectral measure, then its density function is a constant on its support, that is, $\mu$ is essentially the Lebesgue measure restricted on its support [DL00]. However, classification of spectral measures is far from complete. For the study of R-spectral absolutely continuous measures, one can refer to the recent work of Lev et al. with their emphasis on the use of quasicrystals [KN00, Lev12, GL14].

From now on, we concentrate on orthogonally spectral measures. We call $\Lambda$ an *orthogonal set* if $E(\Lambda)$ is a mutually orthogonal sequence for $L^2(\mu)$. Define

$$Q_\Lambda(\xi) = \sum_{\lambda\in\Lambda} |\widehat{\mu}(\xi+\lambda)|^2,$$

where the Fourier transform of $\mu$ is define as usually by

$$\widehat{\mu}(\xi) = \int_{\mathbb{R}^d} e^{-2\pi\xi\cdot x} d\mu(x).$$

$Q_\Lambda$ is crucial in determining whether $E(\Lambda)$ is complete. It is well-known that an orthogonal sequence $E(\Lambda)$ is complete in $L^2(\mu)$ if and only if $Q_\Lambda \equiv 1$ [JP98]. Here, we give a slight generalization of this result and also exploit the analytic property of $Q_\Lambda$.

**Theorem 1.2** *Let $\mu$ be a compactly supported Borel probability measure with compact support in $\mathbb{R}^d$.*

(i) Suppose that $\overline{\text{span}} E(\Gamma) = L^2(\mu)$ and $E(\Lambda)$ is an orthogonal set for $L^2(\mu)$. Then $\Lambda$ is a spectrum of $\mu$ if and only if

$$Q_\Lambda(\gamma) = 1, \quad \text{for } \gamma \in \Gamma.$$

(ii) Suppose that $E(\Lambda)$ is a Bessel sequence for $L^2(\mu)$. Then $Q_\Lambda(\cdot)$ is an entire function in $\mathbb{C}^d$.

The entire property is a simple extension of [JP98, Lemma 4.3]. In our proofs, this property helps us establish the completeness by allowing us to focus on small values of $\xi$.

Our main interest on the spectral measures is when $\mu$ is singularly continuous. The one-fourth contraction Cantor measure was the first example of such spectral measures, which was found by Jorgensen and Pedersen [JP98] in 1998. From that time on, various properties of singular spectral measures are studied extensively [Dai12, DHJ09, DHS09, DHSW11, LaW02, LaW06]. In particular, many exotic spectra were discovered and they do not appear in their absolutely continuous counterpart. Here, we list some of the interesting ones.

(1) There exists a spectrum $\Lambda$ of a singularly continuous measure $\mu$ such that $k\Lambda$ is also a spectrum of $\mu$ for some $k \neq 1$;
(2) There exists a $\Lambda$ so that $E(\Lambda)$ is a maximal orthogonal collection of exponentials for $L^2(\mu)$, but not a basis;
(3) There exists a spectrum $\Lambda$ of a singularly continuous measure so that its Beurling dimension is zero.

Property (1) means that we can sort of dilate a spectrum but preserve its completeness. It was first given by Łaba and Wang [LaW02] and some studies are given in [DJ12].

Property (2) has two types of variants. First, some measures have maximal orthogonal collections of infinite cardinality without being spectral [HuL08, Dai12]. Second, even though the measure is spectral, there still exists some incomplete maximal orthogonal collections. In [DHS09], Dutkay et al. tried to give a classification on maximal orthogonal collection for one-fourth Cantor measures and tried to study which of them are complete. This investigation was generalized and improved in [DaHL13]. Furthermore, we can demonstrate the existence of spectrum satisfying property (3). Beurling dimension is a concept defined in [DHSW11], who tried to generalize Beurling density and the elegant result of Landau [Lan67] on Fourier frame spectra to fractal setting. Their work gave some partial positive results, letting alone a technical assumption on the spectra. In person communication with Wang in 2011, we were told that he can construct an example such that a spectral measure can have a spectrum with zero Beurling dimension. However, he cannot explain why there can be such phenomenon. Our construction gave a better picture of it.

For the rest of our paper, we will prove Theorem 1.1 and 1.2 in Sect. 2. In Sect. 3, we will present a simplified content of [DaHL13] and the examples of zero Beurling dimension spectra will be given. For more results on this issue, reader may refer to [DaHL13, DHS09].

## 2 Law of Pure Types

In this section, we will present a self-contained proof for the law of pure types of $F$-spectral measures. First, we need the following proposition, which was proved in [DHSW11]. This can be viewed as the stability of Bessel sequence under a constant perturbation of a Bessel sequence. It has its origin in the paper of Duffin and Schaeffer [DS52].

**Proposition 2.1** *Let $\{\lambda_n\}_{n=0}^{\infty}$ be a Bessel sequence of $\mu$. If there exists $C$ such that $|\lambda_n - \gamma_n| \le C$ for $n \ge 0$, then $\{\gamma_n\}_{n=0}^{\infty}$ is also a Bessel sequence of $\mu$.*

*Proof* It is sufficient to show that all $\gamma_n = (\gamma_1^{(n)}, \dots, \gamma_d^{(n)})$ differs $\lambda_n = (\lambda_1^{(n)}, \dots, \lambda_d^{(n)})$ only on the first component, and the statement follows by induction on the number of components.

Let $\text{supp}\mu \subseteq [-P, P]^d$ for some $P > 0$. We have that

$$\begin{aligned}
\sum_{n=0}^{\infty} \left| < f(x), e^{-2\pi i \gamma_n \cdot x} > \right|^2 &= \sum_{n=0}^{\infty} \left| < f(x) e^{2\pi i (\gamma_n - \lambda_n)\cdot x}, e^{-2\pi i \lambda_n \cdot x} > \right|^2 \\
&= \sum_{n=0}^{\infty} \left| < f(x) e^{2\pi i (\gamma_1^{(n)} - \lambda_1^{(n)}) x_1}, e^{-2\pi i \lambda_n \cdot x} > \right|^2 \\
&= \sum_{n=0}^{\infty} \left| \sum_{k=0}^{\infty} \frac{(2\pi i (\gamma_1^{(n)} - \lambda_1^{(n)}))^k}{k!} < f(x) x_1^k, e^{-2\pi i \lambda_n \cdot x} > \right|^2 \\
&\le \sum_{n=0}^{\infty} \sum_{k=0}^{\infty} \frac{(2\pi C)^{2k}}{k!} \sum_{k=0}^{\infty} \frac{| < f(x) x_1^k, e^{-2\pi i \lambda_n \cdot x} > |^2}{k!} \\
&\le e^{(2\pi C)^2} \sum_{k=0}^{\infty} \frac{B \| f(x) x_1^k \|^2}{k!} \\
&\le B e^{(2\pi C)^2 + P^2} \| f \|^2.
\end{aligned}$$

Note that the fourth line above uses Cauchy-Schwarz inequality. Hence, the assertion follows. □

In the proof of the pure type property of the F-spectral measures, we need to use the *lower Beurling density* of an infinite discrete set $\Lambda \subset \mathbb{R}^d$:

$$D^-\Lambda := \liminf_{h\to\infty} \inf_{x\in\mathbb{R}^d} \frac{\#(\Lambda \cap Q_h(x))}{h^d},$$

where $Q_h(x)$ is the standard cube of side length $h$ centered at $x$. Intuitively $\Lambda$ is distributed like a lattice if $D^-\Lambda$ is positive. In the seminal paper [Lan67], Landau gave an elegant and useful necessary condition for $\Lambda$ to be an F-spectrum on $L^2(\Omega)$: $D^-\Lambda \ge \mathcal{L}(\Omega)$, where $\mathcal{L}$ is the Lebesgue measure. The following proposition provides some relationships between the lower Beurling density and the types of the measures.

**Proposition 2.2** *Let $\mu$ be a compactly supported probability measure on $\mathbb{R}^d$ and let $\Lambda$ be an F-spectrum of $\mu$. We have*

(i) *If $\mu = \sum_{c\in C} p_c \delta_c$ is discrete, then $\#\Lambda < \infty$ and $\#C < \infty$;*
(ii) *If $\mu$ is singularly continuous, then $D^-\Lambda = 0$;*
(iii) *If $\mu$ is absolutely continuous, then $D^-\Lambda > 0$.*

*Proof* (i) By the definition of Fourier frame, we have for all $f \in L^2(\mu)$,

$$\sum_{\lambda\in\Lambda} |\sum_{c\in C} f(c) e^{2\pi i \langle \lambda, c\rangle} p_c|^2 \le B \sum_{c\in C} |f(c)|^2 p_c.$$

Taking $f = \chi_{c_0}$, where $\chi_{c_0}$ is the indictor function of the set $\{c_0\}$ and $p_{c_0} > 0$, we have $(\#\Lambda) \cdot p_{c_0}^2 \leq B p_{c_0}$. Hence $\#\Lambda \leq B/p_{c_0} < \infty$. This implies $\#\mathcal{C} < \infty$ by the completeness of Fourier frame.

(ii) Suppose on the contrary that $D^-\Lambda \geq c > 0$. We claim that $\mathbb{Z}^d$ is a Bessel sequence of $L^2(\mu)$. By the definition of $D^-\Lambda$, we can choose a large $h \in \mathbb{N}$ such that

$$\inf_{x \in \mathbb{R}^d} (\#(\Lambda \cap Q_h(x))) \geq ch^d > 1.$$

Taking $x = h\mathbf{n}$, where $\mathbf{n} \in \mathbb{Z}^d$, we see that all cubes of the form $h\mathbf{n} + [-h/2, h/2)^d$ contains at least one points of $\Lambda$, say $\lambda_{\mathbf{n}}$. Since $\Lambda$ is an F-spectrum, $\{\lambda_{\mathbf{n}}\}_{\mathbf{n} \in \mathbb{Z}^d}$ is a Bessel sequence. Observing that

$$|\lambda_{\mathbf{n}} - h\mathbf{n}| \leq \operatorname{diam}([-h/2, h/2)^d) = \sqrt{d}\, h,$$

then $h\mathbb{Z}^d$ is also a Bessel sequence of $L^2(\mu)$ by Proposition 2.1. As a Bessel sequence is invariant under translation, we see that the finite union $\mathbb{Z}^d = \bigcup_{\mathbf{k} \in \{0,\ldots,h-1\}^d} (h\mathbb{Z}^d + \mathbf{k})$ is again a Bessel sequence of $L^2(\mu)$, which proves the claim.

Now consider

$$G(x) := \sum_{\mathbf{n} \in \mathbb{Z}^d} |\widehat{\mu}(x + \mathbf{n})|^2.$$

$G$ is a periodic function (mod $\mathbb{Z}^d$). As $\mathbb{Z}^d$ is a Bessel sequence, applying the definition to $e_{-x}$, we see that $G(x) \leq B < \infty$. Hence $G \in L^1([0,1)^d)$ and

$$\int_{\mathbb{R}^d} |\widehat{\mu}(x)|^2\, dx = \sum_{\mathbf{n} \in \mathbb{Z}^d} \int_{[0,1)^d} |\widehat{\mu}(x + \mathbf{n})|^2 dx = \int_{[0,1)^d} |G(x)| dx < \infty.$$

This means that $\widehat{\mu} \in L^2(\mathbb{R}^d)$, which implies that $\mu$ must be absolutely continuous. This is a contradiction.

(iii) If $\mu$ is absolutely continuous, we write $d\mu(x) = \varphi(x)dx$, for some $L^1$ function $\varphi$ and denote by $\Omega$ the support of $\mu$. Let

$$E_N = \left\{ x \in \Omega : \frac{1}{N} \leq \varphi(x) \leq N \right\}.$$

Since $\mu$ is absolutely continuous, the support $\Omega$ must have positive Lebesgue measure and $E_N$ also has positive Lebesgue measure for $N$ large, which we may assume it holds for all $E_N$. Now, we claim that $\Lambda$ is an F-spectrum of $L^2(E_N)$. To see this, let $f \in L^2(E_N)$, then we have $\int_{E_N} |\frac{f(x)}{\varphi(x)}|^2 \varphi(x) dx \leq N \int_{E_N} |f|^2 < \infty$. Hence,

$$\sum_{\lambda\in\Lambda}\left|\int_{E_N} f(x)e^{2\pi i\lambda x}dx\right|^2 = \sum_{\lambda\in\Lambda}\left|\int_{E_N}\frac{f(x)}{\varphi(x)}e^{2\pi i\lambda x}\varphi(x)dx\right|^2$$

$$\leq B\int_{E_N}\left|\frac{f(x)}{\varphi(x)}\right|^2\varphi(x)dx \leq BN\int_{E_N}|f(x)|^2dx.$$

This establishes the upper frame bound. The lower bound can also be established analogously. This justify the claim. By the Landau's density theorem, we have $D^-\Lambda \geq \mathcal{L}(E_N)$. As $E_N$ are increasing sequence of sets and $\bigcup_N E_N = \Omega$ up to a Lebesgue measure zero set, we have

$$D^-\Lambda \geq \mathcal{L}(\Omega) > 0.$$

□

Now it is easy to conclude that an $F$-spectral measure is of pure type.

**Proof of Theorem 1.1** First let us assume that if $\mu$ is decomposed into non-trivial discrete and continuous parts, $\mu = \mu_d + \mu_c$. Let $\Lambda$ be an F-spectrum of $\mu$. As $L^2(\mu_d)$ and $L^2(\mu_s)$ are non-trivial subspaces of $L^2(\mu)$, it is easy to see that $\Lambda$ is also an F-spectrum of both $L^2(\mu_d)$ and $L^2(\mu_c)$. Then $\#\Lambda < \infty$ by Proposition 2.2(i); but $\#\Lambda = \infty$ since $L^2(\mu_c)$ is an infinite dimensional Hilbert space. This contradiction shows that $\mu$ is either discrete or purely continuous.

Suppose $\mu$ is continuous and has non-trivial singular part $\mu_s$ and absolutely continuous part $\mu_a$. By applying the same argument as the above, $\Lambda$ is an F-spectrum of $L^2(\mu_s)$ and $L^2(\mu_a)$. This is impossible in view of the Beurling density of $\Lambda$ in Proposition 2.2(ii) and (iii). □

The following corollary is immediate from Theorem 1.1.

**Corollary 2.3** *A spectral measure or an R-spectral measure must be of pure type.*

In the rest of this section, we will prove Theorem 1.2 and it will be needed in the next section.

**Proof of Theorem 1.2** (i) It is easy to see that the necessity follows by applying Parseval's identity to $e_\gamma$ for $\gamma \in \Gamma$. Now we show the sufficiency. By the hypotheses, it is sufficient to show that $e_\gamma \in \overline{\text{span}}E(\Lambda)$ for each $\gamma \in \Gamma$. Let $\Pi$ be the projection from $L^2(\mu)$ to $\overline{\text{span}}E(\Lambda)$. Then $e_\gamma = \Pi(e_\gamma) + (Id - \Pi)(e_\gamma)$ and thus $1 = \|\Pi(e_\gamma)\|^2 + \|(Id - \Pi)(e_\gamma)\|^2$. Note that

$$\|\Pi(e_\gamma)\|^2 = \sum_{\lambda\in\Lambda}|\langle\Pi(e_\gamma), e_\lambda\rangle|^2 = \sum_{\lambda\in\Lambda}|\langle e_\gamma, e_\lambda\rangle|^2 = \|e_\gamma\|^2 = 1.$$

Then $(Id - \Pi)(e_\gamma) = 0$ and thus $e_\gamma \in \overline{\text{span}}E(\Lambda)$.

(ii) Let $M > 0$ so that $\text{supp}\mu \subseteq B(0, M)$, where $B(0, M)$ is the ball with center at 0 and radius $M$. Denote $\Lambda = \{\lambda_n\}_{n=0}^{\infty}$ and

$$Q_N(w) = \sum_{n=0}^{N}|\widehat{\mu}(w + \lambda_n)|^2, \quad \forall w \in \mathbb{C}^d.$$

Let $B$ be the upper bound of $E(\Lambda)$. Note that

$$\mathcal{Q}_N(w) = \sum_{n=0}^{N} \left|< e_{-w}, e_{\lambda_n} >\right|^2 \le B\|e_{-w}\|^2 \le Be^{4\pi M|\mathcal{I}m(w)|},$$

where $\mathcal{I}m(w)$ is the imaginary part of $w$. This implies that the sequence $\{\mathcal{Q}_N(w)\}_{N=1}^{\infty}$ is uniformly bounded on each compact set of $\mathbb{C}^d$. By Montel theorem (see, e.g., [Gun90] p. 54), we have $\mathcal{Q}_\Lambda$ is an entire function on $\mathbb{C}^d$ and

$$|\mathcal{Q}_\Lambda(w)| \le Be^{4\pi M|\mathcal{I}m(w)|}, \quad \forall w \in \mathbb{C}^d.$$

□

Now the standard Jorgensen-Pedersen Lemma follows as a corollary.

**Corollary 2.4** *An orthogonal sequence $E(\Lambda)$ is complete in $L^2(\mu)$ if and only if $\mathcal{Q}_\Lambda \equiv 1$.*

*Proof* We only need to show that span$E(\mathbb{R}^d)$ is dense in $L^2(\mu)$ by Theorem 1.2. Let $K = \operatorname{supp} \mu$. Since span$E(\mathbb{R}^d)$ is a subalgebra of Banach algebra $C(K)$, the space of all continuous function on $K$, and it separates points $K$. By Stone-Weierstrass theorem, we have that span$E(\mathbb{R}^d)$ is dense in the space $C(K)$. According to Lusin theorem, $C(K)$ is dense in $L^2(\mu)$. This implies the assertion. □

## 3 Spectral Properties of Cantor Measures on $\mathbb{R}$

This section is devoted to a simplified content of [DaHL13]. Our aim is to show the existence of spectra with zero Beurling dimension (Theorem 3.5) when the measures are the Cantor measure with consecutive digits. Let $b, q$ be two integers $> 1$ with $b > q$ and $q \mid b$. Then there exists unique Borel probability measure, denoted by $\mu_{b,q}$, satisfying

$$\mu_{b,q}(\cdot) = \frac{1}{q}\sum_{i=0}^{q-1} \mu_{b,q}(q \cdot -i). \tag{3.1}$$

$\mu_{b,q}$ is called a *Cantor measure* (with consecutive digit). It is well-known that the Hausdorff dimension of the support of $\mu_{b,q}$ is $\ln q/\ln b < 1$ and thus $\mu_{b,q}$ is singularly continuous with respect to Lebesgue measure. We will construct a class of orthogonal set of $\mu_{b,q}$.

Denote $\Sigma_q = \{0, \cdots, q-1\}$, $\Sigma_q^0 = \{\vartheta\}$ and $\Sigma_q^n = \underbrace{\Sigma_q \times \cdots \times \Sigma_q}_{n}$. Let $\Sigma_q^* = \bigcup_{n=0}^{\infty} \Sigma_q^n$ be the set of all finite words. Given $\sigma = \sigma_1\sigma_2\cdots \in \Sigma^*$, we define $\vartheta\sigma = \sigma$, $\sigma|_k = \sigma_1\cdots\sigma_k$ for $k \ge 0$ where $\sigma|_0 = \vartheta$ for any $\sigma$ and adopt the notation $0^k = \underbrace{0\cdots 0}_{k}$ and $\sigma\sigma'$ is the concatenation of $\sigma$ and $\sigma'$. We start with two definitions.

**Definition 3.1** Let $\Sigma_q^*$ be all the finite words defined as above. We say it is a *$q-$adic tree* if we set naturally the root is $\vartheta$, all the $k$-th level nodes are $\Sigma_q^k$ for $k \geq 1$ and all the offsprings of $\sigma \in \Sigma_q^*$ are $\sigma i$ for $i = 0, 1, \ldots, q-1$.

**Definition 3.2** Let $\Sigma_q^*$ be a $q-$adic tree, $\tau$ is called a *regular mapping* from $\Sigma_q^*$ to $\{-1, 0, ..., b-2\}$ if it satisfies

(i) $\tau(\vartheta) = \tau(0^n) = 0$ for all $n \geq 1$.
(ii) For $\sigma_1 \cdots \sigma_k \in \Sigma_q^k$, $\tau(\sigma_1 \cdots \sigma_k) \in (\sigma_k + q\mathbb{Z}) \cap \{-1, 0, ..., b-2\}$.
(iii) For any $\sigma \in \Sigma_q^*$, $\tau(\sigma 0^\ell) = 0$ for $\ell$ large enough.

Let $\tau$ be a regular mapping from $\Sigma_q^*$ to $\{-1, 0, ..., b-2\}$. For any $n \in \mathbb{N}$ with $q^{N-1} \leq n < q^N$, there exists unique $\sigma = \sigma_1 \cdots \sigma_N \in \Sigma_q^N$ such that $\sigma_N \neq 0$ and

$$n = \sigma_1 + \sigma_2 q + \cdots + \sigma_N q^{N-1}.$$

Associated to $\tau$, we define a sequence of integers by $\lambda_0 = 0$ and

$$\lambda_n = \tau(\sigma|_1) + \tau(\sigma|_2)b + \cdots + \tau(\sigma|_N)b^{N-1} + \sum_{k=N}^{\infty} \tau(\sigma 0^{k-N+1})b^k.$$

Note that $\lambda_n$ is uniquely determined by $\tau(\sigma|_1), \tau(\sigma|_2), \ldots, \tau(\sigma|_N) = \tau(\sigma)$. We call $\Lambda = \{\lambda_n\}_{n=0}^\infty$ a *$\tau$-sequence*. Let $\ell_n$ be the number of nonzero terms in the sum $\sum_{k=N}^\infty \tau(\sigma 0^{k-N+1})b^k$, that is

$$\ell_n = \#\{k : \tau(\sigma 0^k) \neq 0 \text{ for } k \geq 1\}. \tag{3.2}$$

We assume that $b, q, r = b/q$ are integers with $b > q$. The following are our main theorems.

**Theorem 3.3** *Let $\tau$ be a regular mapping from $\Sigma_q^*$ to $\{-1, 0, ..., b-2\}$ and let $\Lambda = \{\lambda_n\}_{n=0}^\infty$ be the $\tau$-sequence. Then $E(r\Lambda)$ is a maximal orthogonal collection of exponentials for $L^2(\mu_{b,q})$.*

**Theorem 3.4** *Let $\tau$ be a regular mapping and let $\Lambda = \{\lambda_n\}_{n=0}^\infty$ be the $\tau$-sequence. We have the following:*

(i) *If $\max_{n\geq 1}\{\ell_n\} < \infty$, then $r\Lambda$ is a spectrum of $\mu_{b,q}$;*
(ii) *If $\ell_n \geq \log_q n$ for sufficient large $n$, then $r\Lambda$ is not a spectrum of $\mu_{b,q}$.*

**Theorem 3.5** *Let $g(x)$ be an increasing non-negative function on $[0, \infty)$. Then there exists a spectrum $\Lambda$ of $L^2(\mu_{b,q})$ such that*

$$\lim_{R\to\infty} \sup_{x\in\mathbb{R}} \frac{\#(\Lambda \cap (x-R, x+R))}{g(R)} = 0. \tag{3.3}$$

Let $\mu_{b,q}$ be the Cantor measure given by (3.1) and let

$$M(\xi) = \frac{1}{q}(1 + e^{2\pi i\xi} + \cdots + e^{2\pi i(q-1)\xi}).$$

Then it is easy to obtain that

$$\widehat{\mu}_{b,q}(\xi) = M(b^{-1}\xi)\widehat{\mu}_{b,q}(b^{-1}\xi) = \prod_{k=1}^{\infty} M(b^{-k}\xi). \tag{3.4}$$

Note that $|M(\xi)| = |\sin q\pi\xi|/q|\sin \pi\xi|$. Then

$$\mathcal{Z}_M := \{\xi : M(\xi) = 0\} = \frac{1}{q}(\mathbb{Z} \setminus q\mathbb{Z})$$

and

$$\mathcal{Z}_\mu := \{\xi : \widehat{\mu}_{b,q}(\xi) = 0\} = r\{b^k a : k \geq 0, a \in \mathbb{Z} \setminus q\mathbb{Z}\}.$$

Clearly, $\Theta$ is an orthogonal set of $\mu_{b,q}$ if and only if

$$\Theta - \Theta \subseteq \mathcal{Z}_\mu \cup \{0\}. \tag{3.5}$$

**Proof of Theorem 3.3** We first prove the orthogonal property of $E(r\Lambda)$. Denote $\lambda_{n'} = \tau(\sigma'|_1) + \tau(\sigma'|_2)b + \cdots + \tau(\sigma'|_{N'})b^{N'-1} + \sum_{k=N'}^{\infty} \tau(\sigma 0^{k-N'+1})b^k$ and $n \neq n'$. Let $s$ be the smallest index such that $\tau(\sigma' 0^{|\sigma|}|_s) \neq \tau(\sigma 0^{|\sigma'|}|_s)$, where $|\sigma|$ is the length of $\sigma$. Then

$$\lambda_{n'} - \lambda_n = (\tau(\sigma'|_s) - \tau(\sigma|_s))b^s + b^{s+1}M$$

for some $M \in \mathbb{Z}$. Then $r(\lambda_{n'} - \lambda_n)$ is the zero point of $M(b^{-s+1}\xi)$ by the definition of $\tau$ and thus is a zero point of $\widehat{\mu}_{b,q}$ by (3.4). This implies that $r\Lambda$ is an orthogonal set of $\mu_{b,q}$.

Now we show the maximal property of $E(r\Lambda)$. Suppose that $r\Lambda \cup \{\gamma\}$ is an orthogonal set of $\mu_{b,q}$ with $\gamma \notin r\Lambda$. By (3.5) and $0 \in r\Lambda$, we have $\gamma = rb^k a$ for some $k \geq 0$ and $a \in \mathbb{Z} \setminus q\mathbb{Z}$. Since $a$ can be expressed uniquely as

$$a = a_0 + a_1 b + \cdots + a_m b^m,$$

where all $a_i \in \{-1, 0, 1, \ldots, b-1\}$, $a_m \neq 0$ and $a_0 \in \mathbb{Z} \setminus q\mathbb{Z}$, there exists unique $i_0 \in \{1, 2, \ldots, q-1\}$ such that $a_0 - \tau(0^k i_0) \in q\mathbb{Z}$. By the assumption we have $ab^k - \lambda_{i_0 q^k} \in \mathcal{Z}_\mu$, that is,

$$\frac{ab^k - \lambda_{i_0 q^k}}{b^k} = a_0 - \tau(0^k i_0) + \sum_{s=1}^{m} (a_i - \tau(0^k i_0 0^s)) b^s - \sum_{s=m+1}^{\infty} \tau(0^k i_0 0^s) b^s \in \mathcal{Z}_\mu.$$

Since $q \mid (a_0 - \tau(0^k i_0))$, but $b \nmid (a_0 - \tau(0^k i_0))$ if $a_0 \neq \tau(0^k i_0)$, one has $a_0 = \tau(0^k i_0)$. Similarly, there exists unique $i_1 \in \{0, 1, \ldots, q-1\}$ such that $a_1 - \tau(0^k i_0 i_1) \in q\mathbb{Z}$. From $ab^k - \lambda_{i_0 q^k + i_1 q^{k+1}} \in \mathcal{Z}_\mu$, one has $a_1 = \tau(0^k i_0 i_1)$. By $m$-steps one has $a_s = \tau(0^k i_0 \cdots i_s)$ for $0 \leq s \leq m$.

Let $p = \sum_{s=0}^{m} i_s q^{k+s}$. We claim that $\gamma = r\lambda_p$ and the result follows if the claim holds. In fact,

$$\begin{aligned} \frac{ab^k - \lambda_p}{b^k} &= \sum_{s=0}^{m} (a_s - \tau(0^k i_0 \cdots i_s)) b^s - \sum_{s=m+1}^{\infty} \tau(0^k i_0 \cdots i_m 0^{s-m}) b^s \\ &= -\sum_{s=m+1}^{\infty} \tau(0^k i_0 \cdots i_m 0^{s-m}) b^s. \end{aligned}$$

If $ab^k \neq \lambda_p$, the above implies that $ab^k - \lambda_p \notin \mathcal{Z}_\mu$, which contradicts to the assumption. Hence the claim follows. □

Let $\delta_a$ be the Dirac measure with center $a$. We define

$$\delta_{\mathcal{E}} = \frac{1}{\#\mathcal{E}} \sum_{e \in \mathcal{E}} \delta_e$$

for any finite set $\mathcal{E}$, where $\#\mathcal{E}$ is the cardinality of $\mathcal{E}$. Write $\mathcal{D} = \{0, 1, \ldots, q-1\}$ and $D_N = \frac{1}{b}\mathcal{D} + \cdots + \frac{1}{b^N}\mathcal{D}$ for $N \geq 1$. Let $\mu_N = \delta_{\mathcal{D}_N}$. Then

$$\widehat{\mu_N}(\xi) = \prod_{j=1}^{N} M(b^{-j}\xi).$$

By (3.4) we have

$$\widehat{\mu}_{b,q}(\xi) = \widehat{\mu_N}(\xi) \widehat{\mu}_{b,q}\left(\frac{\xi}{b^N}\right). \tag{3.6}$$

**Lemma 3.6** *Let $\tau$ be a regular mapping and let $\{\lambda_n\}_{n=0}^{\infty}$ be the $\tau$-sequence. Then for all $N \geq 1$,*

$$\sum_{n=0}^{q^N - 1} |\widehat{\mu_N}(\xi + r\lambda_n)|^2 \equiv 1. \tag{3.7}$$

*Proof* Since the dimension of $L^2(\mu_N)$ is $q^N$, the assertion follows by Corollary 2.4 if $\{r\lambda_n\}_{n=0}^{q^N-1}$ is an orthogonal set of $\mu_N$, which can be proved by the same proof of Theorem 3.3. □

For $m \geq 1$, let

$$Q_m(\xi) = \sum_{n=0}^{q^m-1} \left|\widehat{\mu}_{b,q}(\xi + r\lambda_n)\right|^2 \text{ and } Q(\xi) = \sum_{n=0}^{\infty} \left|\widehat{\mu}_{b,q}(\xi + r\lambda_n)\right|^2.$$

Let $\mu = \mu_{b,q}$. For any $m, p > 0$, we have the following identity:

$$\begin{aligned} Q_{m+p}(\xi) &= Q_m(\xi) + \sum_{n=q^m}^{q^{m+p}-1} |\widehat{\mu}(\xi + r\lambda_n)|^2 \\ &= Q_m(\xi) + \sum_{n=q^m}^{q^{m+p}-1} \left|\widehat{\mu_{m+p}}(\xi + r\lambda_n)\right|^2 \left|\widehat{\mu}(\frac{\xi + r\lambda_n}{b^{m+p}})\right|^2. \end{aligned} \tag{3.8}$$

Our goal is see whether $Q(\xi) \equiv 1$. Then by invoking Corollary 2.4, we can determine whether we have a spectrum. As $Q$ is an entire function by Theorem 1.2(ii), we just need to see the value of $Q(\xi)$ for some small values of $\xi$. To do this, we need to make a fine estimation of the terms $\left|\widehat{\mu}(\frac{\xi+r\lambda_n}{b^{m+p}})\right|^2$ in the above. Write

$$\alpha = \min\left\{|M(\xi)\widehat{\mu}(\xi)|^2 : |\xi| \leq \frac{b-1}{qb}\right\} > 0$$

and

$$\beta = \max\left\{|M(\xi)|^2 : \frac{1}{b^2} \leq |\xi| \leq \frac{b-1}{qb}\right\} < 1.$$

where $|M(\xi)| = \frac{|\sin \pi q\xi|}{q|\sin \pi\xi|}$.

**Proposition 3.7** *Let* $|\xi| \leq \frac{r(b-2)}{b-1}$ *and let* $t = \xi + \sum_{k=1}^{N} d_i b^{n_k}$, *where* $d_i \in \{1, 2, \cdots r-1\}$ *and* $1 \leq n_1 < \cdots < n_N$. *Then*

$$\alpha^{N+1} \leq |\widehat{\mu}(t)|^2 \leq \beta^N. \tag{3.9}$$

*Proof* First it is easy to check that, for $|\xi| \leq \frac{r(b-2)}{b-1}$ and all $d_k \in \{0, 1, 2, \ldots, r-1\}$, we have

$$\left|\frac{\xi+\sum_{k=1}^{n}d_kb^k}{b^{n+1}}\right| \le \frac{1}{b^{n+1}}\left(\frac{r(b-2)}{b-1}+(r-1)(b+b^2+\cdots+b^n)\right)$$
$$= \frac{r(b-2)+(r-1)(b^{n+1}-b)}{b^{n+1}(b-1)}$$
$$\le \frac{b-1}{qb} \tag{3.10}$$

for $n \ge 1$. The inequality in the last line follows from a direct comparison of the difference and $q \ge 2$. To simplify notations, we let $n_0 = 0$ and $n_{N+1} = \infty$. Then $|\widehat{\mu}(t)|^2$ equals

$$\prod_{j=1}^{\infty}\left|M\left(b^{-j}t\right)\right|^2 = \prod_{i=0}^{N}\prod_{j=n_i+1}^{n_{i+1}}\left|M\left(b^{-j}t\right)\right|^2. \tag{3.11}$$

We now estimate the products one by one. By (3.10), we have

$$\left|\frac{\xi+\sum_{k=1}^{i}d_kb^{n_k}}{b^{n_i+1}}\right| \le \frac{b-1}{qb}.$$

Hence, together with the integral periodicity of $M(\xi)$ and the definition of $\alpha$, we have for all $i > 0$,

$$\prod_{j=n_i+1}^{n_{i+1}}\left|M(b^{-j}t)\right|^2 = \prod_{j=n_i+1}^{n_{i+1}}\left|M\left(b^{-j}(\xi+\sum_{k=1}^{i}d_kb^{n_k})\right)\right|^2$$
$$\ge \prod_{j=0}^{\infty}\left|M\left(b^{-j}\left(\frac{\xi+\sum_{k=1}^{i}d_kb^{n_k}}{b^{n_i+1}}\right)\right)\right|^2 \ge \alpha. \tag{3.12}$$

For the case $i = 0$, it is easy to see that $\left|\frac{\xi}{b}\right| \le \frac{b-2}{q(b-1)} < \frac{b-1}{qb}$. Hence, $\prod_{j=n_0+1}^{n_1}\left|M(b^{-j}t)\right|^2 \ge \prod_{j=0}^{\infty}\left|M\left(b^{-j}(\xi/b)\right)\right|^2 \ge \alpha$. Putting this fact and (3.12) into (3.11), we have $|\widehat{\mu}_{b,q}(t)|^2 \ge \alpha^{N+1}$.

We next prove the upper bound. From $|M(\xi)| \le 1$, (3.11) and the integral periodicity of $M(\xi)$,

$$|\widehat{\mu}(t)|^2 \le \prod_{i=1}^{N}\left|M\left(b^{-(n_i+1)}t\right)\right|^2 = \prod_{i=1}^{N}\left|M\left(b^{-(n_i+1)}(\xi+\sum_{k=1}^{i}d_kb^{n_k})\right)\right|^2. \tag{3.13}$$

By (3.10) we have

$$|\xi+\sum_{k=1}^{i} d_k b^{n_k}| \geq b^{n_i} - |\xi+\sum_{k=1}^{i-1} d_k b^{n_k}| \geq b^{n_i} - \frac{b^{n_{i-1}}(b-1)}{q} \geq b^{n_i-1}.$$

By (3.10), (3.13), the above and the definition of $\beta$, we obtain that $|\widehat{\mu}(t)|^2 \leq \beta^N$. □

**Proof of Theorem 3.4** (i) Without loss generality we assume that $|\xi| \leq \frac{r(b-2)}{b-1}$. Recall that

$$Q_{m+p}(\xi) = Q_m(\xi) + \sum_{n=q^m}^{q^{m+p}-1} |\widehat{\mu}_{m+p}(\xi+r\lambda_n)|^2 \left|\widehat{\mu}\left(\frac{\xi+r\lambda_n}{q^{m+p}}\right)\right|^2. \qquad (3.14)$$

Let also $L = \max_{n\geq 1} \ell_n$ ($< \infty$ by assumption). For $q^m \leq n < q^{m+p}$, there exists unique $N$, $m < N \leq m+p$, such that $q^{N-1} \leq n < q^N$. By the definition of $\tau$, we have $\tau(\sigma 0^k) \in \{0, q, 2q, \ldots, (r-1)q\}$ for $k \geq 1$. We therefore have

$$\begin{aligned}\xi + r\lambda_n &= \xi + r\tau(\sigma|_1) + r\tau(\sigma|_2)b + \cdots + r\tau(\sigma|_N)b^{N-1} + \sum_{s=N}^{\infty} r\tau(\sigma 0^{s-N+1})b^s \\ &= \xi + r\sum_{i=1}^{N} \tau(\sigma|_i)b^{i-1} + r\tau(\sigma 0)b^N + \cdots + r\tau(\sigma 0^{m+p-N})b^{m+p-1} \\ &\quad + \sum_{s=m+p}^{\infty} d_s b^{s+1}\end{aligned}$$

Hence,

$$\begin{aligned}\frac{\xi+r\lambda_n}{b^{m+p}} &= \frac{1}{b^{m+p}}\left(\xi + r\sum_{i=1}^{N} \tau(\sigma|_i)b^{i-1} + r\tau(\sigma 0)b^N + \cdots + r\tau(\sigma 0^{m+p-N})b^{m+p-1}\right) \\ &\quad + \sum_{s=m+p}^{\infty} d_s b^{s+1-(m+p)} := t + \sum_{s=m+p}^{\infty} d_s b^{s+1-(m+p)}.\end{aligned}$$

Note that, from $|\tau(\sigma)| \leq b-2$ for any multi-indices $\sigma$,

$$|t| \leq \frac{1}{b^{m+p}}\left(|\xi| + r(b-2)(1+b+b^2+\cdots+b^{m+p-1})\right) \leq \frac{r(b-2)}{b-1}.$$

Also, $d_s \in \{0, 1, \ldots, r-1\}$ and there are at most $L$ non-zero terms. By Proposition 3.7, we conclude that $\left|\widehat{\mu}\left(\frac{\xi+r\lambda_n}{q^{m+p}}\right)\right|^2 \geq \alpha^{L+1}$. Using (3.14) and Lemma 3.6. we obtain

$$Q_{m+p}(\xi) \geq Q_m(\xi) + \alpha^{L+1} \sum_{n=q^m}^{q^{m+p}-1} \left|\widehat{\mu}_{m+p}(\xi + r\lambda_n)\right|^2$$
$$= Q_m(\xi) + \alpha^{L+1} \left(1 - \sum_{n=0}^{q^m-1} \left|\widehat{\mu}_{m+p}(\xi + r\lambda_n)\right|^2\right).$$

Fixing $m$, we first let $p$ approaches infinity and obtain

$$Q(\xi) \geq Q_m(\xi) + \alpha^{L+1} \left(1 - \sum_{n=0}^{q^m-1} |\widehat{\mu}(\xi + r\lambda_n)|^2\right).$$

We then finally let $m$ goes to infinity.

$$\alpha^{L+1}\left(1 - \sum_{n=0}^{\infty} |\widehat{\mu}(\xi + r\lambda_n)|^2\right) \leq 0.$$

This means that $Q(\xi) \geq 1$ for $|\xi| \leq r(b-2)/(b-1)$. As $Q(\xi) \leq 1$ for mutually orthogonal sets and by the entire function property of $Q$ on $\mathbb{C}$, we must have $Q(\xi) \equiv 1$ and hence $\Lambda$ is a spectrum for $\mu$.

(ii) With loss of generality we assume that $\ell_n \geq \log_q n$ for $n \geq 1$. Again we begin with

$$Q_m(\xi) = Q_{m-1}(\xi) + \sum_{n=q^{m-1}}^{q^m-1} |\widehat{\mu}_m(\xi + r\lambda_n)|^2 \left|\widehat{\mu}(\frac{\xi + r\lambda_n}{q^m})\right|^2.$$

Note that for $q^{m-1} \leq n < q^m$, $\ell_n \geq \log_q n \geq m-1$. Using it and the same estimate as in (i) so as to apply Proposition 3.7, we have

$$Q_m(\xi) \leq Q_{m-1}(\xi) + \sum_{n=q^{m-1}}^{q^m-1} |\widehat{\mu}_m(\xi + r\lambda_n)|^2 \beta^{\ell_n}$$
$$\leq Q_{m-1}(\xi) + \beta^{m-1} \sum_{n=q^{m-1}}^{q^m-1} |\widehat{\mu}_m(\xi + r\lambda_n)|^2$$
$$= Q_{m-1}(\xi) + \beta^{m-1}(1 - \sum_{n=0}^{q^{m-1}-1} |\widehat{\mu}_m(\xi + r\lambda_n)|^2)$$
$$\leq Q_{m-1}(\xi) + \beta^{m-1}(1 - Q_{m-1}(\xi)).$$

Consequently,

$$1 - Q_m(\xi) \geq (1 - Q_{m-1}(\xi))(1 - \beta^{m-1}) \geq (1 - Q_1(\xi)) \prod_{k=1}^{m-1} (1 - \beta^k).$$

By letting $m$ to infinity, we have

$$1 - Q(\xi) \geq (1 - Q_1(\xi)) \prod_{k=1}^{\infty} (1 - \beta^k).$$

Since $Q_1(\xi) < 1$ for almost all $\xi \in \mathbb{R}$, the second assertion follows by Corollary 2.4. □

**Proof of Theorem 3.5** Let $\{m_k\}_{k=1}^{\infty}$ be a strictly increasing sequence of positive integers with $m_1 \geq 2$. Then $m_k > k$ for $k \geq 1$. We now define a regular mapping in terms of this sequence by induction. Let $\tau(\vartheta) = \tau(0^k) = 0$ for $k \geq 1$. For $\sigma \in \{1, 2, \ldots, q-1\} \subset \Sigma_q^1$, we define $\tau(\sigma) = \sigma$ and $\tau(\sigma 0^l) = 0$ or $q$ according to $l \neq m_\sigma$ or $l = m_\sigma$, respectively. Suppose we have defined all $\tau(\sigma)$, $\sigma = \sigma_1 \cdots \sigma_s$ with $s \leq k$ and $\sigma_s \neq 0$, and $\tau(\sigma 0^l)$ for $l \geq 1$. For $\sigma = \sigma_1 \cdots \sigma_{k+1} \in \Sigma_q^{k+1}$ with $\sigma_{k+1} \neq 0$, we define $\tau(\sigma) = \sigma_{k+1}$ and $\tau(\sigma 0^l) = 0$ or $q$ according to $l \neq m_{p_\sigma}$ or $l = m_{p_\sigma}$, respectively, where $p_\sigma = \sum_{i=1}^{k+1} \sigma_i q^{i-1}$. By induction we have well-defined a regular mapping from the $q$-adic tree to $\{-1, 0, 1, \ldots, b-1\}$.

For any $n \in \mathbb{N}$, there exists unique $k \geq 1$ such that $q^{k-1} \leq n < q^k$. Then $n$ can be expressed by

$$n = \sum_{j=1}^{k} \sigma_j q^{j-1}, \tag{3.15}$$

where all $\sigma_j \in \{0, 1, \ldots, q-1\}$ and $\sigma_k \neq 0$. By the definition of $\tau$-sequence, we have $\lambda_0 = 0$ and

$$\lambda_n = \sum_{j=1}^{k} \tau(\sigma_1 \cdots \sigma_j) b^{j-1} + q b^{m_n},$$

consequently, $\ell_n = 1$ and by Theorem 3.4(i), $\Lambda = \{\lambda_n\}_{n=0}^{\infty}$ is a spectrum of $\mu_{b,q}$.

We now find $\Lambda$ satisfying (3.9) by choosing $m_n$. To do this, we first note that there exists a strictly increasing continuous function $h(t)$ from $[0, \infty)$ onto itself such that $h(t) \leq g(t)$ for $t \geq 0$ and it is sufficient to replace $g(t)$ by $h(t)$ in the proof. In this way, the inverse of $h(t)$ exists, and we denote it by $h^{-1}(t)$.

Now, note that

$$\lambda_n \leq q \frac{b^k - 1}{b - 1} + q b^{m_n} \leq (q+1) b^{m_n}.$$

Hence,

$$\lambda_{n+1} - \lambda_n \geq q b^{m_{n+1}} - (q+1) b^{m_n} \geq b^{m_n + 1}. \tag{3.16}$$

Therefore, we choose $m_n$ so that $b^{m_n} \geq 2h^{-1}(b^{n+1})$ for all $n \geq 1$. For any $h(R) \geq 1$, there exists unique $s \in \mathbb{N}$ such that $b^{s-1} \leq h(R) < b^s$. Then

$$\frac{\sup_{x\in\mathbb{R}} \#(\Lambda \cap (x - R, x + R))}{h(R)} \leq \frac{\sup_{x\in\mathbb{R}} \#(\Lambda \cap (x - h^{-1}(b^s), x + h^{-1}(b^s)))}{b^{s-1}}. \tag{3.17}$$

Note from (3.16) that the length of the open intervals $(x - h^{-1}(b^s), x + h^{-1}(b^s))$ is less than $\lambda_{n+1} - \lambda_n$ whenever $n \geq s$. This implies that the set $\Lambda \cap (x - h^{-1}(b^s), x + h^{-1}(b^s))$ contains at most one $\lambda_n$ where $n \geq s$. We therefore have

$$\sup_{x\in\mathbb{R}} \#(\Lambda \cap (x - h^{-1}(b^s), x + h^{-1}(b^s))) \leq s + 1.$$

Thus the result follows by taking limit in (3.17). □

We conclude the paper with some remarks.

*Remark* (1) When observing the proofs of theorems, the main crux of the proof to spectra of zero Beurling dimension is in Proposition 3.7. The uniform control on the Fourier transform depends only on the number of non-zero digits in the $b$-adic expansion rather than the size of the frequencies.

(2) Indeed, all maximal orthogonal exponentials for $\mu_{b,q}$ can be classified through either *regular* or *irregular* mappings. This note discusses only the regular mappings. For irregular mappings, we can discuss its spectral properties if the number of irregular paths is finite. One can refer the details to [DaHL13].

(3) Much less is known about dilating a spectrum of a spectral measure. A standard example is that if $\Lambda = \{0, 1\} \oplus 4\{0, 1\} \oplus \ldots.$, then $5\Lambda$ is also a spectrum for the standard one-fourth Cantor measure (i.e. $q = 2, b = 4$) [DHSW11]. However, one can prove that the tree mapping corresponding to $5\Lambda$ is irregular with infinitely many irregular paths. To see this, we re-write the following elements $5\Lambda$ into our standard 4-adic expansions.

$$5 \cdot 4^n + 5 \cdot 4^{n+1} + \cdots + 5 \cdot 4^m = 4^n + 2 \cdot 4^{n+1} + 2 \cdot 4^{n+2} + \ldots + 2 \cdot 4^m + 4^{m+1}.$$

This means the paths $0^{n-1}10^\infty$ are irregular paths. Hence, there are infinitely many such paths. This example of spectra cannot be covered by our theory and is also the first example of spectra with infinitely many irregular paths .

## References

[Dai12] Dai, X.-R.: When does a Bernoulli convolution admit a spectrum? Adv. Math. **231**(3–4), 1681–1693 (2012)

[DaHL13] Dai, X.-R., He, X.-G., Lai, C.-K.: Spectral property of Cantor measures with consecutive digits. Adv. Math. **242**, 187–208 (2013)

[DHJ09] Dutkay, D., Han, D., Jorgensen, P.: Orthogonal exponentials, translations and Bohr completions. J. Funct. Anal. **257**, 2999–3019 (2009)

[DHS09] Dutkay, D., Han, D., Sun, Q.: On spectra of a Cantor measure. Adv. Math. **221**, 251–276 (2009)

[DHSW11] Dutkay, D., Han, D., Sun, Q., Weber, E.: On the Beurling dimension of exponential frames. Adv. Math. **226**, 285–297 (2011)

[DJ12] Dutkay, D., Jorgensen, P.: Fourier duality for fractal measures with affine scales. Math. Comp. **81**, 2253–2273 (2012)

[DL00] Dutkay, D., Lai, C.-K.: Uniformity of Measures with Fourier Frames . Adv. Math. **252**, 684–707 (2014)

[DS52] Duffin, R., Schaeffer, A.: A class of nonharmonic Fourier series. Tran. Amer. Math. Soc. **72**, 341–366 (1952)

[Fal90] Falconer, K.J.: Fractal Geometry. Mathematical Foundations and Applications. Wiley, New York (1990)

[Fug74] Fuglede, B.: Commuting self-adjoint partial differential operators and a group theoretic problem. J. Funct. Anal. **16**, 101–121 (1974)

[JP98] Jorgensen, P., Pedersen, S.: Dense analytic subspaces in fractal $L^2$ spaces. J. Anal. Math. **75**, 185–228 (1998)

[GL14] Grepstad, S., Lev, N.: Multi-tiling and Riesz bases. Adv. Math. **252**, 1–6 (2014)

[Gun90] Gunning, R.: intruduction to Holomorphic Functions of Several Variables Volume I, Function Theory. Wadsworth and Brooks , Pacific Grove, Calif (1990)

[HLL13] He, X.-G., Lai, C.-K., Lau, K.-S.: Exponential spectra in $L^2(\mu)$. Appl. Comput. Harmon. Anal. **34**, 327–338 (2013)

[HuL08] Hu, T.-Y., Lau, K.-S.: Spectral property of the Bernoulli convolutions. Adv. Math. **219**, 554–567 (2008)

[KN00] Kozma, G., Nitzan, S.: Combining Riesz bases. arXiv:1210.6383

[Lab01] Łaba, I.: The spectral set conjecture and multiplicative properties of roots of polynomials. J. London Math. Soc. **65**, 661–671 (2001)

[LaW02] Łaba, I., Wang, Y.: On spectral Cantor measures. J. Funct. Anal. **193**, 409–420 (2002)

[LaW06] Łaba, I., Wang, Y.: Some properties of spectral measures. Appl. Comput. Harmon. Anal. **20**, 149–157 (2006)

[Lai11] Lai, C.-K.: On Fourier frame of absolutely continuous measures. J. Funct. Anal. **261**, 2877–2889 (2011)

[Lan67] Landau, H.: Necessary density conditions for sampling and interpolation of certain entire functions. Acta Math. **117**, 37–52 (1967)

[Lev12] Lev, N.: Riesz bases of exponentials on multiband spectra. Proc. Amer. Math. Soc. **140**, 3127–3132 (2012)

# The Role of Transfer Operators and Shifts in the Study of Fractals: Encoding-Models, Analysis and Geometry, Commutative and Non-commutative

**Dorin Ervin Dutkay and Palle E.T. Jorgensen**

**Abstract** We study a class of dynamical systems in $L^2$ spaces of infinite products $X$. Fix a compact Hausdorff space $B$. Our setting encompasses such cases when the dynamics on $X = B^{\mathbb{N}}$ is determined by the one-sided shift in $X$, and by a given transition-operator $R$. Our results apply to any positive operator $R$ in $C(B)$ such that $R1 = 1$. From this we obtain induced measures $\Sigma$ on $X$, and we study spectral theory in the associated $L^2(X, \Sigma)$. For the second class of dynamics, we introduce a fixed endomorphism $r$ in the base space $B$, and specialize to the induced solenoid $\mathrm{Sol}(r)$. The solenoid $\mathrm{Sol}(r)$ is then naturally embedded in $X = B^{\mathbb{N}}$, and $r$ induces an automorphism in $\mathrm{Sol}(r)$. The induced systems will then live in $L^2(\mathrm{Sol}(r), \Sigma)$. The applications include wavelet analysis, both in the classical setting of $\mathbb{R}^n$, and Cantor-wavelets in the setting of fractals induced by affine iterated function systems (IFS). But our solenoid analysis includes such hyperbolic systems as the Smale-Williams attractor, with the endomorphism $r$ there prescribed to preserve a foliation by meridional disks. And our setting includes the study of Julia set-attractors in complex dynamics.

**Keywords** Fractal · Solenoid · Shift-spaces · Transition operators · Ruelle-operators · Attractors · Cantor · Infinite-product measures · Wavelets · Wavelet representations

**2010 Mathematics Subject Classification** 28A80 · 37A30 · 37C30 · 37C85 · 46G25 · 47B65

---

D.E. Dutkay
Department of Mathematics, University of Central Florida, 4000 Central Florida Blvd., P.O. Box 161364, Orlando, FL 32816-1364, USA
e-mail: Dorin.Dutkay@ucf.edu

P.E.T. Jorgensen (✉)
Department of Mathematics, University of Iowa, 14 MacLean Hall, Iowa City, IA 52242-1419, USA
e-mail: jorgen@math.uiowa.edu



D.-J. Feng and K.-S. Lau (eds.), *Geometry and Analysis of Fractals*, Springer Proceedings in Mathematics & Statistics 88,
DOI 10.1007/978-3-662-43920-3_3

# 1 Introduction

The purpose of this paper is to offer a general framework for geometry and analysis of iteration systems. We offer a setting encompassing the kind of infinite product, or solenoid constructions arising in the study of iterated function systems (IFSs). Our aim is to give an operator theoretic construction of infinite product measures in a general setting that includes wavelet analysis of IFSs. To motivate this, recall, that to every affine function system $S$ with fixed scaling matrix, and a fixed set of translation points in $\mathbb{R}^n$, we may associate to $S$ a solenoid. By this we mean a measure space whose $L^2$ space includes $L^2(\mathbb{R}^n)$ in such a way that $\mathbb{R}^n$ embeds densely in the solenoid. (In the more familiar case of $n = 1$, we speak of a dense curve in an infinite-dimensional "torus". The latter being a geometric model of the solenoid.

The need for this generality arose in our earlier investigations, for example in the building of wavelet systems on Cantor systems, of which the affine IFSs are special cases. In these cases (see Theorem 2.6 and Corollary 4.5 below) we found that one must pass to a suitable $L^2$ space of a solenoid. Indeed, we showed that such wavelet bases fail to exist in the usual receptor Hilbert space $L^2(\mathbb{R}^n)$ from wavelet theory.

For reference to earlier papers dealing with measures on infinite products, and their use in harmonic analysis and wavelet theory on fractals; see e.g., [DJ12, DJS12, DLS11, DJ11a, DS11, DJ11b, DHSW11, DJ10, DJP09, LN12, DL10, LW09].

The paper is organized as follows: Starting with a compact Hausdorff space $B$, and a positive operator $R$ in $C(B)$, we pass to a family of induced probability measures $\Sigma$ (depending on $R$) on the infinite product $\Omega = B^{\mathbb{N}}$. Among all probability measures on $\Omega$, we characterize those which are induced. In Sect. 2, we prove a number of theorems about $R$-induced measures on $\Omega$, and we include applications to random walks, and to fractal analysis. In Sect. 3, we then introduce an additional structure: a prescribed endomorphism $r$ in the base space $B$, and we study the corresponding solenoid $\mathrm{Sol}(r)$, contained in $\Omega$, and its harmonic analysis, including applications to generalized wavelets. The latter are studied in detail in Sect. 4 where we introduce wavelet-filters, in the form of certain functions $m$ on $B$.

# 2 Analysis of Infinite Products

**Definition 2.1** Let $B$ be some compact Hausdorff space. $\mathcal{B}$ refes to a $\sigma$-algebra, usually generated by the open sets, so Borel. We will denote by $\mathcal{C}$ the cylinder sets, see below. We denote by $\mathcal{M}(B)$ the set of positive Borel masures on $B$, and by $\mathcal{M}_1(B)$ those that have $\mu(B) = 1$. Let $V$ be some set.

$$B^V = \prod_V B = \text{all functions from } V \text{ to } B.$$

For exàmple $V = \mathbb{N}$ or $\mathbb{Z}$.

For $x \in B^V$ we denote by $\pi_v(x) := x_v$, $v \in V$. If $V = \mathbb{N}$, then we denote by

$$\pi_1^{-1}(x) = \{(x_1, x_2, \dots) \in B^{\mathbb{N}} : x_1 = x\}.$$

Let $r : B \to B$ be some onto mapping, and $\mu$ a Borel probability measure on $B$, $\mu(B) = 1$.

To begin with we do not introduce $\mu$ and $r$, but if $r$ is fixed and

$$1 \leq \#r^{-1}(x) < \infty, \quad \text{for all } x \in B, \tag{2.1}$$

then we introduce two objects

(1) $R = R_W$, the Ruelle operator;
(2) $\mathrm{Sol}(r)$, the solenoid.

For (1), fix $W : B \to [0, \infty)$ such that

$$\sum_{r(y)=0x} W(y) = 1, \text{ for all } x \in B,$$

and set

$$(R_W\varphi)(x) = \sum_{r(y)=x} W(y)\varphi(y). \tag{2.2}$$

For (2),

$$\mathrm{Sol}(r) = \left\{x \in B^{\mathbb{N}} : r(x_{i+1}) = x_i, i = 1, 2, \dots\right\} \tag{2.3}$$

$$\sigma(x)_i = x_{i+1}, \quad (x \in B^{\mathbb{N}}), \qquad \widehat{r}(x) = (r(x_1), x_1, x_2, \dots). \tag{2.4}$$

More generally, consider

$$R : C(B) \to C(B) \text{ or } R : M(B) \to M(B), \tag{2.5}$$

where $M(B)$ is the set of all measurable functions on $B$.

**Definition 2.2** We say that $R$ is *positive* iff

$$\varphi(x) \geq 0 \text{ for all } x \in B \text{ implies } (R\varphi)(x) \geq 0, \text{ for all } x \in B. \tag{2.6}$$

We will always assume $R1 = 1$ where 1 indicates the constant function 1 on $B$. This is satisfied if $R = R_W$ in (2.2), but there are many other positive operators $R$ with these properties.

While what we call "the transfer operator" or a "Ruelle operator" has a host of distinct mathematical incarnations, each dictated by a particular family of applications, they are all examples of positive operators $R$ in the sense of our Definition 2.2. Our paper has two aims: One is to unify, and extend earlier studies; and the other is to prove a number of theorems on measures, dynamical systems, stochastic processes built from infinite products. Indeed there are many positive operators $R$ which might not fall in the class of operators studied as "transfer operators". The earlier literature on transfer operators includes applications to physics [LR69], to the Selberg zeta function [FM12], to dynamical zeta functions [Rue02, Rue96, Nau12, MMS12]; to $C^*$-dynamical systems [Kwa12, ABL11]; to the study of Hausdorff dimension [Hen12]; to spectral theory [ABL12].

These applications are, in addition to the aforementioned, to analysis on fractals, and to generalized wavelets. For book treatments, we refer the reader to [Bal00], and [BJ02]. The literature on positive operators $R$, in the general sense, is much less extensive; but see [Arv86].

**Definition 2.3** A subset $S$ of $B^{\mathbb{N}}$ is said to be *shift-invariant* iff $\sigma(S) \subset S$, where $\sigma$ is as in (2.4), $\sigma(x)_i = x_{i+1}$.

*Remark 2.4* Every solenoid $\mathrm{Sol}(r)$ is shift-invariant.

*Example 2.5* The solenoids introduced in connection with generalized wavelet constructions:

Let $r : B \to B$ as above and let $\mu$ be a strongly invariant measure, i.e.,

$$\int f\,d\mu = \int \frac{1}{\#r^{-1}(x)} \sum_{r(y)=x} f(y)\,d\mu(x)$$

for all $f \in C(B)$.

A *quadrature mirror filter* (*QMF*) for $r$ is a function $m_0$ in $L^\infty(B,\mu)$ with the property that

$$\frac{1}{N} \sum_{r(w)=z} |m_0(w)|^2 = 1, \quad (z \in B) \tag{2.7}$$

As shown by Dutkay and Jorgensen [DJ05, DJ07], every quadrature mirror filter (QMF) gives rise to a wavelet theory. Various extra conditions on the filter $m_0$ will produce wavelets in $L^2(\mathbb{R})$ [Dau92], on Cantor sets [DJ06, MP11], on Sierpinski gaskets [DMP08] and many others.

**Theorem 2.6** [DJ05, DJ07] *Let $m_0$ be a QMF for $r$. Then there exists a Hilbert space $\mathcal{H}$, a representation $\pi$ of $L^\infty(B)$ on $\mathcal{H}$, a unitary operator $U$ on $\mathcal{H}$ and a vector $\varphi$ in $\mathcal{H}$ such that*

(i) **Covariance**

$$U\pi(f)U^* = \pi(f \circ r), \quad (f \in L^\infty(B)) \tag{2.8}$$

(ii) **Scaling equation**

$$U\varphi = \pi(m_0)\varphi \tag{2.9}$$

(iii) **Orthogonality**

$$\langle \pi(f)\varphi\,,\,\varphi\rangle = \int f\,d\mu, \quad (f \in L^{\infty}(B)) \tag{2.10}$$

(iv) **Density**

$$\overline{span}\left\{U^{-n}\pi(f)\varphi : f \in L^{\infty}(B), n \geq 0\right\} = \mathcal{H} \tag{2.11}$$

The system $(\mathcal{H}, U, \pi, \varphi)$ in Theorem 2.6 is called *the wavelet representation associated to the QMF* $m_0$.

While, as we mentioned before, these representations can have incarnations on the real line, or on Cantor sets, they can be also represented using certain random-walk measures on the solenoid (see [DJ05, DJ07, Dut06]).

*Remark 2.7* In examples when the condition (2.1) is not satisfied, the modification of the family of relevant integral operators is as follows.

In the general case when $r : B \to B$ is given, but $\#r^{-1}(x) = \infty$, the modification of the operators $R$, extending those from Example 2.5, is as follows:
Consider

(i) $W : B \to [0, \infty)$ Borel
(ii) $p : B \times \mathcal{B}(B) \to [0, \infty)$ such that for all $x \in B$, $p(x, \cdot) \in \mathcal{M}(r^{-1}(x))$, so is a positive measure such that

$$\int_{r^{-1}(x)} W(y)p(x, dy) = 1, \quad (x \in B).$$

Then set

$$(R\varphi)(x) = \int_{r^{-1}(x)} \varphi(y)W(y)p(x, dy).$$

*Example 2.8* $G = (V, E)$ infinite graph, $V$ are the vertices, $E$ are the edges.

$$i(e) = \text{initial vertex}, \quad t(e) = \text{terminal vertex}. \tag{2.12}$$

$$S^{(G)} = \text{solenoid of } G = \left\{\tilde{e} \in E^{\mathbb{N}} : t(e_j) = i(e_{j+1}) \text{ for all } j \in \mathbb{N}\right\}. \tag{2.13}$$

For example $V = \mathbb{Z}^2$ and the edges are given by $x \sim y$ iff $\|x - y\| = 1$. For details and applications, see [JP10].

**Definition 2.9** Now back to $\mathcal{C}$, the cylinder sets mentioned in Definition 2.1. $C \in \mathcal{C}$ are subsets of $B^V$ indexed by finite sytems $v_1, \ldots, v_n, O_1, \ldots, O_n, v_i \in V, O_i \subset B$ open subsets, $i = 1, \ldots, n, n \in \mathbb{N}$.

$$C_{v_i, O_i} := \left\{ \tilde{x} \in B^V : x_{v_i} \in O_i \text{ for all } i = 1, \ldots, n \right\}. \tag{2.14}$$

Notation: $\mathcal{C}$ generates the topology and the $\sigma$-algebra of subsets in $B^V$ in the usual way, and $B^V$ is compact by Tychonoff's theorem.

If $\varphi$ is a function on $B$, we denote by $M_\varphi$ the multiplication operator $M_\varphi f = \varphi f$, defined on functions $f$ on $B$. In the applications below, we will use $C(B)$, all continuous functions from $B$ to $\mathbb{R}$.

**Lemma 2.10** *Consider $B^{\mathbb{N}}$ and the algebra generated by cylinder functions of the form $f = \varphi_1 \otimes \cdots \otimes \varphi_n$, $\varphi_i \in C(B)$, $n \in \mathbb{N}$, $1 \le i \le n$,*

$$(\varphi_1 \otimes \cdots \otimes \varphi_n)(\tilde{x}) = \varphi_1(x_1)\varphi_2(x_2)\ldots\varphi_n(x_n), \quad (\tilde{x} \in B^{\mathbb{N}}), \tag{2.15}$$

*or*

$$\varphi_1 \otimes \cdots \otimes \varphi_n = (\varphi_1 \circ \pi_1)(\varphi_2 \circ \pi_2)\ldots(\varphi_n \circ \pi_n). \tag{2.16}$$

*Let $\mathcal{A}_\mathcal{C}$ be the algebra of all cylinder functions. Then $\mathcal{A}_\mathcal{C}$ is dense in $C(B^{\mathbb{N}})$.*

*Proof* Easy consequence of Stone-Weierstrass. □

**Theorem 2.11** *Let $R$ be a positive operator as in* (2.5)*, with $R1 = 1$. Then for each $x \in B$ there exists a unique Borel probability measure $\mathbb{P}_x$ on $B^{\mathbb{N}}$ such that*

$$\int_{B^{\mathbb{N}}} \varphi_1 \otimes \cdots \otimes \varphi_n \, d\mathbb{P}_x = \left( M_{\varphi_1} R M_{\varphi_2} \ldots R M_{\varphi_n} 1 \right)(x), \quad (\varphi_i \in C(B), n \in \mathbb{N}). \tag{2.17}$$

*Proof* We only need to check that the right-hand side of (2.17) for $\varphi_1 \otimes \ldots \varphi_n$ equals the right-hand side of (2.17) for $\varphi_1 \otimes \ldots \varphi_n \otimes 1$; but this is immediate from (2.17) and the fact that $R1 = 1$. The existence and uniqueness of $\mathbb{P}_x$ the follows form the inductive method of Kolmogorov. □

**Corollary 2.12** *Let $B$ and $R : C(B) \to C(B)$ be as in* Theorem 2.11*, and let $\mu \in \mathcal{M}_1(B)$ be given. Let $\Sigma = \Sigma^{(\mu)}$ be the measure on $\Omega = B^{\mathbb{N}}$ given by*

$$\int f \, d\Sigma := \int_B \int_{\pi_1^{-1}(x)} f \, d\mathbb{P}_x \, d\mu(x). \tag{2.18}$$

*Then*

(i) *$V_1 : L^2(B, \mu) \to L^2(\Omega, \Sigma)$ given by $V_1 \varphi := \varphi \circ \pi_1$ is isometric.*

(ii) *For its adjoint operator* $V_1^*$, *we have* $V_1^* : L^2(\Omega, \Sigma) \to L^2(B, \mu)$ *with*

$$(V_1^* f)(x) = \int\limits_{\pi_1^{-1}(x)} f \, d\mathbb{P}_x. \tag{2.19}$$

*Proof* The assertion (i) is immediate from Theorem 2.11. To prove (ii) we must show that the following formula holds:

$$\int\limits_B \left( \int\limits_{\pi_1^{-1}(x)} f \, d\mathbb{P}_x \right) \psi(x) \, d\mu(x) = \int\limits_\Omega f \, \psi \circ \pi_1 \, d\Sigma \tag{2.20}$$

for all $f \in L^2(\Omega, \Sigma)$ and all $\psi \in C(B)$.

Recall that $V_1^*$ is determined by

$$\langle V_1^* f \, , \, \psi \rangle_{L^2(\mu)} = \langle f \, , \, V_1 \psi \rangle_{L^2(\Omega, \Sigma)} \, . \tag{2.21}$$

But by Lemma 2.10 (Stone-Weierstrass), to verify (2.20), we may restrict attention to the special case when $f$ has the form given in (2.16). Note that if $f = (\varphi_1 \circ \pi_1)(\varphi_2 \circ \pi_2) \ldots (\varphi_n \circ \pi_n)$ then

$$f(\psi \circ \pi_1) = ((\varphi_1 \psi) \circ \pi_1)(\varphi_2 \circ \pi_2) \ldots (\varphi_n \circ \pi_n),$$

and so the right-hand side of (2.20) is equal to

$$= \int\limits_\Omega ((\varphi_1 \psi) \circ \pi_1)(\varphi_2 \circ \pi_2) \ldots (\varphi_n \circ \pi_n) \, d\Sigma$$

$$= \int\limits_B \varphi_1(x) \psi(x) R(\varphi_2 R(\ldots \varphi_{n-1} R(\varphi_n)) \ldots)))(x) \, d\mu(x) = \int\limits_B \psi(x) \int f \, d\mathbb{P}_x \, d\mu(x)$$

which is the left-hand side of (2.20) and (ii) follows. □

*Remark 2.13* When $R : C(B) \to C(B)$ is a given positive operator, we induce measures on $\Omega = B^{\mathbb{N}}$ by the inductive procedure outlined in the proof of Theorem 2.11; but implicit in this construction is an extension of $\varphi \mapsto R(\varphi)$ from all $\varphi$ continuous to all Borel measurable functions. This extension uses the Riesz theorem in the usual way as follows: Fix $x \in B$ and then apply Riesz' theorem to the positive linear functional $C(B) \ni \varphi \mapsto R(\varphi)(x)$. There is a unique regular Borel measure $\mu_x$ on $B$ such that

$$R(\varphi)(x) = \int_B \varphi(y)\, d\mu_x(y), \quad (\varphi \in C(B)).$$

If $E \subset B$ is Borel, we define

$$\tilde{R}(E)(x) = \tilde{R}(\chi_E)(x) := \mu_x(E);$$

but we shall use this identification without overly burdening our notation with tildes.

**Lemma 2.14** *Let $B$ and $R$ be specified as above. Given $\mu \in \mathcal{M}_1(B)$, let $\Sigma = \Sigma^{(\mu)}$ denote the corresponding measure on $\Omega$, i.e.,*

$$\int_\Omega f\, d\Sigma = \int_B \int_{\pi_1^{-1}(x)} f\, d\mathbb{P}_x^{(R)}\, d\mu(x) \tag{2.22}$$

*We shall consider $V_1 : L^2(B, \mu) \to L^2(\Omega, \Sigma)$ and its adjoint operator $V_1^* : L^2(\Omega, \Sigma) \to L^2(B, \mu)$, where $V_1\varphi = \varphi \circ \pi$, for all $\varphi \in L^2(B, \mu)$. Note that the adjoint operator $V_1^*$ makes reference to the choice of $R$ at the very outset. The following two hold:*

(i) *$R$ naturally extends to $L^2(B, \mu)$; and*

(ii)

$$RV_1^* f = V_1^*(f \circ \sigma), \quad (f \in L^2(\Omega, \Sigma)) \tag{2.23}$$

*Remark 2.15* Given $R$, we say that a function $\varphi \in B$ is *harmonic* iff $R\varphi = \varphi$. It follows that harmonic functions contain the range of $V_1^*$, applied to $\{f : f \circ \sigma = f\}$. For a stronger conclusion, see Corollary 2.21.

*Proof of Lemma 2.14* Using the Stone-Weierstrass theorem, applied to $C(\Omega)$, we note that it is enough for us to check the validity of formula (2.23) on the algebra $\mathcal{A}^{(cyl)}$ spanned by all cylinder functions

$$f = (\varphi_1 \circ \pi_1)(\varphi_2 \circ \pi_2) \dots (\varphi_n \circ \pi_n) \tag{2.24}$$

$n \in \mathbb{N}$, $\varphi_i \in C(B)$. But note that if $f$ is as in (2.24) then

$$f \circ \sigma = (\varphi_1 \circ \pi_2)(\varphi_2 \circ \pi_3) \dots (\varphi_n \circ \pi_{n+1}) \tag{2.25}$$

Using then (2.19) in Corollary 2.12 above, we conclude that

$$(V_1^*(f \circ \sigma))(x) = R(\varphi_1 R(\varphi_2(R \dots \varphi_{n-1} R(\varphi_n)) \dots))(x) = (RV_1^* f)(x);$$

The extension from the cylinder functions $\mathcal{A}^{(cyl)}$ to all of $L^2(\Omega, \Sigma)$ now follows from the usual application of Stone-Weierstrass; recall that $C(\Omega)$ is dense in $L^2(\Omega, \Sigma)$ relative to the $L^2$-norm; and we have the desired conclusion. □

## 2.1 What Measures on $B^{\mathbb{N}}$ have a Transfer Operator?

Below we characterize, among all Borel probability measures $\Sigma$ on $B^{\mathbb{N}}$, precisely those which arise from a pair $\mu$ and $R$ with a transfer operator $R$ and $\mu$ a measure on $B$. The characterization is general and involves only the one-sided shift $\sigma$ on $B^{\mathbb{N}}$.

**Lemma 2.16** *Let $\Sigma \in \mathcal{M}_1(B^{\mathbb{N}})$ and set $\mu := \Sigma \circ \pi_1^{-1} \in \mathcal{M}(B)$; then for $\mu$-almost all $x \in B$ there is a field $\mathbb{P}_x \in \mathcal{M}(\pi_1^{-1}(x))$ such that*

$$d\Sigma = \int_B d\mathbb{P}_x \, d\mu(x) \tag{2.26}$$

*and the following hold*

(i) *The operator $V_1 : L^2(B, \mu) \to L^2(B^{\mathbb{N}}, \Sigma)$ given by*

$$V_1\varphi = \varphi \circ \pi_1 \tag{2.27}$$

*is isometric.*

(ii) *Its adjoint operator $V_1^* : L^2(B^{\mathbb{N}}, \Sigma) \to L^2(B, \mu)$ satisfies*

$$(V_1^* f)(x) = \int_{\pi_1^{-1}(x)} f \, d\mathbb{P}_x =: \mathbb{E}_x(f), \quad (x \in B). \tag{2.28}$$

*Proof* (i) For $\varphi \in C(B)$, we have

$$\begin{aligned} \|V_1\varphi\|^2_{L^2(\Sigma)} &= \int_{B^{\mathbb{N}}} |\varphi \circ \pi_1|^2 \, d\Sigma = \int_{B^{\mathbb{N}}} |\varphi|^2 \circ \pi_1 \, d\Sigma \\ &= \int_B |\varphi|^2 \, d(\Sigma \circ \pi_1^{-1}) = \int_B |\varphi|^2 \, d\mu. \end{aligned}$$

(ii) For $\varphi \in C(B)$ and $f \in L^2(B^{\mathbb{N}}, \Sigma)$, we have

$$\int_{B^{\mathbb{N}}} (V_1\varphi) f \, d\Sigma = \int_B \varphi(x)(V_1^* f)(x) \, d\mu(x), \tag{2.29}$$

where $V_1^* f \in L^2(B, \mu)$. Hence

$$\int_{B^{\mathbb{N}}} (\varphi \circ \pi_1) f \, d\Sigma = \int_B \varphi(x)(V_1^* f)(x) \, d\mu(x) \tag{2.30}$$

and $(V_1^* f)(x)$ is well defined for $\mu$-almost all $x \in B$. Moreover, the mapping

$$C(B^{\mathbb{N}}) \ni f \mapsto (V_1^* f)(x) \tag{2.31}$$

is positive; i.e., $f \geq 0$ implies $(V_1^* f)(x) \geq 0$. This follows from (2.30). For if $E \subset B$, $\mu(E) > 0$, and $V_1^* f < 0$ on $E$ then there exists $\varphi \in C(B)$, $\varphi > 0$ such that $\int_B \varphi(x)(V_1^* f)(x)\, d\mu(x) < 0$, which contradicts (2.30). Now the conclusion in (2.28) follows from an application of Riesz' theorem to (2.31). □

**Proposition 2.17** *Let* $\Sigma \in \mathcal{M}(B^{\mathbb{N}})$, *then* $(\mathbb{P}_x)_{x\in B}$ *from* Lemma 2.16 *has the form* (2.17) *in* Theorem 2.11 *if and only if there is a positive operator* $R$ *such that* $R1 = 1$ *and*

$$\mathbb{E}_x(f \circ \sigma) = (R(\mathbb{E} \bullet f))(x) \tag{2.32}$$

*holds for all* $x \in B$ *and for all* $f \in L^2(B^{\mathbb{N}}, \Sigma)$, *where in* (2.32) *we use the notation*

$$\mathbb{E}_x(\dots) = \int_{\pi_1^{-1}(x)} \dots\, d\mathbb{P}_x = \mathbb{E}^{(\Sigma)}(\dots \mid \pi_1 = x) \tag{2.33}$$

*for the field of conditional expectations, and* $\mathbb{E} \bullet f$ *denotes the map* $x \mapsto \mathbb{E}_x f$.

*Proof* The implication (2.17) ⇒ (2.32) is already established. It is Lemma 2.14(ii). Now assume some positive operator $R$ exists such that (2.32) holds. We will then prove that $\Sigma$ is the measure determined in Theorem 2.11 from $R$ and $\mu = \Sigma \circ \pi_1^{-1}$. It is enough to verify (2.17) on all finite tensors

$$f = (\varphi_1 \circ \pi_1)(\varphi_2 \circ \pi_2) \dots (\varphi_n \circ \pi_n) \tag{2.34}$$

as in (2.16); and we now establish (2.17) by induction, using the assumed (2.32).

The case $n = 1$ is

$$\mathbb{E}_x(\varphi \circ \pi_1) = \varphi(x), \quad (\varphi \in C(B), x \in B);$$

and this follows from Lemma 2.16.

For $n = 2$, we compute as follows

$$\mathbb{E}_x(\varphi_1 \circ \pi_1\, \varphi_2 \circ \pi_2) = \varphi_1(x)(R\varphi_2)(x). \tag{2.35}$$

To do this, we shall prove the following fact, obtained from assumption (2.32):

For $\psi \in C(B)$ and $f \in L^2(B^{\mathbb{N}}, \Sigma)$ we have

$$\mathbb{E}_x((\psi \circ \pi_1) f) = \psi(x)\mathbb{E}_x(f). \tag{2.36}$$

Using (2.28) in Lemma 2.16(ii), note that (2.36) is equivalent to

$$\int_{B^{\mathbb{N}}} (\varphi \circ \pi_1)(\psi \circ \pi_1) f\, d\Sigma = \int_B \varphi\psi V_1^* f\, d\mu,$$

which in turn follows from $V_1(\varphi\psi) = (V_1\varphi)(V_1\psi)$ since $\varphi \mapsto \varphi \circ \pi_1$ is multiplicative.

Returning to (2.35), we then get

$$\begin{aligned}\mathbb{E}_x((\varphi_1 \circ \pi_1)(\varphi_2 \circ \pi_2)) &= \varphi_1(x)\mathbb{E}_x(\varphi_2 \circ \pi_2) = \varphi_1(x)\mathbb{E}_x(\varphi_2 \circ \pi_1 \circ \sigma) \\ &= \varphi_1(x)R(\mathbb{E} \bullet (\varphi_2 \circ \pi_1))(x) = \varphi_1(x)(R\varphi_2)(x).\end{aligned}$$

We shall now be using $\pi_i \circ \sigma = \pi_{i+1}$.

Assume that

$$\mathbb{E}_x(f) = \varphi_1(x)R(\varphi_2 R(\ldots R(\varphi_n)\ldots))(x) \tag{2.37}$$

holds when $f$ in (2.34) has length $n - 1$; then we show it must hold if it has length $n$. We set

$$\mathbb{E}_x(f) = \mathbb{E}_x((\varphi_1 \circ \pi_1)(g \circ \sigma)),$$

where $g$ is a tensor of length $n - 1$. Hence the induction hypothesis yields

$$\mathbb{E}_x(f) = \varphi_1(x)\mathbb{E}_x(g \circ \sigma) = \varphi_1(x)R(\mathbb{E} \bullet (g))$$

which is the right-hand side of (2.37). □

## 2.2 Subalgebras in $L^\infty(\Omega, \Sigma)$ and a Conditional Expectation

Let $B$, $R : C(B) \to C(B)$, $\mu \in \mathcal{M}_1(B)$ and $\Sigma = \Sigma^{(\mu)}$ be as specified. The only assumptions on $R$ are that

(i) it is linear;
(ii) it is positive and
(iii) $R1 = 1$.

We will be using Theorem 2.11 and Corollaries 2.12 and 3.14 referring to the measures

$$\{\mathbb{P}_x^{(R)} : x \in B\} \text{ on } \pi_1^{-1}(x), \quad (x \in B). \tag{2.38}$$

The theorem below is about the operators $\{V_n : n \in \mathbb{N}\}$, $V_n : L^2(B, \mu) \to L^2(\Omega, \Sigma)$ given by

$$V_n\varphi = \varphi \circ \pi_n, \quad (\varphi \in C(B), n \in \mathbb{N}).$$

Since $V_1 : L^2(B, \mu) \to L^2(\Omega, \Sigma)$ is isometric, it follows that

$$Q_1 := V_1 V_1^* \tag{2.39}$$

is a projection in each of the Hilbert spaces $L^2(\Omega, \Sigma^{(\mu)})$.

**Theorem 2.18** *With $B, R, \mu, \Sigma = \Sigma^{(\mu)}$ and $V_n$ specified as above, we have the following formulas:*

(i) $V_1^* V_{n+1} = R^n$ *on* $L^2(B, \mu)$, $n = 0, 1, 2 \ldots$;
(ii) $Q_1 := V_1 V_1^*$ *is a conditional expectation onto*

$$\mathcal{A}_1 := \{\varphi \circ \pi_1 : \varphi \in L^\infty(B, \mu)\}$$

$$Q_1((\varphi \circ \pi_1) f) = (\varphi \circ \pi_1) Q_1(f) \text{ for all } \varphi \in L^\infty(B, \mu),\ f \in L^\infty(\Omega, \Sigma). \tag{2.40}$$

(iii) $Q_1(\varphi \circ \pi_{n+1}) = (R^n \varphi) \circ \pi_1$ *for all* $\varphi \in C(B)$, $n = 0, 1, 2, \ldots$.

*Proof* (i) As a special case of Theorem 2.11, we see that

$$\int_{\pi_1^{-1}(x)} (\varphi \circ \pi_{n+1})\, d\mathbb{P}_x^{(R)} = (R^n \varphi)(x) \tag{2.41}$$

holds for all $\varphi \in C(B)$. We further see that (2.41) extends to both $L^\infty(B, \mu)$ and to $L^2(B, \mu)$. Hence

$$(V_1^* V_{n+1} \varphi)(x) = \int_{\pi_1^{-1}(x)} (\varphi \circ \pi_{n+1})\, d\mathbb{P}_x^{(\mathbb{R})} = (R^n \varphi)(x), \quad (x \in B). \tag{2.42}$$

(ii) By Lemma 2.10, we see that to verify (2.40), it is enough to check it for cylinder functions $f$, i.e.,

$$f = (\psi_1 \circ \pi_1)(\psi_2 \circ \pi_2) \ldots (\psi_n \circ \pi_n), \tag{2.43}$$

$n \in \mathbb{N}$, $\psi_i \in C(B)$. But if $f$ is as in (2.43), then

$$(\varphi \circ \pi_1) f = ((\varphi \psi_1) \circ \pi_1)(\psi_2 \circ \pi_2) \ldots (\psi_n \circ \pi_n), \tag{2.44}$$

and the desired formula (2.40) is immediate.

(iii) Given (i), we may apply $V_1$ to both sides in (2.42), and the desired formula (iii) follows. □

It is important to stress that one obtains a closed-form expression for $V_1^*$ where the operator $V_1 : \varphi \mapsto \varphi \circ \pi_1$ is introduced in Corollary 2.12. Indeed $V_1^*$ is a conditional expectation:

$$(V_1^* f)(x) = \mathbb{E}^{(\Sigma)}(f \mid \pi_1 = x) = \mathbb{E}_x^{(\Sigma)}(f), \quad (x \in B, f \in L^2(\Omega, \Sigma)) \tag{2.45}$$

By contrast, the situation for $V_n^*$, $n > 1$ is more subtle.

**Proposition 2.19** *Let $B$, $R : C(B) \to C(B)$ and $\Sigma = \Sigma^{(\mu)} \in \mathcal{M}_1(\Omega)$ be as above, i.e., $\mu = \Sigma \circ \pi_1^{-1}$. Let $R^*$ be the adjoint of the operator $R$ when considered as a bounded operator in $L^2(B, \mu)$. For $V_2^*$ we have*

$$V_2^*((\varphi_1 \circ \pi_1)(\varphi_2 \circ \pi_2) \ldots (\varphi_n \circ \pi_n)) = R^*(\varphi_1)\varphi_2 R(\varphi_3 \ldots R(\varphi_n) \ldots). \tag{2.46}$$

*Proof* Let $\psi$ be the function given on the right hand side in (2.46). The operator $V_2 : \varphi \mapsto \varphi \circ \pi_2$ maps from $L^2(B, \mu)$ into $L^2(\Omega, \Sigma)$. The assertion in (2.46) follows if we check that, for all $\xi \in C(B)$, we have the following identity:

$$\int_\Omega (\varphi_1 \circ \pi_1)((\xi \varphi_2) \circ \pi_2)(\varphi_3 \circ \pi_3) \ldots (\varphi_n \circ \pi_n)\, d\Sigma = \int_B \xi \psi \, d\mu. \tag{2.47}$$

But we may compare the left-hand side in (2.47) with the use of Theorem 2.11:

$$= \int_B \varphi_1 R((\xi \varphi_2) R(\varphi_3 R(\ldots R(\varphi_n) \ldots)))\, d\mu = \int_B (R^* \varphi_1) \xi \varphi_2 R(\varphi_3 R(\ldots R(\varphi_n) \ldots))\, d\mu,$$

which is the desired conclusion (2.46).

Recall that, by Theorem 2.18, we have $R = V_1^* V_2$ , and so $R^* = V_2^* V_1$. □

The next result is an extension of Lemma 2.14(ii). Note that (2.23) is the assertion that $V_1^*$ intertwines the two operations, $R$ and $f \mapsto f \circ \sigma$. The next result shows that, by contrast, $V_2^*$ acts as a multiplier.

**Corollary 2.20** *Let $B$, $R$, $\Sigma$ and $\mu = \Sigma \circ \pi_1^{-1}$ be as in* Proposition 2.19, *and set $\rho := R^* 1 \in L^2(B, \mu)$; then*

$$(V_2^*(f \circ \sigma))(x) = \rho(x) \mathbb{E}_x(f) = \rho(x)(V_1^* f)(x), \quad (x \in B, f \in L^2(\Omega, \Sigma)).$$

*Proof* This is immediate from Proposition 2.19, see (2.46). Recall that the span of the tensors is dense in $L^2(\Omega, \Sigma)$ and that if $f = (\varphi_1 \circ \pi_1)(\varphi_2 \circ \pi_2) \ldots (\varphi_n \circ \pi_n)$, then $f \circ \sigma = (\varphi_1 \circ \pi_2)(\varphi_2 \circ \pi_3) \ldots (\varphi_n \circ \pi_{n+1})$. □

In Proposition 4.6 we calculate the multiplier $\rho$ for the special case of the wavelet representation from Example 2.5.

**Corollary 2.21** *Let $B$ and $R$ be as in* Theorem 2.18, *and let $\mu \in \mathcal{M}_1(B)$ be given. The induced measure on $\Omega = B^{\mathbb{N}}$ is denoted $\Sigma^{(\mu)}$ and specified as in* (2.18). *We then have the following equivalence:*

(i) $h \in L^2(B, \mu)$ *and* $Rh = h$, *i.e.*, $h$ *is harmonic; and*
(ii) *There exists* $f \in L^2(\Omega, \Sigma^{(\mu)})$ *such that* $f = f \circ \sigma$ *and*

$$h(x) = \int_{\pi_1^{-1}(x)} f \, d\mathbb{P}_x^{(R)}. \tag{2.48}$$

*Proof* The implication (ii)⇒(i) follows from Lemma 2.14 and Remark 2.15. For (i)⇒(ii), let $h$ be given and assume it satisfies (i). An application of (iii) from Theorem 2.18 now yields

$$Q_1(h \circ \pi_{n+1}) = R^n(h) \circ \pi_1 = h \circ \pi_1 = V_1 h.$$

Using (2.39), we get $V_1(h - V_1^*(h \circ \pi_{n+1})) = 0$ for all $n = 0, 1, 2, \ldots$; and therefore

$$h = V_1^*(h \circ \pi_{n+1}), \quad (n \in \mathbb{N}). \tag{2.49}$$

Recalling

$$V_1^*(h \circ \pi_{n+1})(x) = \int_{\pi_1^{-1}(x)} h \circ \pi_{n+1} \, d\mathbb{P}_x^{(R)} \tag{2.50}$$

and using Theorem 2.11, we conclude that $\{h \circ \pi_{n+1}\}_{n \in \mathbb{N}}$ is a bounded $L^2(\Omega, \Sigma^{(\mu)})$-martingale.

By Doob's theorem, there is a $f \in L^2(\Omega, \Sigma^{(\mu)})$ such that

$$\lim_{n \to \infty} \|f - h \circ \pi_{n+1}\|_{L^2(\Sigma^{(\mu)})} = 0.$$

Since $\pi_{n+1} \circ \sigma = \pi_n$, it follows that $f \circ \sigma = f$. Taking the limit in (2.49) and using that the operator norm of $V_1^*$ is one, we get that $h = V_1^* f$ and therefore the desired formula (2.48) holds. □

## 2.3 A Stochastic Process Indexed by $\mathbb{N}$

*Remark 2.22* In the literature one has a number of theorems dealing with the existence of measures $\mu$ satisfying the various conditions; and if $\mu \circ R = \mu$ is satsified, then the measure is called a Ruelle equilibrium measure.

**Theorem 2.23** *Let* $B$ *be compact Hausdorff and* $R : C(B) \to C(B)$ *positive,* $R1 = 1$. *Let* $\mu \in \mathcal{M}_1(B)$ *such that* $\mu(B) = 1$, $\mu \circ R = \mu$. *Set*

$$X_n(\varphi) = \varphi \circ \pi_n, \quad (\varphi \in C(B), n \in \mathbb{N})$$

*and*

$$\int f \, d\Sigma = \int_B \int_{\pi_1^{-1}(x)} f \, d\mathbb{P}_x^{(R)} \, d\mu(x) \tag{2.51}$$

*Then*

$$\mathbb{E}(\dots) = \int \dots \, d\Sigma$$

*satisfies*

$$\mathbb{E}(X_n(\varphi)X_{n+k}(\psi)) = \int_B \varphi(x)(R^k\psi)(x) \, d\mu, \quad (n, k \in \mathbb{N}, \varphi, \psi \in C(B)) \tag{2.52}$$

*i.e., $R^k$ is the transfer operator governing distances k. Asymptotic properties as k goes to infinity govern long-range order.*

*Proof* From the definition of $\mathbb{P}_x^{(R)}$ we have

$$\mathbb{P}_x^{(R)}(X_n(\varphi)) = R^{n-1}(\varphi)(x), \quad \varphi \in C(B) \tag{2.53}$$

Now let $n, k, \varphi, \psi$ as in the statement in (2.52). Let $\Sigma$ be the measure on $B^{\mathbb{N}}$ in (2.51). Then

$$\begin{aligned} \mathbb{E}(X_n(\varphi)X_{n+k}(\psi)) &= \int_{B^{\mathbb{N}}} (\varphi \circ \pi_n)(\psi \circ \pi_{n+k}) \, d\Sigma \\ &= \int_B R^{n-1}(\varphi R^k(\psi))(x) \, d\mu(x) \\ &= \int \varphi(x) R^k(\psi)(x) \, d\mu(x) \end{aligned}$$

which is the desired conclusion. □

**Definition 2.24** We say that $\{X_k(\varphi)\}$ is independent at $\infty$ if

$$\lim_{k\to\infty} \mathbb{E}(X_n(\varphi)X_{n+k}(\psi)) = \left(\int \varphi \, d\mu\right)\left(\int \psi \, d\mu\right), \quad (\varphi, \psi \in C(B), n \in \mathbb{N}). \tag{2.54}$$

**Corollary 2.25** *Suppose for all* $\varphi$ *in* $C(B)$ *we have*

$$\lim_{k\to\infty} R^k(\varphi) = \left(\int \varphi\, d\mu\right) 1,$$

*then* (2.54) *is satisfied.*

*Proof* We proved

$$\mathbb{E}(X_n(\varphi)X_{n+k}(\psi)) = \int \varphi R^k \psi\, d\mu.$$

Now take the limit as $k \to \infty$, the desired conclusion (2.54) follows. □

The next result answers the question: what is the distribution of the random variable $X_n(\varphi)$?

**Corollary 2.26** *Assume* $\mu \circ R = \mu$. *The distribution of* $X_n(\varphi)$ *is* $\mu(\{x \in B : \varphi(x) \leq t\})$ *for all* $n$.

*Proof* Take $\varphi$ real valued for simplicity.

For $t \in \mathbb{R}$,

$$\Sigma(\{\tilde{x} \in B^{\mathbb{N}} : \varphi \circ \pi_n(\tilde{x}) \leq t\}) = \int_B \int_{\pi_1^{-1}(x)} \chi_{\{\varphi\leq t\}} \circ \pi_n\, d\mathbb{P}_x^{(R)}\, d\mu(x)$$

$$= \int_B R^{n-1}\chi_{\{\varphi\leq t\}}\, d\mu = \int_B \chi_{\{\varphi\leq t\}}\, d\mu.$$

In particular, it follows that all the random variables $X_n(\varphi)$ have the same distribution. □

## 2.4 Application to Random Walks

**Corollary 2.27** *Let* $(r, W)$ *be as in* Definition 2.1, *and let* $R_W$ *be the Ruelle operator in* (2.2), $\mathbb{P}_x^{(W)}$*—the random walk measure with transition probability specified as follows*

$$\text{Prob}(x \to y) = \begin{cases} W(y), & if\, r(y) = x \\ 0, & otherwise \end{cases} \tag{2.55}$$

*Then* $\mathbb{P}_x$ *from* Theorem 2.11 *is equal to* $\mathbb{P}_x^{(W)}$.

*Proof* We apply Theorem 2.11 to $R = R_W$ in (2.2) and we compute the right-hand side in (2.15) with induction

$$(M_{\varphi_1} R_W \dots R_W M_{\varphi_{n+1}})(x) = \varphi_1(x) \sum_{y_1} \cdots \sum_{y_n} W(y_1)W(y_2)\dots W(y_n)\varphi_2(y_1)\dots\varphi_{n+1}(y_n),$$

where $r(y_{i+1}) = y_i$, $1 \leq i < n$, $r^{(n)}(y_n) = x$. Further,

$$= \sum \cdots \sum \mathrm{Prob}(x \to y_1)\,\mathrm{Prob}(y_1 \to y_2 \,|, y_1)\dots\mathrm{Prob}(y_{n-1} \to y_n \mid y_{n-1})\varphi(y_1)\dots\varphi(y_n)$$

$$= \int d\mathbb{P}_x^{(W\text{-transition } R_W\text{-measure})} \varphi_1 \otimes \cdots \otimes \varphi_n \tag{2.56}$$

□

*Remark 2.28* The assertion in (2.56) applies to any random walk measure, for example, the one in Example 2.8.

Let $G = (V, E)$ be as in Example 2.8, with $E$ un-directed edges. Let $c : E \to [0, \infty)$ be such that

$$c_{(xy)} = c_{(yx)} \text{ for all } (xy) \in E, \quad c_{(xy)} \neq 0 \text{ if } (xy) \notin E. \tag{2.57}$$

A function as in (2.57) is called conductance.

Set $p = p^c$, where

$$p_{xy} = \frac{c_{xy}}{\sum_{z, z \sim x} c_{xz}} = \frac{c_{xy}}{c(x)}, \tag{2.58}$$

where

$$c(x) = \sum_{z, z \sim x} c_{xz}, \text{ and } z \sim x \text{ means } (zx) \in E.$$

Then there is a unique $\mathbb{P}_x^{(c)}$ such that

$$\int \varphi_1 \otimes \cdots \otimes \varphi_n \, d\mathbb{P}_x^{(c)} = \sum_{y_1} \cdots \sum_{y_n} p_{xy_1} p_{y_1 y_2} \cdots p_{y_{n-1} y_n} \varphi_1(y_1)\dots\varphi_n(y_n),$$

where the sums are over all $y_1, y_2, \dots, y_n$ such that $(y_i y_{i+1}) \in E$.

Note that $\mathbb{P}_x^{(W)}$ is supported on the solenoid, and $\mathbb{P}_x^{(c)}$ is supported on $S^{(G)}$ (see (2.13)).

*Remark 2.29* The last application is useful in the setting of harmonic functions on graphs $G = (V, E)$ with prescribed conductance function $c$ as in (2.57). Set

$$(\Delta\varphi)(x) = \sum_{y \in V, y \sim x} c_{xy}(\varphi(x) - \varphi(y)) \tag{2.59}$$

the graph Laplacian with conductance $c$.

A function $\varphi$ on $V$ satisfies $\Delta\varphi \equiv 0$, iff

$$\varphi(x) = \sum_{y \in V, y \sim x} p_{xy}^{(c)} \varphi(y), \tag{2.60}$$

where $p_{xy}^{(c)} = \frac{c_{xy}}{c(x)}$ as in (2.58).

**Application**. Use $\mathbb{P}_x$ to get harmonic functions. The study of classes of harmonic functions is of interest for infinite networks (see Remarks 2.28 and 2.29), and in Corollary 2.21 is is shown that the harmonic functions $h$ are precisely those that arise from applying $\mathbb{E}_x$ to functions $f$, $f \circ \sigma = f$ on $B^{\mathbb{N}}$, i.e.,

$$h(x) = \int f(x \ldots) \, d\mathbb{P}_x,$$

and conversely a martingale limit constructs $f$ from $h$. For more details on this construction, see Corollary 2.21.

## *2.5 An Application to Integral Operators*

Let $K : B \times B \to [0, \infty)$ be a continuous function and let $\mu$ be a probability measure on $B$ such that

$$\int_B K(x, y) \, d\mu(y) = 1 \text{ for all } x \in B. \tag{2.61}$$

Define

$$R_K f(x) = \int_B K(x, y) f(y) \, d\mu(y), \quad (x \in B, f \in C(B)).$$

Then $R = R_K$ defines a positive operator as in Definition 2.2, $R_K 1 = 1$ and then $\mathbb{P}_x$ in Theorem 2.11 satisfies

$$\int_{B^{\mathbb{N}}} \varphi_1 \otimes \cdots \otimes \varphi_{n+1} \, d\mathbb{P}_x =$$
$$\varphi_1(x) \int \ldots \int K(x, y_1) K(y_1, y_2) \ldots K(y_{n-1}, y_n) \varphi_2(y_1) \ldots \varphi_{n+1}(y_n) \, d\mu(y_1) \ldots d\mu(y_n)$$

We get a measure $\Sigma$ on $B^{\mathbb{N}}$ as follows

$$\int f \, d\Sigma = \int f \, d\mathbb{P}_x \, d\mu(x) \tag{2.62}$$

since the right-hand side in (2.62) is independent of $x$.

# 3 Positive Operators and Endomorphisms

## 3.1 Preliminaries About $r : B \to B$

Given an endomorphism $r$, we form the solenoid $\mathrm{Sol}(r) \subset B^{\mathbb{N}}$. Below we will study $\widehat{r} : \mathrm{Sol}(r) \to \mathrm{Sol}(r)$,

$$\widehat{r}(x_1 x_2 \dots) = (r(x_1) x_1 x_2 \dots)$$

and $\widehat{r} \in \mathrm{Aut}(\mathrm{Sol}(r))$.

Given a positive operator $R : C(B) \to C(B)$, $R1 = 1$ we then form the measure $\mathbb{P}_x^{(R)}$ in the usual way. We will prove the following property $\mathbb{P}_x^{(R)} \circ \widehat{r}^{-1} = \mathbb{P}_x^{(R)}$ on the solenoid but not on $B^{\mathbb{N}}$.

We will impose the conditon (3.10)

$$R((\varphi \circ r)\psi) = \varphi R \psi$$

as the only axiom. It may or may not be satisfied for some examples of positive operators $R$. But it does hold in the following two examples:

$$(R\varphi)(x) = \sum_{r(y)=x} W(y)\varphi(y) \text{ and}$$

$$(R\varphi)(x) = \frac{1}{\# r^{-1}(x)} \sum_{r(y)=x} |m(y)|^2 \varphi(y),$$

where the functions $W$ and $m$ are given subject to the usual conditions.

For reference to earlier papers dealing with measures on infinite products, random walk, and stochastic processes; see e.g., [JP11, JP10, AJ12].

*Example 3.1* Classical wavelet theory on the real line. Let $N = 2$, $B = \mathbb{T} = \{z \in \mathbb{C} : |z| = 1\} \simeq \mathbb{R}/\mathbb{Z} \simeq (-\frac{1}{2}, \frac{1}{2}]$ via $z = e^{2\pi i\theta}$, $\theta \in \mathbb{R}/\mathbb{Z}$; $\mu = d\theta$; $L^2(B, \mu) = L^2((-\frac{1}{2}, \frac{1}{2}], d\theta)$, $r : B \to B$,

$$r(z) = z^2, \text{ or equivalently} r(\theta \mod \mathbb{Z}) = 2\theta \mod \mathbb{Z}. \tag{3.1}$$

Let

$$m_0(\theta) = \sum_{n \in \mathbb{Z}} h_n e^{2\pi i n\theta}, \text{ or equivalently } m_0(z) = \sum_{n \in \mathbb{Z}} h_n z^n, \tag{3.2}$$

where we assume

$$\sum_{n\in\mathbb{Z}} h_n = \sqrt{2}, \quad \sum_{n\in\mathbb{Z}} |h_n|^2 < \infty.$$

**Lemma 3.2** *With $m_0$ as in (3.2), the condition (2.7) is equivalent to*

$$\sum_{k\in\mathbb{Z}} h_k \overline{h}_{k-2n} = \frac{1}{2}\delta_{n,0}. \tag{3.3}$$

**Proposition 3.3** ([BJ02, DJ05]) *Suppose that $m_0$ is as above, and that there is a solution $\varphi \in L^2(\mathbb{R})$ satisfying*

$$\frac{1}{\sqrt{2}}\varphi\left(\frac{x}{2}\right) = \sum_{n\in\mathbb{Z}} h_n \varphi(x-n), \tag{3.4}$$

*and*

$$\text{The translates } \varphi(\cdot - n) \text{ are orthogonal in } L^2(\mathbb{R}),\ n \in \mathbb{Z}. \tag{3.5}$$

*Set $W : L^2(\mathbb{T}) \to L^2(\mathbb{R})$,*

$$(W\xi)(x) = \sum_{n\in\mathbb{Z}} \widehat{\xi}(n)\varphi(x-n) =: \pi(\xi)\varphi, \tag{3.6}$$

*where $\xi \in L^2(\mathbb{T})$, and $\widehat{\xi}(n) = \int_{\mathbb{T}} \overline{e}_n \xi \, d\mu$;*

$$(S_0\xi)(z) = m_0(z)\xi(z^2), \quad (z \in \mathbb{T}); \tag{3.7}$$

*and*

$$(Uf)(x) = \frac{1}{\sqrt{2}} f\left(\frac{x}{2}\right), \quad f \in L^2(\mathbb{R}). \tag{3.8}$$

(i) *Then $S_0$ is isometric, and $(L^2(\mathbb{R}), \varphi, \pi, U)$ is a wavelet representation.*
(ii) *The dilation $W : L^2(\mathbb{T}) \to L^2(\mathbb{R})$ then takes the following form: $W$ is isometric and it intertwines $S_0$ and the unitary operator $U$, i.e., we have*

$$(WS_0\xi)(x) = (UW\xi)(x) = \frac{1}{\sqrt{2}}(W\xi)\left(\frac{x}{2}\right), \quad (\xi \in L^2(\mathbb{T}), x \in \mathbb{R}). \tag{3.9}$$

*Remark 3.4* With $m_0$ as specified in Proposition 3.3, we conclude that the wavelet representation can be realized on $L^2(\mathbb{R})$. On the other hand, we will see in Corollary 4.5 that it can be also realized on the solenoid. The two representations have to be isomorphic. The identifications can be done via the usual embedding of $\mathbb{R}$ into the solenoid $x \mapsto (e^{2\pi i x}, e^{2\pi i x/2}, e^{2\pi i x/2^2}, \ldots)$. The measure $\Sigma$ in this case is supported on the image of $\mathbb{R}$ under this embedding. For details, see [Dut06].

**Axioms.** $B$ compact Hausdorff space, $R : C(B) \to C(B)$ positive linear operator such that $R1 = 1$, $r : B \to B$ onto, continuous.

Assume

$$R((\varphi \circ r)\psi) = \varphi R(\psi), \quad (\varphi, \psi \in C(B)) \tag{3.10}$$

Note that (3.10) is the only property that we assume on the operator $R$.

**Lemma 3.5** *On the solenoid*

$$\mathrm{Sol}(r) = \{(x_1, x_2, \ldots) \in B^{\mathbb{N}} : r(x_{i+1}) = x_i\},$$

$$\pi_i \circ \widehat{r} = r \circ \pi_i.$$

*Proof* For $\tilde{x} = (x_1, x_2, \ldots)$,

$$\widehat{r}(\tilde{x}) = (r(x_1), x_1, x_2, \ldots). \tag{3.11}$$

$$\pi_i \circ \widehat{r}(\tilde{x}) = x_{i-1} = r(x_i) = r \circ \pi_i(\tilde{x}).$$

□

*Remark 3.6* Our initial setup for a given endomorphism $r$ in our present setup is deliberately left open to a variety of possibilities. Indeed, the literature on solenoid analysis is vast, but divides naturally into cases when $r : B \to B$ has only one contractivity degree; as opposed to a mix of non-linear contractive directions. The first case is common in wavelet analysis, such as those studied in [DJ06, DJ07, DJ10, DJ12]. Examples of the second class, often called "hyperbolic" systems, includes the Smale-Williams attractor, with the endomorphism $r$ there prescribed to preserve a foliation by meridional disks; see e.g., [Kuz10, KP07, KP07, Rue04]. Or the study of complex dynamics and Julia sets; see e.g., [BCMN04] .

**Lemma 3.7** *Let* $r : B \to B$ *be given and let* $\widehat{r} \in \mathrm{Aut}(\mathrm{Sol}(r))$ *be the induced automorphism on the solenoid. Then*

$$\widehat{r}(\pi_1^{-1}(x)) = \pi_1^{-1}(r(x)) \cap \pi_2^{-1}(x), \quad (x \in B).$$

*Proof* Use the definition of $\widehat{r}$ in (3.11). □

**Definition 3.8** Given $\mu$ and $R$, they generate the probability measure $\Sigma = \Sigma^{(\mu)}$ on $B^{\mathbb{N}}$. We assume $R1 = 1$ and $\mu(B) = 1$. Define

$$\mathbb{E}(f) = \int_{B^{\mathbb{N}}} f \, d\Sigma \tag{3.12}$$

$$\mathbb{E}_x(f) := \mathbb{E}(f \mid \pi_1 = x) = \int_{\pi_1^{-1}(x)} f \, d\Sigma \tag{3.13}$$

$$\mathbb{E}_{x_1,x_2}(f) := \mathbb{E}(f \mid \pi_1 = x_1, \pi_2 = x_2) = \int_{\pi_1^{-1}(x_1) \cap \pi_2^{-1}(x_2)} f \, d\Sigma. \tag{3.14}$$

for all $x_1, x_2 \in B$. As before we take

$$\mathbb{E}(f) = \int_B \int_{\pi^{-1}(x)} f \, d\mathbb{P}_x \, d\mu(x) \tag{3.15}$$

and we then get

$$\mathbb{E}_x(f) = \int_{\pi^{-1}(x)} f \, d\mathbb{P}_x \tag{3.16}$$

**Lemma 3.9** *Let $B$, $R$ and $r$ be given as above.*

$$R((\varphi \circ r)\psi) = \varphi R(\psi) \tag{3.17}$$

*Then the following two are equivalent for some measure $\mu$ on $B$:*

(i) $\mu \circ R = \mu$
(ii) $\int (\varphi \circ r)\psi \, d\mu = \int \varphi R\psi \, d\mu$.

*Proof* (i)⇒(ii). Assume (i) and (3.17). Then

$$\int \varphi \circ r \cdot \psi \, d\mu = \int R((\varphi \circ r)\psi) \, d\mu = \int \varphi R\psi \, d\mu$$

which is condition (ii).

(ii)⇒(i). Assume (ii). Then set $\varphi = 1$ in (ii) and we get $\int \psi \, d\mu = \int R\psi \, d\mu$ which is the desired property (i). □

**Lemma 3.10** *Assume the basic axiom* (3.17). *For $f \in L^1(\Sigma)$, we denote by $\mathbb{E}\bullet(f)$, the function $x \mapsto \mathbb{E}_x(f)$, $x \in B$. Then*

$$\mathbb{E}_x(f \circ \sigma) = R(\mathbb{E} \bullet (f))(x) \tag{3.18}$$

*Also,*

$$\mathbb{E}_x(f \circ \widehat{r}) = \mathbb{E}_{r(x),x}(f) \tag{3.19}$$

*for all $x \in B$, $f \in L^1(\Sigma)$, or equivalently*

$$\mathbb{E}(f \circ \widehat{r} \mid \pi_1 = x) = \mathbb{E}(f \mid \pi_1 = r(x), \pi_2 = x); \tag{3.20}$$

*see the notations in* Definition 3.8.

*Proof* Equation (3.18) is proved in (2.32). For (3.19), we use the Stone-Weierstrass approximation as before. If

$$f = (\varphi_1 \circ \pi_1)(\varphi_2 \circ \pi_2) \ldots (\varphi_n \circ \pi_n),$$

then

$$f \circ \widehat{r} = (\varphi_1 \circ r \circ \pi_1)(\varphi_2 \circ r \circ \pi_2) \ldots (\varphi_n \circ r \circ \pi_n),$$

and so

$$\begin{aligned}\mathbb{E}_x(f \circ \widehat{r}) &= \varphi_1(r(x)) R(\varphi_2 \circ r R(\varphi_3 \circ r \ldots R(\varphi_n \circ r)) \ldots)(x) \\ &= \varphi_1(r(x)) \varphi_2(x) R(\varphi_3 R(\ldots \varphi_{n-1} R(\varphi_n)) \ldots)(x) = \mathbb{E}_{r(x),x}(f),\end{aligned}$$

or equivalentlly (3.20). □

**Proposition 3.11** *Let $B$ and $R : C(B) \to C(B)$ be as stated in* Theorem 2.11. *For every $\mu \in \mathcal{M}_1(B)$ we denote the induced measure on $\Omega = B^{\mathbb{N}}$ by $\Sigma^{(\mu)}$. If some $r : B \to B$ satisfies*

$$R((\varphi \circ r)\psi) = \varphi R(\psi), \quad (\varphi, \psi \in C(B)) \tag{3.21}$$

*then every one of the induced measures $\Sigma^{(\mu)}$ has its support contained in the solenoid* $\mathrm{Sol}(r)$.

*Proof* Using Lemma 2.14, it is enough to prove that each of the measures $\mathbb{P}_x^{(R)}$ with $x$ fixed (from Corollary 2.12) has its support equal to

$$\pi_1^{-1}(x) \cap \mathrm{Sol}(r) \tag{3.22}$$

For every $n$, consider all infinite words indexed by $y \in r^{-n}(x)$ and specified on the beginning length-$n$ segments as follows $\Omega_n(r, x)$ : $(x, r^{n-1}(y), \ldots, r(y), y,$ free infinite tail) and note that

$$\pi_1^{-1}(x) \cap \mathrm{Sol}(r) = \bigcap_n \Omega_n(r, x) \tag{3.23}$$

For $n = 1$, we have

$$\mathbb{P}_x^{(R)}(\Omega_1(r, x)) = R(\chi_{\{x\}} \circ r)(x) = \chi_{\{x\}}(x) R(1) = 1,$$

where we used assumption (3.21) in the last step in the computation.

The remaining reasoning in the proof is an induction. Indeed, one checks that

$$\begin{aligned}\mathbb{P}_x^{(R)}(\Omega_n(r,x)) &= R((\chi_{\{x\}}\circ r)R((\chi_{\{x\}}\circ r^2)R(\dots(\chi_{\{x\}}\circ r^{n-1})R(\chi_{\{x\}}\circ r^n)\dots)))(x)\\ &= R((\chi_{\{x\}}\circ r)R(\dots R(\chi_{\{x\}}\circ r^{n-1})\dots))(x).\end{aligned}$$

Hence the assertion for $n-1$ implies the next step $n$. By induction, we get

$$\mathbb{P}_x^{(R)}(\Omega_n(r,x)) = 1, \quad (n\in\mathbb{N}, x\in B).$$

Using (3.23), we get

$$\mathbb{P}_x^{(R)}(\pi_1^{-1}(x)\cap \mathrm{Sol}(r)) = \lim_{n\to\infty}\mathbb{P}_x^{(R)}(\Omega_n(r,x)) = 1.$$

As a consequence, the measure $\mathbb{P}_x^{(R)}$ assigns value 1 to the indicator function of $\pi_1^{-1}(x)\cap\mathrm{Sol}(r)$. But

$$\mathrm{Sol}(r) = \bigcup_{x\in B}\pi_1^{-1}(x)\cap\mathrm{Sol}(r).$$

So if $\mu(B)=1$, it follows from (2.51) that

$$\Sigma^{(\mu)}(\mathrm{Sol}(r)) = \int_{B^{\mathbb{N}}}\chi_{\mathrm{Sol}(r)}\,d\Sigma^{(\mu)} = 1;$$

and as a result that

$$\Sigma^{(\mu)}(B^{\mathbb{N}}\backslash\mathrm{Sol}) = 0$$

which is the desired conclusion. □

**Corollary 3.12** *Let $B, r, \mu, R$ be as above and assume* (3.17). *Then $\Sigma$ is supported on* $\mathrm{Sol}(r)$ *and $\widehat{r}$ is invertible on* $\mathrm{Sol}(r)$ *with $\widehat{r}^{-1}=\sigma$. The measure $\Sigma$ is invariant (for $\widehat{r}$) if and only if*

$$\mu\circ R = \mu. \tag{3.24}$$

*Proof* It is enough to prove that

$$\int_{\mathrm{Sol}(r)} f\circ\sigma\,d\Sigma = \int_{\mathrm{Sol}(r)} f\,d\Sigma \tag{3.25}$$

holds for all $f\in L^1(\Sigma)$ if and only if (3.24) holds. But, by (3.18) we have

$$\int_{\mathrm{Sol}(r)} f \circ \sigma \, d\Sigma = \int_B \mathbb{E}_x(f \circ \sigma) \, d\mu(x) = \int_B R(\mathbb{E} \bullet (f))(x) \, d\mu(x),$$

and

$$\int_{\mathrm{Sol}(r)} f \, d\Sigma = \int_B \mathbb{E}_x(f) \, d\mu(x).$$

But the functions $x \mapsto \mathbb{E}_x(f)$ are dense in $L^1(B, \mu)$ as $f$ varies in $L^1(\Sigma)$ (consider for example $f = g \circ \pi_1$ for $g \in C(B)$). Thus the equivalence of (3.24) and (3.25) is immediate from this. □

**Corollary 3.13** *Let $R$ be a positive operator in $C(B)$ satsifying the axioms above, $R1 = 1$, $R((\varphi \circ r)\psi) = \varphi R(\psi)$ for all $\varphi, \psi \in C(B)$. Let $\mu$ be a Borel measure on $B$, and set $\Sigma = \Sigma^{(\mu)}$*

$$\int_{\mathrm{Sol}(r)} f \, d\Sigma := \int_B \int_{\pi_1^{-1}(x)} f \, d\mathbb{P}_x^{(R)} \, d\mu(x);$$

*then*

$$\mathcal{U}^{(R)} f := f \circ \widehat{r} \tag{3.26}$$

*defines a unitary operator on $L^2(\mathrm{Sol}(r), \Sigma)$ if and only if $\mu = \mu \circ R$.*

*Proof* Since $\widehat{r}$ is invertible in $\mathrm{Sol}(r)$, we conclude that $\mathcal{U}^{(R)}$ maps onto $L^2(\mathrm{Sol}(r), \Sigma)$. Recall $\widehat{r}^{-1} = \sigma$

$$\sigma(x_1, x_2, x_3, \ldots) = (x_2, x_3, x_4, \ldots) \tag{3.27}$$

Since, by Proposition 3.11, the measure $\Sigma$ is supported on $\mathrm{Sol}(r)$, the result follows from Corollary 3.12. □

**Corollary 3.14** *Let $B$ and $R : C(B) \to C(B)$ be as in* Corollary 2.12. *Let $\mu \in \mathcal{M}_1(B)$ and consider $\Sigma = \Sigma^{(\mu)}$. Let $r : B \to B$ be an endomorphism.*

(i) *For the operators $V_1 : L^2(B, \mu) \to L^2(\mathrm{Sol}(r), \Sigma)$ and $\mathcal{U}^{(R)} : f \mapsto f \circ \widehat{r}$ acting in $L^2(\mathrm{Sol}(r), \Sigma)$ we have the following covariance relation: $V_1$ is isometric and*

$$(V_1^* \mathcal{U}^{(R)} V_1)(\varphi) = \varphi \circ r, \quad (\varphi \in C(B)).$$

(ii) *Assume that $\mu = \mu \circ R$ and*

$$R((\varphi \circ r)\psi) = \varphi R(\psi), \quad (\varphi, \psi \in C(B)) \tag{3.28}$$

*holds. For functions $F$ on $\Omega$, say $F \in L^{\infty}(\mathrm{Sol}(r))$, let $M_F$ be the multiplication operator defined by $F$. Then $\mathcal{U}^{(R)}$ is unitary and the following covariance relation holds:*

$$(\mathcal{U}^{(R)})^* M_F \mathcal{U}^{(R)} = M_{F\circ\sigma}.$$

*Proof* (i) We will make use of the formula (2.19) for $V_1^*$, see Corollary 2.12(ii). We now compute

$$(V_1^* \mathcal{U}^{(R)} V_1 \varphi)(x) = \mathbb{E}_x \mathcal{U}^{(R)} V_1 \varphi = \mathbb{E}_x(\varphi \circ \pi_1 \circ \widehat{r}) = \mathbb{E}_x(\varphi \circ r \circ \pi_1) = (\varphi \circ r)(x),$$

which is the conclusion in (i). (ii) We proved in Corollary 3.13 that $\mathcal{U}^{(R)}$ is unitary. The covariance relation follows from a simple computation. □

For reference to earlier papers dealing with measures on infinite products, and shift-invariant systems; see e.g., [CGHU12, CH94].

## 3.2 Compact Groups

As a special case of our construction, we mention the compact groups; this will include the case of wavelet theory.

**Proposition 3.15** *Assume $B$ is a compact group with normalized Haar measure $\mu$. Let $r : B \to B$ be a homomorphism $r(xy) = r(x)r(y)$ for all $x, y \in B$, and assume for $N \in \mathbb{N}$, $N > 1$*

$$\#r^{-1}(x) = N, \quad (x \in B).$$

*Set*

$$(R\varphi)(x) = \frac{1}{N} \sum_{r(y)=x} \varphi(y), \quad (\varphi \in C(B)); \tag{3.29}$$

*then*

(i) $\mathrm{Sol}(r)$ *is a compact subgroup of* $B^{\mathbb{N}}$.
(ii) *The induced measure* $\Sigma = \Sigma^{(\mu,R)}$ *is the Haar measure on the group* $\mathrm{Sol}(r)$.

*Proof* (i) If $\tilde{x} = (x_1, x_2, \dots)$, $\tilde{y} = (y_1, y_2, \dots) \in \mathrm{Sol}(r)$ then $r(x_{i+1}y_{i+1}) = r(x_{i+1})r(y_{i+1}) = x_i y_i$, so $\tilde{x}\tilde{y} \in \mathrm{Sol}(r)$.

(ii) From (3.29) we see that the measures $\mathbb{P}_x$ in the decomposition

$$d\Sigma = \int_B \mathbb{P}_x \, d\mu(x) \tag{3.30}$$

$\mathbb{P}_x \in \mathcal{M}(\pi_1^{-1}(x))$, $x \in B$, are random-walk measures with uniform distributions on the points in $r^{-1}(x)$, for all $x \in B$. If $\mathbb{E}$ denotes the $\Sigma$-expectation and $\mathbb{E}_x$ the $\mathbb{P}_x$-expectation, we have

$$\mathbb{E}_x(f) = \mathbb{E}(f \mid \pi_1 = x), \quad (x \in B).$$

For points $\tilde{y} = (y_1, y_2, \dots) \in \mathrm{Sol}(r)$, denote by $f(\cdot\tilde{y})$ the translated function on $\mathrm{Sol}(r)$. Then

$$\mathbb{E}(f(\cdot\tilde{y}) \mid \pi_1 = x) = \mathbb{E}(f \mid \pi_1 = xy_1) \tag{3.31}$$

where we use the terminology in Proposition 2.17.

Now, combining (3.30) and (3.31), we arrive at the formula:

$$\begin{aligned}\int_{\mathrm{Sol}(r)} f(\cdot\tilde{y})\, d\Sigma = \mathbb{E}(f(\cdot\tilde{y})) &= \int_B \mathbb{E}(f(\cdot\tilde{y}) \mid \pi_1 = x)\, d\mu(x) = \int_B \mathbb{E}(f \mid \pi_1 = xy_1)\, d\mu(x) \\ &= \int_B \mathbb{E}(f \mid \pi_1 = x)\, d\mu(x) = \int_{\mathrm{Sol}(r)} f\, d\Sigma.\end{aligned}$$

□

*Remark 3.16* In wavelet theory, one often takes $B = \mathbb{R}^n/\mathbb{Z}^n$, and a fixed $n \times n$ matrix $A$ over $\mathbb{Z}$ such that the eigenvalues $\lambda$ satisfy $|\lambda| > 1$. For $r : B \to B$, then take

$$r(x \bmod \mathbb{Z}^n) = Ax \bmod \mathbb{Z}^n, \quad (x \in \mathbb{R}^n)$$

and it is immediate that $r$ satisfies the multiplicative property in Proposition 3.15.

There are important examples when $r : B \to B$ does not satisfy this property.

*Example 3.17* (Non-group case: the Smale-Williams attractor) Take $B = \mathbb{T} \times \mathbb{D}$, where $\mathbb{T} = \mathbb{R}/\mathbb{Z}$ and $\mathbb{D} = \{z \in \mathbb{C} : |z| \leq 1\}$ the disk. For $(t, z) \in \mathbb{T} \times \mathbb{D}$, set

$$r(t, z) = (2t \bmod \mathbb{Z}, \frac{1}{4}z + \frac{1}{2}e^{2\pi i t}).$$

Then $\mathrm{Sol}(r)$ is the Smale-Williams attractor, see [KP07], a hyperbolic strange attractor.

## 4 Isometries

Below we study condition on functions $m : B \to \mathbb{C}$ which gurantees that $L^2(\mu) \ni f \mapsto m \cdot f \circ r \in L^2(\mu)$ defines an isometry in $L^2(B, \mu)$; and we will study the unitary dilations $L^2(\mathrm{Sol}(r), \Sigma) \ni \tilde{f} \mapsto \tilde{f} \circ \hat{r} \in L^2(\mathrm{Sol}(r), \Sigma)$.

**Setting.** $B$ fixed compact Hausdorff space. We introduce

(i) $r : B \to B$ measurable, onto such that

$$1 \le \# r^{-1}(x) < \infty, \quad (x \in B) \tag{4.1}$$

(ii) $\mu$ Borel measure on $B$, $\mu(B) = 1$.
(iii) $m : B \to \mathbb{C}$ a fixed function on $B$.

Question: Given two of them what are the conditions that the third should satisfy such that

$$L^2(\mu) \ni f \mapsto m \cdot f \circ r \in L^2(\mu) \tag{4.2}$$

is an isometry.

**Definition 4.1** Transformations of measures. Given $\nu$ measure on $B$, $\nu \in \mathcal{M}_1(B)$ and $r : B \to B$, set $\nu \circ r^{-1} \in \mathcal{M}_1(B)$. For $A \in \mathcal{B}(B)$ a Borel set, $(\nu \circ r^{-1})(A) := \nu(r^{-1}(A))$ where $r^{-1}(A) := \{x \in B : r(x) \in A\}$.

Fact: $\nu \circ r^{-1}$ is determined uniquely by the condition

$$\int_B \varphi \circ r \, d\nu = \int \varphi \, d(\nu \circ r^{-1}), \quad (\varphi \in C(B)) \tag{4.3}$$

**Lemma 4.2** *Fix* $r, \mu, m$*; then* (4.2) *is satisfied iff*

$$(|m|^2 \, d\mu) \circ r^{-1} = \mu \tag{4.4}$$

**Definition 4.3** Fix $r$, then we say that $\mu$ is *strongly invariant* iff

$$\int \varphi(x) \, d\mu(x) = \int \frac{1}{\# r^{-1}(x)} \sum_{r(y)=x} \varphi(y) \, d\mu(x) \tag{4.5}$$

**Lemma 4.4** *Given* $r$ *and assume* $\mu$ *is strongly invariant, then the isometry property* (4.2) *holds iff the corresponding positive operator*

$$(R\varphi)(x) := \frac{1}{\# r^{-1}(x)} \sum_{r(y)=x} |m(y)|^2 \varphi(y)$$

*satisfies* $R1 = 1$.

*Proof* Substitute (4.5) into (4.4). Note that then the equation

$$\int_B \varphi(r(x)) |m(x)|^2 \, d\mu(x) = \int_B \varphi(x) \frac{1}{\# r^{-1}(x)} \sum_{r(y)=x} |m(y)|^2 \, \mu(x), \quad (\varphi \in C(B))$$

holds; so $f \mapsto m \cdot f \circ r$ is isometric in $L^2(\mu)$ iff

$$\frac{1}{\# r^{-1}(x)} \sum_{r(y)=x} |m(y)|^2 = 1$$

$\mu$-a.e. $x \in B$. □

The next corollary appears in [DJ07, Theorem 5.5].

**Corollary 4.5** *Let $B$ and $r$ be as specified in* Sect. 3, *let $\mu$ be strongly invariant and let the function $m$ be quadrature mirror filter, as in* Example 2.5. *Assume in addition that $m$ is non-singular, i.e.*

$$\mu(\{x : m(x) = 0\}) = 0.$$

*Then, with $R$ as in* Lemma 4.4, *we get a wavelet representation as in* Theorem 2.6 *to $L^2(\mathrm{Sol}(r), \Sigma^{(\mu)})$ with*

(i) $\mathcal{H} = L^2(\mathrm{Sol}(r), \Sigma^{(\mu)})$;
(ii) $\mathcal{U} f = (m \circ \pi_1)(f \circ \widehat{r})$, *for all* $f \in L^2(\mathrm{Sol}(r), \Sigma^{(\mu)})$;
(iii) $\pi(g) f = (g \circ \pi_1) f$, *for all* $g \in L^\infty(B)$, $f \in L^2(\mathrm{Sol}(r), \Sigma^{(\mu)})$;
(iv) $\varphi = 1$.

*Proof* The details are contained in [DJ07, Theorem 5.5] and require just some simple computations. We only have to check that our measure $\Sigma^{(\mu)}$ coincides with the one defined in [DJ07]. For this, we use [DJ07, Theorem 5.3] and we have to check that

$$\int \varphi \circ \pi_n \, d\Sigma^{(\mu)} = \int R^n(\varphi) \, d\mu, \quad (\varphi \in C(B)).$$

But this follows immediately from the definition of $\Sigma^{(\mu)}$ in (2.17). □

**Proposition 4.6** *Let $B, r, \mu$ and $m_0$ as in* Example 2.5, *i.e., $m_0$ is a QMF and the measure $\mu$ on $B$ is assumed strongly invariant with respect to $r$, and let $R$ as in* Lemma 4.4. *Then the function $\rho = R^* 1$ in Corollary 2.20 is $\rho = |m_0|^2$.*

*Proof* The result follows if we verify the formula for $R^*$; we have

$$(R^* \psi)(x) = |m_0(x)|^2 \psi(r(x)). \tag{4.6}$$

The derivation of (4.6) may be obtained as a consequence of strong invariance as follows: for all $\varphi, \psi \in C(B)$, we have:

$$\int_B |m_0|^2 (\psi \circ r) \varphi \, d\mu = \int_B \psi(x) \frac{1}{N} \sum_{r(y)=x} |m_0(y)|^2 \varphi(y) \, d\mu(x) = \int_B \psi(x) (R\varphi)(x) \, d\mu(x);$$

and the assertion (4.6) follows. □

**Acknowledgments** One of the authors wishes to thank Professors Ka-Sing Lau, De-Jun Feng, and their colleagues, for organizing a wonderful conference in Hong-Kong, "The International Conference on Advances of Fractals and Related Topics", December 2012. Many discussions with participants at the conference inspired this paper. This work was partially supported by a grant from the Simons Foundation (#228539 to Dorin Dutkay).

## References

[AJ12] Alpay, D., Jorgensen, P.E.T.: Stochastic processes induced by singular operators. Numer. Funct. Anal. Optim. **33**(7–9), 708–735 (2012)

[ABL11] Antonevich, A.B., Bakhtin, V.I., Lebedev, A.V.: Crossed product of a $C^*$-algebra by an endomorphism, coefficient algebras, and transfer operators. Mat. Sb. **202**(9), 3–34 (2011)

[ABL12] Antonevich, A.B., Bakhtin, V.I., Lebedev, A.V.: A road to the spectral radius of transfer operators. In: Dynamical Systems and Group Actions, volume 567 of Contemp. Math., pages 17–51. Amer. Math. Soc., Providence, RI, 2012

[Arv86] Arveson, William: Markov operators and OS-positive processes. J. Funct. Anal. **66**(2), 173–234 (1986)

[Bal00] Baladi, V.: Positive Transfer Operators and Decay of Correlations, Advanced Series in Nonlinear Dynamics, vol. 16. World Scientific Publishing Co., Inc., River Edge (2000)

[BCMN04] Beardon, A.F., Carne, T.K., Minda, D., Ng, T.W.: Random iteration of analytic maps. Ergodic Theor. Dyn. Syst. **24**(3), 659–675 (2004)

[BJ02] Bratteli, O., Jorgensen, P.: Wavelets Through a Looking Glass, Applied and Numerical Harmonic Analysis. Birkhäuser Boston Inc., Boston (2002). The world of the spectrum

[CGHU12] Chazottes, J.-R., Gambaudo, J.-M., Hochman, M., Ugalde, E.: On the finite-dimensional marginals of shift-invariant measures. Ergodic Theor. Dyn. Syst. **32**(5), 1485–1500 (2012)

[CH94] Courbage, M., Hamdan, D.: Chapman-Kolmogorov equation for non-Markovian shift-invariant measures. Ann. Prob. **22**(3), 1662–1677 (1994)

[Dau92] Daubechies, I.: Ten lectures on wavelets. In: CBMS-NSF Regional Conference Series in Applied Mathematics. Society for Industrial and Applied Mathematics (SIAM), vol. 61. Philadelphia, PA (1992)

[DMP08] D'Andrea, J., Merrill, K.D., Packer, J.: Fractal wavelets of Dutkay-Jorgensen type for the Sierpinski gasket space. In Frames and operator theory in analysis and signal processing, volume 451 of Contemp. Math., pages 69–88. Amer. Math. Soc., Providence, RI, (2008)

[DHSW11] Dutkay, D.E., Han, D., Sun, Q., Weber, E.: On the Beurling dimension of exponential frames. Adv. Math. **226**(1), 285–297 (2011)

[DJ05] Dutkay D.E., Jorgensen, P.E.T.: Hilbert spaces of martingales supporting certain substitution-dynamical systems. Conform. Geom. Dyn. **9**(electronic), 24–45 (2005)

[DJ06] Dutkay, D.E., Jorgensen, P.E.T.: Wavelets on fractals. Rev. Mat. Iberoam. **22**(1), 131–180 (2006)

[DJ07] Dutkay, D.E., Jorgensen, P.E.T.: Martingales, endomorphisms, and covariant systems of operators in Hilbert space. J. Oper. Theor. **58**(2), 269–310 (2007)

[DJ10] Dutkay, D.E., Jorgensen, P.E.T.: Spectral theory for discrete Laplacians. Complex Anal. Oper. Theory **4**(1), 1–38 (2010)

[DJ11a] Dutkay, D.E., Jorgensen, P.E.T.: Affine fractals as boundaries and their harmonic analysis. Proc. Amer. Math. Soc. **139**(9), 3291–3305 (2011)

[DJ11b] Dutkay, D.E., Jorgensen, P.E.T.: Spectral duality for unbounded operators. J. Oper. Theory **65**(2), 325–353 (2011)

[DJ12] Dutkay, D.E., Jorgensen, P.E.T.: Fourier duality for fractal measures with affine scales. Math. Comp. **81**(280), 2253–2273 (2012)

[DJP09] Dutkay, D.E., Jorgensen, P.E.T, Picioroaga, G.: Unitary representations of wavelet groups and encoding of iterated function systems in solenoids. Ergodic Theor. Dyn. Syst. **29**(6), 1815–1852 (2009)

[DJS12] Dutkay, D.E., Jorgensen, P.E.T., Silvestrov, S.: Decomposition of wavelet representations and Martin boundaries. J. Funct. Anal. **262**(3), 1043–1061 (2012)

[DL10] Dong, X.-H., Lau, K.-S.: Cantor boundary behavior of analytic functions. In Recent developments in fractals and related fields, Appl. Numer. Harmon. Anal., pages 283–294. Birkhäuser Boston Inc., Boston, MA, 2010

[DLS11] Dutkay, D.E., Larson, D.R., Silvestrov, S.: Irreducible wavelet representations and ergodic automorphisms on solenoids. Oper. Matrices **5**(2), 201–219 (2011)

[DS11] Dutkay, D.E., Silvestrov, S.: Reducibility of the wavelet representation associated to the Cantor set. Proc. Amer. Math. Soc. **139**(10), 3657–3664 (2011)

[Dut06] Dutkay, D.E.: Low-pass filters and representations of the Baumslag Solitar group. Trans. Amer. Math. Soc. **358**(12), 5271–5291 (electronic) (2006)

[FM12] Fraczek, M., Mayer, D.: Symmetries of the transfer operator for $\Gamma_0(N)$ and a character deformation of the Selberg zeta function for $\Gamma_0(4)$. Algebra Number Theory **6**(3), 587–610 (2012)

[Hen12] Hensley, D.: Continued fractions, Cantor sets, Hausdorff dimension, and transfer operators and their analytic extension. Discrete Contin. Dyn. Syst. **32**(7), 2417–2436 (2012)

[JP10] Jorgensen, P.E.T., Pearse, E.P.J.: A Hilbert space approach to effective resistance metric. Complex Anal. Oper. Theory **4**(4), 975–1013 (2010)

[JP11] Jorgensen, P.E.T., Pearse, E.P.J.: Ge'fand triples and boundaries of infinite networks. New York J. Math. **17**, 745–781 (2011)

[KP07] Kuznetsov, S.P., Pikovsky, A.: Autonomous coupled oscillators with hyperbolic strange attractors. Phys. D **232**(2), 87–102 (2007)

[Kuz10] Kuznetsov, S.P.: Example of blue sky catastrophe accompanied by a birth of Smale-Williams attractor. Regul. Chaotic Dyn. **15**(2–3), 348–353 (2010)

[Kwa12] Kwaśniewski, B.K.: On transfer operators for $C^*$-dynamical systems. Rocky Mountain J. Math. **42**(3), 919–936 (2012)

[LN12] Lau, K.-S., Ngai, S.-M.: Martin boundary and exit space on the Sierpinski gasket. Sci. China Math. **55**(3), 475–494 (2012)

[LR69] Lanford III, O.E., Ruelle, D.: Observables at infinity and states with short range correlations in statistical mechanics. Comm. Math. Phys. **13**, 194–215 (1969)

[LW09] Lau, K.-S., Wang, X.-Y.: Self-similar sets as hyperbolic boundaries. Indiana Univ. Math. J. **58**(4), 1777–1795 (2009)

[MP11] Marcolli, M., Paolucci, A.M.: Cuntz-Krieger algebras and wavelets on fractals. Complex Anal. Oper. Theory **5**(1), 41–81 (2011)

[MMS12] Mayer, D., Mühlenbruch, T., Strömberg, F.: The transfer operator for the Hecke triangle groups. Discrete Contin. Dyn. Syst. **32**(7), 2453–2484 (2012)

[Nau12] Naud, F.: The Ruelle spectrum of generic transfer operators. Discrete Contin. Dyn. Syst. **32**(7), 2521–2531 (2012)

[Rue96] Ruelle, D.: Sharp zeta functions for smooth interval maps. In: International Conference on Dynamical Systems (Montevideo, 1995), volume 362 of Pitman Res. Notes Math. Ser., pages 188–206. Longman, Harlow, 1996.

[Rue02] Ruelle, D.: Dynamical zeta functions and transfer operators. Notices Amer. Math. Soc. **49**(8), 887–895 (2002)

[Rue04] Ruelle, D.: Thermodynamic Formalism: The Mathematical Structures of Equilibrium Statistical Mechanics. Cambridge Mathematical Library, 2nd edn. Cambridge University Press, Cambridge (2004)

# Generalized Energy Inequalities and Higher Multifractal Moments

**Kenneth Falconer**

**Abstract** We present a class of generalized energy inequalities which may be used to investigate higher multifractal moments, in particular $L^q$-dimensions of images of measures under Brownian-type processes, $L^q$-dimensions of almost self-affine measures, and moments of random cascade measures.

## 1 Introduction

Calculations in fractal geometry often fall into two parts: a geometric part and an analytic part. The geometric part may involve expressing geometric or metric aspects of a problem in mathematical terms leading to an analytic argument to estimate the integrals, sums, etc. so obtained. There are various analytic methods that are applicable to a range of problems across fractal geometry, for example, covering or potential theoretic methods for estimating dimensions. Here we look at an analytic technique which extends the potential theoretic method to higher moments and we indicate several applications.

## 2 $L^q$-Dimensions and Images of Measures

Coarse multifractal analysis reflects the asymptotic behavior of the moment sums of measures over small grid cubes. Let $\mathcal{M}_r$ be the set of mesh cubes of side $r$, that is cubes in $\mathbb{R}^n$ of the form $[j_1 r, (j_1+1)r) \times \cdots \times [j_n r, (j_n+1)r)$ where $j_1, \ldots, j_n \in \mathbb{Z}$. Let $\mu$ be a Borel measure of bounded support on $\mathbb{R}^n$. Define the *$q$-th power moment sum* of $\mu$ by

K. Falconer (✉)
Mathematical Institute, University of St Andrews, North Haugh, St Andrews, Fife KY 16 9SS, Scotland, UK
e-mail: kjf@st-andrews.ac.uk



D.-J. Feng and K.-S. Lau (eds.), *Geometry and Analysis of Fractals*, Springer Proceedings in Mathematics & Statistics 88,
DOI 10.1007/978-3-662-43920-3_4

$$M_r(q) = \sum_{C \in \mathcal{M}_r} \mu(C)^q. \tag{2.1}$$

The $L^q$*-dimension* or *generalized q dimension* of $\mu$ is given by

$$D_q(\mu) = \frac{1}{q-1} \lim_{r \searrow 0} \frac{\log M_r(q)}{\log r} \qquad (q > 0). \tag{2.2}$$

If this limit does not exist we may still take lower or upper limits to get the *lower* and *upper* $L^q$-dimensions:

$$\underline{D}_q(\mu) = \frac{1}{q-1} \underline{\lim}_{r \searrow 0} \frac{\log M_r(q)}{\log r} \quad \text{and} \quad \overline{D}_q(\mu) = \frac{1}{q-1} \overline{\lim}_{r \searrow 0} \frac{\log M_r(q)}{\log r}, \tag{2.3}$$

see, for example, [Fal97] The definitions (2.2) and (2.3) are unchanged if we replace the moment sum by a moment integral

$$M_r(q) = \int \mu(B(x,r))^{q-1} d\mu(x) \qquad (q > 0), \tag{2.4}$$

see [Lau95] for the case of $q > 1$ and [PS00] for $0 < q < 1$.

Often of interest are the dimensions of the image of a set or the generalized dimensions of the image of a measure under a parameterized family of mappings. Let $X$ be a metric space, and let $x_\omega : X \to \mathbb{R}^n$ be a family of continuous mappings where $\omega \in \Omega$ for some parameter space $\Omega$. Let $\mu$ be a Borel measure on $X$ and let $\mu_\omega$ be its image measure under $x_\omega$, so

$$\mu_\omega(A) = \mu(x_\omega^{-1}(A)) \qquad (A \in \mathbb{R}^n)$$

or

$$\int f(x) d\mu_\omega(x) = \int f(x_\omega(t)) d\mu(t) \qquad (f : \mathbb{R}^n \to \mathbb{R}).$$

For a basic example, $x_\omega$ might be orthogonal projection from $\mathbb{R}^m$ onto a line $L_\omega$ (which we may identify with $\mathbb{R}$) in direction $\omega$, with $\mu_\omega$ the corresponding projection of the measure $\mu$ on $\mathbb{R}^m$ onto $L_\omega$.

Now suppose $(\Omega, \mathsf{P}, \mathcal{F})$ is a probability space and write $\mathsf{E}$ for expectation. One way of obtaining lower estimates for $L^q$-dimensions of $\mu_\omega$ valid for almost all $\omega$ is to bound the mean moment integrals. When $q \geq 2$ is an integer:

$$\begin{aligned} &\mathsf{E} \int \mu_\omega(B(x,r))^{q-1} d\mu_\omega(x) \\ &= \mathsf{E} \int \mu_\omega\{y_1 : |x - y_1| \leq r\} \ldots \mu_\omega\{y_{q-1} : |x - y_{q-1}| \leq r\} d\mu_\omega(x) \end{aligned}$$

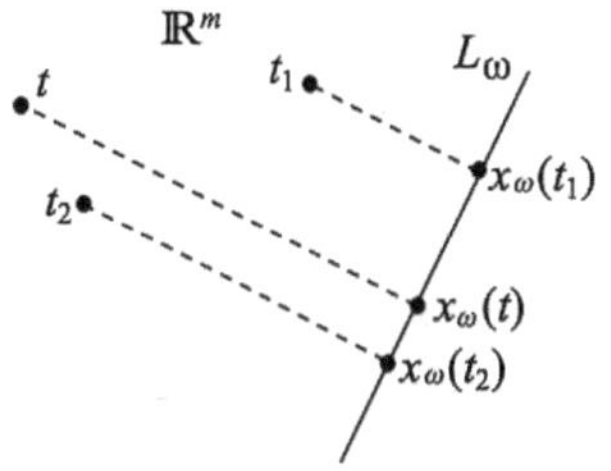


**Fig. 1** Projection of three points onto a line parameterized by $\omega$

$$= \mathsf{E} \int \mu\{t_1 : |x_\omega(t) - x_\omega(t_1)| \le r\} \dots \mu\{t_{q-1} : |x_\omega(t) - x_\omega(t_{q-1})| \le r\} d\mu(t)$$

$$= \mathsf{E} \int \cdots \int \chi_{\{|x_\omega(t) - x_\omega(t_j)| \le r \text{ for all } j\}}(t_1, \dots, t_{q-1}, t) d\mu(t_1) \dots d\mu(t_{q-1}) d\mu(t)$$

$$= \int \cdots \int \mathsf{P}\{|x_\omega(t) - x_\omega(t_j)| \le r \text{ for all } j\} d\mu(t_1) \dots d\mu(t_{q-1}) d\mu(t). \quad (2.5)$$

We may be able to use the geometry of the situation to estimate $\mathsf{P}\{|x_\omega(t) - x_\omega(t_j)| \le r \text{ for all } j\}$, which depends on the relative closeness of the $t_1, \dots, t_{q-1}, t$ in the metric space. For example, with $x_\omega : \mathbb{R}^m \to L_\omega$ as projection onto the line $L_\omega$ where $\omega \in \Omega$ is distributed according to the natural invariant measure on the space of directions $\Omega$, the probability $\mathsf{P}\{|x_\omega(t) - x_\omega(t_j)| \le r \text{ for all } j\}$ is affected more, but not exclusively, by the $t_j$ that are furthest from $t$, see Fig. 1. In particular, bounding (2.5) by const$\cdot r^{s(q-1)}$ may lead to a lower bound of $s$ for the $L^q$-dimension of $\mu_\omega$ for almost all $\omega$.

In the case when $q = 2$ the integral (2.5) may be estimated by

$$\int\int \mathsf{P}\{|x_\omega(t) - x_\omega(t_1)| \le r\} d\mu(t_1) d\mu(t) \le \int\int \mathsf{E}\left(\frac{r^s}{|x_\omega(t) - x_\omega(t_1)|^s}\right) d\mu(t_1) d\mu(t)$$

for all $s > 0$. This expectation can often be estimated using a transversality argument which results in an energy-type integral. The classic case of this is in the projection theorems, see for example [Fal14, Mat99] for the projection case and [PYS00] for a more general setting.

## 3 The Main Inequality

In this section we consider an approach to estimating integrals such as (2.5) for $q > 1$ and present an inequality which may be applied in various settings. It is convenient to take $X$ to be the symbolic space on a set of $m \ge 2$ symbols, $\Lambda \equiv \{1, \dots, m\}$. Thus $\Lambda^k$ consists of the words of length $k$ for $k \ge 0$ and we write $\Lambda^* \equiv \cup_{i=0}^k \Lambda^k$ which we identify with the vertices of the $m$-ary rooted tree in the usual way. The infinite sequences, identified with the boundary of the tree, are denoted by $\Lambda^\infty$.

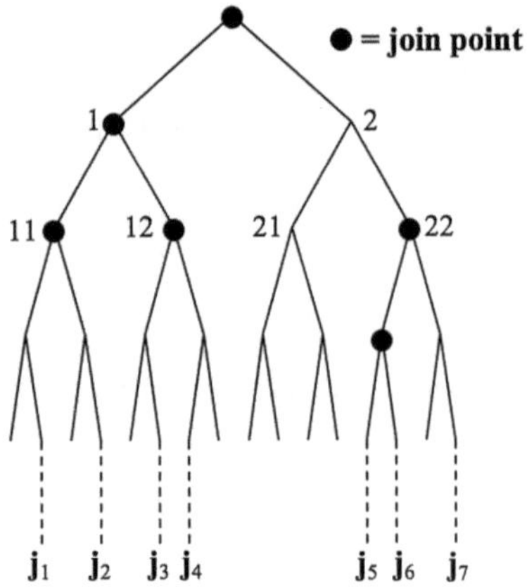


**Fig. 2** A set of 7 points in $\Lambda^\infty$ with their 6 join points in $\Lambda^*$

For $\mathbf{i} = i_1, \ldots, i_k \in \Lambda^*$ we write $|\,\mathbf{i}\,| = k$ for the length of the word $\mathbf{i}$. For $\mathbf{i} \in \Lambda^*$ and $\mathbf{j} \in \Lambda^* \cup \Lambda^\infty$ we write $\mathbf{j} \succ \mathbf{i}$ to mean that $\mathbf{i}$ is an initial segment of $\mathbf{j}$. The *cylinders* are the sets $C_\mathbf{i} = \{\mathbf{j} \in \Lambda^\infty : \mathbf{j} \succ \mathbf{i}\}$ for each $\mathbf{i} \in \Lambda^*$. The cylinders provide a basis for the natural topology on $\Lambda^\infty$.

Write $\mathbf{j}_1 \wedge \mathbf{j}_2 \in \Lambda^*$ for the *join* of $\mathbf{j}_1, \mathbf{j}_2 \in \Lambda^\infty$, that is the longest $\mathbf{i} \in \Lambda^*$ such that $\mathbf{j}_1 \succ \mathbf{i}$ and $\mathbf{j}_2 \succ \mathbf{i}$. For an integer $q \geq 2$ we define the set of *join points* $\mathbf{i}_1, \ldots, \mathbf{i}_{q-1} \in \Lambda^*$ of $\mathbf{j}_1, \mathbf{j}_2, \ldots, \mathbf{j}_q \in \Lambda^\infty$ to be the set

$$J(\mathbf{j}_1, \mathbf{j}_2, \ldots, \mathbf{j}_q) = \{\mathbf{j}_i \wedge \mathbf{j}_j : 1 \leq i < j \leq m\},$$

see Fig. 2. This set will always consist of exactly $q - 1$ points provided that they are counted according to multiplicity, that is if there are $r$ distinct points $\mathbf{j}_{i_1}, \ldots, \mathbf{j}_{i_r}$ such that $\mathbf{i} = \mathbf{j}_{i_p} \wedge \mathbf{j}_{i_q}$ for all $1 \leq p < q \leq r$ then $\mathbf{i}$ is counted as a join point with multiplicity $r - 1$. (If $m = 2$, corresponding to a binary tree, then all join points have multiplicity 1.)

In bounding expressions such as (2.5), where we now take $X = \Lambda^\infty$ so that the $t_i \in X$ are replaced by $\mathbf{j} \in \Lambda^\infty$, a generalised transversality argument may lead to an estimate of the form

$$\mathsf{P}\{|x_\omega(\mathbf{j}_q) - x_\omega(\mathbf{j}_j)| \leq r \text{ for all } j\} \leq F(\mathbf{j}_1, \mathbf{j}_2, \ldots, \mathbf{j}_q) \tag{3.1}$$

where $F$ may be expressed as a product over the join points

$$F(\mathbf{j}_1, \mathbf{j}_2, \ldots, \mathbf{j}_q) = f(\mathbf{i}_1) f(\mathbf{i}_2) \ldots f(\mathbf{i}_{q-1}) \text{ where } \{\mathbf{i}_1, \ldots, \mathbf{i}_{q-1}\} = J(\mathbf{j}_1, \mathbf{j}_2, \ldots, \mathbf{j}_q),$$

for some $f : \Lambda^* \to \mathbb{R}^+$ defined on the vertices of the tree. Then (2.5) takes the form

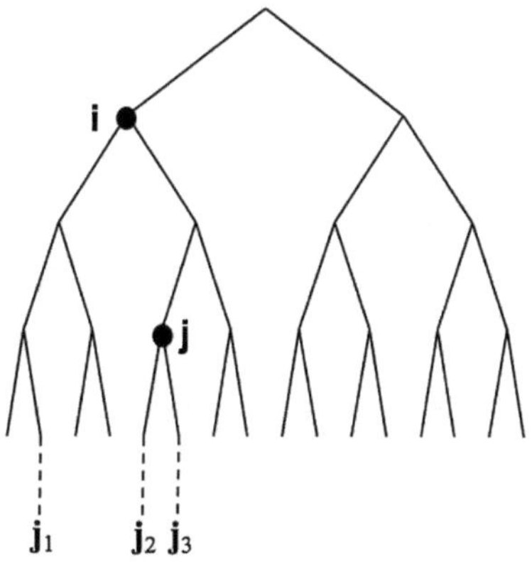


**Fig. 3** The arrangement of three points in $\Lambda^\infty$ the join points used in the proof of Theorem 3.1

$$\mathsf{E}\int \mu_\omega(B(x,r))^{q-1}d\mu_\omega(x) \le \int\cdots\int F(\mathbf{j}_1,\mathbf{j}_2,\ldots,\mathbf{j}_q)d\mu(\mathbf{j}_1)\ldots d\mu(\mathbf{j}_{q-1})d\mu(\mathbf{j}_q). \tag{3.2}$$

The following theorem estimates this integral in terms of $f$ and the cylinder measures $\mu(C_\mathbf{i})$.

**Theorem 3.1** *For each real number $q > 1$ there is a polynomial $p$ such that*

$$\int\cdots\int F(\mathbf{j}_1,\mathbf{j}_2,\ldots,\mathbf{j}_q)d\mu(\mathbf{j}_1)\ldots d\mu(\mathbf{j}_q) \le \Bigl(\sum_{k=0}^{\infty} p(k)\Bigl[\sum_{|\mathbf{i}|=k} f(\mathbf{i})^{q-1}\mu(C_\mathbf{i})^q\Bigr]^{\frac{1}{q-1}}\Bigr)^{q-1}. \tag{3.3}$$

*Proof* We give the proof in the special case when $q = 3$, that is

$$\int\int\int F(\mathbf{j}_1,\mathbf{j}_2,\mathbf{j}_3)d\mu(\mathbf{j}_1)d\mu(\mathbf{j}_2)d\mu(\mathbf{j}_3) \le \Bigl(\sum_{k=0}^{\infty}\Bigl[\sum_{|\mathbf{i}|=k} f(\mathbf{i})^2\mu(C_\mathbf{i})^3\Bigr]^{1/2}\Bigr)^2. \tag{3.4}$$

Splitting this integral into a sum over possible pairs of join points, see Fig. 3,

$$\int\int\int F(\mathbf{j}_1,\mathbf{j}_2,\mathbf{j}_3)d\mu(\mathbf{j}_1)d\mu(\mathbf{j}_2)d\mu(\mathbf{j}_3) \le \sum_{\mathbf{i}\in\Lambda^*}\sum_{\mathbf{j}\in\Lambda^*,\,\mathbf{j}\succ\mathbf{i}} f(\mathbf{i})f(\mathbf{j})\mu(C_\mathbf{i})\mu(C_\mathbf{j})^2. \tag{3.5}$$

We first estimate the restriction of this double sum over vertices of the tree for given levels $|\mathbf{i}| = k$ and $|\mathbf{j}| = l$ where $0 \le k < l$; Cauchy's inequality is used at the places indicated.

$$
\begin{aligned}
&\sum_{|\mathbf{i}|=k}\sum_{|\mathbf{j}|=l,\,\mathbf{j}\succ\mathbf{i}} f(\mathbf{i})f(\mathbf{j})\mu(C_{\mathbf{i}})\mu(C_{\mathbf{j}})^2\\
&\quad=\sum_{|\mathbf{i}|=k}\Big[f(\mathbf{i})\mu(C_{\mathbf{i}})\Big]\Big[\sum_{|\mathbf{j}|=l,\,\mathbf{j}\succ\mathbf{i}}\big(f(\mathbf{j})\mu(C_{\mathbf{j}})^{3/2}\big)\mu(C_{\mathbf{j}})^{1/2}\Big]\\
&\quad\le\sum_{|\mathbf{i}|=k}\Big[f(\mathbf{i})\mu(C_{\mathbf{i}})\Big]\Big[\Big(\sum_{|\mathbf{j}|=l,\,\mathbf{j}\succ\mathbf{i}} f(\mathbf{j})^2\mu(C_{\mathbf{j}})^3\Big)^{1/2}\Big(\sum_{|\mathbf{j}|=l,\,\mathbf{j}\succ\mathbf{i}}\mu(C_{\mathbf{j}})\Big)^{1/2}\Big] \quad \text{(Cauchy)}\\
&\quad=\sum_{|\mathbf{i}|=k}\Big[f(\mathbf{i})\mu(C_{\mathbf{i}})\Big]\Big[\Big(\sum_{|\mathbf{j}|=l,\,\mathbf{j}\succ\mathbf{i}} f(\mathbf{j})^2\mu(C_{\mathbf{j}})^3\Big)^{1/2}\mu(C_{\mathbf{i}})^{1/2}\Big]\\
&\quad=\sum_{|\mathbf{i}|=k}\Big[f(\mathbf{i})\mu(C_{\mathbf{i}})^{3/2}\Big]\Big[\sum_{|\mathbf{j}|=l,\,\mathbf{j}\succ\mathbf{i}} f(\mathbf{j})^2\mu(C_{\mathbf{j}})^3\Big]^{1/2}\\
&\quad\le\Big[\sum_{|\mathbf{i}|=k} f(\mathbf{i})^2\mu(C_{\mathbf{i}})^3\Big]^{1/2}\Big[\sum_{|\mathbf{i}|=k}\sum_{|\mathbf{j}|=l,\,\mathbf{j}\succ\mathbf{i}} f(\mathbf{j})^2\mu(C_{\mathbf{j}})^3\Big]^{1/2} \quad \text{(Cauchy)}\\
&\quad=\Big[\sum_{|\mathbf{i}|=k} f(\mathbf{i})^2\mu(C_{\mathbf{i}})^3\Big]^{1/2}\Big[\sum_{|\mathbf{j}|=l} f(\mathbf{j})^2\mu(C_{\mathbf{j}})^3\Big]^{1/2}
\end{aligned}
$$

Summing over all levels $0 \le k, l$ gives inequality (3.4). □

When $q$ is a larger integer, (3.3) may be established using an induction on configurations of join points, requiring frequent uses of Hölder's inequality rather than Cauchy's inequality. A further extension of the calculation establishes that (3.3) remains valid for any real number $q > 1$, see [Fal10, FX14] for further details.

In applications $f(\mathbf{i}) \equiv f_s(\mathbf{i})$ typically depends on a parameter $s$ such that

$$\sum_{|\mathbf{i}|=k} f_s(\mathbf{i})^{q-1}\mu(C_{\mathbf{i}})^q \asymp (\lambda_s)^k$$

for some $\lambda_s > 0$. Combining (3.3) with (3.2) gives

$$\mathsf{E}\int \mu_\omega(B(x,r))^{q-1}d\mu_\omega(x) \le c\Big(\sum_{k=0}^{\infty} p(k)(\lambda_s)^{k/(q-1)}\Big)^{q-1},$$

so the value of $s$ for which $\lambda_s = 1$ is critical for bounding the mean $L^q$ dimensions of $\mu_\omega$.

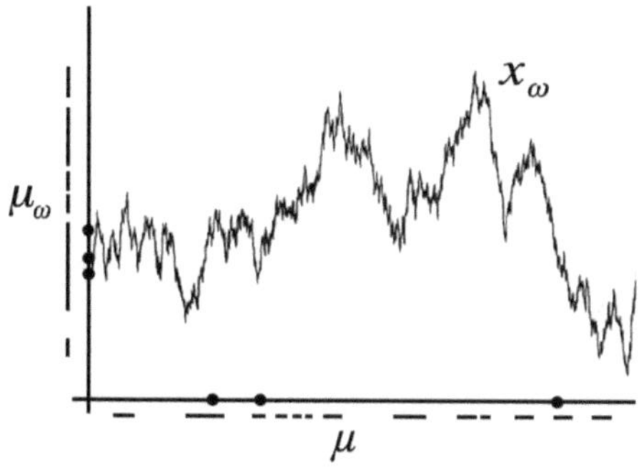


**Fig. 4** A measure $\mu$ and its image $\mu_\omega$ under a process $x_\omega$

# 4 Images of Measures Under Gaussian Processes

For a first application of inequality (3.3), we examine images of measures under certain Gaussian processes. Let $\{x_\omega : [0,1] \to \mathbb{R}, \quad \omega \in \Omega\}$ be index-$\alpha$ fractional Brownian motion ($0 < \alpha \leq 2$) defined on a suitable probability space $\Omega$, see [Adl81, Ka85, MV68]. It was shown by Kahane [Ka85] that for a Borel set $E \subseteq \mathbb{R}$

$$\dim_{\mathrm{H}} X(E) = \min\left\{1, \frac{\dim_{\mathrm{H}} E}{\alpha}\right\} \quad \text{a.s.},$$

where $\dim_{\mathrm{H}}$ denotes Hausdorff dimension. It is natural to seek similar relationships between the generalized dimensions of measures and their images under such processes (Fig. 4). Full details of the following result are in [FX14].

**Theorem 4.1** *Let $x_\omega : [0,1] \to \mathbb{R}$ be index-$\alpha$ fractional Brownian motion, let $\mu$ be a finite measure on $[0,1]$ and let $\mu_\omega$ be the image of $\mu$ under $x_\omega$. Let $q > 1$. Assuming that $D_q(\mu)$ exists then $D_q(\mu_\omega)$ exists almost surely and*

$$D_q(\mu_\omega) = \min\left\{1, \frac{D_q(\mu)}{\alpha}\right\} \quad \textit{a.s.}.$$

*Sketch of proof* Since index-$\alpha$ fractional Brownian motion almost surely satisfies an $(\alpha - \epsilon)$-Hölder condition for all $\epsilon > 0$, it follows easily from the definition of $L^q$-dimensions that $D_q(\mu_\omega) \leq D_q(\mu)/\alpha$.

For the opposite inequality we use the local nondeterminism (LND) of fBm. Roughly this states that the variance of $x_\omega(t_1)$ conditional on $x_\omega(t_2), \ldots, x_\omega(t_q)$ is comparable with the variance of $x_\omega(t_1) - x_\omega(t_j)$ for the $j$ for which $|t_1 - t_j|$ is least, see [Ber73, Xia06, Xia11]. It may be shown that the calculations are essentially unaffected if, for a suitably large $m$, we consider the numbers in $[0,1]$ to base $m$ and identify the base $m$ number $0.a_1a_2a_3\ldots$ with $(a_1+1, a_2+1, a_3+1, \ldots) \in \Lambda^\infty$, so that the hierarchy of $m$-ary subintervals of $[0,1]$ are the cylinders $C_{\mathbf{i}}$ in symbolic space. Using LND inductively we obtain, in symbolic space notation,

$$\mathsf{P}\{|x_\omega(\mathbf{j}_q) - x_\omega(\mathbf{j}_j)| \leq r \text{ for all } j\} \leq cF(\mathbf{j}_1, \mathbf{j}_2, \ldots, \mathbf{j}_q)$$

where $F$ is a product over the join points $\mathbf{i}_1, \ldots, \mathbf{i}_{q-1} \in J(\mathbf{j}_1, \mathbf{j}_2, \ldots, \mathbf{j}_q)$ of the form

$$F(\mathbf{j}_1, \mathbf{j}_2, \ldots, \mathbf{j}_q) = cr^{s(q-1)} m^{|\mathbf{i}_1|\alpha s} m^{|\mathbf{i}_2|\alpha s} \ldots m^{|\mathbf{i}_{q-1}|\alpha s}$$

for any $s > 0$, where we have replaced Euclidean distance on [0, 1] by the $m$-ary ultrametric $d(\mathbf{j}_1, \mathbf{j}_2) = m^{-|\mathbf{j}_1 \wedge \mathbf{j}_2|}$. In this notation (2.5) becomes

$$\begin{aligned}
&\mathsf{E} \int \mu_\omega(B(x,r))^{q-1} d\mu_\omega(x) \\
&\quad \leq cr^{s(q-1)} \int \cdots \int_{\mathbf{i}_1, \ldots, \mathbf{i}_{q-1} \in J(\mathbf{j}_1, \ldots, \mathbf{j}_q)} m^{|\mathbf{i}_1|\alpha s} m^{|\mathbf{i}_2|\alpha s} \ldots m^{|\mathbf{i}_{q-1}|\alpha s} d\mu(\mathbf{j}_1) \ldots d\mu(\mathbf{j}_q).
\end{aligned}$$

Inequality (3.3) with $f(\mathbf{i}) = f_s(\mathbf{i}) \equiv m^{|\mathbf{i}|\alpha s}$ now gives

$$\mathsf{E} \int \mu_\omega(B(x,r))^{q-1} d\mu_\omega(x) \leq cr^{s(q-1)} \Big( \sum_{k=0}^{\infty} p(k) \Big[ \sum_{|\mathbf{i}|=k} \lambda_{s,k} \Big]^{1/(q-1)} \Big)^{q-1} \tag{4.1}$$

where

$$\lambda_{s,k} \equiv \sum_{|\mathbf{i}|=k} f_s(\mathbf{i})^{q-1} \mu(C_\mathbf{i})^q = m^{k\alpha s(q-1)} \sum_{|\mathbf{i}|=k} \mu(C_\mathbf{i})^q. \tag{4.2}$$

The sum in (4.1) is finite if $\limsup_{k\to\infty}(\lambda_{s,k})^{1/k} < 1$, that is if $\alpha s < D_q(\mu)$ using (2.2) and noting that the sum in (4.2) is a sum over the $m$-ary mesh intervals of lengths $m^{-k}$ that are identified with the cylinders $C_\mathbf{i}$ where $| i | = k$. It follows that if $s_1 < s < D_q(\mu)/\alpha$ then

$$\begin{aligned}
&\mathsf{E} \sum_{k=1}^{\infty} 2^{-s_1(q-1)} \int \mu_\omega(B(x, 2^{-k}))^{q-1} d\mu_\omega(x) \\
&\qquad = \mathsf{E} \int \Big( \sum_{k=1}^{\infty} 2^{-s_1(q-1)} \mu_\omega(B(x, 2^{-k}))^{q-1} \Big) d\mu_\omega(x) < \infty,
\end{aligned}$$

which implies that $\overline{D}_q(\mu_\omega) > s_1$ almost surely, since the generalized dimensions are determined by the sequence of $r = 2^{-k}$. □

This method yields similar conclusions for the $L^q$-dimensions of the images of measures under other classes of Gaussian process such as fractional Riesz-Bessel motion and infinity scale fractional Brownian motion, see [FX14].

## 5 Measures on Almost Self-affine Sets

Next we consider $L^q$-dimensions of measures on self-affine and almost self-affine sets. For $i = 1, \ldots, m$ let $T_i$ be linear contractions on $\mathbb{R}^n$ and let $\omega_i$ be translation vectors. The iterated function system $\{T_j(x) + \omega_j\}_{j=1}^m$ has an non-empty compact attractor $E$ satisfying $E = \cup_{j=1}^m (T_j(E) + \omega_j)$; such a set is termed a *self-affine set*. Writing $\boldsymbol{\omega} = (\omega_1, \ldots, \omega_m)$ for the set of translations, the attractor $E$ may be characterised in terms of $m$-ary words: $E_{\boldsymbol{\omega}} = \bigcup_{\mathbf{j} \in \Lambda^\infty} x_{\boldsymbol{\omega}}(\mathbf{j})$, where $x_{\boldsymbol{\omega}} : \Lambda^\infty \to \mathbb{R}^n$ is given by the single point in the decreasing intersection

$$x_{\boldsymbol{\omega}}(\mathbf{j}) \equiv x_{\boldsymbol{\omega}}(j_1, j_2, \ldots) = \bigcap_{k=1}^{\infty} (T_{j_1} + \omega_{j_1})(T_{j_2} + \omega_{j_2}) \cdots (T_{j_k} + \omega_{j_k})(B), \quad (5.1)$$

where $B$ is any ball such that $T_j(B) + \omega_i \subseteq B$ for all $j$.

Let $p_1, \ldots, p_m$ be probabilities, so that $0 < p_j < 1$ and $\sum_{j=1}^m p_j = 1$. Let $\mu$ be the Bernoulli probability measure on $\Lambda^\infty$ defined on cylinders by

$$\mu(C_{\mathbf{j}}) = p_{j_1} p_{j_2} \cdots p_{j_k} \quad \text{where} \quad \mathbf{j} = (j_1, \ldots, j_k) \in \Lambda^*, \quad (5.2)$$

and extended to a Borel measure on $\Lambda^\infty$. For each $\boldsymbol{\omega} \in \Omega$ let $\mu_{\boldsymbol{\omega}}$ be the image measure of $\mu$ under $x_{\boldsymbol{\omega}}$, which is supported by $E_{\boldsymbol{\omega}}$.

We wish to find the generalized dimensions $D_q(\mu_{\boldsymbol{\omega}})$. This is well-known in the case where the $T_j + \omega_j$ are similarities and $E_{\boldsymbol{\omega}}$ is a self-similar set. Provided the open set condition is satisfied (that is, there exists a non-empty open set $U$ such that $\cup_{j=1}^m (T_j(U) + \omega_j) \subset U$ with this union disjoint), then the generalized dimension $D_q(\mu_{\boldsymbol{\omega}}) = d_0$ where $d_0$ satisfies the equation $\sum_{j=1}^m r_j^{(1-q)d_0} p_j^q = 1$, see [CM92, Fal14]. Closed formulae have also been obtained for the generalized dimensions or self-affine 'carpets' and 'sponges', where the $T_j$ are all equal and the affine transformations $T_{j_i} + \omega_{j_i}$ map a given cube onto similarly-aligned rectangles or rectangular parallelepipeds [Kin95, Ols98].

In general it is difficult to obtain formulae for $L^q$-dimensions of measures on self-affine sets, or even for the Hausdorff dimension of the supporting self-affine sets, not least because they need not be continuous in $\boldsymbol{\omega}$. Nevertheless, using a potential-theoretic approach, one may obtain formulae that are valid for almost all $\boldsymbol{\omega} = (\omega_1, \ldots, \omega_m)$ in the sense of $mn$-dimensional Lebesgue measure in the case that $1 < q \leq 2$, see [Fal99]. However, in general there is 'not enough transversality' as $\boldsymbol{\omega}$ varies for the estimates to extend to $q > 2$.

One way of circumventing this difficulty is to introduce more randomness by allowing a random perturbation in the translation component at each stage of the construction. We let

$$\boldsymbol{\omega} = \{\omega_{j_1, j_2, \ldots, j_k} : (j_1, j_2, \ldots, j_k) \in \Lambda^*\} \in (\mathbb{R}^n)^{\Lambda^*} \quad (5.3)$$

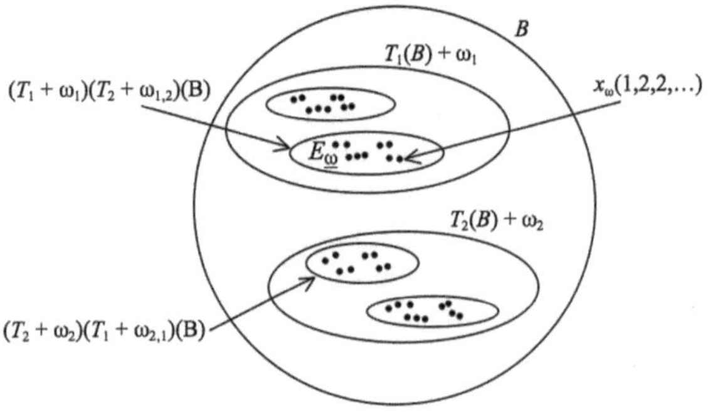


**Fig. 5** Hierarchical construction of an almost self-affine set $E_\omega$

be a family of translation vectors in $\mathbb{R}^n$ which we assume to be bounded. Analogously to (5.1) we let

$$\begin{aligned} x_\omega(\mathbf{j}) &= \bigcap_{k=1}^{\infty}(T_{j_1} + \omega_{j_1})(T_{j_2} + \omega_{j_1,j_2})(T_{j_3} + \omega_{j_1,j_2,j_3})\cdots(T_{j_k} + \omega_{j_1,j_2,\ldots j_k})(B) \\ &= \lim_{k\to\infty}(T_{j_1} + \omega_{j_1})(T_{j_2} + \omega_{j_1,j_2})(T_{j_3} + \omega_{j_1,j_2,j_3})\cdots(T_{j_k} + \omega_{j_1,j_2,\ldots j_k})(0) \\ &= \omega_{j_1} + T_{j_1}\omega_{j_1,j_2} + T_{j_1}T_{j_2}\omega_{j_1,j_2,j_3} + \cdots \end{aligned} \tag{5.4}$$

for each $\mathbf{j} \equiv (j_1, j_2, \ldots) \in \Lambda^\infty$, for some ball $B$ large enough to ensure that $T_j(B) + \omega_{j_1,j_2,\ldots,j_k} \subseteq B$ for all $j_1, j_2, \ldots, j_k \in \Lambda^*$. We call

$$E_\omega = \bigcup_{\mathbf{j}\in\Lambda^\infty} x_\omega(\mathbf{j})$$

an *almost self-affine set*, see Fig. 5.

We may randomize the translation vectors in the self-affine construction. Assume now that $w_{j_1,j_2,\ldots,j_k}$ in (5.3) are independent identically distributed (i.i.d.) random vectors for $j_1, j_2, \ldots, j_k \in \Lambda^*$ with absolutely continuous density with respect to $n$-dimensional Lebesgue measure. We put the product probability measure on $(\mathbb{R}^n)^{\Lambda^*}$. We then term $E_\omega$ a *random almost self-affine set* [JPS07], see Fig. 6.

To analyse self-affine and almost self-affine sets we utilize the singular values of the mappings which control the geometry of the components in the construction. The *singular values* $\alpha_1 \geq \alpha_2 \geq \cdots \geq \alpha_n \geq 0$ of a linear mapping $T : \mathbb{R}^n \to \mathbb{R}^n$ are the positive square roots of the eigenvalues of $TT^*$ or equivalently are the semi-axis lengths of the ellipsoid $T(B)$ where $B$ is the unit ball. The *singular value function* of $T$ is then defined by

$$\phi^s(T) = \alpha_1 \ldots \alpha_{p-1}\alpha_p^{s-p+1} \tag{5.5}$$

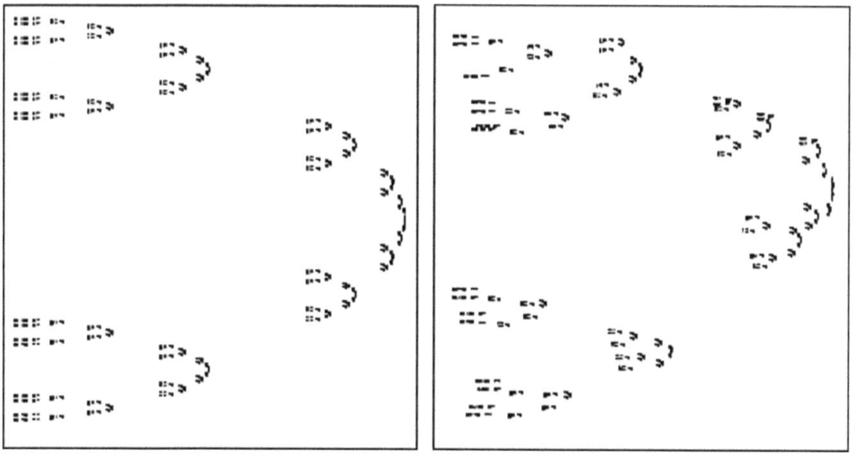

**Fig. 6** A self-affine set and a random almost self-affine set with the same linear components in the defining transformations

where $p$ is the integer such that $p - 1 \le s \le p$. (If $T$ is a similarity then $\phi^s(T)$ is just the $s$th power of the scaling ratio of $T$.)

There are two important properties of $\phi^s$. Firstly it is *submultiplicative*, that is

$$\phi^s(T_1 T_2) \le \phi^s(T_1)\phi^s(T_2), \tag{5.6}$$

and secondly, if $T$ is a contracting linear map, then $\phi^s(T)$ is continuous and strictly decreasing in $s$, see [Fal88]. It follows, writing

$$\Phi_k^s := \sum_{i_1,\dots,i_k \in \Lambda^k} \phi^s(T_{i_1} \circ \cdots \circ T_{i_k}),$$

that $\Phi_k^s$ itself is also submultiplicative, that is $\Phi_{k+l}^s \le \Phi_k^s \Phi_l^s$, so, by the standard property of submultiplicative sequences, the limit

$$\Phi^s := \lim_{k\to\infty} (\Phi_k^s)^{1/k}$$

exists and is decreasing in $s$.

The positive number $d_0$ that satisfies $\Phi^{d_0} = 1$ is called the *affinity dimension* $d_0 \equiv d_0(T_1, \dots, T_m)$ of the self-affine set $E_\omega$ that is the attractor of the IFS of affine maps $\{T_i + \omega_i\}_{i=1}^m$. In other words $d_0$ is given by

$$\Phi^{d_0}(T_1, \dots, T_m) \equiv \Phi^{d_0} = \lim_{k\to\infty} \Big( \sum_{i_1 \dots i_k \in \Lambda^k} \phi^{d_0}(T_{i_1} \circ \cdots \circ T_{i_k}) \Big)^{1/k} = 1; \tag{5.7}$$

notice that the affinity dimension depends only on the linear parts of the IFS functions. Affinity dimensions provide 'generic' values for the Hausdorff and box-counting dimensions of self-affine sets. We write $\dim_{\rm H}$ and $\overline{\dim}_{\rm B}$ for Hausdorff and upper box-counting dimensions respectively.

**Proposition 5.1** *Let $E_\omega$ be a self-affine or almost self-affine subset of $\mathbb{R}^n$. Then*

$$\dim_{\rm H} E_\omega \leq \overline{\dim}_{\rm B} E_\omega \leq d_0(T_1, \ldots, T_m) \tag{5.8}$$

*where $(T_1, \ldots, T_m)$ are the linear parts of the affine contractions in the construction of $E_\omega$. If $E_\omega$ is self-affine with $\|T_j\| < \frac{1}{2}$ for all $j$ then there is equality in (5.8) for almost all translation vectors $\boldsymbol{\omega} \in (\mathbb{R}^n)^m$. If $E_\omega$ is a random almost self-affine set then there is equality in (5.8) for almost all $\boldsymbol{\omega} \in (\mathbb{R}^n)^{\Lambda^*}$ with no restriction on $\|T_j\|$.*

*Proof* Inequality (5.8) is obtained by a covering method. Almost sure equality for self-affine sets and random almost self-affine sets may be derived from energy estimates for measures supported on the sets, see [Fal88] and [JPS07] for the two settings. □

To obtain generic formulae for $L^q$-dimensions, we adapt the definition of affinity dimension. With $\mu$ the Bernoulli measure on $\Lambda^\infty$ defined by (5.2), let

$$\Phi_q^s = \lim_{k\to\infty}\Big(\sum_{i_1,\ldots,i_k\in\Lambda^k} \phi^s(T_{i_1}\circ T_{i_2}\circ\cdots\circ T_{i_k})^{1-q}\mu(C_{i_1,i_2,\ldots,i_k})^q\Big)^{1/k} \tag{5.9}$$

$$= \lim_{k\to\infty}\Big(\sum_{i_1,\ldots,i_k\in\Lambda^k} \phi^s(T_{i_1}\circ T_{i_2}\circ\cdots\circ T_{i_k})^{1-q}(p_{i_1}p_{i_2}\cdots p_{i_k})^q\Big)^{1/k}.$$

Again the limits exist as a consequence of supermultiplicativity, and if $q > 1$ then $\Phi_q^s$ is strictly increasing and continuous in $s$. Thus we may define positive numbers $d_q$ by the requirement that

$$\Phi_q^{d_q} = 1. \tag{5.10}$$

As before we write $\mu_\omega$ for the image of the Bernoulli measure $\mu$ under $x_\omega$. We refer to $\mu_\omega$ as a *self-affine measure* when $\boldsymbol{\omega} \in (\mathbb{R}^n)^m$ and the support $E_\omega$ is a self-affine set, and as a *(random) almost self-affine measure* when $\boldsymbol{\omega} \in (\mathbb{R}^n)^{\Lambda^*}$ and the support is a (random) almost self-affine set.

**Proposition 5.2** *Let $1 < q \leq 2$. Let $\mu$ be a Bernoulli measure on $\Lambda^\infty$. For every self-affine or almost self-affine measure $\mu_\omega$ on $\mathbb{R}^n$*

$$\overline{D}_q(\mu_\omega) \leq \min\{d_q, n\} \tag{5.11}$$

*where $d_q$ is given by (5.10). Moreover, $D_q(\mu_\omega)$ exists and*

$$D_q(\mu_\omega) = \min\{d_q, n\}$$

*in the self-affine case provided* $\|T_j\| < \frac{1}{2}$ *for all* $j$, *for almost all* $\boldsymbol{\omega} \in (\mathbb{R}^n)^m$, *and also in the random almost self-affine case for almost all* $\boldsymbol{\omega} \in (\mathbb{R}^n)^{\Lambda^*}$ *(with no restriction on the* $\|T_j\|$*).*

*Note on proof* This is proved in [Fal99] using a potential-theoretic method; the proof adapts easily to give equality in the random almost self-affine case. □

It is natural to ask whether the conclusion of Proposition 5.2 is valid for $q > 2$ when the basic potential-theoretic method is inadequate. However, for self-affine measures $\mu_\omega$ there is, in general, not enough randomness or transversality to get an adequate estimate in (3.1) to lead to equality for almost all $\boldsymbol{\omega} \in (\mathbb{R}^n)^m$. Thus we consider the lower bound for random almost self-affine measures using the inequality of Sect. 3.

**Theorem 5.1** *Let* $q > 1$. *Let* $\mu$ *be a Bernoulli measure on* $\Lambda^\infty$. *For every self-affine or almost self-affine measure* $\mu_\omega$ *on* $\mathbb{R}^n$

$$\overline{D}_q(\mu_\omega) \leq \min\{d_q, n\} \tag{5.12}$$

*where* $d_q$ *is given by* (5.10). *If* $\mu_\omega$ *is a random almost self-affine measure then* $D_q(\mu_\omega)$ *exists and*

$$D_q(\mu_\omega) = \min\{d_q, n\}$$

*for almost all* $\boldsymbol{\omega} \in (\mathbb{R}^n)^{\Lambda^*}$.

*Sketch of proof* The upper bound (5.12) comes from splitting ellipses of the form that occur in the intersections in (5.4) into appropriate pieces and summing the powers of the measures, see [Fal99, Fal10].

For the case where $q \geq 2$ is an integer and $\mu_\omega$ a random almost self-affine measure, let $\mathbf{j}_1, \ldots, \mathbf{j}_q \in \Lambda^\infty$. Using the geometry and randomness or higher transversality available in the construction, we may obtain an estimate

$$\mathsf{P}\{|x_\omega(\mathbf{j}_q) - x_\omega(\mathbf{j}_j)| \leq r \text{ for all } j\} \leq c r^{s(q-1)} \phi^s(T_{\mathbf{i}_1})^{-1} \phi^s(T_{\mathbf{i}_2})^{-1} \ldots \phi^s(T_{\mathbf{i}_{q-1}})^{-1} \tag{5.13}$$

where $\mathbf{i}_1, \ldots, \mathbf{i}_{q-1}$ are the join points of $\mathbf{j}_1, \ldots, \mathbf{j}_q$. Using (2.5) we get, for all $s > 0$,

$$\begin{aligned}
&\mathsf{E} \int \mu_\omega(B(x,r))^{q-1} d\mu_\omega(x) \\
&\leq c r^{s(q-1)} \int \cdots \int \phi^s(T_{\mathbf{i}_1})^{-1} \phi^s(T_{\mathbf{i}_2})^{-1} \ldots \phi^s(T_{\mathbf{i}_{q-1}})^{-1} d\mu(\mathbf{j}_1) \ldots d\mu(\mathbf{j}_q) \\
&\leq c r^{s(q-1)} \bigg( \sum_{k=0}^{\infty} p(k) \Big[ \sum_{|\mathbf{i}|=k} \phi^s(T_{\mathbf{i}})^{1-q} \mu(C_{\mathbf{i}})^q \Big]^{\frac{1}{q-1}} \bigg)^{q-1},
\end{aligned}$$

for some polynomial $p$, taking $f(\mathbf{i}) = \phi^s(T_{\mathbf{i}})^{-1}$ in inequality (3.3). From the definition (5.9),(5.10) of $\Phi_q^s$, this series converges if $0 < s < d_q$, in which case

$$\mathsf{E}\sum_{k=1}^{\infty} 2^{-s_1(q-1)} \int \mu_\omega(B(x,2^{-k}))^{q-1} d\mu_\omega(x)$$
$$= \mathsf{E}\int \Big(\sum_{k=1}^{\infty} 2^{-s_1(q-1)} \mu_\omega(B(x,2^{-k}))^{q-1}\Big) d\mu_\omega(x) < \infty,$$

for all $0 < s_1 < s$, giving $D_q(\mu_\omega) > s_1$ for almost all $\omega \in (\mathbb{R}^n)^{\Lambda^*}$ for all $s_1 < d_q$, as required.

For full details of this argument and the case of non-integer $q > 1$ see [Fal00]. □

# 6 Random Multiplicative Cascade Measures

Let $\Lambda = \{1, 2, \ldots, m\}$, let $W_{\mathbf{i}}$ be independent positive random variables indexed by $\mathbf{i} = i_1, i_2, \ldots, i_k \in \Lambda^*$ and let

$$X_{\mathbf{i}} = W_{i_1} W_{i_1,i_2} \cdots W_{i_1,i_2,\ldots,i_k}.$$

We may identify the cylinders in symbolic space with the hierarchy of $m$-ary subintervals of $[0, 1]$ in the obvious way, see Fig. 7. We assume that $\mathsf{E}(W_{\mathbf{i}}) = 1$ for all $\mathbf{i} \in \Lambda^*$ in which case $(X_{\mathbf{j}|k}, \mathcal{F}_k)$ is a martingale for each $\mathbf{j} \in \Lambda^\infty$, where $\mathbf{j}|k$ denotes the curtailment of $\mathbf{j}$ after $k$ terms and $\mathcal{F}_k$ is the $\sigma$-field generated by $\{W_{\mathbf{i}} : \mathbf{i} \in \cup_{l=1}^{k} \Lambda^l\}$.

These martingales, termed *random multiplicative cascade measures*, were introduced and studied in the 1970 s by Mandelbrot [Man74] and Kahane and Peyrière [Kah85, KP76] who obtained many properties in the 'self-similar' case, that is when the $W_{\mathbf{i}}$ are independent and identically distributed. Let $\mu$ be a Borel probability measure on $\Lambda^\infty$. Of particular interest are the $k$-th level sums

$$\sum_{|\mathbf{i}|=k} X_{\mathbf{i}}\, \mu(C_{\mathbf{i}}) \equiv \int X_{\mathbf{j}|k} d\mu(\mathbf{j}),$$

which moments of the sums remain bounded as $k \to \infty$ and in what setting the integral converges. It follows from Minkowski's inequality that if $\mathsf{E}\big(\big(\sum_{|\mathbf{i}|=k} X_{\mathbf{i}}\, \mu(C_{\mathbf{i}})\big)^q\big)$ is bounded in $k$ then so is $\sum_{|\mathbf{i}|=k} \mathsf{E}\big((X_{\mathbf{i}}\, \mu(C_{\mathbf{i}}))^q\big)$, but the opposite implications are more subtle. Using the inequality from Sect. 3 we get the following result.

**Theorem 6.1** *Let $q > 1$ be a real number. If*

$$\limsup_{k\to\infty} \Big(\sum_{|\mathbf{i}|=k} \mathsf{E}(X_{\mathbf{i}}^q)\mu(C_{\mathbf{i}})^q\Big)^{1/k} < 1 \tag{6.1}$$

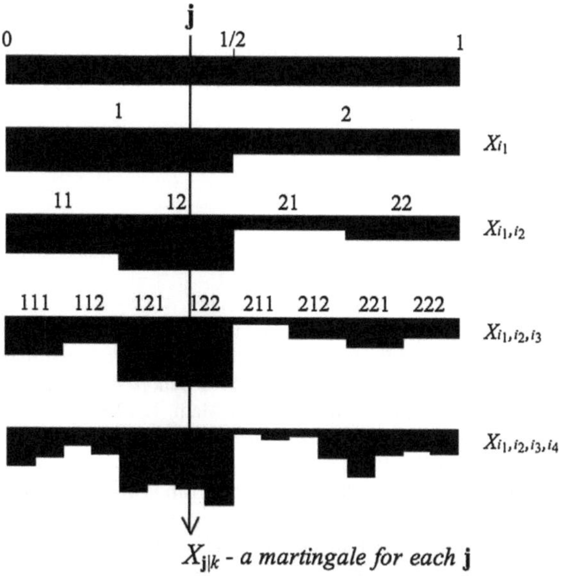


**Fig. 7** A random multiplicative binary cascade measure represented on the interval [0, 1], with $X_{\mathbf{j}|k}$ a martingale for each $\mathbf{j} \in [0, 1]$

*then*

$$\limsup_{k\to\infty} \mathsf{E}\Big(\Big(\sum_{|\mathbf{i}|=k} X_{\mathbf{i}}\, \mu(C_{\mathbf{i}})\Big)^q\Big) < \infty \tag{6.2}$$

*and* $\int X_{\mathbf{j}|k} d\mu(\mathbf{j})$ *converges a.s. and in* $L^q$. *Note that we require the underlying* $W_{\mathbf{i}}$ *to be independent but not necessarily identically distributed.*

This result, along with many other properties of these martingales, was obtained by Kahane and Peyrière [Kah85, KP76] when the random cascade is 'self-similar', that is when the $W_{\mathbf{i}}$ are identically distributed, utilizing the self-similarity to show that the sums satisfy a random difference equation. There have been many subsequent extensions and variants, see [BM04, Liu00] which contain many further references. Barral [BM04] proved this result without the i.i.d. requirement on the $W_{\mathbf{i}}$ in the case $1 < q \leq 2$, also with the martingales defined in a more general continuous setting.
*Note on the proof of Theorem* 6.1. When $q > 1$ is an integer we may expand

$$\mathsf{E}\Big(\Big(\sum_{|\mathbf{i}|=k} X_{\mathbf{i}}\, \mu(C_{\mathbf{i}})\Big)^q\Big) = \sum_{|\mathbf{i}_1|,|\mathbf{i}_2|,\ldots,|\mathbf{i}_q|=k} \mathsf{E}(X_{\mathbf{i}_1} X_{\mathbf{i}_2} \ldots X_{\mathbf{i}_q}) \mu(C_{\mathbf{i}_1}) \mu(C_{\mathbf{i}_2}) \ldots \mu(C_{\mathbf{i}_q})$$
$$\leq \Big(\sum_{k=0}^{\infty} p(k) \Big(\sum_{|\mathbf{i}|=k} \mathsf{E}(X_{\mathbf{i}}^q) \mu(C_{\mathbf{i}})^q\Big)^{\frac{1}{q-1}}\Big)^{q-1}.$$

for a polynomial $p$, where this inequality may be established using induction in a manner akin to that of Theorem 3.1 by relating the expectations of products of the $X_{\mathbf{i}_j}$ to expectations of powers of the $X_{\mathbf{i}}$ at the join points of $\mathbf{i}_1, \ldots, \mathbf{i}_q$. The conclusion (6.2) then follows from (6.1).

As with Theorem 3.1 the argument for non-integer $q$ requires a more involved induction argument. □

## References


[Adl81] Adler, R.J.: The Geometry of Random Fields. Wiley, New York (1981)

[BM04] Barral, J., Mandelbrot, B.B.: Random multiplicative multifractal measures I, II, III, in Fractal Geometry and Applications: A Jubilee of Benoît Mandelbrot (M. Lapidus and M. V. Frankenkuijsen, eds.), Proc. Sympos. Pure Math. 72 (2004) 3–90

[Ber73] Berman, S.M.: Local nondeterminism and local times of Gaussian processes. Indiana Univ. Math. J. **23**, 69–94 (1973)

[CM92] Cawley, R., Mauldin, R.D.: Multifractal decomposition of Moran fractals. Adv. Math. **92**, 196–236 (1992)

[Fal88] Falconer, K.J.: The Hausdorff dimension of self-affine fractals. Math. Proc. Cambridge Philos. Soc. **103**, 339–350 (1988)

[Fal97] Falconer, K.J.: Techniques in Fractal Geometry. Wiley, New York (1997)

[Fal99] Falconer, K.J.: Generalized dimensions of measures on self-affine sets. Nonlinearity **12**, 877–891 (1999)

[Fal00] Falconer, K.J.: Higher moments of random multiplicative cascade measures: to appear. J. Fractal geometry. arXiv:1311.0160

[Fal10] Falconer, K.J.: Generalised dimensions of measures on almost self-affine sets. Nonlinearity **23**, 1047–1069 (2010)

[Fal14] Falconer, K.J.: Fractal Geometry-Mathematical Foundations and Applications, 3rd Edn. Wiley, New York (2014)

[FX14] Falconer, K.J., Xiao, Y.: Generalized Dimensions of Images of Measures Under Gaussian Processes. Adv. Math. **252**, 492–517 (2014)

[JPS07] Jordan, T., Pollicott, M., Simon, K.: Hausdorff dimension for randomly perturbed self affine attractors. Commun. Math. Phys. **270**, 519–544 (2007)

[Kah85] Kahane, J.-P.: Sur le chaos multiplicatif. Ann. Sci. Math. Québec **9**, 105–150 (1985)

[Ka85] Kahane, J.-P.: Some Random Series of Functions, 2nd Edn. Cambridge University Press, Cambridge (1985)

[KP76] Kahane, J.-P., Peyrière, J.: Sur certaines martingales de Benoit Mandelbrot. Adv. Math. **22**, 131–145 (1976)

[Kin95] King, J.: The singularity spectrum for general Sierpinski carpets. Adv. Math. **116**, 1–8 (1995)

[Lau95] Lau, K.-S.: Self-similarity, $L^p$-spectrum and multifractal formalism. In: Bandt, C., Graf, S., Zähle, M. (eds.) Fractal Geometry and Stochastics, Progress in Probability 37, pp. 55–90. Birkhäuser, Basel (1995)

[Liu00] Liu, Quansheng: On generalized multiplicative cascades. Stoch. Process. Appl. **86**, 263–286 (2000)

[Man74] Mandelbrot, B.: Intermittent turbulence in self-similar cascades: divergence of higher moments and dimension of the carrier. J. Fluid. Mech. **62**, 331–358 (1974)

[MV68] Mandelbrot, B.B., Van Ness, J.W.: Fractional Brownian motions, fractional noises and applications. SIAM Rev. **10**, 422–437 (1968)

[Mat99] Mattila, P.: Geometry of Sets and Measures in Euclidean Spaces. Cambridge University Press, Cambridge (1999)

[Ols98] Olsen, L.: Self-affine multifractal Sierpinski sponges in $\mathbb{R}^d$. Pacific J. Math. **183**, 143–199 (1998)

[PYS00] Peres, Y., Schlag, B.: Smoothness of projections, Bernoulli convolutions, and the dimension of exceptions. Duke Math. J. **102**, 193–251 (2000)

[PS00] Peres, Y., Solomyak, B.: Existence of $L^q$ dimensions and entropy dimension for self-conformal measures. Indiana Univ. Math. J. **49**, 1603–1621 (2000)

[Xia06] Xiao, Y.: Properties of local nondeterminism of Gaussian and stable random fields and their applications. Ann. Fac. Sci. Toulouse Math. **15**, 157–193 (2006)

[Xia11] Xiao, Y.: Properties of strong local nondeterminism and local times of stable random fields. In: Dalang, R.C., Dozzi, M., Russo, F. (eds.) Seminar on Stochastic Analysis, Random Fields and Applications VI, Progress in Probability, vol. 63, pp. 279–310. Birkhäuser, Basel (2011)

# Some Aspects of Multifractal Analysis

**Ai-Hua Fan**

**Abstract** The aim of this survey is to present some aspects of multifractal analysis around the recently developed subject of multiple ergodic averages. Related topics include dimensions of measures, oriented walks, Riesz products etc. The exposition on the multifractal analysis of multiple ergodic averages is mainly based on [FLM12, KPS12, FSW00].

**Keywords** Dynamical systems · Ergodic averages · Multifractal analysis · Hausdorff dimension

**2010 Mathematics Subject Classification** Primary 37C45

## 1 Introduction

Multifractal problems can be put into the following frame. Let $(X, d)$ be a metric space and $\mathcal{P}(x)$ a property (quantitative or qualitative) depending on a point $x$ of the space $X$. For any prescribed property $\mathbf{P}$, we look at the set of those points $x$ which have the property $\mathbf{P}$:

$$E(\mathbf{P}) = \{x \in X : \mathcal{P}(x) = \mathbf{P}\}.$$

The size of the sets $E(\mathbf{P})$ for different $\mathbf{P}$'s is problematic in the multifractal analysis. According to popular folklore, the function $x \mapsto \mathcal{P}(x)$ is multifractal if $E(\mathbf{P})$ is not empty for uncountably many properties $\mathbf{P}$. Usually the size of a set $A$ in $X$ is described by its Hausdorff dimension $\dim_H A$ or its packing dimension $\dim_P A$, or its topological entropy in a dynamical setting. See [Fal03, Mat95] for the dimension theory and [Bow73, Pes97] for the notion of topological entropy.

Seeds of multifractals were sown in Mandelbrot's works on multiplicative chaos in 1970's [Man74, Man74]. First rigorous results are due to Kahane and Peyrère [KP76]. The concept of multifractality came from geophysics and theoretical physics.

A.-H. Fan (✉)
LAMFA, UMR, 7352 CNRS, Université de Picardie, 33 rue Saint Leu, 80039 Amiens, France
e-mail: ai-hua.fan@u-picardie.fr



D.-J. Feng and K.-S. Lau (eds.), *Geometry and Analysis of Fractals*,
Springer Proceedings in Mathematics & Statistics 88,
DOI 10.1007/978-3-662-43920-3_5

At the beginning, Frisch and Parisi [FP85], Hentschel and Procaccia [HP83] had the rather vague idea of mixture of subsets of different dimensions each of which has a given Hölder singularity exponent. The multifractal formalism became clearer in the 1980–1990's in the works of Halsey et al. [HJKPS86], of Collet et al. [CLP87], and of Grown et al. [BMP92]. The multifractal formalism is tightly related to the thermodynamics and Ruelle [Rue78] was the first to use the thermodynamical formalism to compute the Hausdorff dimensions of some Julia sets.

The research on the subject has been very active and very fruitful since four decades. The first most studied multifractal quantity is the local dimension of a Borel measure $\mu$ on $X$. Recall that the local lower dimension of $\mu$ at $x \in X$ is defined by

$$\underline{D}_{\mu}(x) = \liminf_{r \to 0} \frac{\log \mu(B(x,r))}{\log r}$$

where $B(x,r)$ denotes the ball centered at $x$ with radius $r$. The local upper dimension $\overline{D}_{\mu}(x)$ is similarly defined. There is a huge literature on this subject. Let us mention another example, Hölder exponent of a function (see [Jaf97]). Let $\alpha > 0$. Let $f : \mathbb{R}^d \to \mathbb{R}$ be a function and $x \in \mathbb{R}^d$ be a fixed point. We say $f$ is $\alpha$-Hölder at $x$ and we write $f \in C^{\alpha}(x)$ if there exist two constants $\delta > 0$ and $C > 0$ and a polynomial $P(x)$ of degree strictly smaller than $\alpha$ such that

$$|y - x| < \delta \Rightarrow |f(y) - f(x) - P(y - x)| \le C|y - x|^{\alpha}.$$

The Hölder exponent of $f$ at $x$ is then defined by

$$h_f(x) = \sup\{\alpha > 0 : f \in C^{\alpha}(x)\}.$$

The multifractal analysis has now become a set of tools applicable in analysis, probability and stochastic processus, number theory, ergodic theory and dynamical systems etc if we don't account applications in physics and other sciences.

The main goal of this paper is to present some problems in the multifractal analysis of Birkhoff ergodic averages, especially of multiple Birkhoff ergodic averages, and some related topics like dimensions of measures and Riesz products as tools, and oriented walks as similar subject.

Let $T : X \to X$ be a map from $X$ into $X$. We consider the dynamical system $(X, T)$. The main concern about the system is the behavior of an orbit $\{T^n x\}$ of a given point $x \in X$. Some aspects of the behavior of the orbit may be described by the so-called Birkhoff averages

$$A_n f(x) = \frac{1}{n} \sum_{k=0}^{n-1} f(T^k x) \qquad (1)$$

where $f : \mathbb{X}^d \to \mathbb{R}$ is a given function, called observable. We refer to [Wal82] for basic facts in the theory of dynamical system and ergodic theory.

The famous and fundamental Birkhoff Ergodic Theorem states that for any $T$-invariant ergodic probability measure $\mu$ on $X$ and for any integrable function $f \in L^1(\mu)$, the limit

$$\lim_{n\to+\infty} A_n f(x) = \int_X f \, d\mu$$

holds for $\mu$-almost all points $x \in X$. Even if $\mu$ is only $T$-invariant but not ergodic, the limit still exists for $\mu$-almost all points $x \in X$. For many dynamical systems, there is a rich class of invariant measures and so that the limit of Birkhoff averages $A_n f(x)$ may vary for different points $x$. This variety reflects the chaotic feature of the dynamical system. A typical example is the doubling dynamics $x \mapsto 2x (\bmod 1)$ on the unit circle $\mathbb{T} := \mathbb{R}/\mathbb{Z}$.

The multifractal analysis of Birkhoff ergodic averages provides a way to study the chaotic feature of the dynamics. Let $\alpha \in \mathbb{R}$. We define the $\alpha$-level set

$$E_f(\alpha) = \left\{ x \in X : \lim_{n\to+\infty} A_n f(x) = \alpha \right\}.$$

The purpose of the multifractal analysis is the determination of the size of sets $E_f(\alpha)$. If the Hausdorff dimension is used as measuring device, we are led to the Haudorff multifractal spectrum of $f$:

$$\mathbb{R} \ni \alpha \mapsto d_H(\alpha) := \dim_H E_f(\alpha).$$

The existing works show that in many cases it is possible to compute the spectrum $d_H(\cdot)$ and it is also possible to distinguish a nice invariant measure sitting on the set $F_f(\alpha)$ for each $\alpha$. By "nice" we mean that the measure is supported by $E_f(\alpha)$ and its dimension is equal to that of $E_f(\alpha)$. The dimension of a measure is defined to be the dimension of the "smallest" Borel support of the measure. Therefore the nice measure is a maximal measure in the sense that it attains the maximum among all measures supported by $E_f(\alpha)$. This maximal measure may be invariant, ergodic and even mixing. Some other nice properties are also shared by the maximal measure. For this well studied classic ergodic averages, see for example [BSS02, FF99, FFW01, FLP08, FS03, FLW02, Oli99, TV03]. Let us also mention some useful tools for dimension estimation [BS07, BV06, Dur08, FST00].

The multiple ergodic theory started almost at the same time of the development of multifractal analysis. It started with Fürstenberg's proof of Szemerédi theorem on the existence of arbitrary long arithmetic sequence in a set of integers of positive density [Für77]. This theory involves several dynamics rather than one dynamics. Let $T_1$, $T_2$,..., $T_d$ be $d$ transformations on a space $X$. We assume that they are commuting each other and preserving a given probability measure $\mu$. For $d$ measurable functions $F = (f_1, \ldots, f_d)$ we define the multiple Birkhoff averages by

$$A_n F(x) := \frac{1}{n}\sum_{k=0}^{n-1} f_1(T_1^k x) f_2(T_2^k x) \cdots f_d(T_d^k x). \tag{2}$$

An inspiring example is the couple $(\tau_2, \tau_3)$ on the circle $\mathbb{T}$ where

$$\tau_2 x = 2x \ (\bmod\ 1), \quad \tau_3 x = 3x \ (\bmod\ 1).$$

The mixture of the dynamics $T_1, T_2, \ldots, T_d$ is much more difficult to understand. After the first works of Fürstenberg-Weiss [FW96] and of Conze-Lesigne [LT84], Host and Kra [HK05] proved the $L^2$-convergence of $A_n F$ when $T_j = T^j$ (powers of a fixed dynamics) and $f_j \in L^\infty(\mu)$. For the almost every convergence, results are sparse. Bourgain proved the almost everywhere convergence when $d = 2$. We should point out that even if the limit exists, it is not easy to recognize the limit. In particular, the limit may not be constant for some ergodic measures. For nilsystems, explicit formula for the limit was known to Lesigne [Les89] and Ziegler [Zie05]. Anyway, not like the "simple" ergodic theory, the multiple ergodic theory has not yet reached its maturity.

Although the multiple ergodic theory has been still developing, this situation doesn't prevent us from investigating the multifractal feature of multiple systems. Let us consider the following general set-up. For a given observable $\Phi : X^d \to \mathbb{R}$, we consider the Multiple Ergodic Averages

$$A_n \Phi(x) := \frac{1}{n}\sum_{k=0}^{n-1} \Phi(T_1^k x, T_2^k x, \ldots, T_d^k x). \tag{3}$$

The Birkhoff averages (2) correspond to the special case of tensor product $\Phi = f_1 \otimes f_2 \otimes \cdots \otimes f_d$.

It is natural to introduce the following generalization of multifractal spectrum. In this paper, we will only consider the case where $T_j = T^j$ for $1 \le j \le d$. So

$$A_n \Phi(x) = \frac{1}{n}\sum_{k=0}^{n-1} \Phi(T^k x, T^{2k} x, \ldots, T^{dk} x). \tag{4}$$

The multiple Hausdorff spectrum of the observable $\Phi$ is defined by

$$d_H(\alpha) = \dim_H E_\Phi(\alpha), \quad (\alpha \in \mathbb{R})$$

where

$$E_\Phi(\alpha) = \left\{ x \in X : \lim_{n\to+\infty} \frac{1}{n}\sum_{k=0}^{n-1} \Phi(T^k x, T^{2k} x, \ldots, T^{dk} x) = \alpha \right\}. \tag{5}$$

As we said, the classical theory ($d = 1$) is well developed. When $d > 1$, there are several results on the multifractal analysis of the limit of the averages $A_n \Phi$ in some special cases; but questions remain largely unanswered. In Sect. 2, we will present the first results obtained in [FLM12] in a very special case which give us a feeling of the problem and illustrate the difficulty of the problem. One result is the Hausdorff spectrum obtained by using Riesz products, a tool borrowed from Fourier analysis. Another result concerns the box dimension of a multiplicatively invariant set. These two kinds of result are respectively generalized in [FSW00] and [KPS12, PSSS00] in some general setting. In Sect. 5, we will present these results. As we mentioned, the local dimension of a measure was the first study object of multifractal analysis. In Sect. 3, we will give an account of dimensions of measures which are related to the local dimension and have their own interests. The idea of using Riesz products is inspired by a work on oriented walks [Fan00] to which Sect. 4 will be devoted. In the last section, we will collect some remarks and open problems.

## 2 First Multifractal Results on the Multiple Ergodic Averages

The question of computing the dimension of $E_\Phi(\alpha)$ was raised by Fan et al. in [FLM12], where the following special case was studied: $X =: \mathbb{M}_2 := \{-1, 1\}^{\mathbb{N}^*}$, $T$ is the shift, $d \geq 1$ and $\Phi$ is the function

$$\Phi(x^{(1)}, x^{(2)}, \ldots, x^{(d)}) = x_1^{(1)} x_1^{(2)} \ldots x_1^{(d)}, \quad (x^{(1)}, x^{(2)}, \ldots, x^{(d)}) \in \mathbb{M}_2^d.$$

Note that $x_1^{(j)}$ is the first coordinate of $x^{(j)}$. We consider $\mathbb{M}_2$ as the infinite product of the multiplicative group $\{-1, 1\}$. Then the function $x \mapsto x_j$ is a group character of $\mathbb{M}_2$, called Rademacher function and

$$\Phi(T^k x, T^{2d} x, \ldots, T^{dk} x) = x_k x_{2k} \ldots x_{dk}$$

is also a group character, called Walsh function. Recall that $\mathbb{M}_2 = \{-1, 1\}^{\mathbb{N}^*}$, considered as a symbolic space, is endowed with the metric

$$d(x, y) = 2^{-\min\{k : x_k \neq y_k\}}, \quad \text{for } x, y \in \mathbb{M}_2.$$

### *2.1 Multifractal Spectrum of a Sequence of Walsh Functions*

Under the above assumption, we have the following result.

**Theorem 1** [FLM12] *For every $\alpha \in [-1, 1]$, the set*

$$B_\alpha := \left\{ x \in \mathbb{M}_2 : \lim_{n \to +\infty} \frac{1}{n} \sum_{k=1}^{n} x_k x_{2k} \cdots x_{dk} = \alpha \right\}$$

*has the Hausdorff dimension*

$$\dim_H B_\alpha = 1 - \frac{1}{d} + \frac{1}{d \log 2} H\left(\frac{1+\alpha}{2}\right),$$

*where* $H(t) := -t \log t - (1-t)\log(1-t)$.

A key observation is that all these Walsh functions constitute a dissociated system in the sense of Hewitt-Zuckermann [HZ66]. This allows us to define probability measures called Riesz products on the group $\mathbb{M}_2$:

$$\mu_b := \prod_{k=1}^{\infty}(1 + bx_k x_{2k} \cdots x_{dk}) := w^* - \lim_{N\to\infty} \prod_{k=1}^{N}(1 + bx_k x_{2k} \cdots x_{dk})dx$$

for $b \in [-1, 1]$, where $dx$ denote the Haar measure on $\mathbb{M}_2$. Fortunately, the Riesz product $\mu_\alpha$ $(b = \alpha)$ is a maximizing measure on $B_\alpha$. So, by computing the dimension of the measure $\mu_\alpha$, we get the stated formula. We will go back to Riesz products in Sect. 3 and to dimensions of measures in Sect. 4.

Note that the case $d = 1$ is nothing but the well-known Besicovich-Eggleston theorem dated back to 1940's, which would be considered as the first result of multifractal analysis.

## 2.2 Box Dimension of Some Multiplicatively Invariant Set

The very motivation of [FLM12] was the multiple ergodic averages in the following case where $X := \mathbb{D}_2 := \{0, 1\}^{\mathbb{N}^*}$, $T$ is the shift on $\mathbb{D}_2$ and $\Phi(x, y) = x_1 y_1$ for $x = x_1 x_2 \ldots$ and $y = y_1 y_2 \ldots \in \mathbb{D}_2$. The space $\mathbb{D}_2$ may also be considered as the infinite product group of $\mathbb{Z}/2\mathbb{Z}$. But the function $x \mapsto x_1$ is no longer group character and the Fourier method fails. Then the authors of [FLM12] proposed to look at a subset of the 0-level set

$$E_\Phi(0) = \left\{ x \in \Sigma_2 : \lim_{n\to+\infty} \frac{1}{n} \sum_{k=1}^{n} x_k x_{2k} = 0 \right\}. \tag{6}$$

The proposed subset is

$$X_2 = \{x \in \Sigma_2 : \forall k \geq 1, \quad x_k x_{2k} = 0\}. \tag{7}$$

This set $X_2$ has a nicer structure than $E_\Phi(0)$. The condition $x_k x_{2k} = 0$ is imposed to all integers $k$ without exception for all points $x$ in $X_2$, while the same condition is imposed to "most" integers $k$ for points $x$ in $E_\Phi(0)$.

**Theorem 2** [FLM12] *The box dimension of* $X_2$ *is equal to*

$$\dim_B X_2 = \frac{1}{2\log 2}\sum_{n=1}^{+\infty} \frac{\log F_n}{2^n},$$

*where* $F_n$ *is the Fibonacci sequence:* $F_0 = 1$, $F_1 = 2$ *and* $F_{n+2} = F_{n+1} + F_n$ *for* $n \geq 0$.

The key idea to prove the formula is the following observation, which is also one of the key points for all the obtained results up to now in different cases. Look at the definition (7) of $X_2$. The value of the digit $x_1$ of an element $x = (x_k) \in X_2$ has an impact on the value of $x_2$, which in turn on the value of $x_4$,...and so forth on the values of $x_{2^k}$ for all $k \geq 1$. But it has no influence on $x_3$, $x_5$,...Similarly, the value of $x_i$ for an odd integer $i$ only has influence on $x_{i2^k}$. This suggests us the following partition

$$\mathbb{N}^* = \bigsqcup_{i \text{ odd}} \Lambda_i, \quad \text{with } \Lambda_i := \{i\, 2^n : n \geq 0\}.$$

We could say that the defining conditions of $X_2$ restricted to different $\Lambda_i$ are independent. We are then led to investigate, for each odd number $i$, the restriction of $x$ to $\Lambda_i$ which will be denoted by

$$x_{|\Lambda_i} = x_i x_{i2} x_{i2^2} \ldots x_{i2^n} \ldots.$$

If we rewrite $x_{|\Lambda_i} = z_1 z_2 \ldots$, which is considered as a point in $\mathbb{D}_2$, then $(z_n)$ belongs to the subshift of finite type subjected to $z_k z_{k+1} = 0$.

It is clear that

$$\dim_B X_2 = \lim_{n\to\infty} \frac{\log_2 N_n}{n}$$

where $N_n$ is the cardinality of the set

$$\{(x_1 x_2 \cdots x_n) : x_k x_{2k} = 0 \ \text{ for } k \geq 1 \text{ such that } \ 2k \leq n\}.$$

Let us decompose the set of the first $n$ integers as follows $\{1, \ldots, n\} = C_0 \sqcup C_1 \sqcup \cdots \sqcup C_m$ with

$$\begin{aligned}
C_0 &:= \{1, 3, 5, \ldots, 2n_0 - 1\},\\
C_1 &:= \left\{1\cdot 2, 3\cdot 2, 5\cdot 2, \ldots, 2\cdot (2n_1 - 1)\right\},\\
&\ldots\\
C_k &:= \left\{1\cdot 2^k, 3\cdot 2^k, 5\cdot 2^k, \ldots, 2^k\cdot (2n_k - 1)\right\},\\
&\ldots\\
C_m &:= \left\{1\cdot 2^m\right\}.
\end{aligned}$$

These finite sequences have different length $n_k$ $(0 \le k \le m)$. Actually $n_k$ is the biggest integer such that $2^k(2n_k - 1) \le n$, i.e. $n_k = \left\lfloor \frac{n}{2^{k+1}} + \frac{1}{2} \right\rfloor$. The number $m$ is the biggest integer such that $2^m \le n$, i.e. $m = \lfloor \log_2 n \rfloor$. It clear that $n_0 > n_1 > \cdots > n_{m-1} > n_m = 1$. The conditions $x_\ell x_{2\ell} = 0$ with $\ell$ in different columns in the table defining $C_0, \ldots, C_m$ are independent. This independence allows us to count the number of possible choices for $(x_1, \ldots, x_n)$ by the multiplication principle. First, we have $n_m (= 1)$ column which has $m + 1$ elements. We have $F_{m+1}$ choices for those $x_\ell$ with $\ell$ in the first column because $x_\ell x_{2\ell}$ is conditioned to be different from the word 11. We repeat this argument for other columns. Each of the next $n_{m-1} - n_m$ columns has $m$ elements, so we have $F_m^{n_{m-1}-n_m}$ choices for those $x_\ell$ with $\ell$ in these columns. By induction, we get

$$N_n = F_{m+1}^{n_m} F_m^{n_{m-1}-n_m} F_{m-1}^{n_{m-2}-n_{m-1}} \cdots F_1^{n_0-n_1}.$$

To finish the computation we need to note that $\frac{n_k}{n}$ tends to $2^{-(k+1)}$ as $n$ tends to the infinity.

The set $X_2$ is not invariant under the shift. But as observed by Kenyon et al. [KPS12], it is multiplicatively invariant in the sense that $M_r X_2 \subset X_2$ for all integers $r \ge 1$ where

$$M_r(x_n) = (x_{rn}).$$

The Hausdorff dimension of $X_2$ was obtained in [KPS12] and the gauge function of $X_2$ was obtained in [PS13]. We will discuss the work done in [KPS12] in Sect. 5.

## 3 Dimensions of Measures

Multifractal properties were first investigated for measures. The multifractal analysis of a measure is the analysis of the local dimension on the whole space, while the dimensions of a measure concern with what happens on a Borel support of the measure. In [Fan89], lower and upper Hausdorff dimensions of a measure were introduced and systematically studied, inspired by Peyrière [Pey75] and Kahane [Kah87]. The lower and upper packing dimensions were later studied independently in [Tam95, Heu98]. Some aspects were also considered in [Haa92]. A fundamental theorem in the theory of fractals is Frostman theorem. Howroyd [How94] and Kaufman [Kau94] generalized it from Euclidean spaces to complete separable metric spaces. This fundamental theorem allows us to employ the potential theory.

### *3.1 Potential Theory*

Let $(X, d)$ be a complete separable metric space, called Polish space. Let $0 < \alpha < \infty$. For any locally finite Borel measure $\mu$ on $X$, we define its potential of order $\alpha$ by

$$U_\alpha^\mu(x) := \int_X \frac{d\mu(y)}{(d(x,y))^\alpha} \qquad (x \in X)$$

and its energy of order $\alpha$ by

$$I_\alpha^\mu := \int_X U_\alpha^\mu(x) d\mu(x) = \int_X \int_X \frac{d\mu(x) d\mu(y)}{(d(x,y))^\alpha}.$$

The capacity of order $\alpha$ of a compact set $K$ in $X$ is defined by

$$\mathrm{Cap}_\alpha K = \left( \inf_{\mu \in \mathcal{M}_1^+(K)} I_\alpha^\mu \right)^{-1}.$$

For an arbitrary set $E$ of $X$, we define its capacity of order $\alpha$ by

$$\mathrm{Cap}_\alpha E = \sup\{\mathrm{Cap}_\alpha K : K \text{ compact contained in } E\}.$$

For a set $E$ in $X$, we define its capacity dimension by

$$\dim_C E = \inf\{\alpha > 0 : \mathrm{Cap}_\alpha(E) = 0\} = \sup\{\alpha > 0 : \mathrm{Cap}_\alpha(E) > 0\}.$$

The following is the theorem of Frostman-Kaufman-Howroyd. Frostman initially dealt with the Euclidean space. Kaufman proved the result by generalizing a min-max theorem on quadratic function to the mutual potential energy functional, while Howroyd used the technique of weighted Hausdorff measures.

**Theorem 3** [Kau94, How94] *Let $(X, d)$ be a complete metric space. For any Borel $E \subset X$, we have* $\dim_C E = \dim_H E$.

## *3.2 Hausdorff Dimensions of a Measures*

An important tool to study a measure is its dimensions, which attempt to estimate the size of the "supports" of the measure. The idea finds its origin in Peyrière's works on Riesz products [Pey75] and also in that of Kahane on Dvoretzky covering [Kah87]. The following definitions were introduced in [Fan89] (see also [Fan94a, Fal03]).

Let $(X, d)$ be a complete separable metric space. Let $\mu$ be a Borel measure on $X$. The *lower Hausdorff dimension* and the *upper Hausdorff dimension* of a measure $\mu$ are respectively defined by

$$\dim_* \mu = \inf\{\dim_H A : \mu(A) > 0\}, \qquad \dim^* \mu = \inf\{\dim_H A : \mu(A^c) = 0\}.$$

It is evident that $\dim_* \mu \le \dim^* \mu$. When the equality holds, $\mu$ is said to be *unidimensional* or *$\alpha$-dimensional* where $\alpha$ is the common value of $\dim_* \mu$ and $\dim^* \mu$. The Hausdorff dimensions $\dim_* \mu$ and $\dim^* \mu$ are described by the lower local dimension function $\underline{D}_\mu(x)$ in the following way.

**Theorem 4** [Fan89, Fan94a]

$$\dim_* \mu = \operatorname{ess\,inf}_\mu \underline{D}_\mu(x), \quad \dim^* \mu = \operatorname{ess\,sup}_\mu \underline{D}_\mu(x).$$

There are also a continuity-singularity criterion using Hausdorff measures and a energy-potential criterion. Sometimes these criteria are more practical.

**Theorem 5** [Fan89, Fan94a]

$$\begin{aligned}\dim_* \mu &= \sup\{\alpha > 0 : \mu \ll H^\alpha\} = \sup\{\alpha > 0 : \mu = \sum \mu_k, I_\alpha^{\mu_k} < \infty\}\\ \dim^* \mu &= \inf\{\alpha > 0 :\ \mu \perp H^\alpha\} \ = \inf\{\alpha > 0 : U_\alpha^\mu(x) = \infty,\ \mu\text{−p.p.}\}.\end{aligned}$$

Theorem 4 holds for lower and upper packing dimensions of a measures, similarly defined, if we replace $\underline{D}(\mu, x)$ by $\overline{D}(\mu, x)$ (see [Tam95, Heu98]). They are denoted by $\mathrm{Dim}_*\mu$ and $\mathrm{Dim}^*\mu$.

We say $\mu$ is exact if $\dim_* \mu = \dim^* \mu = \mathrm{Dim}_*\mu = \mathrm{Dim}^*\mu$.

## *3.3 Sums, Products, Convolutions, Projections of Measures*

What we present in this section was in the first unpublished version of [Fan94a]. Some part was restated by in [Tam95] and some part was used in [Fan94b, Fan94c].

*Sum* The absolute continuity $\nu \ll \mu$ is a partial order on the space of positive Borel measures $M^+(X)$ on the space $X$. Using the continuity-singularity criterion, it is easy to see that

$$0 < \nu \ll \mu \Rightarrow \dim_* \mu \le \dim_* \nu \le \dim^* \nu \le \dim^* \mu. \tag{8}$$

The relation $\nu \sim \mu$ (meaning $\nu \ll \mu \ll \nu$) is an equivalent relation. Since two equivalent measures have the same lower and upper Hausdorff dimensions, both $\dim_* \mu$ and $\dim^*$ are well defined for equivalent classes.

Given a family of positive measures $\{\mu_i\}_{i\in I}$ which is bounded under the order $\ll$, we denote its supremum by $\bigvee_{i\in I} \mu_i$. If the family is finite, we have $\bigvee_{i\in I} \mu_i \sim \sum_{i\in I} \mu_i$. Such equivalence also holds when the family is countable. In general, there is a countable sub-family $\mu_{i_k}$ such that $\bigvee_{i\in I} \mu_i \sim \sum_k \mu_{i_k}$. This is what we mean by sum of measures.

**Theorem 6** *If $\{\mu_i\}_{i\in I}$ is a bounded family of measures in $M^+(X)$, we have*

$$\dim_* \bigvee_{i\in I} \mu_i = \inf_{i\in I} \dim_* \mu_i, \quad \dim^* \bigvee_{i\in I} \mu_i = \sup_{i\in I} \dim^* \mu_i.$$

Let us see how to prove the first formula for a family of two measures $\mu$ and $\nu$. For any $\alpha < \dim_*(\mu+\nu)$, we have $U_\alpha^\mu(x) + U_\alpha^\mu(x) = U_\alpha^{\mu+\nu}(x) < \infty \quad (\mu+\nu)-a.e.$ This implies $\dim_*(\mu+\nu) \leq \min\{\dim_* \mu, \dim_* \nu\}$. The inverse inequality follows from the fact that if $\beta < \min\{\dim_* \mu, \dim_* \nu\}$, then $\mu \ll H^\beta$ and $\nu \ll H^\beta$ which implies $\mu + \nu \ll H^\beta$.

Let us consider now the infimum $\bigwedge_{i\in I}$ of a family of measure $\{\mu_i\}_{i\in I}$. Recall that by definition we have

$$\bigwedge_{i\in I} \mu_i \sim \bigvee_{\mu:\forall i\in I, \mu\ll\mu_i} \mu.$$

For a family of two measures $\mu$ and $\nu$, we have $\mu \bigwedge \nu \sim \frac{d\mu}{d\nu} \bigvee \frac{d\nu}{d\mu}$.

**Theorem 7** *If $\{\mu_i\}_{i\in I}$ is family of measures in $M^+(X)$ such that $\bigwedge_{i\in I} \mu_i \neq 0$, we have*

$$\sup_{i\in I} \dim_* \mu_i \leq \dim_* \bigwedge_{i\in I} \mu_i \leq \dim^* \bigwedge_{i\in I} \mu_i \leq \inf_{i\in I} \dim^* \mu_i.$$

Consequently, $\bigvee_{i\in I} \mu_i$ (resp. $\bigwedge_{i\in I} \mu_i$) is unidimensional if and only if all measures $\mu_i$ are unidimensional and have the same dimension.

*Product* Let $(X, \delta_X)$ and $(Y, \delta_Y)$ be two Polish spaces. Then $d := \delta_X \vee \delta_Y$ is a compatible metric on the product space. Let $\mu \in M^+(X)$ and $\nu \in M^+(Y)$. Let us consider the product measure $\mu \otimes \nu$.

**Theorem 8** *For $\mu \in M^+(X)$ and $\nu \in M^+(Y)$, we have*

$$\dim_* \mu \otimes \nu \geq \dim_* \mu + \dim_* \nu, \qquad \dim^* \mu \otimes \nu \geq \dim^* \mu + \dim^* \nu.$$

We say a measure $\mu \in M^+(X)$ is regular if $\underline{D}(\mu, x) = \overline{D}(\mu, x)$ $\mu$-a.e.

**Theorem 9** *If $\mu \in M^+(X)$ or $\nu \in M^+(Y)$ is regular, we have*

$$\dim_* \mu \otimes \nu = \dim_* \mu + \dim_* \nu, \qquad \dim^* \mu \otimes \nu = \dim^* \mu + \dim^* \nu.$$

*Convolution* Assume that the Polish space $X$ is a locally compact abelian group $G$. Assume further that $G$ satisfies the following hypothesis

$$I_\alpha^{\mu*\nu} \leq C(\alpha, \nu) I_\alpha^\mu$$

for all measures $\mu, \nu \in M^+(X)$, where $C(\alpha, \nu)$ is a constant independent of $\mu$. For example, if $G = \mathbb{R}^d$, we have

$$I_\alpha^\mu = \int |\xi|^{-(d-\alpha)} |\widehat{\mu}(\xi)|^2 d\xi$$

which implies that the hypothesis is satisfied by $\mathbb{R}^d$. The hypothesis is also satisfied by the group $\prod_{n=1}^{\infty} \mathbb{Z}/m_n\mathbb{Z}$ [Fan89].

**Theorem 10** *For any measures $\mu, \nu \in M^+(G)$, we have*

$$\dim_* \mu * \nu \geq \max\{\dim_* \mu, \dim_* \nu\}, \qquad \dim^* \mu * \nu \geq \max\{\dim^* \mu, \dim^* \nu\}.$$

For a given measure $\mu \in M^+(G)$, we consider the following two subgroups

$$H_- := \{t \in G : \mu \ll \mu * \delta_t\}, \qquad H_+ := \{t \in G : \mu * \delta_t \ll \mu\}.$$

We could call $H = H_- \cap H_+$ the quasi-invariance group of $\mu$.

**Theorem 11** *Under the above assumption, if $\nu(H_-) > 0$, we have the equality* $\dim_* \mu * \nu = \dim_* \mu$ *and consequently* $\dim_* \nu \leq \dim_* \mu$ *and* $\dim_H H_- \leq \dim_* \mu$; *if $\nu(H_+) > 0$, we have the equality* $\dim^* \mu * \nu = \dim^* \mu$ *and consequently* $\dim^* \nu \leq \dim^* \mu$ *and* $\dim_H H_+ \leq \dim^* \mu$.

*Projection* Let $\mu \in M^+(\mathbb{R}^2)$ with $\mathbb{R}^2 = \mathbb{C}$. Let $L_\theta$ $(0 \leq \theta < 2\pi)$ be the line passing the origin and having angle $\theta$ with the line of abscissa. The orthogonal projection $P_\theta$ on $L_\theta$ is defined by $P_\theta(x) = \langle x, e^{i\theta} \rangle$. Let $\mu_\theta := \mu \circ P_\theta^{-1}$ be the projection of $\mu$ on $L_\theta$.

**Theorem 12** *Let $\mu \in M^+(\mathbb{R}^2)$. For almost all $\theta$, we have* $\dim_* \mu_\theta = \dim_* \mu \wedge 1$ *and* $\dim^* \mu_\theta = \dim^* \mu \wedge 1$.

## 3.4 Ergodicity and Dimension

Young [You82] considered diffeomorphisms of surfaces leaving invariant an ergodic Borel probability measure $\mu$. She proved that $\mu$ is exact and found a formula relating $\dim \mu$ to the entropy and Lyapunov exponents of $\mu$. One of the main problems in the interface of dimension theory and dynamical systems is the Eckmann-Ruelle conjecture on the dimension of hyperbolic ergodic measures: the local dimension of every hyperbolic measure invariant under a $C^{1+\alpha}$-diffeomorphism exists almost everywhere. This conjecture was proved by Barreira et al. [BPS99] based on the fundamental fact that such a measure possesses asymptotically "almost" local product structure. But, in general, the ergodicity of the measure doesn't imply that the measure is exact [Cut95]. In [Fan94b], $D$-ergodicity and unidimensionality were studied.

# 4 Oriented Walks and Riesz Products

## 4.1 Oriented Walks

Let $(\epsilon_n)_{n\geq 1} \subset [0, 2\pi)^{\mathbb{N}}$ be a sequence of angles. For $n \geq 1$, define

$$S_n(\epsilon) = \sum_{k=1}^{n} e^{i(\epsilon_1+\epsilon_2+\cdots+\epsilon_k)}.$$

We call $(S_n(\epsilon))_{n\geq 1}$ an oriented walk on the plan $\mathbb{C}$. In his book [Fel68] (vol. 1, pp. 240–241), Feller mentioned a model describing the length of long polymer molecules in chemistry. It is a random chain consisting of $n$ links, each of unit length, and the angle between two consecutive links is $\pm\alpha$ where $\alpha$ is a positive constant. Then the distance $L_n$ from the beginning to the end of the chain can be expressed by

$$L_n = |S_n(\epsilon)|$$

where $(\epsilon_n)$ is an i.i.d. sequence of random variables taking values in $\{-\alpha, \alpha\}$. If $\alpha = 0$, $L_n = n$ is deterministic. If $0 < \alpha < 2\pi$, the random variable $L_n$ is not expressed as sums of independent variables. However Feller succeeded in computing the second order moment of $L_n$. It is actually proved in [Fel68] that $\|L_n\|_2$ is of order $\sqrt{n}$. More precisely, for $0 < \alpha < 2\pi$ we have

$$\mathbb{E}L_n(\alpha)^2 = n\frac{1+\cos\alpha}{1-\cos\alpha} - 2\cos\alpha\frac{1-\cos^n\alpha}{(1-\cos\alpha)^2}.$$

Observe that

$$\mathbb{E}L_n^2 = \frac{1-(-1)^n}{2} \text{ if } \alpha = \pi; \qquad \mathbb{E}L_n^2 \sim n\frac{1+\cos\alpha}{1-\cos\alpha} \text{ if } 0 < \alpha < 2\pi, \alpha \neq \pi.$$

What is the behavior of $S_n(\epsilon)$ as $n \to \infty$ for individuals $\epsilon$? We could study the behavior from the multifractal point of view. Let us consider a more general setting. Fix $d \in \mathbb{N}^*$. Let $\tau \in GL(\mathbb{R}^d)$, $v \in \mathbb{R}^d$ and $A$ a finite subset of $\mathbb{Z}$. For any $x = (x_n) \in \mathbb{D} := A^{\mathbb{N}}$, we define the oriented walk

$$S_0(x) = v, \qquad S_n(x) = \sum_{k=1}^{n} \tau^{x_1+x_2+\cdots+x_k} v.$$

For $\alpha \in \mathbb{R}^d$, we define the *$\alpha$-level set*

$$E_\tau(\alpha) := \left\{ x \in \mathbb{D} : \lim_{n\to\infty} \frac{1}{n} S_n(x) = \alpha \right\}.$$

Let $L_\tau := \{\alpha \in \mathbb{R}^d : E_\tau(\alpha) \neq \emptyset\}$.

The following two cases were first studied in [Fan00]. Case 1: $d = 2$, $\tau = -1$ and $A = \{0, 1\}$; Case 2: $d = 1$, $\tau = e^{i\pi/2}$ and $A = \{-1, 1\}$.

**Theorem 13** [Fan00] *In the first case, we have* $L_\tau = [-1, 1]$ *and for* $\alpha \in L_\tau$ *we have*

$$\dim_H E_\tau(\alpha) = \dim_P E_\tau(\alpha) = H\left(\frac{1+\alpha}{2}\right).$$

*In the seconde case, we have* $L_\tau = \{z = a + ib : |a| \le 1/2, |b| \le 1/2\}$ *and for* $\alpha = a + bi \in L_\tau$ *we have*

$$\dim_H E_\tau(\alpha) = \dim_P E_\tau(\alpha) = \frac{1}{2\log 2}\left[H\left(\frac{1}{2}+a\right) + H\left(\frac{1}{2}+b\right)\right].$$

This theorem was proved by using Riesz products which will be described in the following section.

A new construction of measures allows us to deal with a class of oriented walks. We assume that $\tau \in GL(\mathbb{R}^d)$ is idempotent. That is to say $\tau^p = Id$ for some integer $p > 1$ (the case $p = 1$ is trivial). The least $p$ is called the order of $\tau$. The above two cases are special cases. In fact, $\theta = -1$ is idempotent with order $p = 2$ and $\theta = e^{i\pi/2}$ is idempotent with order $p = 4$. The following rotations in $\mathbb{R}^3$

$$\tau_1 = \begin{pmatrix} 0 & 0 & 1 \\ 1 & 0 & 0 \\ 0 & 1 & 0 \end{pmatrix}, \qquad \tau_2 = \begin{pmatrix} 0 & 0 & -1 \\ 1 & 0 & 0 \\ 0 & 1 & 0 \end{pmatrix}$$

are idempotent with order respectively equal to 3 and 6. Also remark that $\tau_1 \in SO_3(\mathbb{R})$ and $\tau_2 \in O_3(\mathbb{R}) \setminus SO_3(\mathbb{R})$.

Since $\tau^p = Id$, the sum $x_1 + x_2 + \cdots + x_k$ in the definition of $S_n(x)$ can be made modulo $p$. For $s \in \mathbb{R}^d$, we define a $p \times p$-matrix $M_s = (M_s(i, j))$: for $(i, j) \in \mathbb{Z}/p\mathbb{Z} \times \mathbb{Z}/p\mathbb{Z}$ define

$$M_s(i, j) = 1_A(j - i)\exp[\langle s, \tau^j\rangle v].$$

where $\langle \cdot, \cdot \rangle$ denotes the scalar product in $\mathbb{R}^d$. It is clear that $M_s$ is irreducible iff $M_0$ is so. We consider $A$ as a subset (modulo $p$) of $\mathbb{Z}/p\mathbb{Z}$. It is easy to see that $M_0$ is irreducible iff $A$ generates the group $\mathbb{Z}/p\mathbb{Z}$.

Assume that $A$ generates the group $\mathbb{Z}/p\mathbb{Z}$. Then $M_s$ is irreducible and by the Perron-Frobenius theorem, the spectral radius $\lambda(s)$ of $M_s$ is a simple eigenvalue and there is a unique corresponding probability eigenvector $t(s) = (t_s(0), t_s(1) \ldots, t_s(p-1)$. Let

$$P(s) = \log \lambda(s).$$

It is real analytic and strictly convex function on $\mathbb{R}^d$. We call it the pressure function associated to the oriented walk.

**Theorem 14** [FW00] *Assume $\tau$ is idempotent with order $p$ and $A$ generates the group $\mathbb{Z}/p\mathbb{Z}$. Then $L_\tau = \overline{\{\nabla P(s) : s \in \mathbb{R}^d\}}$ and for $\alpha \in \Delta$, we have*

$$\dim_H E_\tau(\alpha) = \dim_P E_\tau(\alpha) = \frac{1}{\log p} \inf_{s \in \mathbb{R}^d} \{P(s) - \langle s, \alpha \rangle\} = \frac{P(s_\alpha) - \langle s_\alpha, \alpha \rangle}{\log p},$$

*where $s_\alpha$ is the unique $s \in \mathbb{R}^d$ such that $\nabla P(s) = \alpha$.*

## 4.2 Riesz Products

Theorem 1 was proved by using Riesz products. While Hausdorff introduced the Hausdorff dimension (1919), Riesz constructed a class of continuous but singular measures on the circle (1918), called Riesz products. Riesz products are used as tool in harmonic analysis and some of them are Gibbs measures in the sense of dynamical systems.

Let us recall the definition of Riesz product on a compact abelian group $G$, due to Hewitt-Zuckerman [HZ66] (1966). Let $\widehat{G}$ be the dual group of $G$. A sequence of characters $\Lambda = (\gamma_n)_{n \geq 1} \subset \widehat{G}$ is said to be *dissociated* if for any $n \geq 1$, the following characters are all distinct:

$$\gamma_1^{\epsilon_1} \gamma_2^{\epsilon_2} \cdots \gamma_n^{\epsilon_n}$$

where $\epsilon_j \in \{-1, 0, 1\}$ if $\gamma_j$ is not of order 2, or $\epsilon_j \in \{0, 1\}$ otherwise. Given such a dissociated sequence $\Lambda = (\gamma_n)_{n \geq 1}$ and a sequence of complex numbers $a = (a_n)_{n \geq 1}$ such that $|a_n| \leq 1$, we can define a probability measure on $G$, called *Riesz product*,

$$\mu_a = \prod_{n=1}^{\infty} \big(1 + \mathrm{Re}\ a_n \gamma_n(t)\big) \tag{9}$$

as the weak* limit of $\prod_{n=1}^{N} \big(1 + \mathrm{Re}\ a_n \gamma_n(t)\big) \mathrm{d}t$ where $\mathrm{d}t$ denotes the Haar measure on $G$.

A very useful fact is that the Fourier coefficients of the Riesz product $\mu_a$ can be explicitly expressed in term of the coefficients $a_n$'s:

$$\widehat{\mu}_a(\gamma) = \prod_{k=1}^{n} a_k^{(\epsilon_k)} \text{ if } \gamma = \gamma_1^{\epsilon_1} \gamma_2^{\epsilon_2} \cdots \gamma_n^{\epsilon_n}, \quad \widehat{\mu}_a(\gamma) = 0 \text{ otherwise.}$$

where $a_n^{(\epsilon)} = 1, a_n/2$ or $\bar{a}_n/2$ according to $\epsilon = 0, 1$ or $-1$. Consequently the sequence $\{\gamma_n - \bar{a}_n/2\}_{n \geq 1}$ is an orthogonal system in $L^2(\mu_a)$. Here are some properties of $\mu_a$.

**Theorem 15** [Zyg68] *The measure $\mu_a$ is either absolutely continuous or singular (with respect to the Haar measure) according to $\sum_{n=1}^{\infty} |a_n|^2 < \infty$ or $= \infty$.*

**Theorem 16** [Fan93, Pey90] *Let $\{\alpha_n\}$ be a sequence of complex numbers. The orthogonal series $\sum_{n=1}^{\infty} \alpha_n(\gamma_n(t) - \bar{a}_n/2)$ converges $\mu_a$-everywhere iff $\sum_{n=1}^{\infty} |\alpha_n|^2 < \infty$.*

The proof of Theorem 16 in [Fan93] involved the following Riesz product with $\omega = (\omega_n) \in G^{\mathbb{N}}$ as phase translation:

$$\mu_{a,\omega} = \prod_{n=1}^{\infty} (1 + \mathrm{Re}\, a_n \gamma_n(t + \omega_n)). \tag{10}$$

Actually $\mu_{a,\omega}$ was considered as a random measure and $(\omega_n)$ was considered as an i.i.d. random sequence with Haar measure as common probability law.

When two Riesz products $\mu_a$ and $\mu_b$ are singular or mutually absolutely continuous? It is a unsolved problem. Bernoulli infinite product measures can be viewed as Riesz products on the group $(\mathbb{Z}/m\mathbb{Z})^{\mathbb{N}}$. For these Bernoulli infinite measures, the Kakutani dichotomy theorem [Kak48] applies and there is a complete solution. But there is no complete solution for other groups.

The classical Riesz products are of the form

$$\mu_a = \prod_{k=1}^{\infty} (1 + \mathrm{Re}\, a_n e^{i\lambda_k x}) \tag{11}$$

where $\{\lambda_n\} \subset \mathbb{N}$ is a lacunary sequence in the sense that $\lambda_{n+1} \geq 3\lambda_n$. Peyrière has first studied the lower and upper dimensions of $\mu_a$, without introducing the notion of dimension of measures. Let us mention an estimation for the energy integrals of $\mu_a$ [Fan89]:

$$\int\int \frac{d\mu_a(x)d\mu_a(y)}{|x-y|^\alpha} \approx \lambda_1^{\alpha-1}|a_1|^2 + \sum_{n=2}^{\infty} \lambda_n^{\alpha-1}|a_n|^2 \prod_{k=1}^{n-1}\left(1 + \frac{|a_n|^2}{2}\right).$$

## 4.3 Evolution Measures

The key for the proof of Theorem 14 is the construction of the following measures on $A^{\mathbb{N}}$, which describe the evolution of the oriented walk. It is similar to Markov measure but it is not. It plays the role of Gibbs measure but it is not Gibbs measure either.

Recall that $M_s t(s) = \lambda(s)t(s)$. In other words, for every $i \in \mathbb{Z}/p\mathbb{Z}$ we have

$$\lambda(s)t_i(s) = \sum_j 1_A(j-i)t_j(s)\exp[\langle s, \tau^j v\rangle].$$

Denote, for $a \in A$ and for $(x_1, \ldots, x_{k+1}) \in A^{k+1}$,

$$\pi(a) = \frac{t_a(s)}{\sum_{b \in A} t_b(s)};$$

$$Q_k(x_1, x_2, \ldots, x_{k+1}) = \frac{t_{x_1+x_2+\cdots+x_{k+1}}(s) \exp[\langle s, \tau^{x_1+x_2+\cdots+x_{k+1}} v\rangle]}{\lambda(s) t_{x_1+x_2+\cdots+x_k}(s)}.$$

Then we define a probability measure $\mu_s$ on $A^{\mathbb{N}}$ as follows. For any word $x_1 x_2 \cdots x_n \in A^n$, let

$$\mu_s([x_1 x_2 \cdots x_n]) = \pi(x_1) Q_1(x_1, x_2) Q_2(x_1, x_2, x_3) \cdots Q_{n-1}(x_1, x_2, \ldots, x_n).$$

For $x = (x_n) \in A^{\mathbb{N}}$, let

$$w_n(x) = x_1 + x_2 + \cdots + x_n \pmod{p}.$$

The mass $\mu_s([x_1 x_2 \cdots x_n])$ and the partial sum $S_n(x)$ are directly related as follows.

$$\log \mu_s([x_1 x_2 \cdots x_n]) = \langle s, S_n(x)\rangle - (n-1)\log \lambda(s) - \log \sum_{a \in A} t_a(s) + \log t_{w_n(x)}(s).$$

As the $t_i(s)$'s are bounded, we deduce the following relation between the measure $\mu_s$ and the oriented walk $S_n$.

**Proposition 1** *For any $x \in \mathbb{D}$, we have*

$$\log \mu_s([x_1 x_2 \cdots x_n]) - \langle s, S_n(x)\rangle = -n \log \lambda(s) + O(1).$$

# 5 Multiple Birkhoff Averages

Let $\mathcal{A} = \{0, 1, 2, \ldots, m-1\}$ be a set of $m$ symbols ($m \geq 2$). Denote $\Sigma_m = \mathcal{A}^{\mathbb{N}^*}$. Let $q \geq 2$ be a integer. Fan, Schmeling and Wu made a forward step in [FSW00] by obtaining a Hausdorff spectrum of multiple ergodic averages for a class of potentials. They consider an arbitrary function $\varphi : \mathcal{A}^d \to \mathbb{R}$ and study the sets

$$E(\alpha) = \left\{x \in \Sigma_m : \lim_{n\to\infty} A_n \varphi(x) = \alpha\right\}$$

for $\alpha \in \mathbb{R}$, where

$$A_n \varphi(x) = \frac{1}{n} \sum_{k=1}^{n} \varphi(x_k, x_{qk}, \ldots, x_{q^{d-1}k}). \tag{12}$$

Let

$$\alpha_{\min} = \min_{a_1,\ldots,a_d \in \mathcal{A}} \varphi(a_1,\ldots,a_d), \quad \alpha_{\max} = \max_{a_1,\ldots,a_d \in \mathcal{A}} \varphi(a_1,\ldots,a_d).$$

It is assumed that $\alpha_{\min} < \alpha_{\max}$ (otherwise $\varphi$ is constant and the problem is trivial). A key ingredient of the proof is a class of measures constructed by Kenyon, Peres and Solomyak [KPS12] that we call telescopic product measures. In [FSW00], a nonlinear thermodynamic formalism was developed.

## 5.1 Thermodynamic Formalism

The Hausdorff dimension of $E(\alpha)$ is determined through the following thermodynamic formalism. Let $\mathcal{F}(\mathcal{A}^{d-1}, \mathbb{R}^+)$ be the cone of functions defined on $\mathcal{A}^{d-1}$ taking non-negative real values. For any $s \in \mathbb{R}$, consider the transfer operator $\mathcal{L}_s$ defined on $\mathcal{F}(\mathcal{A}^{d-1}, \mathbb{R}^+)$ by

$$\mathcal{L}_s \psi(a) = \sum_{j \in \mathcal{A}} e^{s\varphi(a,j)} \psi(Ta, j) \tag{13}$$

where $T : \mathcal{A}^{d-1} \to \mathcal{A}^{d-2}$ is defined by $T(a_1,\ldots,a_{d-1}) = (a_2,\ldots,a_{d-1})$. Then define the non-linear operator $\mathcal{N}_s$ on $\mathcal{F}(\mathcal{A}^{d-1}, \mathbb{R}^+)$ by $\mathcal{N}_s\psi(a) = (\mathcal{L}_s\psi(a))^{1/q}$. It is proved in [FSW00] that the equation

$$\mathcal{N}_s \psi_s = \psi_s \tag{14}$$

admits a unique strictly positive solution $\psi_s = \psi_s^{(d-1)} : \mathcal{A}^{d-1} \to \mathbb{R}_+^*$. Extend the function $\psi_s$ onto $\mathcal{A}^k$ for all $1 \le k \le d-2$ by induction:

$$\psi_s^{(k)}(a) = \left( \sum_{j \in \mathcal{A}} \psi_s^{(k+1)}(a, j) \right)^{\frac{1}{q}}, \quad (a \in \mathcal{A}^k). \tag{15}$$

For simplicity, we write $\psi_s(a) = \psi_s^{(k)}(a)$ for $a \in \mathcal{A}^k$ with $1 \le k \le d-1$. Then the pressure function is defined by

$$P_\varphi(s) = (q-1)q^{d-2} \log \sum_{j \in \mathcal{A}} \psi_s(j). \tag{16}$$

It is proved [FSW00] that $P_\varphi(s)$ is an analytic and convex function of $s \in \mathbb{R}$ and even strictly convex when $\alpha_{\min} < \alpha_{\max}$. The Legendre transform of $P_\varphi$ is defined as

$$P_\varphi^*(\alpha) = \inf_{s \in \mathbb{R}} (P_\varphi(s) - s\alpha).$$

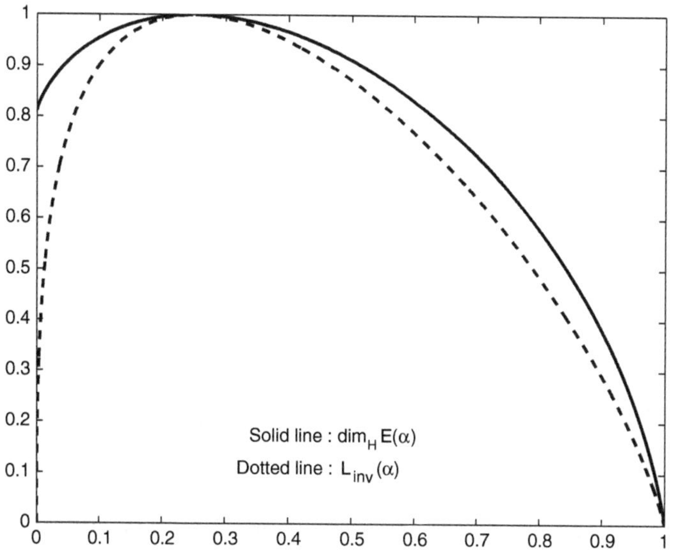


**Fig. 1** Spectra $\alpha \mapsto \dim_H E(\alpha)$ and $\alpha \mapsto F_{\text{inv}}(\alpha)$ for $\varphi_1$

We denote by $L_\varphi$ the set of levels $\alpha \in \mathbb{R}$ such that $E(\alpha) \neq \emptyset$.

**Theorem 17** [FSW00] *We have* $L_\varphi = [P'_\varphi(-\infty), P'_\varphi(+\infty)]$. *If* $\alpha = P'_\varphi(s_\alpha)$ *for some* $s_\alpha \in \mathbb{R} \cup \{-\infty, +\infty\}$, *then* $E(\alpha) \neq \emptyset$ *and the Hausdorff dimension of* $E(\alpha)$ *is equal to*

$$\dim_H E(\alpha) = \frac{P^*_\varphi(\alpha)}{q^{d-1} \log m}.$$

Similar results hold for vector valued functions $\varphi$ [FSW00]. Peres and Solomyak [PS13] have obtained a result for the special case $\varphi(x, y) = x_1 y_1$ on $\Sigma_2$. Kifer [Kif12] has obtained a result on the multiple recurrence sets for some frequency of product form.

Let us consider two examples. Let $q = 2$ and $\ell = 2$ and let $\varphi_1(x, y) = x_1 y_1$ (See Fig. 1) and $\varphi_2(x, y) = (2x_1 - 1)(2y_1 - 1)$ (see Fig. 2) be two potentials on $\Sigma_2$. The invariant spectra (see Sect. 6.2) are also shown in the figures.

## *5.2 Telescopic Product Measures*

One of the key points in the proof of the Hausdorff spectrum (Theorem 17) is the observation that the coordinates $x_1, \ldots, x_n, \ldots$ of $x$ appearing in the definition of $A_n\varphi(x)$ share the following independence. Consider the partition of $\mathbb{N}^*$:

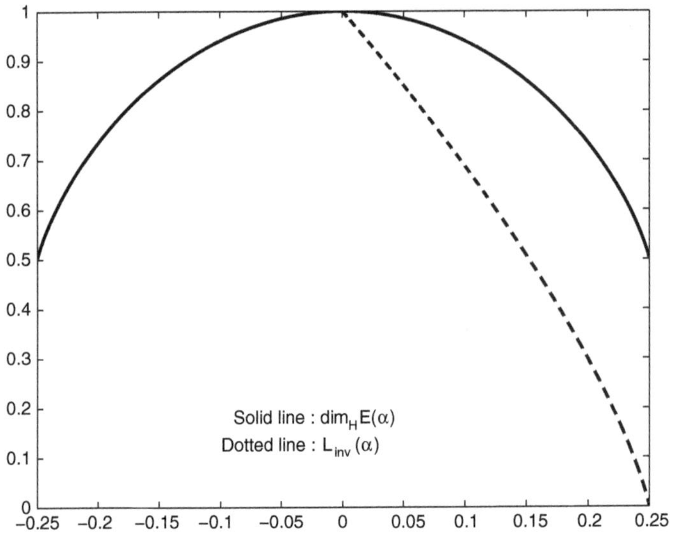


**Fig. 2** Spectra $\alpha \mapsto \dim_H E(\alpha)$ and $\alpha \mapsto F_{\text{inv}}(\alpha)$ for $\varphi_2$

$$\mathbb{N}^* = \bigsqcup_{i\geq 1, q\nmid i} \Lambda_i \quad \text{with } \Lambda_i = \{iq^j\}_{j\geq 0}.$$

Observe that if $k = iq^j$ with $q \nmid i$, then $\varphi(x_k, x_{kq}, \ldots, x_{kq^{d-1}})$ depends only on $x_{|\Lambda_i}$, the restriction of $x$ on $\Lambda_i$. So the summands in the definition of $A_n\varphi(x)$ can be put into different groups, each of which depends on one restriction $x_{|\Lambda_i}$. For this reason, we decompose $\Sigma_m$ as follows:

$$\Sigma_m = \prod_{i\geq 1, q\nmid i} \mathcal{A}^{\Lambda_i}.$$

Telescopic product measures are now constructed as follows. Let $\mu$ be a probability measure on $\Sigma_m$. Notice that $\mathcal{A}^{\Lambda_i}$ is nothing but a copy of $\Sigma_m$. We consider $\mu$ as a measure on $\mathcal{A}^{\Lambda_i}$ for every $i$ with $q \nmid i$. Then we define the infinite product measure $\mathbb{P}_\mu$ on $\prod_{i\geq 1, q\nmid i} \mathcal{A}^{\Lambda_i}$ of the copies of $\mu$. More precisely, for any word $u$ of length $n$ we define

$$\mathbb{P}_\mu([u]) = \prod_{i\leq n, q\nmid i} \mu([u_{|\Lambda_i}]),$$

where $[u]$ denotes the cylinder of all sequences starting with $u$. The probability measure $\mathbb{P}_\mu$ is called *telescopic product measure*. Kenyon, Peres and Solomyak [KPS12] have first constructed these measures.

The Hausdorff dimension of every telescopic product measure is computable.

**Theorem 18** [KPS12, FSW00] *For any given measure $\mu$, the telescopic product measure $\mathbb{P}_\mu$ is exact and its dimension is equal to*

$$\dim_H \mathbb{P}_\mu = \dim_P \mathbb{P}_\mu = \frac{(q-1)^2}{\log m} \sum_{k=1}^{\infty} \frac{H_k(\mu)}{q^{k+1}}$$

*where $H_k(\mu) = -\sum_{a_1,\dots,a_k \in S} \mu([a_1 \cdots a_k]) \log \mu([a_1 \cdots a_k])$.*

## 5.3 Dimension Formula of Ruelle-Type

The function $\psi_s$ defined by (14) and (15) determine a special telescopic product measure which plays the role of Gibbs measure in the proof of the Hausdorff spectrum.

First we define a $(d-1)$-step Markov measure on $\Sigma_m$, which will be denoted by $\mu_s$, with the initial law

$$\pi_s([a_1, \dots, a_{d-1}]) = \prod_{j=1}^{d-1} \frac{\psi_s(a_1, \dots, a_j)}{\psi_s^q(a_1, \dots, a_{j-1})} \tag{17}$$

and the transition probability

$$Q_s([a_1, \dots, a_{d-1}], [a_2, \dots, a_d]) = e^{s\varphi(a_1,\dots,a_d)} \frac{\psi_s(a_2, \dots, a_d)}{\psi_s^q(a_1, \dots, a_{d-1})}. \tag{18}$$

The corresponding telescopic product measure $\mathbb{P}_{\mu_s}$ is proved to be a dimension maximizing measure of $E(\alpha)$ if $s$ is chosen to be the solution of $P_\varphi'(s) = \alpha$. The dimension of $\mathbb{P}_{\mu_s}$ is simply expressed by the pressure function. In other words, we have the following formula of Ruelle-type.

**Theorem 19** [FSW00] *For any $s \in \mathbb{R}$, we have*

$$\dim_H \mathbb{P}_{\mu_s} = \frac{1}{q^{d-1}} [P_\varphi(s) - s P_\varphi'(s)].$$

## 5.4 Multiplicatively Invariant Sets

Kenyon et al. [KPS12] were able to compute both the Hausdorff dimension and the box dimension of $X_2$, already considered in Sect. 2.2, and of a class of generalizations of $X_2$. Peres et al. [PSSS00] generalized the results to a more general class of sets.

Recall that $\Sigma_m = \{0, 1, \dots, m-1\}^{\mathbb{N}^*}$, $q$ is an integer greater than 2 and $\Lambda_i := \{i\, q^n : n \geq 0\}$. For any subset $\Omega \subset \Sigma_m$, define

$$X_\Omega := \{x = (x_k)_{k\ge 1} \in \Sigma_m : \forall i \ne 0 \mod q,\ x_{|\Lambda_i} \in \Omega\}. \tag{19}$$

We get $X_\Omega = X_2$ when $q = 2$ and $\Omega$ is the Fibonacci set $\{y \in \Sigma_m : \forall k \ge 1,\ y_k y_{k+1} = 0\}$.

The set $X_\Omega$ is not shift-invariant but is multiplicatively invariant in the sense that $M_r X_\Omega \subset X_\Omega$ for every integer $r \in \mathbb{N}^*$ where $M_r$ maps $(x_n)$ to $(x_{rn})$.

The generating set $\Omega$ has a tree of prefixes, which is a directed graph $\Gamma$. The set $V(\Gamma)$ of vertices consists of all possible prefixes of finite length in $\Omega$, i.e.

$$V(\Gamma) := \bigcup_{k\ge 0} \mathrm{Pref}_k(\Omega),$$

where $\mathrm{Pref}_0(\Omega) = \{\emptyset\}$, $\mathrm{Pref}_k(\Omega) := \{u \in \{0, 1, \dots, m-1\}^k : \Omega \cap [u] \ne \emptyset\}$. There is a directed edge from a vertex $u$ to another $v$ if and only if $v = ui$ for some $i \in \{0, 1, \dots, m-1\}$.

**Theorem 20** [KPS12] *There exists a unique vector $\bar{t} = (t_v)_{v\in\Gamma} \in [1, m^{\frac{1}{q-1}}]^{V(\Gamma)}$ defined on the tree such that*

$$\forall v \in V(\Gamma), \quad (t_v)^q = \sum_{i\in\{0,1,\dots,m-1\}:\ vi\in\Omega} t_{vi}. \tag{20}$$

*The Hausdorff dimension and the box dimension of $X_\Omega$ are respectively equal to*

$$\dim_H(X_\Omega) = (q-1)\log_m t_\emptyset \tag{21}$$

$$\dim_B(X_\Omega) = (q-1)^2 \sum_{k=1}^{+\infty} \frac{\log_m |\mathrm{Pref}_k(\Omega)|}{q^{k+1}}. \tag{22}$$

*The two dimensions coincide if and only if the tree $\Gamma$ is spherically symmetric, i.e. all prefixes of length $k$ in $\Omega$ have the same number of continuations of length $k+1$ in $\Omega$.*

The vector $\bar{t}$ defines a measure $\mu$ on $\Omega \subset \{0, 1, \dots, m-1\}^{\mathbb{N}^*}$. Then a telescopic product measure can be built on $X_\Omega$. It is proved in [KPS12] that there is a maximizing measure on $X_\Omega$ of this form.

A typical example of the class of sets studied by Peres et al. [PSSS00] is

$$X_{2,3} = \{x \in \Sigma_m : x_k x_{2k} x_{3k} = 0\}.$$

The construction of the sets is as follows. Let $\kappa \ge 1$ be an integer and let $p_1, \dots, p_\kappa$ be $\kappa$ primes, which generates a semigroup $S$ of $\mathbb{N}^*$:

$$S = \langle p_1, p_2, \dots, p_\kappa\rangle = \{p_1^{\alpha_1} p_2^{\alpha_2} \cdots p_\kappa^{\alpha_\kappa} : \alpha_1, \dots, \alpha_\kappa \in \mathbb{N}\}.$$

The elements of $S$ are arranged in increasing order and denoted by $\ell_k$ the $k$-th element of $S = \{\ell_k\}_{k=1}^{\infty} : 1 = \ell_1 < \ell_2 < \cdots$ . Define

$$\gamma(S) := \sum_{k=1}^{\infty} \frac{1}{\ell_k} . \tag{23}$$

Write $(i, S) = 1$ when $(i, p_j) = 1$ for all $j \leq \kappa$. We have the following partition of $\mathbb{N}^*$:

$$\mathbb{N}^* = \bigsqcup_{(i,S)=1} iS. \tag{24}$$

For each element $x = (x_k)_{k=1}^{\infty}$, $x|_{iS}$ denotes the restriction $x|_{iS} := (x_{i\ell_k})_{k=1}^{\infty}$, which is also viewed as an element of $\Sigma_m$. Given a closed subset $\Omega \subset \Sigma_m$, we define a new subset of $\Sigma_m$:

$$X_{\Omega}^{(S)} := \left\{ x = (x_k)_{k=1}^{\infty} \in \Sigma_m : \ x|_{iS} \in \Omega \ \text{ for all } i, \ (i, S) = 1 \right\}. \tag{25}$$

**Theorem 21** [PSSS00] *There exists a vector* $\overline{t} = (t(u))_{u \in \mathrm{Pref}(\Omega)} \in [1, +\infty)^{\mathrm{Pref}(\Omega)}$ *defined on the tree of prefixes of* $\Omega$ *such that*

$$t(\varnothing) \in [1, m], \ \ t(u) \in [1, m^{\ell_k(\ell_{k+1}^{-1} + \ell_{k+2}^{-1} + \cdots)}], \ \ |u| = k, \ \ k \geq 1,$$

*which is the solution of the system*

$$t(\varnothing)^{\gamma(S)} = \sum_{j=0}^{m-1} t(j),$$
$$t(u)^{\ell_{k+1}/\ell_k} = \sum_{j:\ uj \in \mathrm{Pref}_{k+1}(\Omega)} t(uj), \quad \forall\, u \in \mathrm{Pref}_k(\Omega), \ \forall k \geq 1.$$

*The Hausdorff dimension and the box dimension of* $X_{\Omega}^{(S)}$ *are respectively equal to*

$$\dim_H \left( X_{\Omega}^{(S)} \right) = \log_m t(\varnothing)$$
$$\dim_B \left( X_{\Omega}^{(S)} \right) = \gamma(S)^{-1} \sum_{k=1}^{\infty} \Big( \frac{1}{\ell_k} - \frac{1}{\ell_{k+1}} \Big) \log_m |\mathrm{Pref}_k(\Omega)|.$$

*We have* $\dim_H \left( X_{\Omega}^{(S)} \right) = \dim_B \left( X_{\Omega}^{(S)} \right)$ *if and only if the tree of prefixes of* $\Omega$ *is spherically symmetric.*

Ban et al. [BHL00] studied the Minkowski dimension of $X_{2,3}$ and of some other multiplicative sets as pattern generating problem.

# 6 Remarks and Problems

## 6.1 Vector Valued Potential

The non-linear thermodynamic formalism can be generalized to vectorial potentials. Let $\varphi, \gamma$ be two functions defined on $\mathcal{A}^\ell$ taking real values. Instead of considering the transfer operator $\mathcal{L}_s$ as defined in (13), we consider the following one:

$$\mathcal{L}_s\psi(a) = \sum_{j\in S} e^{s\varphi(a,j)+\gamma(a,j)}\psi(Ta, j), \ a \in S^{\ell-1}, \ s \in \mathbb{R}.$$

There exists a unique solution to the equation

$$(\mathcal{L}_s\psi)^{\frac{1}{q}} = \psi.$$

Then we define the pressure function $P_{\varphi,\gamma}(s)$ as indicated in $P_{\varphi,\gamma}(s)$. The function $s \mapsto P_{\varphi,\gamma}(s)$ is convex and analytic. Now, let $\underline{\varphi} = (\varphi_1, \ldots, \varphi_d)$ be a function defined on $S^\ell$ taking values in $\mathbb{R}^d$. For $\underline{s} = (s_1, \ldots, s_d) \in \mathbb{R}^d$, we consider the following transfer operator.

$$\mathcal{L}_{\underline{s}}\psi(a) = \sum_{j\in S} e^{\langle \underline{s}, \underline{\varphi}\rangle}\psi(Ta, j), \ a \in S^{\ell-1},$$

where $\langle \cdot, \cdot \rangle$ denotes the scalar product in $\mathbb{R}^d$. We denote the associated pressure function by $P(\underline{\varphi})(\underline{s})$. Then for any vectors $u, v \in \mathbb{R}^d$ the function

$$\mathbb{R} \ni s \longmapsto P(\underline{\varphi})(us + v)$$

is analytic and convex. We deduce from this that the function $\underline{s} \mapsto P(\underline{\varphi})(\underline{s})$ is infinitely differentiable and convex on $\mathbb{R}^d$.

Similarly, we define the level sets $E(\underline{\alpha})$ ($\underline{\alpha} \in \mathbb{R}^d$) of $\underline{\varphi}$. A vector version of Theorem 17 is stated by just replacing the derivative of the pressure function by the gradient.

## 6.2 Invariant Spectrum and Mixing Spectrum

The set $E_\Phi(\alpha)$ defined by (4) is not invariant. The size of the invariant part of $E_\Phi(\alpha)$ could be considered to be

$$d_{\text{inv}}(\alpha) = \sup\{\dim^* \mu : \mu \text{ invariant}, \ \mu(E_\Phi(\alpha)) = 1\}.$$

The function $\alpha \mapsto d_{\text{inv}}(\alpha)$ is called the invariant spectrum of $\Phi$. Similarly we define the mixing spectrum of $\Phi$ by

$$d_{\text{mix}}(\alpha) = \sup\{\dim^* \mu : \mu \text{ mixing}, \ \mu(E_\Phi(\alpha)) = 1\}.$$

Examples in [FSW00] show that it is possible to have

$$d_{\text{mix}}(\alpha) < d_{\text{inv}}(\alpha) < d_H(\alpha).$$

## 6.3 Semigroups

The semigroup $\{q^n\}_{n\geq 0}$ of $\mathbb{N}^*$ appeared in [FSW00, KPS12]. Other semigroup structures appeared in [PSSS00]. Combining the ideas in [PSSS00, FSW00], averages like

$$\lim_{n\to\infty} \frac{1}{n} \sum_{k=1}^{n} \varphi(x_k, x_{2k}, x_{3k})$$

can be treated [Wu13]. The Riesz product method used in [FLM12] is well adapted to the study of the special limit on $\Sigma_2$:

$$\lim_{n\to\infty} \frac{1}{n} \sum_{k=1}^{n} (2x_k - 1)(2x_{2k} - 1) \cdots (2x_{\ell k} - 1)$$

where $\ell \geq 2$ is any integer.

## 6.4 Subshifts of Finite Type

What we have presented is strictly restricted to the full shift dynamics. It is a challenging problem to study the dynamics of subshift of finite type and the dynamics with Markov property. New ideas are needed to deal with these dynamics. It is also a challenging problem to deal with potential depending more than one coordinates.

The doubling dynamics $Tx = 2x \bmod 1$ on the interval $[0, 1)$ is essentially a shift dynamics. Cookie cutters are the first interval maps coming into the mind after the doubling map. If the cookie cutter maps are not linear, it is a difficult problem. A cookie-cutter can be coded, but the non-linearity means that the derivative is a potential depending more than one codes.

Based on the computation made in [PS13], Liao and Rams [LR00] considered a special piecewise linear map of two branches defined on two intervals $I_0$ and $I_1$ and studied the following limit

$$\lim_{n\to\infty}\frac{1}{n}\sum_{k=1}^{n}1_{I_1}(T^k x)1_{I_1}(T^{2k}x).$$

The techniques presented in [FSW00] can be used to treat the problem for general piecewise *linear* cookie cutter dynamics [FLW00, Wu13].

## 6.5 Discontinuity of Spectrum for V-Statistics

The limit of V-statistics

$$\lim_{n\to\infty} n^{-r}\sum_{1\le i_1,\ldots,i_r\le n}\Phi(T^{i_1}x,\ldots,T^{i_r}x).$$

was studied in [FSW13] where it is proved that the multifractal spectrum of topological entropy of the above limit is expressed by an variational principle when the system satisfies the specification property. Unlike the classical case ($r = 1$) where the spectrum is an analytic function when $\Phi$ is Hölder continuous, the spectrum of the limit of higher order V-statistics ($r \ge 2$) may be discontinuous even for very regular kernel $\Phi$. It is an interesting problem to determine the number of discontinuities. Rauch [Rau13] has recently established a variational principe relative to $V$-statistics.

## 6.6 Mutual Absolute Continuity of Two Riesz Products

Let us state two conjectures. See [BFP10] for the discussion on these conjectures.

*Conjecture 1. Let $\mu_a$ and $\mu_b$ be two Riesz products and let $\omega := (\omega_n) \in G^{\mathbb{N}}$. Then $\mu_a \ll \mu_b \Rightarrow \mu_{a,\omega} \ll \mu_{b,\omega}$, and $\mu_a \perp \mu_b \Rightarrow \mu_{a,\omega} \perp \mu_{b,\omega}$.*

For a function $f$ defined on $G$, we use $\mathbb{E}f$ to denote the integral of $f$ with respect to the Haar measure. The truthfulness is that the preceding conjecture implies the following one.

*Conjecture 2. Let $\mu_a$ and $\mu_b$ be two Riesz products. Then*

$$\prod_{n=1}^{\infty}\mathbb{E}\sqrt{(1+\mathrm{Re}\ a_n\gamma_n)(1+\mathrm{Re}\ b_n\gamma_n)} > 0 \Longrightarrow \mu_a \ll \mu_b;$$

$$\prod_{n=1}^{\infty}\mathbb{E}\sqrt{(1+\mathrm{Re}\ a_n\gamma_n)(1+\mathrm{Re}\ b_n\gamma_n)} = 0 \Longrightarrow \mu_a \perp \mu_b.$$

## 6.7 Doubling and Tripling

For any integer $m \geq 2$, we define the dynamics $\tau_m x = mx \pmod 1$ on $[0, 1)$. A typical couple of commuting transformations is the couple $(\tau_2, \tau_3)$. Let us take, for example, $\Phi(x, y) = e^{2\pi i(ax+by)}$ with $a, b$ being two fixed integers. We are then led to the multiple ergodic averages, a special case of (3),

$$A_n^{(2,3)}(x) := \frac{1}{n}\sum_{k=1}^{n} e^{2\pi i(a2^k+b3^k)x}. \tag{26}$$

This is an object not yet well studied in the literature (but if $a = 0$, we get a classical Birkhoff average). We propose to develop a thermodynamic formalism by studying Gibbs type measures which are weak limits $\mu_{s,t}$ $(s, t \in \mathbb{R})$ of

$$Z_n(s,t)^{-1} Q_n(x)dx$$

where

$$Q_n(x) := \prod_{k=1}^{n} e^{s\cos(2\pi(a2^k+b3^k)x)+t\sin(2\pi(a2^k+b3^k)x)}.$$

The pressure function defined by

$$P(s,t) := \lim_{n\to\infty} \frac{\log Z_n(s,t)}{n}$$

would be differentiable. But first we have to prove the existence of the limit defining $P(s, t)$.

More generally, let $(c_n)$ be a sequence of complex numbers and $(\lambda_n)$ a lacunary sequence of positive integers (by lacunary we mean $\inf_n \frac{\lambda_{n+1}}{\lambda_n} > 1$). We can consider the following weighted lacunary trigonometric averages

$$\frac{1}{n}\sum_{k=1}^{n} c_k e^{2\pi i\lambda_k x}.$$

Under the divisibility condition $\lambda_n | \lambda_{n+1}$, such averages and more general averages were studied in [Fan97]. For example, if $c_k = e^{2\pi i\omega_k}$ with $(\omega_k)$ being an i.i.d. sequence of Lebesgue distributed random variables, from the results obtained in [Fan97] we deduce that almost surely the pressure is well defined and equal to the following deterministic function

$$P(s,t) = \log \int_0^{2\pi} e^{\sqrt{t^2+s^2}\cos x} \frac{dx}{2\pi}.$$

Recall that

$$J_0(r) = \frac{1}{2\pi}\int_0^{2\pi} e^{r\cos x}dx = \sum_{n=0}^{\infty}\frac{r^{2n}}{(n!)^2 2^{2n}}.$$

is the Bessel function.

But, the condition $\lambda_n|\lambda_{n+1}$ for $\lambda_n = 2^n + 3^n$ is not satisfied. Neither the condition is satisfied for $\lambda_n = 2^n + 4^n$. No rigorous results are known for the multifractal analysis of the averages defined by (26).

As conjectured by Fürstenberg, the Lebesgue measure is the unique continuous probability measure which is both $\tau_2$-invariant and $\tau_3$-invariant. However, common $\tau_2$- and $\tau_3$-periodic points (different from the trivial one 0) do exists. Given two integers $n \geq 1$ and $m \geq 1$. We can prove that there is a point $x(\neq 0)$ which is $n$-periodic with respect to $\tau_2$ and $m$-periodic with respect to $\tau_3$ if and only if

$$(2^n - 1, 3^m - 1) > 1. \tag{27}$$

Let $d = (2^n - 1, 3^m - 1)$. When the above condition on GCD is satisfied, there are $d - 1$ such common periodic points. These common periodic points $x(\neq 0)$ are of the form $x = \frac{k}{2^n-1} = \frac{j}{3^m-1}$ for some $1 \leq k < 2^n - 2$ and $1 \leq j < 3^m - 2$. Actually choices for $k$ are

$$1 \cdot \frac{2^n - 1}{d}, \quad 2 \cdot \frac{2^n - 1}{d}, \ldots, \quad (d-1) \cdot \frac{2^n - 1}{d}.$$

Choices for $j$ are $1 \cdot \frac{3^m-1}{d}$, $2 \cdot \frac{3^m-1}{d}, \ldots, (d-1) \cdot \frac{3^m-1}{d}$. Thus the $d - 1$ common periodic points are $\frac{1}{d}, \frac{2}{d}, \ldots, \frac{d-1}{d}$. For such a point $x$, the following limit exists

$$\lim_{N\to\infty} A_N^{(2,3)}(x) = A_{nm}^{(2,3)}(x).$$

Note that there is an infinite number of such couples $n$ and $m$ such that (27) holds. There would be some relation between these common periodic points and the multifractal behavior of $A_N^{(2,3)}(x)$.

## References

[BHL00] Ban, J.C., Hu, W.G., Lin, S.S.: Pattern Generating Problems Arising in Multiplicative Interger Systems (preprint)

[BS07] Barral, J., Seuret, S.: Heterogeneous ubiquitous systems in $\mathbb{R}^d$ and Hausdorff dimension. Bull. Braz. Math. Soc. **38**(3), 467–515 (2007)

[BFP10] Barral, J., Fan, A.H., Peyrière, J.: Mesures engendrées par multiplication, Quelques Interactions Entre Analyse, Probabilités et Fractales, pp. 57–189. Panorama et Synthèse/SMF, No. 32 (2010)

[BPS99] Barreira, L., Pesin, Y., Schmeling, J.: Dimension and product structure of hyperbolic measures. Ann. Math. **149**(3), 755–783 (1999) (2nd ser)

[BSS02] Barreira, L., Saussol, B., Schmeling, J.: Higher dimensional multifractal analysis. J. Math. Pure. Appl. **81**, 67–91 (2002)

[BV06] Beresnevich, V., Velani, S.: A mass transference principle and the Duffin-Schaeffer conjecture for Hausdorff measures. Ann. Math. **164**, 971–992 (2006)

[Bou90] Bourgain, J.: Double recurrence and almost sure convergence. J. Reine. Angew. Math. **404**, 140–161 (1990)

[Bow73] Bowen, R.: Topological entropy for noncompact sets. Trans. Amer. Math. Soc. **184**, 125–136 (1973)

[BMP92] Brown, G., Michon, G., Peyrière, J.: On the multifractal analysis of measures. J. Statist. Phys. **66**, 775–790 (1992)

[CLP87] Collet, P., Lebowitz, J.L., Porzio, A.: The dimension spectrum of some dynamical systems. J. Stat. Phys. **47**, 609–644 (1987)

[LT84] Conze, J.P., Lesigne, E.: Théorèmes ergodiques pour des mesures diagonales. Bull. Soc. Math. France **112**(2), 143–175 (1984)

[Cut95] Cutler, C.: Connecting ergodicity and dimension in dynamical systems. Ergodic. Theor. Dynam. Syst. **118**, 393–410 (1995)

[Dur08] Durand, A.: Ubiquitous systems and metric number theory. Adv. Math. **218**, 368–394 (2008)

[Fal03] Falconer, K.J.: Fractal Geometry: Mathematical Foundations and Applications, 2nd ed. Wiley, Chichester (2003)

[Fan89] Fan, A.H.: Recouvrement aléatoire et Décompositions de Mesures. Publication d'Orsay, Orsay (1989)

[Fan93] Fan, A.H.: Quelques propriétés de produits de Riesz. Bull. Sci. Math. **117**(3), 421–439 (1993)

[Fan94a] Fan, A.H.: Sur les dimensions de mesures. Studia Math. **111**(1), 1–17 (1994)

[Fan94b] Fan, A.H.: On ergodicity and unidimensionality. Kyushu J. Math. **48**(2), 249–255 (1994)

[Fan94c] Fan, A.H.: Une formule approximative de dimension pour certains produits de Riesz. Monatsh. für Math. **118**, 83–89 (1994)

[Fan97] Fan, A.H.: Multifractal analysis of infinite products. J. Stat. Phys. **86**(5/6), 1313–1336 (1997)

[Fan00] Fan, A.H.: Individual behaviors of oriented walks. Stoc. Proc. Appl. Stoc. Proc. Appl. **90**, 263–275 (2000)

[FF99] Fan, A.H., Feng, D.J.: On the distribution of long-term time averages on symbolic space. J. Statist. Phys. **99**, 813–856 (1999)

[FFW01] Fan, A.H., Feng, D.J., Wu, J.: Recurrence, dimension and entropy. J. Lond. Math. Soc. **64**, 229–244 (2001)

[FLR02] Fan, A.H., Lau, K.S., Rao, H.: Relationship between different dimension of a measure. Monatsch. Math. **135**, 191–201 (2002)

[FLP08] Fan, A.H., Liao, L.M., Peyrière, J.: Generic points in systems of specification and Banach valued Birkhoff ergodic average. Discr. Cont. Dynam. Syst. **21**(4), 1103–1128 (2008)

[FLM12] Fan, A.-H., Liao, L., Ma, J.-H.: Level sets of multiple ergodic averages. Monatsh. Math. **168**(1), 17–26 (2012)

[FLW00] Fan, A.H., Liao, L.M., Wu, M.: Multifractal Analysis of Some Multiple Ergodic Averages in Linear Cookie-Cutter Dynamical Systems (preprint)

[FS03] Fan, A.H., Schmeling, J.: On fast Birkhoff averaging. Math. Proc. Camb. Phil. Soc. **135**, 443–467 (2003)

[FSW00] Fan, A.-H., Schmeling, J., Wu, M.: Multifractal Analysis of Some Multiple Ergodic Average (preprint)

[FSW13] Fan, A.-H., Schmeling, J., Wu, M.: Multifractal analysis of V-statistics, in further developments in fractals and related fields. Barral, J., Seuret, S (Eds.) Trends in Mathematics, pp. 135–151 (2013)

[FST00] Fan, A.H., Schmeling, J., Troubetzkoy, S.: A multifractal mass transference principle for Gibbs measures with applications to dynamical Diophantine approximation. Proc. London Math. Soc. **107**(3), 1173–1219 (2013)

[FW00] Fan, A.H., Wu, M.: Oriented Walks on Euclidean Spaces (preprint)

[Fel68] Feller, W.: An Introduction to Probability Theory and Its Applications. 3rd Edn, vol. 1. Wiley, New York (1968)

[FLW02] Feng, D.J., Lau, K.-S., Wu, J.: Ergodic limits on the conformal repellers. Adv. Math. **169**, 58–91 (2002)

[FP85] Frisch, U., Parisi, G.: Fully developped turbulence and intermittency in turbulence, and predictability in geophysical fluid dynamics and climate dymnamics. In: Ghil, M. (ed.) International School of Physics "Enrico Fermi": Course 88, p. 84. North Holland (1985)

[Für77] Fürstenberg, H.: Ergodic behavior of diagonal measures and a theorem of Szemerédi on arithmetic progressions. J. Anal. Math. **31**, 204–256 (1977)

[FW96] Fürstenberg, H., Weiss, B.: A mean ergodic theorem for $(1/N)\sum_{n=1}^{N} f(T^n x)g(T^{n^2} x)$. Convergence in ergodic theory and probability, Ohio State University, Mathematical Sciences Reseach Institute Publication, vol 5, pp. 193–227. de Gruyter (1996)

[Haa92] Haase, H.: A survey on the dimension of measures, topology, measures and fractals. Warnemünde Math. Res. **66**(1992), 66–75 (1991)

[HJKPS86] Halsey, T.C., Jensen, M.H., Kadanoff, L.P., Procaccia, I., Shraiman, B.I.: Fractal measures and their singularities: the characterisation of strange sets. Phys. Rev. A **33**, 1141 (1986)

[HP83] Hentschel, H.G., Procaccia, I.: The infinite number of generalized dimensions of fractals and strange attractors. Physica **8D**, 435 (1983)

[Heu98] Heurteaux, Y.: Estimations de la dimension inférieure et de la dimension supérieure des mesures. Ann. Inst. H. Poincaré Probab. Statist. **34**, 309–338 (1998)

[HZ66] Hewitt, E., Zuckerman, H.S.: Singular measures with absolutely continuous convolution squares. Proc. Camb. Phil. Soc. **62**, 399–420 (1966)

[HK05] Host, B., Kra, B.: Nonconventional ergodci averages and nilmanifolds. Ann. Math. **161**, 397–488 (2005)

[How94] Howroyd, J.D.: On the theory of Hausdorff measures in metric spaces. Thesis, University of London (1994)

[Jaf97] Jaffard, S.: Multifractal formalism for functions I. Results valid for all functions II. Self-similar functions. SIAM J. Math. Anal. **28**, 944–970, 971–998 (1997)

[Kah87] Kahane, J.-P.: Intervalles aléatoires et décomposition des mesures. C. R. Acad. Sci. Paris Sér. I **304**, 551–554 (1987)

[KP76] Kahane, J.-P., Peyrière, J.: Sur certaines martingales de Mandelbrot. Adv. Math. **22**, 131–145 (1976)

[Kak48] Kakutani, S.: On equivalence of infinite product measures. Ann. Math. **49**, 214–224 (1948)

[Kau94] Kaufman, R.: A "min-max" theorem on potentials. C. R. Acad. Sci. Paris Sér. I Math. **319**(8), 799–800 (1994)

[KPS12] Kenyon, R., Peres, Y., Solomyak, B.: Hausdorff dimension for fractals invariant under the multiplicative integers. Erg. Th. Dyn. Syst. **32**(5), 1567–1584 (2012)

[Kif12] Kifer, Y.: A nonconventional strong law of large numbers and fractal dimensions of some multiple recurrence sets. Stochast. Dyn. **12**(3), 21 (2012)

[Les89] Lesigne, E.: Theorems ergodiques pour une translation sur une nilvariete. Erg. Thm. Dyn. Syst. **9**(1), 115–126 (1989)

[LR00] Liao, L.M., Rams, M.: Multifractal Analysis of Some Multiple Ergodic Averages for the Systems with Non-constant Lyapunov Exponents Real Analysis Exchange, to appear

[Man72] Mandelbrot, B.: Possible refinement of the log-normal hypothesis concerning the distribution of energy dissipation in intermittent trubulence in statistical models and trubulence. In: Abstracts of the Symposium at University of California, San Diego 1972. Lecture Notes in Physics, pp. 333–351 (1971)

[Man74] Mandelbrot, B.: Multiplications aléatoires itérées et distributions invariantes par moyennes pondérée aléatoire. C. R. Acad. Sci. Paris **278**, 289–292 (1974)

[Man74] Mandelbrot, B.: Multiplications aléatoires itérées et distributions invariantes par moyennes pondérée aléatoire: quelques extensions. C. R. Acad. Sci. Paris **278**, 355–358 (1974)

[Mat95] Mattila, P.: Geometry of sets and measures in Euclidean spaces. Fractals and rectifiability. Cambridge Studies in Advanced Mathematics, vol. 44. Cambridge University Press, Cambridge (1995)

[Oli99] Olivier, E.: Multifractal analysis in symbolic dynamics and distribution of pointwise dimension for $g$-measures. Nonlinearity **12**, 1571–1585 (1999)

[Ols95] Olsen, L.: A multifractal formalism. Adv. Math. **116**, 92–195 (1995)

[PSSS00] Peres, Y., Schmeling, J., Seuret, S., Solomyak, B.: Hausdorff Dimension for Fractals Generated by Multiplication by 2 and 3 Israel Journal Math., to appear

[PS11-12] Peres, Y., Solomyak, B.: Dimension spectrum for a nonconventional erdofic average. Real Anal. Exch. **37**(2), 375–388 (2011/2012)

[PS13] Peres, Y., Solomyak, B.: The multiplicative golden mean shift has infinite Hausdorff measure: further developments in fractals and related fields. Barral, J., Seuret, S. (eds.) Trends in Mathematics, pp. 193–212 (2013)

[Pes97] Pesin, Y.: Dimension Theory in Dynamical Systems: Contemporary Views and Applications. The University of Chicago Press, Chicago (1997). Chicago lectures in Mathematics

[Pey75] Peyrière, J.: Étude de quelques propriétés des produits de Riesz. Ann. Inst. Fourier **25**, 127–169 (1975)

[Pey90] Peyrière, J.: Almost everywhere convergence of lacunary trigonometric series with respect to Riesz products. Aust. J. Math. Ser. A **48**, 376–383 (1990)

[Ran89] Rand, D.A.: The singularity spectrum $f(\alpha)$ for cookie-cutters. Ergod. Th. Dynam. Syst. **9**, 527–541 (1989)

[Rau13] Rauch, M.: A multivariate variational principle. Diploma Thesis, University of Jena (2013)

[Rue78] Ruelle, D.: Thermodynamic Formalism, Encyclopedia of Mathematics and its Applications, vol. 5. Addison-Weysley, Reading (1978)

[TV03] Takens, F., Verbitskiy, E.: On the variational principle for the topological entropy of certain non-compact sets. Ergod. Th. Dyn. Syst. **23**(1), 317–348 (2003)

[Tam95] Tamashira, M.: Dimensions in a separable metric space. Kyushu J. Math. **49**(1), 143–162 (1995)

[Wal82] Walters, P.: An Introduction to Ergodic Theory, GTM 79. Springer, New York (1982)

[Wu13] Wu, M.: Multifractal Analysis of Some Nonconventional Ergodic Averages. Ph. D. Thesis, University of Picardie, Jules Verne (2013)

[You82] Young, L.S.: Dimension, entropy and Lyapounov exponents. Erg. Th. Dyn. Syst. **2**, 109–124 (1982)

[Zie05] Ziegler, T.: A non-conventional ergodic theorem for a nilsystem. Erg. Th. Dyn. Syst. **25**(4), 1357–1370 (2005)

[Zyg68] Zygmund, A.: Trigonometric Series. Cambridge University Press, Cambridge (1968)

# Heat Kernels on Metric Measure Spaces

**Alexander Grigor'yan, Jiaxin Hu and Ka-Sing Lau**

## Contents



Supported by SFB 701 of the German Research Council and the Grants from the Department of Mathematics and IMS of CUHK.
Corresponding author. Supported by NSFC (Grant No. 11271122), SFB 701 and the HKRGC Grant of CUHK.
Supported by the HKRGC Grant and the Focus Investment Scheme of CUHK

A. Grigor'yan
Department of Mathematics, University of Bielefeld, 33501 Bielefeld, Germany

J. Hu (✉)
Department of Mathematical Sciences, Tsinghua University, Beijing 100084, China
e-mail: hujiaxin@mail.tsinghua.edu.cn

K.-S. Lau
Department of Mathematics, The Chinese University of Hong Kong, Shatin, N.T., Hong Kong



D.-J. Feng and K.-S. Lau (eds.), *Geometry and Analysis of Fractals*,
Springer Proceedings in Mathematics & Statistics 88,
DOI 10.1007/978-3-662-43920-3_6

# 1 What is the Heat Kernel

In this section we shall discuss the notion of the heat kernel on a metric measure space $(M, d, \mu)$. Loosely speaking, a heat kernel $p_t(x, y)$ is a family of measurable functions in $x, y \in M$ for each $t > 0$ that is *symmetric*, *Markovian* and satisfies the *semigroup* property and the *approximation of identity* property. It turns out that the heat kernel coincides with the *integral kernel* of the *heat semigroup* associated with the *Dirichlet form* in $L^2(M, \mu)$.

Let us start with some basic examples of the heat kernels.

## *1.1 Examples of Heat Kernels*

### 1.1.1 Heat Kernel in Euclidean Spaces

The classical Gauss-Weierstrass heat kernel is the following function

$$p_t(x, y) = \frac{1}{(4\pi t)^{n/2}} \exp\left(-\frac{|x-y|^2}{4t}\right), \tag{1.1}$$

where $x, y \in \mathbb{R}^n$ and $t > 0$. This function is a fundamental solution of the heat equation

$$\frac{\partial u}{\partial t} = \Delta u,$$

where $\Delta = \sum_{i=1}^n \frac{\partial^2}{\partial x_i^2}$ is the Laplace operator. Moreover, if $f$ is a continuous bounded function on $\mathbb{R}^n$, then the function

$$u(t, x) = \int_{\mathbb{R}^n} p_t(x, y) f(y)\, dy$$

solves the Cauchy problem

$$\begin{cases} \frac{\partial u}{\partial t} = \Delta u, \\ u(0, x) = f(x). \end{cases}$$

This can be also written in the form

$$u(t, \cdot) = \exp(-t\mathcal{L}) f,$$

where $\mathcal{L}$ here is a self-adjoint extension of $-\Delta$ in $L^2(\mathbb{R}^n)$ and $\exp(-t\mathcal{L})$ is understood in the sense of the functional calculus of self-adjoint operators. That means that $p_t(x, y)$ is the integral kernel of the operator $\exp(-t\mathcal{L})$.

The function $p_t(x, y)$ has also a probabilistic meaning: it is the transition density of Brownian motion $\{X_t\}_{t \geq 0}$ in $\mathbb{R}^n$ (Fig. 1). The graph of $p_t(x, 0)$ as a function of $x$ is shown here:

The term $\frac{|x-y|^2}{t}$ determines the *space/time scaling*: if $|x - y|^2 \leq Ct$, then $p_t(x, y)$ is comparable with $p_t(x, x)$, that is, the probability density in the $C\sqrt{t}$-neighborhood of $x$ is nearly constant.

### 1.1.2 Heat Kernels on Riemannian Manifolds

Let $(M, g)$ be a connected Riemannian manifold, and $\Delta$ be the Laplace-Beltrami operator on $M$. Then the heat kernel $p_t(x, y)$ can be defined as the integral kernel of the heat semigroup $\{\exp(-t\mathcal{L})\}_{t \geq 0}$, where $\mathcal{L}$ is the Dirichlet Laplace operator, that is, the minimal self-adjoint extension of $-\Delta$ in $L^2(M, \mu)$, and $\mu$ is the Riemannian volume. Alternatively, $p_t(x, y)$ is the minimal positive fundamental solution of the heat equation

$$\frac{\partial u}{\partial t} = \Delta u.$$

The function $p_t(x, y)$ can be used to define Brownian motion $\{X_t\}_{t \geq 0}$ on $M$. Namely, $\{X_t\}_{t \geq 0}$ is a *diffusion process* (that is, a Markov process with continuous trajectories), such that

$$\mathbb{P}_x(X_t \in A) = \int_A p_t(x, y)\, d\mu(y)$$

for any Borel set $A \subset M$ (Fig. 2).

Let $d(x, y)$ be the geodesic distance on $M$. It turns out that the *Gaussian* type space/time scaling $\frac{d^2(x,y)}{t}$ appears in heat kernel estimates on general Riemannian manifolds:

1. (*Varadhan*) For an arbitrary Riemannian manifold,

$$\log p_t(x, y) \sim -\frac{d^2(x, y)}{4t} \quad \text{as } t \to 0.$$

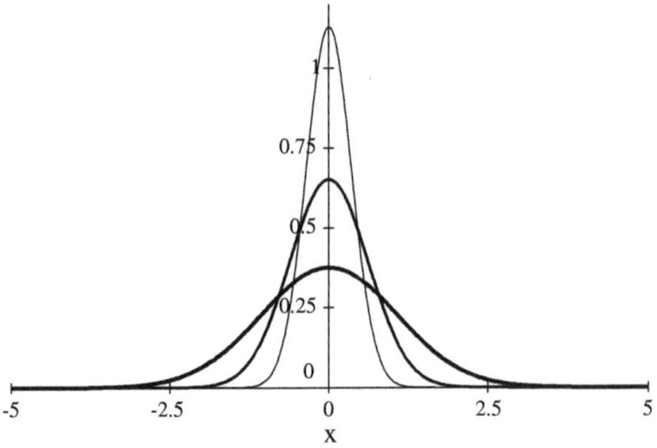


**Fig. 1** The Gauss-Weierstrass heat kernel at different values of $t$

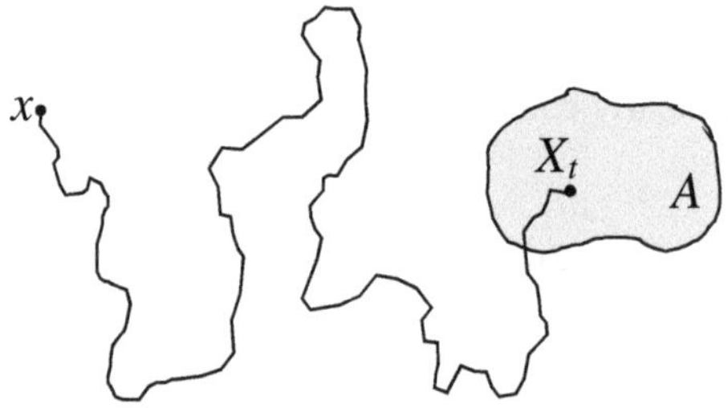


**Fig. 2** The Brownian motion $X_t$ hits a set $A$

2. (*Davies*) For an arbitrary manifold $M$, for any two measurable sets $A, B \subset M$

$$\int_A \int_B p_t(x,y)\, d\mu(x)\, d\mu(y) \le \sqrt{\mu(A)\,\mu(B)} \exp\left(-\frac{d^2(A,B)}{4t}\right).$$

Technically, all these results depend upon the property of the geodesic distance: $|\nabla d| \le 1$.

It is natural to ask the following question:

*Are there settings where the space/time scaling is different from Gaussian?*

### 1.1.3 Heat Kernels of Fractional Powers of Laplacian

Easy examples can be constructed using another operator instead of the Laplacian. As above, let $\mathcal{L}$ be the Dirichlet Laplace operator on a Riemannian manifold $M$, and consider the evolution equation

$$\frac{\partial u}{\partial t} + \mathcal{L}^{\beta/2} u = 0,$$

where $\beta \in (0, 2)$. The operator $\mathcal{L}^{\beta/2}$ is understood in the sense of the functional calculus in $L^2(M, \mu)$. Let $p_t(x, y)$ be now the heat kernel of $\mathcal{L}^{\beta/2}$, that is, the integral kernel of $\exp\left(-t\mathcal{L}^{\beta/2}\right)$.

The condition $\beta < 2$ leads to the fact that the semigroup $\exp\left(-t\mathcal{L}^{\beta/2}\right)$ is *Markovian*, which, in particular, means that $p_t(x, y) > 0$ (if $\beta > 2$ then $p_t(x, y)$ may be signed). Using the techniques of subordinators, one obtains the following estimate for the heat kernel of $\mathcal{L}^{\beta/2}$ in $\mathbb{R}^n$:

$$p_t(x, y) \asymp \frac{C}{t^{n/\beta}} \left(1 + \frac{|x - y|}{t^{1/\beta}}\right)^{-(n+\beta)} \asymp \frac{C}{t^{n/\beta}} \left(1 + \frac{|x - y|^\beta}{t}\right)^{-\frac{n+\beta}{\beta}}. \quad (1.2)$$

(the symbol $\asymp$ means that both $\leq$ and $\geq$ are valid but with different values of the constant $C$).

The heat kernel of $\sqrt{\mathcal{L}} = (-\Delta)^{1/2}$ in $\mathbb{R}^n$ (that is, the case $\beta = 1$) is known explicitly:

$$p_t(x, y) = \frac{c_n}{t^n} \left(1 + \frac{|x - y|^2}{t^2}\right)^{-\frac{n+1}{2}} = \frac{c_n t}{\left(t^2 + |x - y|^2\right)^{\frac{n+1}{2}}},$$

where $c_n = \Gamma\left(\frac{n+1}{2}\right)/\pi^{(n+1)/2}$. This function coincides with the Poisson kernel in the half-space $\mathbb{R}^{n+1}_+$ and with the density of the Cauchy distribution in $\mathbb{R}^n$ with the parameter $t$.

As we have seen, the space/time scaling is given by the term $\frac{d^\beta(x,y)}{t}$ where $\beta < 2$. The heat kernel of the operator $\mathcal{L}^{\beta/2}$ is the transition density of a *symmetric stable process of index* $\beta$ that belongs to the family of Lévy processes. The trajectories of this process are discontinuous, thus allowing jumps. The heat kernel $p_t(x, y)$ of such process is nearly constant in some $Ct^{1/\beta}$-neighborhood of $y$. If $t$ is large, then

$$t^{1/\beta} \gg t^{1/2},$$

that is, this neighborhood is much larger than that for the diffusion process, which is not surprising because of the presence of jumps. The space/time scaling with $\beta < 2$ is called *super-Gaussian*.

### 1.1.4 Heat Kernels on Fractal Spaces

A rich family of heat kernels for diffusion processes has come from Analysis on *fractals*. Loosely speaking, fractals are subsets of $\mathbb{R}^n$ with certain self-similarity

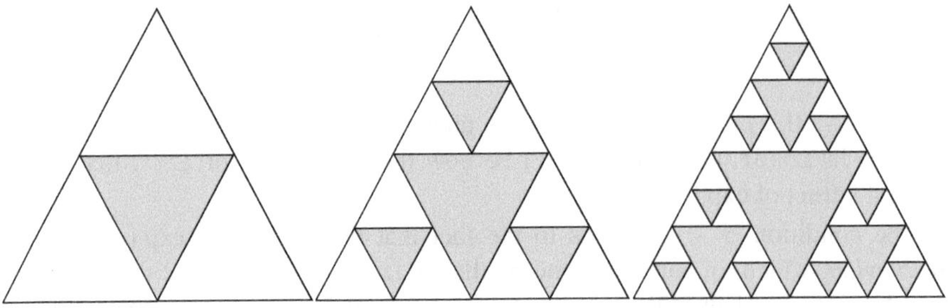

**Fig. 3** Construction of the Sierpinski gasket

**Fig. 4** The unbounded SG is obtained from SG by merging the latter (at the *left lower corner* of the diagram) with two shifted copies and then by repeating this procedure at larger scales

properties. One of the best understood fractals is *the Sierpinski gasket* (SG). The construction of the Sierpinski gasket is similar to the Cantor set: one starts with a triangle as a closed subset of $\mathbb{R}^2$, then eliminates the open middle triangle (shaded on the diagram), then repeats this procedure for the remaining triangles, and so on (Fig. 3).

Hence, SG is a compact connected subset of $\mathbb{R}^2$. The *unbounded* SG is obtained from SG by merging the latter (at the left lower corner of the next diagram) with two shifted copies and then by repeating this procedure at larger scales (Fig. 4).

Barlow and Perkins [BP88], Goldstein [Gol87] and Kusuoka [Kus87] have independently constructed by different methods a natural diffusion process on SG that has the same self-similarity as SG. Barlow and Perkins considered random walks on the graph approximations of SG and showed that, with an appropriate scaling, the random walks converge to a diffusion process. Moreover, they proved that this process has a transition density $p_t(x, y)$ with respect to a proper Hausdorff measure $\mu$ of SG, and that $p_t$ satisfies the following elegant estimate:

$$p_t(x, y) \asymp \frac{C}{t^{\alpha/\beta}} \exp\left(-c\left(\frac{d^\beta(x, y)}{t}\right)^{\frac{1}{\beta-1}}\right), \tag{1.3}$$

where $d(x, y) = |x - y|$ and

$$\alpha = \dim_H SG = \frac{\log 3}{\log 2}, \quad \beta = \frac{\log 5}{\log 2} > 2.$$

Similar estimates were proved by Barlow and Bass for other families of fractals, including Sierpinski carpets, and the parameters $\alpha$ and $\beta$ in (1.3) are determined by the intrinsic properties of the fractal. In all cases, $\alpha$ is the Hausdorff dimension (which is also called the *fractal dimension*). The parameter $\beta$, that is called the *walk dimension*, is larger than 2 in all interesting examples.

The heat kernel $p_t(x, y)$, satisfying (1.3) is nearly constant in some $Ct^{1/\beta}$-neighborhood of $y$. If $t$ is large, then

$$t^{1/\beta} \ll t^{1/2},$$

that is, this neighborhood is much smaller than that for the diffusion process, which is due to the presence of numerous holes-obstacles that the Brownian particle must bypass. The space/time scaling with $\beta > 2$ is called *sub-Gaussian*.

### 1.1.5 Summary of Examples

Observe now that in all the above examples, the heat kernel estimates can be unified as follows:

$$p_t(x, y) \asymp \frac{C}{t^{\alpha/\beta}} \Phi\left(c\frac{d(x, y)}{t^{1/\beta}}\right), \tag{1.4}$$

where $\alpha, \beta$ are positive parameters and $\Phi(s)$ is a positive decreasing function on $[0, +\infty)$. For example, the Gauss-Weierstrass function (1.1) satisfies (1.4) with $\alpha = n, \beta = 2$ and

$$\Phi(s) = \exp\left(-s^2\right)$$

(Gaussian estimate).

The heat kernel (1.2) of the symmetric stable process in $\mathbb{R}^n$ satisfies (1.4) with $\alpha = n, 0 < \beta < 2$, and

$$\Phi(s) = (1 + s)^{-(\alpha+\beta)}$$

(super-Gaussian estimate).

The heat kernel (1.3) of diffusions on fractals satisfies (1.4) with $\beta > 2$ and

$$\Phi(s) = \exp\left(-s^{\frac{\beta}{\beta-1}}\right)$$

(sub-Gaussian estimate).

There are at least two questions related to the estimates of the type (1.4):

1. What values of the parameters $\alpha, \beta$ and what functions $\Phi$ can actually occur in the estimate (1.4)?
2. How to obtain estimates of the type (1.4)?

To give these questions a precise meaning, we must define what is a heat kernel.

## 1.2 Abstract Heat Kernels

Let $(M, d)$ be a locally compact, separable metric space and let $\mu$ be a Radon measure on $M$ with full support. The triple $(M, d, \mu)$ will be called a *metric measure space*.

**Definition 1.1** (*heat kernel*) A family $\{p_t\}_{t>0}$ of measurable functions $p_t(x, y)$ on $M \times M$ is called a *heat kernel* if the following conditions are satisfied, for $\mu$-almost all $x, y \in M$ and all $s, t > 0$:

(i) Positivity: $p_t(x, y) \geq 0$.
(ii) The total mass inequality:

$$\int_M p_t(x, y) d\mu(y) \leq 1.$$

(iii) Symmetry: $p_t(x, y) = p_t(y, x)$.
(iv) The semigroup property:

$$p_{s+t}(x, y) = \int_M p_s(x, z) p_t(z, y) d\mu(z).$$

(v) Approximation of identity: for any $f \in L^2 := L^2(M, \mu)$,

$$\int_M p_t(x, y) f(y) d\mu(y) \xrightarrow{L^2} f(x) \text{ as } t \to 0+.$$

If in addition we have, for all $t > 0$ and almost all $x \in M$,

$$\int_M p_t(x, y) d\mu(y) = 1,$$

then the heat kernel $p_t$ is called *stochastically complete* (or *conservative*).

## 1.3 Heat Semigroups

Any heat kernel gives rise to the family of operators $\{P_t\}_{t\geq 0}$ where $P_0 = \text{id}$ and $P_t$ for $t > 0$ is defined by

$$P_t f(x) = \int_M p_t(x, y) f(y) d\mu(y),$$

where $f$ is a measurable function on $M$. It follows from (i) –(ii) that the operator $P_t$ is *Markovian*, that is, $f \geq 0$ implies $P_t f \geq 0$ and $f \leq 1$ implies $P_t f \leq 1$. It follows that $P_t$ is a bounded operator in $L^2$ and, moreover, is a contraction, that is, $\|P_t f\|_2 \leq \|f\|_2$.

The symmetry property (iii) implies that the operator $P_t$ is *symmetric* and, hence, self-adjoint. The semigroup property (iv) implies that $P_t P_s = P_{t+s}$, that is, the family $\{P_t\}_{t\geq 0}$ is a *semigroup* of operators. It follows from $(v)$ that

$$s\text{-}\lim_{t\to 0} P_t = \text{id} = P_0$$

where $s$-lim stands for the *strong* limit. Hence, $\{P_t\}_{t\geq 0}$ is a strongly continuous, symmetric, Markovian semigroup in $L^2$. In short, we call that $\{P_t\}$ is a *heat semigroup.*

Conversely, if $\{P_t\}$ is a heat semigroup and if it has an integral kernel $p_t(x, y)$, then the latter is a heat kernel in the sense of the above Definition.

Given a heat semigroup $P_t$ in $L^2$, define the *infinitesimal generator* $\mathcal{L}$ of the semigroup by

$$\mathcal{L} f := \lim_{t\to 0} \frac{f - P_t f}{t},$$

where the limit is understood in the $L^2$-norm. The *domain* $\text{dom}(\mathcal{L})$ of the generator $\mathcal{L}$ is the space of functions $f \in L^2$ for which the above limit exists. By the Hille–Yosida theorem, $\text{dom}(\mathcal{L})$ is dense in $L^2$. Furthermore, $\mathcal{L}$ is a self-adjoint, positive definite operator, which immediately follows from the fact that the semigroup $\{P_t\}$ is self-adjoint and contractive. Moreover, $P_t$ can be recovered from $\mathcal{L}$ as follows

$$P_t = \exp(-t\mathcal{L}),$$

where the right hand side is understood in the sense of spectral theory.

Heat kernels and heat semigroups arise naturally from Markov processes. Let $(\{X_t\}_{t\geq 0}, \{\mathbb{P}_x\}_{x\in M})$ be a Markov process on $M$, that is reversible with respect to measure $\mu$. Assume that it has the transition density $p_t(x, y)$, that is, a function such that, for all $x \in M, t > 0$, and all Borel sets $A \subset M$,

$$\mathbb{P}_x (X_t \in A) = \int_M p_t (x, y) \, d\mu (y) .$$

Then $p_t (x, y)$ is a heat kernel in the sense of the above Definition.

## 1.4 Dirichlet Forms

Given a heat semigroup $\{P_t\}$ on a metric measure space $(M, d, \mu)$, for any $t > 0$, we define a bilinear form $\mathcal{E}_t$ on $L^2$ by

$$\mathcal{E}_t (u, v) := \left( \frac{u - P_t u}{t}, v \right) = \frac{1}{t} ((u, v) - (P_t u, v)),$$

where $(\cdot, \cdot)$ is the inner product in $L^2$. Since $P_t$ is symmetric, the form $\mathcal{E}_t$ is also symmetric. Since $P_t$ is a contraction, it follows that

$$\mathcal{E}_t (u) := \mathcal{E}_t (u, u) = \frac{1}{t} ((u, u) - (P_t u, u)) \geq 0,$$

that is, $\mathcal{E}_t$ is a positive definite form.

In terms of the spectral resolution $\{E_\lambda\}$ of the generator $\mathcal{L}$, $\mathcal{E}_t$ can be expressed as follows

$$\mathcal{E}_t (u) = \frac{1}{t} ((u, u) - (P_t u, u)) = \frac{1}{t} \left( \int_0^\infty d\|E_\lambda u\|_2^2 - \int_0^\infty e^{-t\lambda} d\|E_\lambda u\|_2^2 \right)$$

$$= \int_0^\infty \frac{1 - e^{-t\lambda}}{t} d\|E_\lambda u\|_2^2,$$

which implies that $\mathcal{E}_t (u)$ is decreasing in $t$, since the function $t \mapsto \frac{1-e^{-t\lambda}}{t}$ is decreasing. Define for any $u \in L^2$

$$\mathcal{E} (u) = \lim_{t \downarrow 0} \mathcal{E}_t (u)$$

where the limit (finite or infinite) exists by the monotonicity, so that $\mathcal{E} (u) \geq \mathcal{E}_t (u)$. Since $\frac{1-e^{-t\lambda}}{t} \to \lambda$ as $t \to 0$, we have

$$\mathcal{E}(u) = \int_0^\infty \lambda d\|E_\lambda u\|_2^2.$$

Set

$$\mathcal{F} := \{u \in L^2 : \mathcal{E}(u) < \infty\} = \operatorname{dom}\left(\mathcal{L}^{1/2}\right) \supset \operatorname{dom}(\mathcal{L})$$

and define a bilinear form $\mathcal{E}(u, v)$ on $\mathcal{F}$ by the polarization identity

$$\mathcal{E}(u, v) := \frac{1}{4}\left(\mathcal{E}(u+v) - \mathcal{E}(u-v)\right),$$

which is equivalent to

$$\mathcal{E}(u, v) = \lim_{t \to 0} \mathcal{E}_t(u, v).$$

Note that $\mathcal{F}$ contains $\operatorname{dom}(\mathcal{L})$. Indeed, if $u \in \operatorname{dom}(\mathcal{L})$, then we have for all $v \in L^2$

$$\lim_{t \to 0} \mathcal{E}_t(u, v) = \left(\lim_{t \to 0} \frac{u - P_t u}{t}, v\right) = (\mathcal{L}u, v).$$

Setting $v = u$ we obtain $u \in \mathcal{F}$. Then choosing any $v \in \mathcal{F}$ we obtain the identity

$$\mathcal{E}(u, v) = (\mathcal{L}u, v) \text{ for all } u \in \operatorname{dom}(\mathcal{L}) \text{ and } v \in \mathcal{F}.$$

The space $\mathcal{F}$ is naturally endowed with the inner product

$$[u, v] := (u, v) + \mathcal{E}(u, v).$$

It is possible to show that the form $\mathcal{E}$ is *closed*, that is, the space $\mathcal{F}$ is *Hilbert*. Furthermore, $\operatorname{dom}(\mathcal{L})$ is dense in $\mathcal{F}$.

The fact that $P_t$ is Markovian implies that the form $\mathcal{E}$ is also *Markovian*, that is

$$u \in \mathcal{F} \Rightarrow \widetilde{u} := \min(u_+, 1) \in \mathcal{F} \text{ and } \mathcal{E}(\widetilde{u}) \le \mathcal{E}(u).$$

Indeed, let us first show that for any $u \in L^2$

$$\mathcal{E}_t(u_+) \le \mathcal{E}_t(u).$$

We have

$$\mathcal{E}_t(u) = \mathcal{E}_t(u_+ - u_-) = \mathcal{E}_t(u_+) + \mathcal{E}_t(u_-) - 2\mathcal{E}_t(u_+, u_-) \ge \mathcal{E}_t(u_+)$$

because $\mathcal{E}_t(u_-) \ge 0$ and

$$\mathcal{E}_t(u_+, u_-) = \frac{1}{t}(u_+, u_-) - \frac{1}{t}(P_t u_+, u_-) \le 0.$$

Assuming $u \in \mathcal{F}$ and letting $t \to 0$, we obtain

$$\mathcal{E}(u_+) = \lim_{t\to 0} \mathcal{E}_t(u_+) \le \lim_{t\to 0} \mathcal{E}_t(u) = \mathcal{E}(u) < \infty$$

whence $\mathcal{E}(u_+) \le \mathcal{E}(u)$ and, hence, $u_+ \in \mathcal{F}$.

Similarly one proves that $\widetilde{u} = \min(u_+, 1)$ belongs to $\mathcal{F}$ and $\mathcal{E}(\widetilde{u}) \le \mathcal{E}(u_+)$.

**Conclusion** *Hence, $(\mathcal{E}, \mathcal{F})$ is a* Dirichlet form, *that is, a bilinear, symmetric, positive definite, closed, densely defined form in $L^2$ with Markovian property.*

If the heat semigroup is defined by means of a heat kernel $p_t$, then $\mathcal{E}_t$ can be equivalently defined by

$$\mathcal{E}_t(u) = \frac{1}{2t}\int_M\int_M (u(x) - u(y))^2\, p_t(x, y)d\mu(y)d\mu(x) + \frac{1}{t}\int_M (1 - P_t 1(x))\, u^2(x)d\mu(x). \tag{1.5}$$

Indeed, we have

$$\begin{aligned} u(x) - P_t u(x) &= u(x)\, P_t 1(x) - P_t u(x) + (1 - P_t 1(x))\, u(x) \\ &= \int_M (u(x) - u(y))\, p_t(x, y)d\mu(y) + (1 - P_t 1(x))\, u(x), \end{aligned}$$

whence

$$\mathcal{E}_t(u) = \frac{1}{t}\int_M\int_M (u(x) - u(y))\, u(x) p_t(x, y)d\mu(y)d\mu(x) + \frac{1}{t}\int_M (1 - P_t 1(x))\, u^2(x)d\mu(x).$$

Interchanging the variables $x$ and $y$ in the first integral and using the symmetry of the heat kernel, we obtain also

$$\mathcal{E}_t(u) = \frac{1}{t}\int_M\int_M (u(y) - u(x))\, u(y) p_t(x, y)d\mu(y)d\mu(x) + \frac{1}{t}\int_M (1 - P_t 1(x))\, u^2(x)d\mu(x),$$

and (1.5) follows by adding up the two previous lines.

Since $P_t 1 \leq 1$, the second term in the right hand side of (1.5) is non-negative. If the heat kernel is stochastically complete, that is, $P_t 1 = 1$, then that term vanishes and we obtain

$$\mathcal{E}_t(u) = \frac{1}{2t} \int_M \int_M (u(x) - u(y))^2 \, p_t(x, y) d\mu(y) d\mu(x). \tag{1.6}$$

**Definition 1.2** The form $(\mathcal{E}, \mathcal{F})$ is called *local* if $\mathcal{E}(u, v) = 0$ whenever the functions $u, v \in \mathcal{F}$ have compact disjoint supports. The form $(\mathcal{E}, \mathcal{F})$ is called *strongly local* if $\mathcal{E}(u, v) = 0$ whenever the functions $u, v \in \mathcal{F}$ have compact supports and $u \equiv \text{const}$ in an open neighborhood of $\operatorname{supp} v$.

For example, if $p_t(x, y)$ is the heat kernel of the Laplace-Beltrami operator on a complete Riemannian manifold, then the associated Dirichlet form is given by

$$\mathcal{E}(u, v) = \int_M \langle \nabla u, \nabla v \rangle d\mu, \tag{1.7}$$

and $\mathcal{F}$ is the Sobolev space $W_2^1(M)$. Note that this Dirichlet form is strongly local because $u = \text{const}$ on $\operatorname{supp} v$ implies $\nabla u = 0$ on $\operatorname{supp} v$ and, hence, $\mathcal{E}(u, v) = 0$.

If $p_t(x, y)$ is the heat kernel of the symmetric stable process of index $\beta$ in $\mathbb{R}^n$, that is, $\mathcal{L} = (-\Delta)^{\beta/2}$, then

$$\mathcal{E}(u, v) = c_{n,\beta} \int_{\mathbb{R}^n} \int_{\mathbb{R}^n} \frac{(u(x) - u(y))(v(x) - v(y))}{|x - y|^{n+\beta}} dx dy,$$

and $\mathcal{F}$ is the Besov space $B_{2,2}^{\beta/2}(\mathbb{R}^n) = \{u \in L^2 : \mathcal{E}(u, u) < \infty\}$. This form is clearly non-local.

Denote by $C_0(M)$ the space of continuous functions on $M$ with compact supports, endowed with sup-norm.

**Definition 1.3** The form $(\mathcal{E}, \mathcal{F})$ is called *regular* if $\mathcal{F} \cap C_0(M)$ is dense both in $\mathcal{F}$ (with $[\cdot, \cdot]$-norm) and in $C_0(M)$ (with sup-norm).

All the Dirichlet forms in the above examples are regular.

Assume that we are given a Dirichlet form $(\mathcal{E}, \mathcal{F})$ in $L^2(M, \mu)$. Then one can define the generator $\mathcal{L}$ of $(\mathcal{E}, \mathcal{F})$ by the identity

$$(\mathcal{L}u, v) = \mathcal{E}(u, v) \quad \text{for all } u \in \operatorname{dom}(\mathcal{L}),\ v \in \mathcal{F}, \tag{1.8}$$

where $\operatorname{dom}(\mathcal{L}) \subset \mathcal{F}$ must satisfy one of the following two equivalent requirements:

1. $\operatorname{dom}(\mathcal{L})$ is a maximal possible subspace of $\mathcal{F}$ such that (1.8) holds
2. $\mathcal{L}$ is a densely defined self-adjoint operator.

Clearly, $\mathcal{L}$ is positive definite so that spec$\mathcal{L} \subset [0, +\infty)$. Hence, the family of operators $P_t = e^{-t\mathcal{L}}$, $t \geq 0$, forms a strongly continuous, symmetric, contraction semigroup in $L^2$. Moreover, using the Markovian property of the Dirichlet form $(\mathcal{E}, \mathcal{F})$, it is possible to prove that $\{P_t\}$ is Markovian, that is, $\{P_t\}$ is a heat semigroup. The question whether and when $P_t$ has the heat kernel requires a further investigation.

## 1.5 More Examples of Heat Kernels

Let us give some examples of stochastically complete heat kernels that do not satisfy (1.4).

*Example 1.4* (*A frozen heat kernel*) Let $M$ be a countable set and let $\{x_k\}_{k=1}^{\infty}$ be the sequence of all distinct points from $M$. Let $\{\mu_k\}_{k=1}^{\infty}$ be a sequence of positive reals and define measure $\mu$ on $M$ by $\mu(\{x_k\}) = \mu_k$. Define a function $p_t(x, y)$ on $M \times M$ by

$$p_t(x, y) = \begin{cases} \frac{1}{\mu_k}, & x = y = x_k \\ 0, & \text{otherwise.} \end{cases}$$

It is easy to check that $p_t(x, y)$ is a heat kernel. For example, let us check the approximation of identity: for any function $f \in L^2(M, \mu)$, we have

$$P_t f(x) = \int_M p_t(x, y) f(y)\, d\mu(y) = p_t(x, x) f(x) \mu(\{x\}) = f(x).$$

This identity also implies the stochastic completeness. The Dirichlet form is

$$\mathcal{E}(f) = \lim_{t \to 0} \left( \frac{f - P_t f}{t}, f \right) = 0.$$

The Markov process associated with the frozen heat kernel is very simple: $X_t = X_0$ for all $t \geq 0$ so that it is a frozen diffusion.

*Example 1.5* (*The heat kernel in* $\mathbb{H}^3$) The heat kernel of the Laplace-Beltrami operator on the 3-dimensional hyperbolic space $\mathbb{H}^3$ is given by the formula

$$p_t(x, y) = \frac{1}{(4\pi t)^{3/2}} \frac{r}{\sinh r} \exp\left( -\frac{r^2}{4t} - t \right),$$

where $r = d(x, y)$ is the geodesic distance between $x, y$. The Dirichlet form is given by (1.7).

*Example 1.6* (*The Mehler heat kernel*) Let $M = \mathbb{R}$, measure $\mu$ be defined by

$$d\mu = e^{x^2} dx,$$

and let $\mathcal{L}$ be given by

$$\mathcal{L} = -e^{-x^2}\frac{d}{dx}\left(e^{x^2}\frac{d}{dx}\right) = -\frac{d^2}{dx^2} - 2x\frac{d}{dx}.$$

Then the heat kernel of $\mathcal{L}$ is given by the formula

$$p_t(x,y) = \frac{1}{(2\pi \sinh 2t)^{1/2}} \exp\left(\frac{2xye^{-2t} - x^2 - y^2}{1 - e^{-4t}} - t\right).$$

The associated Dirichlet form is also given by (1.7).

Similarly, for the measure

$$d\mu = e^{-x^2}dx$$

and for the operator

$$\mathcal{L} = -e^{x^2}\frac{d}{dx}\left(e^{-x^2}\frac{d}{dx}\right) = -\frac{d^2}{dx^2} + 2x\frac{d}{dx},$$

we have

$$p_t(x,y) = \frac{1}{(2\pi \sinh 2t)^{1/2}} \exp\left(\frac{2xye^{-2t} - (x^2+y^2)\,e^{-4t}}{1 - e^{-4t}} + t\right).$$

## 2 Necessary Conditions for Heat Kernel Bounds

In this Chapter we assume that $p_t(x,y)$ is a heat kernel on a metric measure space $(M,d,\mu)$ that satisfies certain upper and/or lower estimates, and state the consequences of these estimates. The reader may consult [GK08, GHL03] or [GHL09] for the proofs.

Fix two positive parameters $\alpha$ and $\beta$, and let $\Phi : [0,+\infty) \to [0,+\infty)$ be a monotone decreasing function. We will consider the bounds of the heat kernel via the function $\frac{1}{t^{\alpha/\beta}}\Phi\left(\frac{d(x,y)}{t^{1/\beta}}\right)$.

### *2.1 Identifying $\Phi$ in the Non-local Case*

**Theorem 2.1** (Grigor'yan and Kumagai [GK08]) *Let $p_t(x,y)$ be a heat kernel on $(M,d,\mu)$.*

*(a) If the heat kernel satisfies the estimate*

$$p_t(x,y) \le \frac{1}{t^{\alpha/\beta}} \Phi\left(\frac{d(x,y)}{t^{1/\beta}}\right),$$

*for all $t > 0$ and almost all $x, y \in M$, then either the associated Dirichlet form $(\mathcal{E}, \mathcal{F})$ is local or*

$$\Phi(s) \ge c(1+s)^{-(\alpha+\beta)}$$

*for all $s > 0$ and some $c > 0$.*

(b) *If the heat kernel satisfies the estimate*

$$p_t(x,y) \ge \frac{1}{t^{\alpha/\beta}} \Phi\left(\frac{d(x,y)}{t^{1/\beta}}\right),$$

*then we have*

$$\Phi(s) \le C(1+s)^{-(\alpha+\beta)}$$

*for all $s > 0$ and some $C > 0$.*

(c) *Consequently, if the heat kernel satisfies the estimate*

$$p_t(x,y) \asymp \frac{C}{t^{\alpha/\beta}} \Phi\left(c\frac{d(x,y)}{t^{1/\beta}}\right),$$

*then either the Dirichlet form $\mathcal{E}$ is local or*

$$\Phi(s) \asymp (1+s)^{-(\alpha+\beta)}.$$

## 2.2 Volume of Balls

Denote by $B(x,r)$ a metric ball in $(M,d)$, that is

$$B(x,r) := \{y \in M : d(x,y) < r\}.$$

**Theorem 2.2** (Grigor'yan et al. [GHL03]) *Let $p_t$ be a heat kernel on $(M,d,\mu)$. Assume that it is stochastically complete and that it satisfies the two-sided estimate*

$$p_t(x,y) \asymp \frac{C}{t^{\alpha/\beta}} \Phi\left(c\frac{d(x,y)}{t^{1/\beta}}\right). \tag{2.1}$$

*Then, for all $x \in M$ and $r > 0$,*

$$\mu(B(x,r)) \asymp r^{\alpha},$$

*that is, $\mu$ is $\alpha$-regular.*

*Consequently,* $\dim_H(M,d) = \alpha$ *and* $\mu \asymp H^\alpha$ *on all Borel subsets of* $M$*, where* $H^\alpha$ *is the* $\alpha$*-dimensional Hausdorff measure in* $M$.

In particular, the parameter $\alpha$ is the invariant of the metric space $(M,d)$, and measure $\mu$ is determined (up to a factor $\asymp 1$) by the metric space $(M,d)$.

## 2.3 Besov Spaces

Fix $\alpha > 0$, $\sigma > 0$. We introduce the following seminorms on $L^2 = L^2(M,\mu)$:

$$N_{2,\infty}^{\alpha,\sigma}(u) = \sup_{0<r\le 1} \frac{1}{r^{\alpha+2\sigma}} \iint_{\{x,y\in M:d(x,y)<r\}} |u(x)-u(y)|^2\, d\mu(x)d\mu(y), \tag{2.2}$$

and

$$N_{2,2}^{\alpha,\sigma}(u) = \int_0^\infty \frac{dr}{r}\frac{1}{r^{\alpha+2\sigma}} \iint_{\{x,y\in M:d(x,y)<r\}} |u(x)-u(y)|^2\, d\mu(x)d\mu(y). \tag{2.3}$$

Define the space $\Lambda_{2,\infty}^{\alpha,\sigma}$ by

$$\Lambda_{2,\infty}^{\alpha,\sigma} = \left\{u \in L^2 : N_{2,\infty}^{\alpha,\sigma}(u) < \infty\right\},$$

and the norm by

$$\|u\|^2_{\Lambda_{2,\infty}^{\alpha,\sigma}} = \|u\|_2^2 + N_{2,\infty}^{\alpha,\sigma}(u).$$

Similarly, one defines the space $\Lambda_{2,2}^{\alpha,\sigma}$. More generally one can define $\Lambda_{p,q}^{\alpha,\sigma}$ for $p \in [1,+\infty)$ and $q \in [1,+\infty]$.

In the case of $\mathbb{R}^n$, we have the following relations

$$\begin{aligned}
\Lambda_{p,q}^{n,\sigma}\left(\mathbb{R}^n\right) &= B_{p,q}^{\sigma}\left(\mathbb{R}^n\right),\ 0<\sigma<1,\\
\Lambda_{2,\infty}^{n,1}\left(\mathbb{R}^n\right) &= W_p^1\left(\mathbb{R}^n\right),\\
\Lambda_{2,2}^{n,1}\left(\mathbb{R}^n\right) &= \{0\},\\
\Lambda_{p,q}^{n,\sigma}\left(\mathbb{R}^n\right) &= \{0\},\ \sigma>1.
\end{aligned}$$

where $B_{p,q}^{\sigma}$ is the Besov space and $W_p^1$ is the Sobolev space. The spaces $\Lambda_{p,q}^{\alpha,\sigma}$ will also be called Besov spaces.

**Theorem 2.3** (Jonsson [Jon96], Pietruska-Pałuba [Pie00], Grigor'yan et al. [GHL03]) *Let* $p_t$ *be a heat kernel on* $(M,d,\mu)$*. Assume that it is stochastically* *complete and that it satisfies the following estimate: for all* $t > 0$ *and almost all* $x, y \in M$,

$$\frac{1}{t^{\alpha/\beta}}\Phi_1\left(\frac{d(x,y)}{t^{1/\beta}}\right) \le p_t(x,y) \le \frac{1}{t^{\alpha/\beta}}\Phi_2\left(\frac{d(x,y)}{t^{1/\beta}}\right), \tag{2.4}$$

*where* $\alpha, \beta$ *be positive constants, and* $\Phi_1, \Phi_2$ *are monotone decreasing functions from* $[0, +\infty)$ *to* $[0, +\infty)$ *such that* $\Phi_1(s) > 0$ *for some* $s > 0$ *and*

$$\int_0^\infty s^{\alpha+\beta}\Phi_2(s)\frac{ds}{s} < \infty. \tag{2.5}$$

*Then, for any* $u \in L^2$,

$$\mathcal{E}(u) \asymp N_{2,\infty}^{\alpha,\beta/2}(u),$$

*and, consequently,* $\mathcal{F} = \Lambda_{2,\infty}^{\alpha,\beta/2}$.

By Theorem 2.1, the upper bound in (2.4) implies that either $(\mathcal{E}, \mathcal{F})$ is local or

$$\Phi_2(s) \ge c(1+s)^{-(\alpha+\beta)}.$$

Since the latter contradicts condition (2.5), the form $(\mathcal{E}, \mathcal{F})$ must be local. For non-local forms the statement is not true. For example, for the operator $(-\Delta)^{\beta/2}$ in $\mathbb{R}^n$, we have $\mathcal{F} = B_{2,2}^{\beta/2} = \Lambda_{2,2}^{n,\beta/2}$ that is strictly smaller than $B_{2,\infty}^{\beta/2} = \Lambda_{2,\infty}^{n,\beta/2}$. This case will be covered by the following theorem.

**Theorem 2.4** (Stós [Sto00]) *Let* $p_t$ *be a stochastically complete heat kernel on* $(M, d, \mu)$ *satisfying estimate* (2.4) *with functions*

$$\Phi_1(s) \asymp \Phi_2(s) \asymp (1+s)^{-(\alpha+\beta)}.$$

*Then, for any* $u \in L^2$,

$$\mathcal{E}(u) \asymp N_{2,2}^{\alpha,\beta/2}(u).$$

*Consequently, we have* $\mathcal{F} = \Lambda_{2,2}^{\alpha,\beta/2}$.

## 2.4 Subordinated Semigroups

Let $\mathcal{L}$ be the generator of a heat semigroup $\{P_t\}$. Then, for any $\delta \in (0,1)$, the operator $\mathcal{L}^\delta$ is also a generator of a heat semigroup, that is, the semigroup $\{e^{-t\mathcal{L}^\sigma}\}$

is a heat semigroup. Furthermore, there is the following relation between the two semigroups

$$e^{-t\mathcal{L}^{\delta}} = \int_{0}^{\infty} e^{-s\mathcal{L}} \eta_t(s)\, ds,$$

where $\eta_t(s)$ is a *subordinator* whose Laplace transform is given by

$$e^{-t\lambda^{\delta}} = \int_{0}^{\infty} e^{-s\lambda} \eta_t(s)\, ds, \ \lambda > 0.$$

Using the known estimates for $\eta_t(s)$, one can obtain the following result.

**Theorem 2.5** *Let a heat kernel $p_t$ satisfy the estimate* (2.4) *where $\Phi_1(s) > 0$ for some $s > 0$ and*

$$\int_{0}^{\infty} s^{\alpha+\beta'} \Phi_2(s) \frac{ds}{s} < \infty,$$

*where $\beta' = \delta\beta$, $0 < \delta < 1$. Then the heat kernel $q_t(x, y)$ of operator $\mathcal{L}^{\delta}$ satisfies the estimate*

$$q_t(x, y) \asymp \frac{1}{t^{\alpha/\beta'}} \left(1 + \frac{d(x, y)}{t^{1/\beta'}}\right)^{-(\alpha+\beta')} \asymp \min\left(t^{-\alpha/\beta'}, \frac{t}{d(x, y)^{\alpha+\beta'}}\right),$$

*for all $t > 0$ and almost all $x, y \in M$.*

## 2.5 The Walk Dimension

It follows from definition that the Besov seminorm

$$N_{2,\infty}^{\alpha,\sigma}(u) = \sup_{0<r\leq 1} \frac{1}{r^{\alpha+2\sigma}} \iint_{\{x,y\in M:d(x,y)<r\}} |u(x) - u(y)|^2 \, d\mu(x) d\mu(y)$$

increases when $\sigma$ increases, which implies that the space

$$\Lambda_{2,\infty}^{\alpha,\sigma} := \left\{u \in L^2 : N_{2,\infty}^{\alpha,\sigma}(u) < \infty\right\}$$

shrinks. For a certain value of $\sigma$, this space may become trivial. For example, as was already mentioned, $\Lambda_{2,\infty}^{n,\sigma}(\mathbb{R}^n) = \{0\}$ for $\sigma > 1$, while $\Lambda_{2,\infty}^{n,\sigma}(\mathbb{R}^n)$ is non-trivial for $\sigma \leq 1$.

**Definition 2.6** Fix $\alpha > 0$ and set

$$\beta^* := \sup\left\{\beta > 0 : \Lambda_{2,\infty}^{\alpha,\beta/2} \text{ is dense in } L^2(M,\mu)\right\}. \tag{2.6}$$

The number $\beta^* \in [0, +\infty]$ is called the *critical exponent* of the family $\left\{\Lambda_{2,\infty}^{\alpha,\beta/2}\right\}_{\beta>0}$ of Besov spaces.

Note that the value of $\beta^*$ is an intrinsic property of the space $(M, d, \mu)$, which is defined independently of any heat kernel. For example, for $\mathbb{R}^n$ with $\alpha = n$ we have $\beta^* = 2$.

**Theorem 2.7** (Jonsson [Jon96], Pietruska-Pałuba [Pie00], Grigor'yan et al. [GHL03]) *Let $p_t$ be a heat kernel on a metric measure space $(M, d, \mu)$. If the heat kernel is stochastically complete and satisfies* (2.4), *where $\Phi_1(s) > 0$ for some $s > 0$ and*

$$\int_0^\infty s^{\alpha+\beta+\varepsilon}\Phi_2(s)\frac{ds}{s} < \infty \tag{2.7}$$

*for some $\varepsilon > 0$, then $\beta = \beta^*$.*

By Theorem 2.1, condition (2.7) implies that the Dirichlet form $(\mathcal{E}, \mathcal{F})$ is local. For non-local forms the statement is not true: for example, in $\mathbb{R}^n$ for symmetric stable processes we have $\beta < 2 = \beta^*$.

**Theorem 2.8** *Under the hypotheses of Theorem 2.7, the values of the parameters $\alpha$ and $\beta$ are the invariants of the metric space $(M, d)$ alone. Moreover, we have*

$$\mu \asymp H^\alpha \text{ and } \mathcal{E} \asymp N_{2,\infty}^{\alpha,\beta/2}.$$

*Consequently, both the measure $\mu$ and the energy form $\mathcal{E}$ are determined (up to a factor $\asymp 1$) by the metric space $(M, d)$ alone.*

*Example 2.9* Consider in $\mathbb{R}^n$ the Gauss-Weierstrass heat kernel

$$p_t(x, y) = \frac{1}{(4\pi t)^{n/2}}\exp\left(-\frac{|x-y|^2}{4t}\right)$$

and its generator $\mathcal{L} = -\Delta$ in $L^2(\mathbb{R}^n)$ with the Lebesgue measure. Then $\alpha = n$, $\beta = 2$, and

$$\mathcal{E}(u) = \int_{\mathbb{R}^n} |\nabla u|^2 \, dx.$$

Consider now another elliptic operator in $\mathbb{R}^n$:

$$\mathcal{L} = -\frac{1}{m(x)} \sum_{i,j=1}^{n} \frac{\partial}{\partial x_i} \left( a_{ij}(x) \frac{\partial}{\partial x_j} \right),$$

where $m(x)$ and $a_{ij}(x)$ are continuous functions, $m(x) > 0$ and the matrix $(a_{ij}(x))$ is positive definite. The operator $\mathcal{L}$ is symmetric with respect to measure

$$d\mu = m(x) \, dx,$$

and its Dirichlet form is

$$\mathcal{E}(u) = \int_{\mathbb{R}^n} a_{ij}(x) \frac{\partial u}{\partial x_i} \frac{\partial u}{\partial x_j} dx.$$

Let $d(x, y) = |x - y|$ and assume that the heat kernel $p_t(x, y)$ of $\mathcal{L}$ satisfies the conditions of Theorem 2.7. Then we conclude by Corollary 2.8 that $\alpha$ and $\beta$ must be the same as in the Gauss-Weierstrass heat kernel, that is, $\alpha = n$ and $\beta = 2$; moreover, measure $\mu$ must be comparable to the Lebesgue measure, which implies that $m \asymp 1$, and the energy form must admit the estimate

$$\mathcal{E}(u) \asymp \int_{\mathbb{R}^n} |\nabla u|^2 \, dx,$$

which implies that the matrix $(a_{ij}(x))$ is uniformly elliptic. Hence, the operator $\mathcal{L}$ is uniformly elliptic.

By Aronson's theorem [Aro67, Aro68] the heat kernel for uniformly elliptic operators satisfies the estimate

$$p_t(x, y) \asymp \frac{C}{t^{n/2}} \exp\left( -c \frac{|x - y|^2}{t} \right).$$

What we have proved here implies the converse to Aronson's theorem: if the Aronson estimate holds for the operator $\mathcal{L}$, then $\mathcal{L}$ is uniformly elliptic.

The next theorem handles the non-local case.

**Theorem 2.10** *Let $p_t$ be a heat kernel on a metric measure space $(M, d, \mu)$. If the heat kernel satisfies the lower bound*

$$p_t(x, y) \geq \frac{1}{t^{\alpha/\beta}} \Phi_1 \left( \frac{d(x, y)}{t^{1/\beta}} \right),$$

*where* $\Phi_1(s) > 0$ *for some* $s > 0$, *then* $\beta \leq \beta^*$.

*Proof* In the proof of Theorem 2.3 one shows that the lower bound of the heat kernel implies $\mathcal{F} \subset \Lambda_{2,\infty}^{\alpha,\beta/2}$ (and the opposite inclusion follows from the upper bound and the stochastic completeness). Since $\mathcal{F}$ is dense in $L^2$, it follows that $\beta \leq \beta^*$. ∎

As a conclusion of this part, we briefly explain the walk dimension from three different points of view. As we have seen, there is a parameter appears in three different places:

- A parameter $\beta$ in heat kernel bounds (2.4).
- A parameter $\theta$ in Markov processes: for a process $X_t$, one may have (cf. [Bar98, formula (1.1)])

$$\mathbb{E}_x \left( |X_t - x|^2 \right) \asymp t^{2/\theta}.$$

  Then $\theta$ is a parameter that measures how fast the process $X_t$ goes away from the starting point $x$ in time $t$. Alternatively, one may have that, for any ball $B(x, r) \subset M$,

$$\mathbb{E}_x \left( \tau_{B(x,r)} \right) \asymp r^{\theta},$$

  where $\tau_{B(x,r)}$ is the *first exit time* of $X_t$ from the ball

$$\tau_B = \inf \{ t > 0 : X_t \notin B(x, r) \}.$$

- A parameter $\sigma$ in function spaces $N_{2,\infty}^{\alpha,\sigma}$ or $N_{2,2}^{\alpha,\sigma}$. By (2.2) or by (2.3), it is not hard to see that $\sigma$ measures how much smooth of the functions in the space $N_{2,\infty}^{\alpha,\sigma}$ or $N_{2,2}^{\alpha,\sigma}$.

In general the three parameters $\beta, \theta, 2\sigma$ may be different. However, it turns out that, under some certain conditions, all these parameters are the same:

$$\beta = \theta = 2\sigma. \tag{2.8}$$

For examples, by Theorems 2.3 and 2.4, we see that $\sigma = \frac{\beta}{2}$, whilst by Theorems 3.8 and 4.3 below, we will see that $\beta = \theta$.

## 2.6 Inequalities for the Walk Dimension

**Definition 2.11** We say that a metric space $(M, d)$ satisfies the *chain condition* if there exists a (large) constant $C$ such that for any two points $x, y \in M$ and for any positive integer $n$ there exists a sequence $\{x_i\}_{i=0}^n$ of points in $M$ such that $x_0 = x$, $x_n = y$, and

$$d(x_i, x_{i+1}) \le C\frac{d(x, y)}{n}, \quad \text{for all } i = 0, 1, \dots, n-1. \tag{2.9}$$

The sequence $\{x_i\}_{i=0}^n$ is referred to as a *chain* connecting $x$ and $y$.

**Theorem 2.12** (Grigor'yan et al. [GHL03]) *Let $(M, d, \mu)$ be a metric measure space. Assume that*

$$\mu(B(x, r)) \asymp r^\alpha \tag{2.10}$$

*for all $x \in M$ and $0 < r \le 1$. Then*

$$\beta^* \ge 2.$$

*If in addition $(M, d)$ satisfies the chain condition, then*

$$\beta^* \le \alpha + 1.$$

Observe that the chain condition is essential for the inequality $\beta^* \le \alpha + 1$ to be true. Indeed, assume for a moment that the claim of Theorem 2.12 holds without the chain condition, and consider a new metric $d'$ on $M$ given by $d' = d^{1/\gamma}$ where $\gamma > 1$. Let us mark by a dash all notions related to the space $(M, d', \mu)$ as opposed to those of $(M, d, \mu)$. It is easy to see that $\alpha' = \alpha\gamma$ and $\beta^{*\prime} = \beta^*\gamma$. Hence, if Theorem 2.12 could apply to the space $(M, d', \mu)$ it would yield $\beta^*\gamma \le \alpha\gamma + 1$ which implies $\beta^* \le \alpha$ because $\gamma$ may be taken arbitrarily large. However, there are spaces with $\beta^* > \alpha$, for example on SG.

Clearly, the metric $d'$ does not satisfy the chain condition; indeed the inequality (2.9) implies

$$d'(x_i, x_{i+1}) \le C\frac{d'(x, y)}{n^{1/\gamma}},$$

which is not good enough. Note that if in the inequality (2.9) we replace $n$ by $n^{1/\gamma}$, then the proof of Theorem 2.12 will give that $\beta^* \le \alpha + \gamma$ instead of $\beta^* \le \alpha + 1$.

**Theorem 2.13** (Grigor'yan et al. [GHL03]) *Let $p_t$ be a stochastically complete heat kernel on $(M, d, \mu)$ such that*

$$p_t(x,y) \asymp \frac{C}{t^{\alpha/\beta}} \Phi\left(c\frac{d(x,y)}{t^{1/\beta}}\right).$$

*(a) If for some* $\varepsilon > 0$

$$\int_0^\infty s^{\alpha+\beta+\varepsilon}\Phi(s)\frac{ds}{s} < \infty, \tag{2.11}$$

*then* $\beta \geq 2$.

*(b) If* $(M,d)$ *satisfies the chain condition, then* $\beta \leq \alpha + 1$.

*Proof* By Theorem 2.2 $\mu$ is $\alpha$-regular so that Theorem 2.12 applies.

$(a)$ By Theorem 2.12, $\beta^* \geq 2$, and by Theorem 2.12, $\beta = \beta^*$, whence $\beta \geq 2$.

$(b)$ By Theorem 2.12, $\beta^* \leq \alpha + 1$, and by Theorem 2.10, $\beta \leq \beta^*$, whence $\beta \leq \alpha + 1$. ∎

Note that the condition (2.11) can occur only for a local Dirichlet form $\mathcal{E}$. If both (2.11) and the chain condition are satisfied, then we obtain

$$2 \leq \beta \leq \alpha + 1. \tag{2.12}$$

This inequality was stated by Barlow [Bar98] without proof.

The set of couples $(\alpha, \beta)$ satisfying (2.12) is shown on the diagram (Fig. 5):

Barlow [Bar04] proved that any couple of $\alpha, \beta$ satisfying (2.12) can be realized for the heat kernel estimate

$$p_t(x,y) \asymp \frac{C}{t^{\alpha/\beta}} \exp\left(-c\left(\frac{d^\beta(x,y)}{t}\right)^{\frac{1}{\beta-1}}\right) \tag{2.13}$$

For a non-local form, we can only claim that

$$0 < \beta \leq \alpha + 1$$

(under the chain condition). In fact, any couple $\alpha, \beta$ in the range

$$0 < \beta < \alpha + 1$$

can be realized for the estimate

$$p_t(x,y) \asymp \frac{1}{t^{\alpha/\beta'}}\left(1 + \frac{d(x,y)}{t^{1/\beta'}}\right)^{-(\alpha+\beta')}.$$

Indeed, if $\mathcal{L}$ is the generator of a diffusion with parameters $\alpha$ and $\beta$ satisfying (2.13), then the operator $\mathcal{L}^\delta$, $\delta \in (0,1)$, generates a jump process with the walk dimension

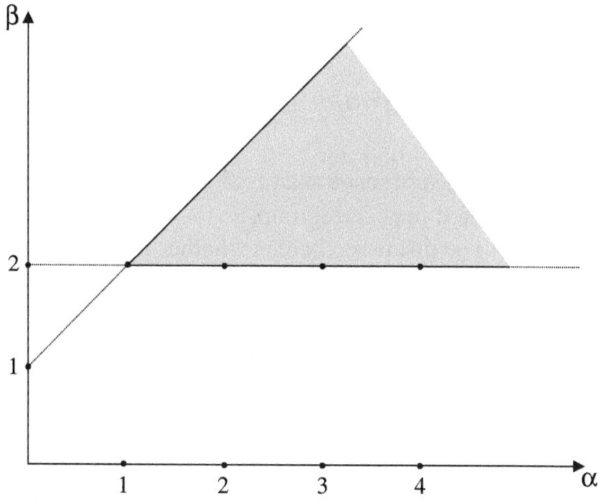


**Fig. 5** The set $2 \le \beta \le \alpha + 1$

$\beta' = \delta\beta$ and the same $\alpha$ (cf. Theorem 2.5). Clearly, $\beta'$ can take any value from $(0, \alpha + 1)$.

It is not known whether the walk dimension for a non-local form can be equal to $\alpha + 1$.

## 2.7 Identifying Φ in the Local Case

**Theorem 2.14** (Grigor'yan and Kumagai [GK08]) *Assume that the metric space $(M, d)$ satisfies the chain condition and all metric balls are precompact. Let $p_t$ be a stochastically complete heat kernel in $(M, d, \mu)$. Assume that the associated Dirichlet form $(\mathcal{E}, \mathcal{F})$ is regular, and the following estimate holds with some $\alpha$, $\beta > 0$ and $\Phi : [0, +\infty) \to [0, +\infty)$:*

$$p_t(x, y) \asymp \frac{C}{t^{\alpha/\beta}} \Phi\left(c \frac{d(x, y)}{t^{1/\beta}}\right).$$

*Then the following dichotomy holds:*

- *either the Dirichlet form $\mathcal{E}$ is local, $2 \le \beta \le \alpha + 1$, and $\Phi(s) \asymp C \exp(-cs^{\frac{\beta}{\beta-1}})$.*
- *or the Dirichlet form $\mathcal{E}$ is non-local, $\beta \le \alpha + 1$, and $\Phi(s) \asymp (1 + s)^{-(\alpha+\beta)}$.*

# 3 Sub-Gaussian Upper Bounds

## *3.1 Ultracontractive Semigroups*

Let $(M, d, \mu)$ be a metric measure space and $(\mathcal{E}, \mathcal{F})$ be a Dirichlet form in $L^2(M, \mu)$, and let $\{P_t\}$ be the associated heat semigroup, $P_t = e^{-t\mathcal{L}}$ where $\mathcal{L}$ is the generator of $(\mathcal{E}, \mathcal{F})$. The question to be discussed here is whether $P_t$ possesses the heat kernel, that is, a function $p_t(x, y)$ that is non-negative, jointly measurable in $(x, y)$, and satisfies the identity

$$P_t f(x) = \int_M p_t(x, y) f(y) d\mu(y)$$

for all $f \in L^2$, $t > 0$, and almost all $x \in M$. Usually the conditions that ensure the existence of the heat kernel give at the same token some upper bounds.

Given two parameters $p, q \in [0, +\infty]$, define the $L^p \to L^q$ norm of $P_t$ by

$$\|P_t\|_{L^p \to L^q} = \sup_{f \in L^p \cap L^2 \setminus \{0\}} \frac{\|P_t f\|_q}{\|f\|_p}.$$

In fact, the Markovian property allows to extend $P_t$ to an operator in $L^p$ so that the range $L^p \cap L^2$ of $f$ can be replaced by $L^p$. Also, it follows from the Markovian property that $\|P_t\|_{L^p \to L^p} \leq 1$ for any $p$.

**Definition 3.1** The semigroup $\{P_t\}$ is said to be $L^p \to L^q$ *ultracontractive* if there exists a positive decreasing function $\gamma$ on $(0, +\infty)$, called the *rate* function, such that, for each $t > 0$

$$\|P_t\|_{L^p \to L^q} \leq \gamma(t).$$

By the symmetry of $P_t$, if $P_t$ is $L^p \to L^q$ ultracontractive, then $P_t$ is also $L^{q^*} \to L^{p^*}$ ultracontractive with the same rate function, where $p^*$ and $q^*$ are the Hölder conjugates to $p$ and $q$, respectively. In particular, $P_t$ is $L^1 \to L^2$ ultracontractive if and only if it is $L^2 \to L^\infty$ ultracontractive.

**Theorem 3.2** *(a) The heat semigroup $\{P_t\}$ is $L^1 \to L^2$ ultracontractive with a rate function $\gamma$, if and only if $\{P_t\}$ has the heat kernel $p_t$ satisfying the estimate*

$$\operatorname*{esup}_{x,y \in M} p_t(x, y) \leq \gamma(t/2)^2 \quad \textit{for all } t > 0.$$

*(b) The heat semigroup $\{P_t\}$ is $L^1 \to L^\infty$ ultracontractive with a rate function $\gamma$, if and only if $\{P_t\}$ has the heat kernel $p_t$ satisfying the estimate*

$$\operatorname*{esup}_{x,y \in M} p_t(x, y) \leq \gamma(t) \quad \textit{for all } t > 0.$$

This result is "well-known" and can be found in many sources. However, there are hardly complete proofs of the measurability of the function $p_t(x, y)$ in $(x, y)$, which is necessary for many applications, for example, to use Fubini. Normally the existence of the heat kernel is proved in some specific setting where $p_t(x, y)$ is continuous in $(x, y)$, or one just proves the existence of a family of functions $p_{t,x} \in L^2$ so that

$$P_t f(x) = \left(p_{t,x}, f\right) = \int_M p_{t,x}(y) f(y) \, d\mu(y)$$

for all $t > 0$ and almost all $x$. However, if one defines $p_t(x, y) = p_{t,x}(y)$, then this function does not have to be jointly measurable. The proof of the existence of a jointly measurable version can be found in [GH10]. Most of the material of this section can also be found there.

## 3.2 Restriction of the Dirichlet Form

Let $\Omega$ be an open subset of $M$. Define the function space $\mathcal{F}(\Omega)$ by

$$\mathcal{F}(\Omega) = \overline{\{f \in \mathcal{F} : \operatorname{supp} f \subset \Omega\}}^{\mathcal{F}}.$$

Clearly, $\mathcal{F}(\Omega)$ is a closed subspace of $\mathcal{F}$ and a subspace of $L^2(\Omega)$.

**Theorem 3.3** *If $(\mathcal{E}, \mathcal{F})$ is a regular Dirichlet form in $L^2(M)$, then $(\mathcal{E}, \mathcal{F}(\Omega))$ is a regular Dirichlet form in $L^2(\Omega)$. If $(\mathcal{E}, \mathcal{F})$ is (strongly) local then so is $(\mathcal{E}, \mathcal{F}(\Omega))$.*

The regularity is used, in particular, to ensure that $\mathcal{F}(\Omega)$ is dense in $L^2(\Omega)$. From now on let us assume that $(\mathcal{E}, \mathcal{F})$ is a regular Dirichlet form. Other consequences of this assumptions are as follows (cf. [FOT11]):

1. The existence of cutoff functions: for any compact set $K$ and any open set $U \supset K$, there is a function $\varphi \in \mathcal{F} \cap C_0(U)$ such that $0 \le \varphi \le 1$ and $\varphi \equiv 1$ in an open neighborhood of $K$.
2. The existence of a Hunt process $\left(\{X_t\}_{t \ge 0}, \{\mathbb{P}_x\}_{x \in M}\right)$ associated with $(\mathcal{E}, \mathcal{F})$.

Hence, for any open subset $\Omega \subset M$, we have the Dirichlet form $(\mathcal{E}, \mathcal{F}(\Omega))$ that is called a *restriction* of $(\mathcal{E}, \mathcal{F})$ to $\Omega$.

*Example 3.4* Consider in $\mathbb{R}^n$ the canonical Dirichlet form

$$\mathcal{E}(u) = \int_{\mathbb{R}^n} |\nabla u|^2 \, dx$$

in $\mathcal{F} = W_2^1(\mathbb{R}^n)$. Then $\mathcal{F}(\Omega) = \overline{C_0^1(\Omega)}^{W_2^1} =: H_0^1(\Omega)$.

Using the restricted form $(\mathcal{E}, \mathcal{F}(\Omega))$ corresponds to imposing the Dirichlet boundary conditions on $\partial\Omega$ (or on $\Omega^c$), so that the form $(\mathcal{E}, \mathcal{F}(\Omega))$ could be called the Dirichlet form with the Dirichlet boundary condition.

Denote by $\mathcal{L}_\Omega$ the generator of $(\mathcal{E}, \mathcal{F}(\Omega))$ and set

$$\lambda_{\min}(\Omega) := \inf \operatorname{spec} \mathcal{L}_\Omega = \inf_{u \in \mathcal{F}(\Omega) \setminus \{0\}} \frac{\mathcal{E}(u)}{\|u\|_2^2}. \tag{3.1}$$

Clearly, $\lambda_{\min}(\Omega) \geq 0$ and $\lambda_{\min}(\Omega)$ is decreasing when $\Omega$ expands.

*Example 3.5* If $(\mathcal{E}, \mathcal{F})$ is the canonical Dirichlet form in $\mathbb{R}^n$ and $\Omega$ is the bounded domain in $\mathbb{R}^n$, then the operator $\mathcal{L}_\Omega$ has the discrete spectrum $\lambda_1(\Omega) \leq \lambda_2(\Omega) \leq \lambda_3(\Omega) \leq \ldots$ that coincides with the eigenvalues of the Dirichlet problem

$$\begin{cases} \Delta u + \lambda u = 0, \\ u|_{\partial\Omega} = 0, \end{cases}$$

so that $\lambda_1(\Omega) = \lambda_{\min}(\Omega)$.

## 3.3 Faber-Krahn and Nash Inequalities

Continuing the above example, we have by a theorem of Faber-Krahn

$$\lambda_1(\Omega) \geq \lambda_1(\Omega^*),$$

where $\Omega^*$ is the ball of the same volume as $\Omega$. If $r$ is the radius of $\Omega^*$, then we have

$$\lambda_1(\Omega^*) = \frac{c'}{r^2} = \frac{c}{|\Omega^*|^{2/n}} = \frac{c}{|\Omega|^{2/n}},$$

whence

$$\lambda_1(\Omega) \geq c_n |\Omega|^{-2/n}.$$

It turns out that this inequality, that we call *the Faber-Krahn inequality*, is intimately related to the existence of the heat kernel and its upper bound.

**Theorem 3.6** *Let $(\mathcal{E}, \mathcal{F})$ be a regular Dirichlet form in $L^2(M, \mu)$. Fix some constant $\nu > 0$. Then the following conditions are equivalent:*

*(i)* (The Faber-Krahn inequality) *There is a constant $a > 0$ such that, for all non-empty open sets $\Omega \subset M$,*

$$\lambda_{\min}(\Omega) \geq a\mu(\Omega)^{-\nu}. \tag{3.2}$$

*(ii)* (The Nash inequality) *There exists a constant $b > 0$ such that*

$$\mathcal{E}(u) \geq b\|u\|_2^{2+2\nu}\|u\|_1^{-2\nu}, \tag{3.3}$$

*for any function $u \in \mathcal{F} \setminus \{0\}$.*

*(iii)* (On-diagonal estimate of the heat kernel) *The heat kernel exists and satisfies the upper bound*

$$\operatorname*{esup}_{x,y\in M} p_t(x,y) \leq ct^{-1/\nu} \tag{3.4}$$

*for some constant $c$ and for all $t > 0$.*

*The relation between the parameters $a, b, c$ is as follows:*

$$a \asymp b \asymp c^{-\nu}$$

*where the ratio of any two of these parameters is bounded by constants depending only on $\nu$.*

In $\mathbb{R}^n$, we see that $\nu = 2/n$.

The implication (ii) $\Rightarrow$ (iii) was proved by Nash [Nas58], and (iii) $\Rightarrow$ (ii) by Carlen-Kusuoka-Stroock [CKS87], and (i) $\Leftrightarrow$ (iii) by

Grigor'yan [Gri94] and Carron [Car96].

**Proof of** (i) $\Rightarrow$ (ii) $\Rightarrow$ (iii) Observe first that (ii) $\Rightarrow$ (i) is trivial: choosing in (3.3) a function $u \in \mathcal{F}(\Omega) \setminus \{0\}$ and applying the Cauchy-Schwarz inequality

$$\|u\|_1 \leq \mu(\Omega)^{1/2}\|u\|_2,$$

we obtain

$$\mathcal{E}(u) \geq b\mu(\Omega)^{-\nu}\|u\|_2^2,$$

whence (3.2) follow by the variational principle (3.1).

The opposite inequality (i) $\Rightarrow$ (ii) is a bit more involved, and we prove it for functions $0 \leq u \in \mathcal{F} \cap C_0(M)$ (a general $u \in \mathcal{F}$ requires some approximation argument). By the Markovian property, we have $(u-t)_+ \in \mathcal{F} \cap C_0(M)$ for any $t > 0$ and

$$\mathcal{E}(u) \geq \mathcal{E}\left((u-t)_+\right). \tag{3.5}$$

For any $s > 0$, consider the set

$$U_s := \{x \in M : u(x) > s\},$$

which is clearly open and precompact. If $t > s$, then $(u-t)_+$ is supported in $U_s$, and whence, $(u-t)_+ \in \mathcal{F}(U_s)$. It follows from (3.1)

$$\mathcal{E}\left((u-t)_+\right) \geq \lambda_{\min}(U_s) \int_{U_s} (u-t)_+^2 \, d\mu. \tag{3.6}$$

For simplicity, set $A = \|u\|_1$ and $B = \|u\|_2^2$. Since $u \geq 0$, we have

$$(u-t)_+^2 \geq u^2 - 2tu,$$

which implies that

$$\int_{U_s} (u-t)_+^2 d\mu = \int_M (u-t)_+^2 d\mu \geq B - 2tA. \tag{3.7}$$

On the other hand, we have

$$\mu(U_s) \leq \frac{1}{s} \int_{U_s} u \, d\mu \leq \frac{A}{s},$$

which together with the Faber-Krahn inequality implies

$$\lambda_{\min}(U_s) \geq a\mu(U_s)^{-\nu} \geq a\left(\frac{s}{A}\right)^{\nu}. \tag{3.8}$$

Combining (3.5)–(3.8), we obtain

$$\mathcal{E}(u) \geq \lambda_{\min}(U_s) \int_{U_s} (u-t)_+^2 \, d\mu \geq a\left(\frac{s}{A}\right)^{\nu} (B - 2tA).$$

Letting $t \to s+$ and then choosing $s = \frac{B}{4A}$, we obtain

$$\mathcal{E}(u) \geq a\left(\frac{s}{A}\right)^{\nu} (B - 2sA) = a\left(\frac{B}{4A^2}\right)^{\nu} \frac{B}{2} = \frac{a}{4^{\nu}2} B^{\nu+1} A^{-2\nu},$$

which is exactly (3.3).

To prove (ii)$\Rightarrow$(iii), choose $f \in L^2 \cap L^1$, and consider $u = P_t f$. Since $u = e^{-t\mathcal{L}} f$ and $\frac{d}{dt} u = -\mathcal{L}u$, we have

$$\begin{aligned} \frac{d}{dt} \|u\|_2^2 &= \frac{d}{dt}(u,u) = -2(\mathcal{L}u, u) = -2\mathcal{E}(u,u) \\ &\leq -2b\|u\|_2^{2+2\nu} \|u\|_1^{-2\nu} \leq -2b\|u\|_2^{2+2\nu} \|f\|_1^{-2\nu}, \end{aligned}$$

since $\|u\|_1 \leq \|f\|_1$. Solving this differential inequality, we obtain

$$\|P_t f\|_2^2 \le ct^{-1/v} \|f\|_1^2,$$

that is, the semigroup $P_t$ is $L^1 \to L^2$ ultracontractive with the rate function $\gamma(t) = \sqrt{ct^{-1/v}}$. By Theorem 3.2 we conclude that the heat kernel exists and satisfies (3.4). ■

Let $M$ be a Riemannian manifold with the geodesic distance $d$ and the Riemannian volume $\mu$. Let $(\mathcal{E}, \mathcal{F})$ be the canonical Dirichlet form on $M$. The heat kernel on manifolds always exists and is a smooth function. In this case the estimate (3.4) is equivalent to the on-diagonal upper bound

$$\sup_{x \in M} p_t(x, x) \le ct^{-1/\nu}.$$

It is known (but non-trivial) that the on-diagonal estimate implies the Gaussian upper bound

$$p_t(x, y) \le Ct^{-1/\nu} \exp\left(-\frac{d^2(x, y)}{(4 + \varepsilon)t}\right),$$

for all $t > 0$ and $x, y \in M$, which is due to the specific property of the geodesic distance function that $|\nabla d| \le 1$.

In the context of abstract metric measure space, the distance function does not have to satisfy this property, and typically it does not (say, on fractals). Consequently, one needs some additional conditions that would relate the distance function to the Dirichlet form and imply the off-diagonal bounds.

## *3.4 Off-diagonal Upper Bounds*

From now on, let $(\mathcal{E}, \mathcal{F})$ be a regular *local* Dirichlet form, so that the associated Hunt process $\left(\{X_t\}_{t \ge 0}, \{\mathbb{P}_x\}_{x \in M}\right)$ is a diffusion. Recall that it is related to the heat semigroup $\{P_t\}$ of $(\mathcal{E}, \mathcal{F})$ by means of the identity

$$\mathbb{E}_x(f(X_t)) = P_t f(x)$$

for all $f \in \mathcal{B}_b(M)$, $t > 0$ and almost all $x \in M$ (Fig. 6).

Fix two parameters $\alpha > 0$ and $\beta > 1$ and introduce some conditions.

$(V_\alpha)$ (Volume regularity) For all $x \in M$ and $r > 0$,

$$\mu(B(x, r)) \asymp r^\alpha.$$

*(FK)* (The Faber-Krahn inequality) For any open set $\Omega \subset M$,

$$\lambda_{\min}(\Omega) \ge c\mu(\Omega)^{-\beta/\alpha}.$$

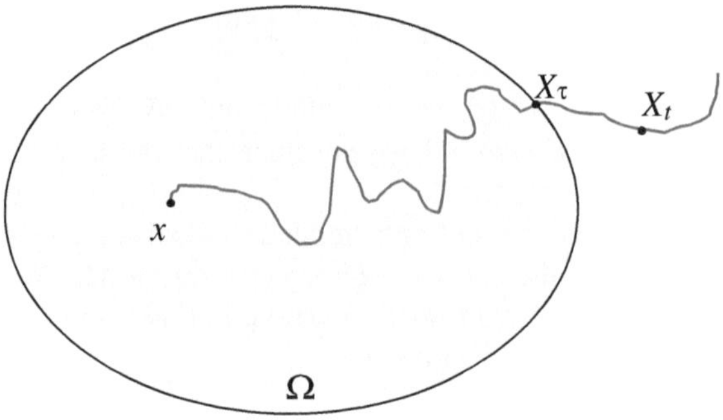


**Fig. 6** First exit time $\tau$

For any open set $\Omega \subset M$, define the *first exist time* from $\Omega$ by

$$\tau_\Omega = \inf\{t > 0 : X_t \notin \Omega\}.$$

A set $N \subset M$ is called *properly exceptional*, if it is a Borel set of measure 0 that is almost never hit by the process $X_t$ starting outside $N$. In the next conditions $N$ denotes some properly exceptional set.

$(E_\beta)$ (*An estimate for the mean exit time from balls*) For all $x \in M \setminus N$ and $r > 0$

$$\mathbb{E}_x\left[\tau_{B(x,r)}\right] \asymp r^\beta$$

(the parameter $\beta$ is called the walk dimension of the process).

$(P_\beta)$ (*The exit probability estimate*) There exist constants $\varepsilon \in (0, 1)$, $\delta > 0$ such that, for all $x \in M \setminus N$ and $r > 0$,

$$\mathbb{P}_x\left(\tau_{B(x,r)} \leq \delta r^\beta\right) \leq \varepsilon.$$

$(E\Omega)$ (*An isoperimetric estimate for the mean exit time*) For any open subset $\Omega \subset M$,

$$\sup_{x\in\Omega\setminus N} \mathbb{E}_x\left(\tau_\Omega\right) \leq C\mu\left(\Omega\right)^{\beta/\alpha}.$$

If both $(V_\alpha)$ and $(E_\beta)$ are satisfied, then we obtain for any ball $B \subset M$

$$\sup_{x\in B\setminus N} \mathbb{E}_x\left(\tau_B\right) \asymp r^\beta \asymp \mu\left(B\right)^{\beta/\alpha}.$$

It follows that the balls are in some sense optimal sets for the condition $(E\Omega)$.

*Example 3.7* If $X_t$ is Brownian motion in $\mathbb{R}^n$, then it is known that

$$\mathbb{E}_x \tau_{B(x,r)} = c_n r^2,$$

so that $\left(E_\beta\right)$ holds with $\beta = 2$. This can also be rewritten in the form

$$\mathbb{E}_x \tau_B = c_n \left|B\right|^{2/n},$$

where $B = B\left(x, r\right)$.

It is also known that for any open set $\Omega \subset \mathbb{R}^n$ with finite volume and for any $x \in \Omega$,

$$\mathbb{E}_x \left(\tau_\Omega\right) \leq \mathbb{E}_x \left(\tau_{B(x,r)}\right),$$

provided that ball $B\left(x, r\right)$ has the same volume as $\Omega$; that is, for a fixed value of $|\Omega|$, the mean exist time is maximal when $\Omega$ is a ball and $x$ is the center. It follows that

$$\mathbb{E}_x \left(\tau_\Omega\right) \leq c_n \left|\Omega\right|^{2/n}$$

so that $(E\Omega)$ is satisfied with $\beta = 2$ and $\alpha = n$.

Finally, introduce notation for the following estimates of the heat kernel:

$(UE_{loc})$ (*Sub-Gaussian upper estimate*) The heat kernel exists and satisfies the estimate

$$p_t\left(x, y\right) \leq \frac{C}{t^{\alpha/\beta}} \exp\left(-c\left(\frac{d^\beta(x, y)}{t}\right)^{\frac{1}{\beta-1}}\right)$$

for all $t > 0$ and almost all $x, y \in M$.

$(\Phi UE)$ ($\Phi$-*upper estimate*) The heat kernel exists and satisfies the estimate

$$p_t\left(x, y\right) \leq \frac{1}{t^{\alpha/\beta}} \Phi\left(\frac{d\left(x, y\right)}{t^{1/\beta}}\right)$$

for all $t > 0$ and almost all $x, y \in M$, where $\Phi$ is a decreasing positive function on $[0, +\infty)$ such that

$$\int_0^\infty s^\alpha \Phi\left(s\right) \frac{ds}{s} < \infty.$$

(*DUE*) (*On-diagonal upper estimate*) The heat kernel exists and satisfies the estimate

$$p_t\left(x, y\right) \leq \frac{C}{t^{\alpha/\beta}}$$

for all $t > 0$ and almost all $x, y \in M$.

$(T_{\exp})$ *(The exponential tail estimate)* The heat kernel $p_t$ exists and satisfies the estimate

$$\int_{B(x,r)^c} p_t(x, y)\, d\mu(y) \leq C \exp\left(-c \left(\frac{r}{t^{1/\beta}}\right)^{\frac{\beta}{\beta-1}}\right), \tag{3.9}$$

for some constants $C, c > 0$, all $t > 0$, $r > 0$ and $\mu$-almost all $x \in M$.
Note that it is easy to show that (3.9) is equivalent to the following inequality: for any ball $B = B(x_0, r)$ and $t > 0$,

$$P_t \mathbf{1}_{B^c}(x) \leq C \exp\left(-c \left(\frac{r}{t^{1/\beta}}\right)^{\frac{\beta}{\beta-1}}\right) \text{ for } \mu\text{-almost all } x \in \frac{1}{4}B$$

(see [GH08, Remark 3.3]).
$(T_\beta)$ *(The tail estimate)* There exist $0 < \varepsilon < \frac{1}{2}$ and $C > 0$ such that, for all $t > 0$ and all balls $B = B(x_0, r)$ with $r \geq Ct^{1/\beta}$,

$$P_t \mathbf{1}_{B^c}(x) \leq \varepsilon \quad \text{for } \mu\text{-almost all } x \in \frac{1}{4}B.$$

$(S_\beta)$ *(The survival estimate)* There exist $0 < \varepsilon < 1$ and $C > 0$ such that, for all $t > 0$ and all balls $B = B(x_0, r)$ with $r \geq Ct^{1/\beta}$,

$$1 - P_t^B \mathbf{1}_B(x) \leq \varepsilon \quad \text{for } \mu\text{-almost all } x \in \frac{1}{4}B.$$

Clearly, we have

$$(UE_{loc}) \Rightarrow (\Phi UE) \Rightarrow (DUE).$$

**Theorem 3.8** (Grigor'yan and Hu [GH10]) *Let $(M, d, \mu)$ be a metric measure space and let $(V_\alpha)$ hold. Let $(\mathcal{E}, \mathcal{F})$ be a regular, local, conservative Dirichlet form in $L^2(M, \mu)$. Then, the following equivalences are true:*

$$\begin{aligned}(UE_{loc}) &\Leftrightarrow (FK) + (E_\beta) \Leftrightarrow (E\Omega) + (E_\beta)\\ &\Leftrightarrow (FK) + (P_\beta) \Leftrightarrow (E\Omega) + (P_\beta)\\ &\Leftrightarrow (DUE) + (E_\beta) \Leftrightarrow (DUE) + (P_\beta),\\ &\Leftrightarrow (\Phi UE)\\ &\Leftrightarrow (FK) + (S_\beta) \Leftrightarrow (FK) + (T_\beta)\\ &\Leftrightarrow (DUE) + (S_\beta) \Leftrightarrow (DUE) + (T_\beta)\\ &\Leftrightarrow (DUE) + (T_{\exp}).\end{aligned}$$

Let us emphasize the equivalence

$$(UE_{loc}) \Leftrightarrow (E\Omega) + (E_\beta)$$

where the right hand side means the following: the mean exit time from all sets $\Omega$ satisfies the isoperimetric inequality, and this inequality is optimal for balls (up to a constant multiple). Note that the latter condition relates the properties of the diffusion (and, hence, of the Dirichlet form) to the distance function.

**Conjecture 3.9** *Under the hypotheses of Theorem 3.8,*

$$(UE_{loc}) \Leftrightarrow (FK) + \left\{\lambda_{\min}(B_r) \asymp r^{-\beta}\right\}$$

Indeed, the Faber-Krahn inequality $(FK)$ can be regarded as an isoperimetric inequality for $\lambda_{\min}(\Omega)$, and the condition

$$\lambda_{\min}(B_r) \asymp r^{-\beta}$$

means that $(FK)$ is optimal for balls (up to a constant multiple).

Theorem 3.8 is an oversimplified version of a result of [GH10], where instead of $(V_\alpha)$ one uses the volume doubling condition, and other hypotheses must be appropriately changed.

The following lemma is used in the proof of Theorem 3.8.

**Lemma 3.10** *For any open set $\Omega \subset M$*

$$\lambda_{\min}(\Omega) \geq \frac{1}{\operatorname{esup}_{x\in\Omega} \mathbb{E}_x(\tau_\Omega)}.$$

*Proof* Let $G_\Omega$ be the Green operator in $\Omega$, that is,

$$G_\Omega = \mathcal{L}_\Omega^{-1} = \int_0^\infty e^{-t\mathcal{L}_\Omega} dt.$$

We claim that

$$\mathbb{E}_x(\tau_\Omega) = G_\Omega 1(x)$$

for almost all $x \in \Omega$. We have

$$G_\Omega 1(x) = \int_0^\infty e^{-t\mathcal{L}_\Omega} 1_\Omega(x)\, dt = \int_0^\infty \mathbb{E}_x\left(1_\Omega\left(X_t^\Omega\right)\right)$$

$$= \int_0^\infty \mathbb{E}_x\left(\mathbf{1}_{\{t<\tau_\Omega\}}\right) dt = \mathbb{E}_x \int_0^\infty \left(\mathbf{1}_{\{t<\tau_\Omega\}}\right) dt = \mathbb{E}_x(\tau_\Omega).$$

Setting

$$m = \operatorname*{esup}_{x\in\Omega} \mathbb{E}_x\left(\tau_\Omega\right),$$

we obtain that $G_\Omega 1 \leq m$, so that $m^{-1}G_\Omega$ is a Markovian operator. Therefore, $\left\|m^{-1}G_\Omega\right\|_{L^2\to L^2} \leq 1$ whence spec$G_\Omega \in [0,m]$. It follows that spec $\mathcal{L}_\Omega \subset [m^{-1},\infty)$ and $\lambda_{\min}(\Omega) \geq m^{-1}$. ■

A new analytical approach is developed in [GH10] to prove Theorem 3.8, which is different from the Davies-Gaffney approach [Dav92] . The difficult part in proving Theorem 3.8 is to deduce $(UE_{loc})$ from various conditions.

**Sketch of proof for Theorem 3.8** We sketch the main steps.

- By a direct integration, we have

$$(\Phi UE) \Rightarrow \left(T_\beta\right).$$

Indeed, for any $x \in \frac{1}{4}B$, we see that $B(x,\frac{1}{2}r) \subset B$ . Thus, setting $r_k = 2^k(r/2)$ and using condition $(\Phi UE)$ and the monotonicity of $\Phi$, we obtain that

$$\begin{aligned}
\int_{M\setminus B} p_t(x,y)d\mu(y) &\leq \int_{M\setminus B(x,r/2)} p_t(x,y)d\mu(y) \qquad (3.10)\\
&= \sum_{k=0}^{\infty} \int_{B(x,r_{k+1})\setminus B(x,r_k)} p_t\left(x,y\right)d\mu(y)\\
&\leq \sum_{k=0}^{\infty} \int_{B(x,r_{k+1})\setminus B(x,r_k)} Ct^{-\alpha/\beta}\Phi\left(\frac{r_k}{t^{1/\beta}}\right)d\mu(y)\\
&\leq \sum_{k=0}^{\infty} Cr_{k+1}^{\alpha}t^{-\alpha/\beta}\Phi\left(\frac{r_k}{t^{1/\beta}}\right)\\
&= C'\sum_{k=0}^{\infty}\left(\frac{2^{k-1}r}{t^{1/\beta}}\right)^{\alpha}\Phi\left(\frac{2^{k-1}r}{t^{1/\beta}}\right)\\
&\leq C'\int_{\frac{1}{4}r/t^{1/\beta}}^{\infty} s^{\alpha}\Phi(s)\frac{ds}{s}.
\end{aligned}$$

The integral (3.10) converges, and its value can be made arbitrarily small provided that $r^\beta/t$ is large enough. Hence, condition $\left(T_\beta\right)$ follows.

- The following implications hold:

$$(E\Omega) \overset{\text{L. 3.10}}{\Rightarrow} (FK) \overset{\text{T. 3.6}}{\Rightarrow} (DUE).$$

In particular, we see that the heat kernel exists under any of the hypotheses of Theorem 3.8.

- We can also show that

$$\left(E_\beta\right) \Rightarrow \left(P_\beta\right) \Longrightarrow \left(T_\beta\right)$$

(the implication $\left(E_\beta\right) \Rightarrow \left(P_\beta\right)$ was pointed out in [Bar98]).

- By a bootstrapping technique, we obtain (hard!) the implication

$$\left(T_\beta\right) \Longrightarrow \left(T_{\exp}\right)$$

(see also [GH08]). Hence, any set of the hypothesis of Theorem 3.8 imply both $(DUE)$ and $\left(T_{\exp}\right)$.

- Finally, it is easy to check the implication

$$(DUE) + \left(T_{\exp}\right) \Rightarrow \left(UE_{loc}\right). \tag{3.11}$$

Indeed, using the semigroup identity, we have that, for all $t > 0$, almost all $x, y \in M$, and $r := \frac{1}{2} d(x, y)$,

$$\begin{aligned} p_t(x, y) &= \int_M p_{\frac{t}{2}}(x, z)\, p_{\frac{t}{2}}(z, y)\, d\mu(z) \\ &\leq \left( \int_{B(x,r)^c} + \int_{B(y,r)^c} \right) p_{\frac{t}{2}}(x, z)\, p_{\frac{t}{2}}(z, y)\, d\mu(z) \\ &\leq \operatorname*{esup}_{z\in M} p_{\frac{t}{2}}(z, y) \int_{B(x,r)^c} p_{\frac{t}{2}}(x, z)\, d\mu(z) \\ &\quad + \operatorname*{esup}_{z\in M} p_{\frac{t}{2}}(x, z) \int_{B(y,r)^c} p_{\frac{t}{2}}(y, z)\, d\mu(z). \end{aligned} \tag{3.12}$$

On the other hand, by condition $(DUE)$,

$$\operatorname{esup} p_t \leq C t^{-\alpha/\beta},$$

whilst by condition $\left(T_\beta\right)$,

$$\int_{B(x,r)^c} p_{\frac{t}{2}}(x, z)\, d\mu(z) \leq C \exp\left( -c \left( \frac{d^\beta(x, y)}{t} \right)^{\frac{1}{\beta-1}} \right).$$

Therefore, it follows from (3.12) that, for almost all $x, y \in M$,

$$p_t(x,y) \le \frac{C}{t^{\alpha/\beta}} \exp\left(-c\left(\frac{d^\beta(x,y)}{t}\right)^{\frac{1}{\beta-1}}\right),$$

proving the implication (3.11). ■

Recently, Andres and Barlow [AB] gave a new equivalence condition for $(UE_{loc})$. Consider the following functional inequality.

$(CSA_\beta)$ (*The cutoff Sobolev annulus inequality*) There exists a constant $C > 0$ such that, for all two concentric balls $B(x,R)$, $B(x,R+r)$, there exists a cutoff function $\varphi$ satisfying

$$\int_U f^2 d\mu_{\langle\varphi\rangle} \le \frac{1}{8}\int_U \varphi^2 d\mu_{\langle f\rangle} + Cr^{-\beta}\int_U f^2 d\mu$$

for any $f \in \mathcal{F}$, where $U = B(x,R+r) \setminus B(x,R)$ is the annulus and $\mu_{\langle\varphi\rangle}$ is the energy measure associated with $\varphi$:

$$\int_M u d\mu_{\langle\varphi\rangle} = 2\mathcal{E}(u\varphi,\varphi) - \mathcal{E}(\varphi^2,u) \text{ for any } u \in \mathcal{F} \cap C_0(M).$$

We remark here that constant $C$ is universal that is independent of two concentric balls $B(x,R)$, $B(x,R+r)$ and function $f$, whilst the cutoff function $\varphi$ may depend on the balls but is independent of function $f$. The coefficient $\frac{1}{8}$ is not essential and is chosen for technical reasons.

**Theorem 3.11** (Andres, Barlow [AB]) *Let $(M,d,\mu)$ be an unbounded metric measure space and let $(V_\alpha)$ hold. Let $(\mathcal{E},\mathcal{F})$ be a regular, local Dirichlet form in $L^2(M,\mu)$. Then, the following equivalence is true:*

$$(UE_{loc}) \Leftrightarrow (FK) + (CSA_\beta).$$

We mention that here the Dirichlet form is not required to be conservative as in Theorem 3.8.

The key point in proving Theorem 3.11 is to derive the "Davies-Gaffney" bound [Dav92], and then use the technique developed in [Gri92, CG98] to show a mean value inequality for weak solutions of the heat equation. It is quite surprising that the Davies-Gaffney method still works when the walk dimension $\beta$ may be greater than 2.

# 4 Two-Sided Sub-Gaussian Bounds

## 4.1 Using Elliptic Harnack Inequality

Now we would like to extend the results of Theorems 3.8, 3.11, and obtain also the lower estimates and the Hölder continuity of the heat kernel. As before, $(M, d, \mu)$ is a metric measure space, and assume in addition that all metric balls are precompact. Let $(\mathcal{E}, \mathcal{F})$ is a local regular conservative Dirichlet form in $L^2(M, \mu)$.

**Definition 4.1** We say that a function $u \in \mathcal{F}$ is *harmonic* in an open set $\Omega \subset M$ if

$$\mathcal{E}(u, v) = 0 \text{ for all } v \in \mathcal{F}(\Omega).$$

For example, if $M = \mathbb{R}^n$ and $(\mathcal{E}, \mathcal{F})$ is the canonical Dirichlet form in $\mathbb{R}^n$, then a function $u \in W_2^1(\mathbb{R}^n)$ is harmonic in an open set $\Omega \subset \mathbb{R}^n$ if

$$\int_{\mathbb{R}^n} \langle \nabla u, \nabla v \rangle dx = 0$$

for all $v \in H_0^1(\Omega)$ or for $v \in C_0^\infty(\Omega)$. This of course implies that $\Delta u = 0$ in a weak sense in $\Omega$ and, hence, $u$ is harmonic in $\Omega$ in the classical sense. However, unlike the classical definition, we a priori require $u \in W_2^1(\mathbb{R}^n)$.

**Definition 4.2** (*Elliptic Harnack inequality* $(H)$) We say that $M$ satisfies the *elliptic Harnack inequality* $(H)$ if there exist constants $C > 1$ and $\delta \in (0, 1)$ such that for any ball $B(x, r)$ and for any function $u \in \mathcal{F}$ that is non-negative and harmonic in $B(x, r)$,

$$\operatorname*{esup}_{B(x,\delta r)} u \le C \operatorname*{einf}_{B(x,\delta r)} u.$$

We remark that constants $C$ and $\delta$ are independent of ball $B(x, r)$ and function $u$. We introduce the near-diagonal lower estimate of heat kernel.

$(NLE)$ (*Near-diagonal lower estimate*) The heat kernel $p_t(x, y)$ exists, and satisfies

$$p_t(x, y) \ge \frac{c}{t^{\alpha/\beta}}$$

for all $t > 0$ and $\mu \times \mu$-almost all $x, y \in M$ such that $d(x, y) \le \delta t^{1/\beta}$, where $\delta > 0$ is a sufficiently small constant.

Denote by $(UE_{strong})$ a modification of condition $(UE_{loc})$ that is obtained by adding the Hölder continuity of $p_t(x, y)$ and by restricting inequality in $(UE_{loc})$ to all $x, y \in M$. In a similar way, we can define *condition* $(NLE_{strong})$.

**Theorem 4.3** (Grigor'yan, Telcs [GT12, Theorem 7.4]) *Let $(M, d, \mu)$ be a metric measure space and let $(V_\alpha)$ hold. Let $(\mathcal{E}, \mathcal{F})$ be a regular, strongly local Dirichlet form in $L^2(M, \mu)$. Then, the following equivalences are true:*

$$\begin{aligned}(H) + \left(E_\beta\right) &\Leftrightarrow (UE_{loc}) + (NLE) \\ &\Leftrightarrow \left(UE_{strong}\right) + \left(NLE_{strong}\right).\end{aligned}$$

This theorem is proved in [GT12] for a more general setting of volume doubling instead of $(V_\alpha)$.

Observe that the following implications hold [GT12, Lemma 7.3]:

$$\begin{aligned}(H) &\Rightarrow (M, d) \text{ is connected,} \\ \left(E_\beta\right) &\Rightarrow (\mathcal{E}, \mathcal{F}) \text{ is conservative,} \\ \left(E_\beta\right) &\Rightarrow diam\,(M) = \infty.\end{aligned}$$

*Proof* Sketch of proof for Theorem 4.3 First one shows that

$$(V_\alpha) + \left(E_\beta\right) + (H) \Rightarrow (FK),$$

which is quite involved and uses, in particular, Lemma 3.10. Once having $(V_\alpha) + \left(E_\beta\right) + (FK)$, we obtain $(UE_{loc})$ by Theorem 3.8.

Using the elliptic Harnack inequality, one obtains in a standard way the oscillating inequality for harmonic functions and then for functions of the form $u = G_\Omega f$ (that solves the equation $\mathcal{L}_\Omega u = f$) in terms of $\|f\|_\infty$.

If now $u = P_t^\Omega f$ then $u$ satisfies the equation

$$\frac{d}{dt}u = -\mathcal{L}_\Omega u,$$

and whence

$$u = -G_\Omega\left(\frac{d}{dt}u\right).$$

Knowing an upper bound for $u$, which follows from the upper bound of the heat kernel, one obtains also an upper bound for $\frac{d}{dt}u$ in terms of $u$. Applying the oscillation inequality one obtains the Hölder continuity of $u$ and, hence, of the heat kernel.

Let us prove the on-diagonal lower bound

$$p_t\,(x, x) \geq ct^{-\alpha/\beta}.$$

Note that $(UE_{loc})$ and $(V_\alpha)$ imply that

$$\int_{B(x,r)} p_t(x,y)\,d\mu(y) \geq \frac{1}{2}$$

provided $r \geq Kt^{1/\beta}$ (cf. [GHL03, formula (3.8)]). Choosing $r = Kt^{1/\beta}$, we obtain

$$\begin{aligned} p_{2t}(x,x) &= \int_M p_t^2(x,y)\,d\mu(y) \\ &\geq \frac{1}{\mu(B(x,r))}\left(\int_{B(x,r)} p_t(x,y)\,d\mu(y)\right)^2 \\ &\geq \frac{c}{r^\alpha} = \frac{c'}{t^{\alpha/\beta}}. \end{aligned}$$

Then $(NLE)$ follows from the upper estimate for

$$|p_t(x,x) - p_t(x,y)|$$

when $y$ close to $x$, which follows from the oscillation inequality. ■

We next characterize $(UE_{loc}) + (NLE)$ by using the estimates of the capacity and of the Green function.

**Definition 4.4** *(capacity)* Let $\Omega$ be an open set in $M$ and $A \Subset \Omega$ be a Borel set. Define the *capacity* $\mathrm{cap}(A,\Omega)$ by

$$\mathrm{cap}(A,\Omega) := \inf\{\mathcal{E}(\varphi) : \varphi \text{ is a cutoff function of } (A,\Omega)\}. \tag{4.1}$$

It follows from the definition that the capacity $\mathrm{cap}(A,\Omega)$ is increasing in $A$, and decreasing in $\Omega$, namely, if $A_1 \subset A_2, \Omega_1 \supset \Omega_2$, then $\mathrm{cap}(A_1,\Omega_1) \leq \mathrm{cap}(A_2,\Omega_2)$. Using the latter property, let us extend the definition of capacity when $A \subset \Omega$ as follows:

$$\mathrm{cap}(A,\Omega) = \lim_{n\to\infty} \mathrm{cap}(A \cap \Omega_n, \Omega) \tag{4.2}$$

where $\{\Omega_n\}$ is any increasing sequence of precompact open subsets of $\Omega$ exhausting $\Omega$ (in particular, $A \cap \Omega_n \Subset \Omega$).

Note that by the monotonicity property of the capacity, the limit in the right hand side of (4.2) exists (finite or infinite) and is independent of the choice of the exhausting sequence $\{\Omega_n\}$.

Next, define the *resistance* $\mathrm{res}(A,\Omega)$ by

$$\mathrm{res}(A,\Omega) = \frac{1}{\mathrm{cap}(A,\Omega)}. \tag{4.3}$$

We introduce the notions of the Green operator and the Green function.

**Definition 4.5** For an open $\Omega \subset M$, a linear operator $G^{\Omega} : L^2(\Omega) \to \mathcal{F}(\Omega)$ is called a *Green operator* if, for any $\varphi \in \mathcal{F}(\Omega)$ and any $f \in L^2(\Omega)$,

$$\mathcal{E}(G^{\Omega} f, \varphi) = (f, \varphi) . \tag{4.4}$$

If $G^{\Omega}$ admits an integral kernel $g^{\Omega}$, that is,

$$G^{\Omega} f(x) = \int_{\Omega} g^{\Omega}(x, y) f(y) d\mu(y) \text{ for any } f \in L^2(\Omega), \tag{4.5}$$

then $g^{\Omega}$ is called a *Green function.*

It is known (cf. [GH00, Lemma 5.1]) that if $(\mathcal{E}, \mathcal{F})$ is regular and if $\Omega \subset M$ is open such that $\lambda_{\min}(\Omega) > 0$, then the Green operator $G^{\Omega}$ exists, and in fact, $G^{\Omega} = (-\mathcal{L}^{\Omega})^{-1}$, the inverse of $-\mathcal{L}^{\Omega}$, where $\mathcal{L}^{\Omega}$ is the generator of $(\mathcal{E}, \mathcal{F}(\Omega))$. However, the issue of the Green function $g^{\Omega}$ is much more involved, and is one of the key topics in [GH00].

For an open set $\Omega \subset M$, function $E^{\Omega}$ is defined by

$$E^{\Omega}(x) := G^{\Omega}\mathbf{1}(x) \ (x \in M) , \tag{4.6}$$

namely, the function $E^{\Omega}$ is a unique weak solution of the following Poisson-type equation

$$-\mathcal{L}^{\Omega} E^{\Omega} = 1, \tag{4.7}$$

provided that $\lambda_{\min}(\Omega) > 0$.

It is known that

$$E^{\Omega}(x) = \mathbb{E}_x(\tau_{\Omega}) \text{ for } \mu\text{-a.a. } x \in M. \tag{4.8}$$

Clearly, if the Green function $g^{\Omega}$ exists, then

$$E^{\Omega}(x) = G^{\Omega}\mathbf{1}(x) = \int_{\Omega} g^{\Omega}(x, y)\, d\mu(y) \tag{4.9}$$

for $\mu$-almost all $x \in M$.

We introduce the following hypothesis.

$(R_\beta)$ (*Resistance condition* $(R_\beta)$) We say that the *resistance condition* $(R_\beta)$ is satisfied if, there exist constants $K, C > 1$ such that, for any ball $B$ of radius $r > 0$,

$$C^{-1}\frac{r^\beta}{\mu(B)} \le \operatorname{res}(B, KB) \le C\frac{r^\beta}{\mu(B)}, \tag{4.10}$$

where constants $K$ and $C$ are independent of the ball $B$. Equivalently, (4.10) can be written in the form

$$\operatorname{res}(B, KB) \asymp \frac{r^{\beta}}{\mu(B)}.$$

$\left(E'_{\beta}\right)$ *(Condition* $\left(E'_{\beta}\right)$*)* We say that *condition* $\left(E'_{\beta}\right)$ holds if, there exist two constants $C > 1$ and $\delta_1 \in (0,1)$ such that, for any ball $B$ of radius $r > 0$,

$$\begin{aligned} \operatorname*{esup}_{B} E^B &\leq C r^{\beta}, \\ \operatorname*{einf}_{\delta_1 B} E^B &\geq C^{-1} r^{\beta}. \end{aligned}$$

$(G_{\beta})$ *(Condition* $(G_{\beta})$*)* We say that *condition* $(G_{\beta})$ holds if, there exist constants $K > 1$ and $\dot{C} > 0$ such that, for any ball $B := B(x_0, R)$, the Green kernel $g^B$ exists and is jointly continuous off the diagonal, and satisfies

$$g^B(x_0, y) \leq C \int_{K^{-1}d(x_0,y)}^{R} \frac{s^{\beta} ds}{sV(x,s)} \text{ for all } y \in B \setminus \{x_0\},$$

$$g^B(x_0, y) \geq C^{-1} \int_{K^{-1}d(x_0,y)}^{R} \frac{s^{\beta} ds}{sV(x,s)} \text{ for all } y \in K^{-1}B \setminus \{x_0\},$$

where $V(x,r) = \mu(B(x,r))$ as before.

**Theorem 4.6** (Grigor'yan and Hu) [GH00, Theorem 3.14]) *Let* $(M, d, \mu)$ *be a metric measure space and let* $(V_{\alpha})$ *hold. Let* $(\mathcal{E}, \mathcal{F})$ *be a regular, strongly local Dirichlet form in* $L^2(M, \mu)$*. Then, the following equivalences are true:*

$$\begin{aligned} (H) + \left(E'_{\beta}\right) \Leftrightarrow (G_{\beta}) &\Leftrightarrow (H) + (R_{\beta}) \\ &\Leftrightarrow (UE_{loc}) + (NLE) \\ &\Leftrightarrow \left(UE_{strong}\right) + \left(NLE_{strong}\right). \end{aligned}$$

We mention that condition $(V_{\alpha})$ can be replaced by conditions (*VD*) and (*RVD*), the latter refers to the reverse doubling condition (cf. [GH00]).

**Sketch of proof for Theorem 4.6** The proofs of Theorem 4.6 consists of two parts.

- *Part One*. Firstly, the following implications hold:

$$\boxed{(UE_{strong}) + (NLE_{strong})}$$
$$\Updownarrow$$
$$\boxed{(UE_{loc}) + (NLE)}$$
$$\Updownarrow \qquad \Uparrow$$
$$\boxed{(H) + (E_\beta)} \Rightarrow \boxed{(H) + \left(E'_\beta\right)}$$

In fact, by Theorem 4.3, we only need to show that

$$\left(E_\beta\right) \Rightarrow \left(E'_\beta\right), \tag{4.11}$$

$$(H) + \left(E'_\beta\right) \Longrightarrow (UE_{loc}) + (NLE). \tag{4.12}$$

The implication (4.11) can be proved directly by using the probability argument, see [GH00, Theorem 3.14]. And the implication (4.12) can be done by showing the following

$$
\begin{aligned}
(H) + \left(E'_\beta\right) &\Rightarrow (FK) \ \ \text{([GT12, formula (3.17) and T.3.11])}\\
\left(E'_\beta\right) &\Rightarrow \left(S_\beta\right) \ \ \text{(by [GHL00, formula (6.34)])}\\
(FK) + \left(S_\beta\right) &\Rightarrow (UE_{loc}) \ \ \text{(by Theorem 3.8)}\\
(H) + \left(E'_\beta\right) &\Rightarrow (NLE) \ \ \text{(by [GT12, Section 5.4]).}
\end{aligned}
$$

- *Part Two*. Secondly, we need to show that

$$(H) + \left(E'_\beta\right) \Leftrightarrow \left(G_\beta\right) \Leftrightarrow (H) + \left(R_\beta\right).$$

This is the hard part. The cycle implications are obtained in [GH00, Section 8 ] as follows:

$$(H) + \left(R_\beta\right) \Longrightarrow \left(G_\beta\right) \Longrightarrow (H) + \left(E'_\beta\right) \Longrightarrow (H) + \left(R_\beta\right).$$

One of the most challenging results (cf. [GH00, Lemma 5.7]) is to obtain an *annulus Harnack inequality* for the Green function, without assuming any specific properties of the metric $d$, unlike previously known similar results in [Bar05], [GT02] where the geodesic property of the distance function was used. ■

## 4.2 Matching Upper and Lower Bounds

The purpose of this subsection is to improve both $(UE_{loc})$ and $(NLE)$ in order to obtain matching upper and lower bounds for the heat kernel. The reason why $(UE_{loc})$ and $(NLE)$ do not match, in particular, why $(NLE)$ contains no information about lower bound of $p_t(x, y)$ for distant $x, y$ is the lack of *chaining properties* of the distance function, that is an ability to connect any two points $x, y \in M$ by a chain of balls of controllable radii so that the number of balls in this chain is also under control.

For example, the chain condition considered above is one of such properties. If $(M, d)$ satisfies the chain condition, then as we have already mentioned, $(NLE)$ implies the full sun-Gaussian lower estimate by the chain argument and the semigroup property (see for example [GHL03, Corollary 3.5]).

Here we consider a setting with weaker chaining properties. For any $\varepsilon > 0$, we introduce a modified distance $d_\varepsilon(x, y)$ by

$$d_\varepsilon(x, y) = \inf_{\{x_i\} \text{ is } \varepsilon\text{-chain}} \sum_{i=1}^{N} d(x_i, x_{i-1}), \tag{4.13}$$

where an *ε-chain* is a sequence $\{x_i\}_{i=0}^{N}$ of points in $M$ such that

$$x_0 = x, \quad x_N = y, \quad \text{and } d(x_i, x_{i-1}) < \varepsilon \text{ for all } i = 1, 2, ..., N.$$

Clearly, $d_\varepsilon(x, y)$ is decreases as $\varepsilon$ increases and $d_\varepsilon(x, y) = d(x, y)$ if $\varepsilon > d(x, y)$. As $\varepsilon \downarrow 0$, $d_\varepsilon(x, y)$ increases and can go to $\infty$ or even become equal to $\infty$. It is easy to see that $d_\varepsilon(x, y)$ satisfies all properties of a distance function except for finiteness, so that it is a distance function with possible value $+\infty$.

It is easy to show that

$$d_\varepsilon(x, y) \asymp \varepsilon N_\varepsilon(x, y),$$

where $N_\varepsilon(x, y)$ is the smallest number of balls in a chain of balls of radius $\varepsilon$ connecting $x$ and $y$ (Fig. 7):

$N_\varepsilon$ can be regarded as the graph distance on a graph approximation of $M$ by an $\varepsilon$-net.

If $d$ is geodesic, then the points $\{x_i\}$ of an $\varepsilon$-chain can be chosen on the shortest geodesic, whence $d_\varepsilon(x, y) = d(x, y)$ for any $\varepsilon > 0$. If the distance function $d$ satisfies the chain condition, then one can choose in (4.13) an $\varepsilon$-chain so that $d(x_i, x_{i+1}) \le C\frac{d(x,y)}{N}$, whence $d_\varepsilon(x, y) \le Cd(x, y)$. In general, $d_\varepsilon(x, y)$ may go to $\infty$ as $\varepsilon \to 0$, and the rate of growth of $d_\varepsilon(x, y)$ as $\varepsilon \to 0$ can be regarded as a quantitative description of the chaining properties of $d$.

We need the following hypothesis

$C_\beta$ (*Chaining property*) For all $x, y \in M$,

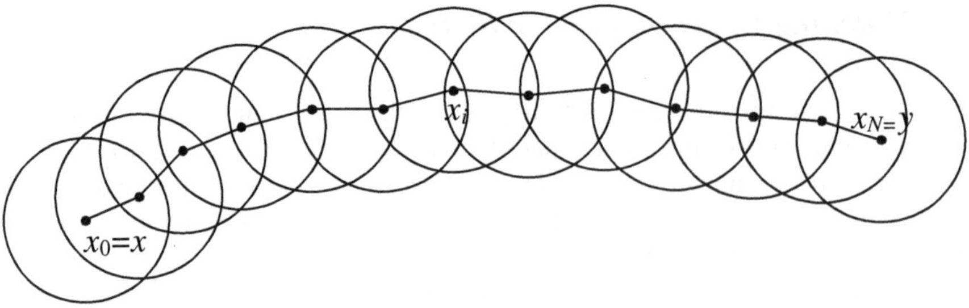


**Fig. 7** Chain of balls connecting $x$ and $y$

$$\varepsilon^{\beta-1} d_\varepsilon(x,y) \to 0 \text{ as } \varepsilon \to 0,$$

or equivalently,

$$\varepsilon^{\beta} N_\varepsilon(x,y) \to 0 \text{ as } \varepsilon \to 0.$$

For $x \neq y$ we have $\varepsilon^{\beta-1} d_\varepsilon(x,y) \to \infty$ as $\varepsilon \to \infty$, which implies under $(C_\beta)$ that for any $t > 0$, there is $\varepsilon = \varepsilon(t,x,y)$ satisfying the identity

$$\varepsilon^{\beta-1} d_\varepsilon(x,y) = t \tag{4.14}$$

(always take the maximal possible value of $\varepsilon$). If $x = y$, then set $\varepsilon(t,x,x) = \infty$.

**Theorem 4.7** (Grigor'yan, Telcs [GT12, Section 6]) *Assume that all the hypothesis of Theorem 4.6 hold. If $(E_\beta) + (H)$ and $(C_\beta)$ are satisfied, then*

$$p_t(x,y) \asymp \frac{C}{t^{\alpha/\beta}} \exp\left(-c\left(\frac{d_\varepsilon^\beta(x,y)}{t}\right)^{\frac{1}{\beta-1}}\right) \tag{4.15}$$

$$\asymp \frac{C}{t^{\alpha/\beta}} \exp\left(-cN_\varepsilon(x,y)\right), \tag{4.16}$$

*where $\varepsilon = \varepsilon(t,x,y)$.*

Since $d_\varepsilon(x,y) \geq d(x,y)$, the upper bound in (4.15) is an improvement of $(UE_{loc})$; similarly the lower bound in (4.15) is an improvement of $(NLE)$. The proof of the upper bound in (4.15) follows the same line as the proof of $(UE_{loc})$ with careful tracing all places where the distance $d(x,y)$ is used and making sure that it can be replaced by $d_\varepsilon(x,y)$. The proof of the lower bound in (4.16) uses $(NLE)$ and the semigroup identity along the chain with $N_\varepsilon$ balls connecting $x$ and $y$ . Finally, observe that (4.15) and (4.16) are equivalent, that is

$$N_\varepsilon \asymp \left(\frac{d_\varepsilon^\beta(x,y)}{t}\right)^{\frac{1}{\beta-1}},$$

which follows by substituting here $N_\varepsilon \asymp d_\varepsilon/\varepsilon$ and $t = \varepsilon^{\beta-1} d_\varepsilon(x,y)$.

By Theorem 4.6, the same conclusion in Theorem 4.7 is true if $(E_\beta) + (H)$ is instead replaced by the one of conditions $(H) + \left(E'_\beta\right)$, $(G_\beta)$ and $(H) + (R_\beta)$.

*Example 4.8* A good example to illustrate Theorem 4.7 is the class of post critically finite (p.c.f.) fractals. For connected p.c.f. fractals with regular harmonic structure, the heat kernel estimate (4.16) was proved by Hambly and Kumagai [HK99], see also [KS05, Theorem 5.2]. In this setting $d(x,y)$ is the resistance metric of the fractal $M$ and $\mu$ is the Hausdorff measure of $M$ of dimension $\alpha := \dim_H M$. Hambly and Kumagai proved that $(V_\alpha)$ and $(E_\beta)$ are satisfied with $\beta = \alpha + 1$. The condition $(C_\beta)$ follows from their estimate

$$N_\varepsilon(x,y) \le C\left(\frac{d(x,y)}{\varepsilon}\right)^{\beta/2},$$

because

$$\varepsilon^\beta N_\varepsilon(x,y) \le C d(x,y)^{\beta/2} \varepsilon^{\beta/2} \to 0 \text{ as } \varepsilon \to 0.$$

The Harnack inequality $(H)$ on p.c.f. fractals was proved by Kigami [Kig01, Proposition 3.2.7, p.78]. Hence, Theorem 4.7 applies and gives the estimates (4.15) and (4.16).

The estimate (4.16) means that the diffusion process goes from $x$ to $y$ in time $t$ in the following way. The process firstly "computes" the value $\varepsilon(t,x,y)$, secondly "detects" a shortest chain of $\varepsilon$-balls connecting $x$ and $y$, and then goes along that chain (Fig. 8).

This phenomenon was first observed by Hambly and Kumagai on p.c.f. fractals, but it seems to be generic. Hence, to obtain matching upper and lower bounds, one needs in addition to the usual hypotheses also the following information, encoded in the function $N_\varepsilon(x,y)$: the graph distance between $x$ and $y$ on any $\varepsilon$-net approximation of $M$.

*Example 4.9* (*Computation of* $\varepsilon$) Assume that the following bound is known for all $x, y \in M$ and $\varepsilon > 0$

$$N_\varepsilon(x,y) \le C\left(\frac{d(x,y)}{\varepsilon}\right)^{\gamma},$$

where $0 < \gamma < \beta$, so that $(C_\beta)$ is satisfied (since $N_\varepsilon \ge d(x,y)/\varepsilon$, one must have $\gamma \ge 1$). Since by (4.14) we have $\varepsilon^\beta N_\varepsilon \asymp t$, it follows that

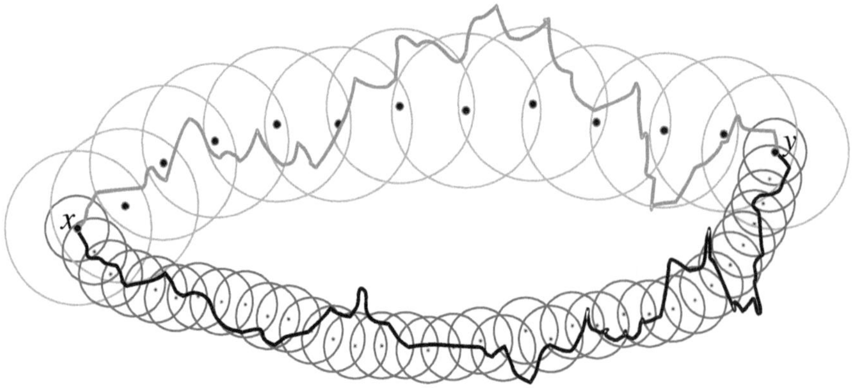


**Fig. 8** Two shortest chains of $\varepsilon$-ball for two distinct values of $\varepsilon$ provide different routes for the diffusion from $x$ to $y$ for two distinct values of $t$

$$\varepsilon^{\beta}\left(\frac{d(x,y)}{\varepsilon}\right)^{\gamma}\geq ct,$$

whence

$$\varepsilon\geq c\left(\frac{t}{d(x,y)^{\gamma}}\right)^{\frac{1}{\beta-\gamma}}.$$

Consequently, we obtain

$$N_{\varepsilon}(x,y)\leq Cd(x,y)^{\gamma}\varepsilon^{-\gamma}\leq Cd(x,y)^{\gamma}\left(\frac{d(x,y)^{\gamma}}{t}\right)^{\frac{\gamma}{\beta-\gamma}}=C\left(\frac{d(x,y)^{\beta}}{t}\right)^{\frac{\gamma}{\beta-\gamma}},$$

and so

$$p_t(x,y)\geq\frac{c}{t^{\alpha/\beta}}\exp\left(-\left(\frac{d(x,y)^{\beta}}{ct}\right)^{\frac{\gamma}{\beta-\gamma}}\right).$$

Similarly, the lower estimate of $N_{\varepsilon}$

$$N_{\varepsilon}(x,y)\geq c\left(\frac{d(x,y)}{\varepsilon}\right)^{\gamma}$$

implies an upper bound for the heat kernel

$$p_t(x,y)\leq\frac{C}{t^{\alpha/\beta}}\exp\left(-\left(\frac{d(x,y)^{\beta}}{Ct}\right)^{\frac{\gamma}{\beta-\gamma}}\right).$$

*Remark 4.10* Assume that $(V_\alpha)$ holds and all balls in $M$ of radius $\geq r_0$ are connected, for some $r_0 > 0$. We claim that $(C_\beta)$ holds with any $\beta > \alpha$. The $\alpha$-regularity of measure $\mu$ implies, by the classical ball covering argument, that any ball $B_r$ of radius $r$ can be covered by at most $C\left(\frac{r}{\varepsilon}\right)^\alpha$ balls of radii $\varepsilon \in (0, r)$. Consequently, if $B_r$ is connected then any two points $x, y \in B_r$ can be connected by a chain of $\varepsilon$-balls containing at most $C\left(\frac{r}{\varepsilon}\right)^\alpha$ balls, so that

$$N_\varepsilon(x, y) \leq C\left(\frac{r}{\varepsilon}\right)^\alpha.$$

Since any two points $x, y \in M$ are contained in a connected ball $B_r$ (say, with $r = r_0 + d(x, y)$), we obtain

$$\varepsilon^\beta N_\varepsilon(x, y) \leq C\varepsilon^{\beta-\alpha} r^\alpha \to 0$$

as $\varepsilon \to 0$, which was claimed.

## 4.3 Further Results

We discuss here some consequences and extensions of the above results. For this, we introduce two-sided estimates of the heat kernel.

$(ULE_{loc})$ (*Upper and lower estimates*) The heat kernel $p_t(x, y)$ exists and satisfies

$$p_t(x, y) \asymp \frac{C}{t^{\alpha/\beta}} \exp\left(-c\left(\frac{d^\beta(x, y)}{t}\right)^{\frac{1}{\beta-1}}\right). \tag{4.17}$$

**Theorem 4.11** *Let $(M, d, \mu)$ be a metric measure space, and let $(\mathcal{E}, \mathcal{F})$ be a regular, conservative Dirichlet form in $L^2(M, \mu)$. If $(M, d)$ satisfies the chain condition, then the following equivalences take place:*

$$(V_\alpha) + \begin{cases} (E_\beta) + (H) \\ (E'_\beta) + (H) \\ (R_\beta) + (H) \\ (G_\beta) \end{cases} + (\textit{locality}) \Longleftrightarrow (ULE_{loc}),$$

*where condition (locality) means that $(\mathcal{E}, \mathcal{F})$ is local.*

*Remark 4.12* Observe that if $(\mathcal{E}, \mathcal{F})$ is regular, conservative and local, then $(\mathcal{E}, \mathcal{F})$ is strongly local; this is easily seen by using the Beuling-Deny decomposition [FOT11, Theorem 3.2.1, p. 120] and by noting that both killing and jump measures disappear.

*Remark 4.13* Observe also that $(V_\alpha) + (NLE)+$ (chain condition) implies that the off-diagonal lower estimate

$$p_t(x, y) \geq \frac{C'}{t^{\alpha/\beta}} \exp\left(-c'\left(\frac{d^\beta(x, y)}{t}\right)^{\frac{1}{\beta-1}}\right) \tag{4.18}$$

for $\mu$-almost all $x, y \in M$ and all $t > 0$, see for example [GHL08, Proposition 3.1] or [Bar98], [GHL03, Corollary 3.5].

**Sketch of proof for Theorem 4.11** (1) "$\Rightarrow$".
Let us show the implication

$$(V_\alpha) + (E_\beta) + (H) + (\text{ locality}) \Rightarrow (ULE_{loc}). \tag{4.19}$$

Indeed, by Remark 4.12, we have that $(\mathcal{E}, \mathcal{F})$ is strongly local. Now, using Theorem 4.3, we obtain $(UE_{loc}) + (NLE)$. Using Remark 4.13, we see that (4.18) holds, showing that $(ULE_{loc})$ is true.

Similarly, using Theorem 4.6, we obtain the other three implications "$\Rightarrow$".
(2) "$\Leftarrow$".
Let us show the opposite implication

$$(ULE_{loc}) \Rightarrow (V_\alpha) + (E_\beta) + (H) + (\text{locality}). \tag{4.20}$$

Indeed, note that

$$\begin{aligned}(ULE_{loc}) &\Rightarrow (V_\alpha) \ \text{(by Theorem 2.2)}\\ (UE_{loc}) &\Rightarrow \text{(locality) (by Theorem 2.14)}\\ (UE_{loc}) + (NLE) &\Rightarrow (E_\beta) + (H) \ \text{(by Theorem 4.3)}\end{aligned}$$

showing that the implication (4.20) holds.
Similarly, all the other three implications "$\Longleftarrow$" also hold. ■

*Remark 4.14* The implication (4.19) can also be proved by using Theorem 4.7 and the fact that $d_\mathcal{E} \asymp d$.

**Conjecture 4.15** *The condition $(E_\beta)$ above may be replaced by*

$$\lambda_{\min}(B(x, r)) \asymp r^{-\beta}. \tag{$\lambda_\beta$}$$

In fact, $(E_\beta)$ in all statements can be replaced by the resistance condition:

$$\text{res}(B_r, B_{2r}) \asymp r^{\beta-\alpha} \tag{$R_\beta$}$$

where $B_r = B(x, r)$. In the strongly recurrent case $\alpha < \beta$, it alone implies the elliptic Harnack inequality $(H)$ so that two sided heat kernel estimates are equivalent to $(V_\alpha) + (R_\beta)$ as was proved by Barlow, Coulhon, Kumagai [BCK05] (in a setting of graphs) and was discussed in M. Barlow's lectures.

An interesting (and obviously hard) question is the characterization of the elliptic Harnack inequality $(H)$ in more geometric terms—so far nothing is known, not even a conjecture.

One can consider also a *parabolic* Harnack inequality (*PHI*), which uses caloric functions instead of harmonic functions. Then in a general setting and assuming the volume doubling condition (*VD*) (instead of $(V_\alpha)$), the following holds (cf. [BGK12]):

$$(PHI) \Leftrightarrow (UE_{loc}) + (NLE)\,.$$

On the other hand, (*PHI*) is equivalent to

Poincaré inequality + cutoff Sobolev inequality,

see [BBK06].

**Conjecture 4.16** *The cutoff Sobolev inequality here can be replaced by* $(\lambda_\beta)$ *and/or* $(R_\beta)$.

# 5 Upper Bounds for Jump Processes

We have investigated above the heat kernel for the *local* Dirichlet form. In this section we shall study the *non-local* Dirichlet form and present the equivalence conditions for upper bounds of the associated heat kernel. As an interesting example, we discuss the heat kernel estimates for effective metric spaces.

A non-local Dirichlet form will give arise to a jump process, that is, the trajectories of this process are discontinuous, as we have already seen for a symmetric stable process of index $\beta$ (Lévy process). And the heat kernel decays at a polynomial rate (cf. 1.2), instead of an exponential rate as for a local Dirichlet form.

Jump process have found various applications in science. For instance, a Lévy flight is a jump process and can be used to describe animal foraging patterns, the distribution of human travel and some aspects of earthquake behavior (cf. [BBW08]).

## *5.1 Upper Bounds for Non-local Dirichlet Forms*

The techniques for obtaining heat kernel bounds for non-local Dirichlet forms has been developed by a number of authors, see for example [BBCK09, BGK09, BL02, CK03, CK08] and the references therein. The basic approach to obtaining heat kernel upper estimates used in these papers consists of the two steps. The first step is to obtain the heat kernel upper bounds for a *truncated* Dirichlet form, that is, in the case when the jump density $J(x, y)$ has a bounded range. In this case one uses the Davies method as it was presented in the seminal work [CKS87] and where the cut-off functions of form $(\lambda - d(x_0, x))_+$ were used (where $\lambda$ is a positive constant). This method can be used as long as the cut-off functions belong to the domain of the Dirichlet form, which is the case only when $\beta < 2$ (hence, if $\beta \geq 2$ then this method does not work).

The second step is to obtain heat kernel estimates for the original Dirichlet form by comparing the heat semigroup of the truncated Dirichlet form with the original heat semigroup. We remark that while the first step was done by purely analytic means, the second step in the above-mentioned papers used a probabilistic argument.

Here we describe an alternative new approach of [GHL00] for obtaining upper bounds.

Recall that by a theorem of Beurling and Deny, any regular conservative Dirichlet form admits a decomposition

$$\mathcal{E}(u, v) = \mathcal{E}^{(L)}(u, v) + \mathcal{E}^{(J)}(u, v), \tag{5.1}$$

where $\mathcal{E}^{(L)}$ is a *local part* and

$$\mathcal{E}^{(J)}(u, v) = \iint_{M \times M \setminus \mathrm{diag}} (u(x) - u(y))(v(x) - v(y))\, dj(x, y) \tag{5.2}$$

is a *jump part* with a jump measure $j$ defined on $M \times M \setminus$ diag. In our setting the jump measure $j$ will have a density with respect to $\mu \times \mu$, which will be denoted by $J(x, y)$, and so the jump part $\mathcal{E}^{(J)}$ becomes

$$\mathcal{E}^{(J)}(u, v) = \iint_{M \times M} (u(x) - u(y))(v(x) - v(y))\, J(x, y) d\mu(y) d\mu(x). \tag{5.3}$$

We introduce the following hypothesis.

$(\mathbf{V}_{\leq})$ (*Upper $\alpha$- regularity*) For all $x \in M$ and all $r > 0$,

$$V(x, r) \leq C r^{\alpha}.$$

**(UE)** (*Upper estimate of non-local type*) The heat kernel $p_t$ exists and satisfies the off-diagonal upper estimate

$$p_t(x, y) \leq \frac{C}{t^{\alpha/\beta}} \left(1 + \frac{d(x, y)}{t^{1/\beta}}\right)^{-(\alpha+\beta)}$$

for all $t > 0$ and $\mu$-almost all $x, y \in M$.

$(\mathbf{J}_{\leq})$ (*Upper bound of jump density*) The jump density exists and admits the estimate

$$J(x, y) \leq Cd(x, y)^{-(\alpha+\beta)},$$

for $\mu$-almost all $x, y \in M$.
($\mathbf{T}_{strong}$) (*Strong tail estimate*) There exist constants $c > 0$ and $\beta > 0$ such that, for all balls $B = B(x_0, r)$ and for all $t > 0$,

$$P_t \mathbf{1}_{B^c}(x) \leq \frac{ct}{r^\beta} \text{ for } \mu \text{ -almost all } x \in \frac{1}{4}B.$$

Clearly, we have that $\left(T_{\exp}\right) \Rightarrow \left(T_{strong}\right) \Rightarrow \left(T_\beta\right)$.

We now state the main technical result of [GHL00].

**Theorem 5.1** (Grigor'yan et al. [GHL00]) *Let $(M, d, \mu)$ be a metric measure space with precompact balls, and let $(\mathcal{E}, \mathcal{F})$ be a regular conservative Dirichlet form in $L^2(M, \mu)$ with jump density $J$. Then the following implication holds:*

$$\left(V_{\leq}\right) + (DUE) + (J_{\leq}) + \left(S_\beta\right) \Rightarrow (UE). \tag{5.4}$$

We remark that by [GHL03, Theorem 3.2], if $(\mathcal{E}, \mathcal{F})$ is conservative then

$$\left(V_{\leq}\right) + (UE) \Rightarrow (V_\alpha).$$

Hence, the hypotheses of Theorem 5.1 imply that $\mu$ is $\alpha$-regular.

**Sketch of proof for Theorem 5.1** We sketch the ideas of the proof.

- *Step* 1. We decompose $\mathcal{E}(u)$ into two parts:

$$\mathcal{E}(u) = \mathcal{E}^{(\rho)}(u) + \int_M \int_{M \setminus B(x,\rho)} (u(x) - u(y))^2 J(x, y) d\mu(y) d\mu(x),$$

where $\rho \in (0, \infty)$ is any fixed number. Then the form $\left(\mathcal{E}^{(\rho)}, \mathcal{F}\right)$ can be extended to a regular Dirichlet form $\left(\mathcal{E}^{(\rho)}, \mathcal{F}^{(\rho)}\right)$. Indeed, since using condition $(J_{\leq})$,

$$\operatorname*{esup}_{x \in M} \int_{B(x,\rho)^c} J(x, y) d\mu(y) < \infty,$$

the form $\left(\mathcal{E}^{(\rho)}, \mathcal{F}\right)$ is closable, and its closure $\left(\mathcal{E}^{(\rho)}, \mathcal{F}^{(\rho)}\right)$ in $L^2$ is a regular Dirichlet form in $L^2$. Note that $\left(\mathcal{E}^{(\rho)}, \mathcal{F}^{(\rho)}\right)$ is $\rho$-*local* (non-local): $\mathcal{E}^{(\rho)}(f, g) = 0$ for any two functions $f, g \in \mathcal{F}^{(\rho)}$ with compact supports such that

$$\text{dist}\,(\text{supp}\, f, \text{supp}\, g) > \rho.$$

- *Step* 2. We need to obtain upper estimates of the heat kernel $q_t(x, y)$ of the *truncated* Dirichlet form $\left(\mathcal{E}^{(\rho)}, \mathcal{F}^{(\rho)}\right)$. Indeed, conditions $(DUE)$, $\left(J_{\leq}\right)$, $(S_\beta)$ and $\left(V_{\leq}\right)$

imply the following estimate of $q_t(x, y)$ :

$$q_t(x, y) \leq \frac{C}{t^{\alpha/\beta}} \exp\left(4\rho^{-\beta}t\right) \exp\left(-c\left(\frac{d(x, y)}{\rho} \wedge \frac{\rho}{t^{1/\beta}}\right)\right) \tag{5.5}$$

for all $t > 0$ and $\mu$-almost all $x, y \in M$, where constants $C, c > 0$ depend on the constants in the hypotheses but are independent of $\rho$. This can be done with a certain amount of effort, by using the bootstrapping technique where the comparison inequality [GHL10, Corollary 4.8, Remark 4.10] for heat semigroups play an important rôle.

- *Step* 3. Next we apply the following useful inequality between two heat kernels:

$$p_t(x, y) \leq q_t(x, y) + 2t \operatorname*{esup}_{x\in M, y\in B(x,\rho)^c} J(x, y) \tag{5.6}$$

for all $t > 0$ and almost all $x, y \in M$; this inequality follows from the parabolic maximum principle alone. Therefore, by choosing an appropriate $\rho$, it follows from (5.5), (5.6) that, for any real $n \geq 0$,

$$p_t(x, y) \leq \frac{c(n)}{t^{\alpha/\beta}} \left(1 + \frac{d(x, y)}{t^{1/\beta}}\right)^{-\frac{(\alpha+\beta)n}{n+\alpha+\beta}} \tag{5.7}$$

for almost all $x, y \in M$ and all $t > 0$.
Note that (5.7) is nearly close to our desired estimate $(UE)$. However, one can not just obtain $(UE)$ by directly taking the limit as $n \to \infty$, since we do not know whether the coefficient $c(n)$ is bounded uniformly in $n$. We need the second iteration.

- *Step* 4. Finally, we will obtain $(UE)$ by a self-improvement of (5.7). Indeed, one can use (5.7) to obtain

$$\int_{B(x,r)^c} p_t(x, y)d\mu(y) \leq C(n)\left(rt^{-1/\beta}\right)^{-\theta},$$

where $\theta = \frac{n\beta-\alpha(\alpha+\beta)}{n+\alpha+\beta} \in (0, \beta)$ (note that this estimate is sharper than condition $(S_\beta)$), and then repeating the above procedure, we arrive at $(UE)$. ■

Now we can state some equivalences for $(UE)$.

**Theorem 5.2** (Grigor'yan et al. [GHL00]) *Let $(M, d, \mu)$ be a metric measure space with precompact balls, and let $(\mathcal{E}, \mathcal{F})$ be a regular conservative Dirichlet form in $L^2(M, \mu)$ with jump density $J$. If $(V_{\leq})$ holds, then the following equivalences are true:*

$$\begin{aligned}(UE) &\Leftrightarrow (UE\Phi) + (J_{\leq}) \\ &\Leftrightarrow (DUE) + (J_{\leq}) + (T_\beta) \\ &\Leftrightarrow (DUE) + (J_{\leq}) + (S_\beta) \\ &\Leftrightarrow (DUE) + (J_{\leq}) + (T_{strong}) .\end{aligned} \tag{5.8}$$

*Proof* Observe that the implication $(UE) \Rightarrow (J_{\leq})$ holds by [BGK09, p. 150], and $(UE) \Rightarrow (UE\Phi)$ is trivial by taking $\Phi(s) = (1+s)^{-(\alpha+\beta)}$. The implication $(UE\Phi) \Rightarrow (DUE)$ is obvious. The implication $(UE\Phi) \Rightarrow (T_\beta)$ was proved in (3.10) (see also [GHL03, formula (3.6), p. 2072]). Since $(\mathcal{E}, \mathcal{F})$ is conservative, the equivalence $(T_\beta) \Leftrightarrow (S_\beta)$ holds by [GH08, Theorem 3.1, p. 96]. By Theorem 5.1 we have

$$(DUE) + (J_{\leq}) + (S_\beta) \Rightarrow (UE),$$

which closes the cycle of implications, thus proving the first three equivalences.

Finally, the implication $(UE) \Rightarrow (T_{strong})$ is true by using (3.10), and hence

$$\begin{aligned}(UE) &\Rightarrow (DUE) + (J_{\leq}) + (T_{strong}) \\ &\Rightarrow (DUE) + (J_{\leq}) + (T_\beta) \Rightarrow (UE),\end{aligned}$$

which finishes the proof. ■

*Remark 5.3* The upper estimate $(UE)$ is best possible for non-local forms in the following sense: if the heat kernel $p_t$ satisfies the estimate

$$p_t(x, y) \leq \frac{1}{t^{\alpha/\beta}} \Phi\left(\frac{d(x, y)}{t^{1/\beta}}\right)$$

for all $t > 0$ and $\mu$-almost all $x, y \in M$, where $\Phi$ is a continuous decreasing function on $[0, +\infty)$, then necessarily

$$\Phi(s) \geq c(1+s)^{-(\alpha+\beta)}$$

for some $c > 0$ (see Theorem 2.14).

*Remark 5.4* Under the standing assumptions of Theorem 5.2, the following equivalence is true

$$(UE_{loc}) \Leftrightarrow (DUE) + (\text{"locality"}) + (S_\beta).$$

Indeed, since $(UE_{loc})$ is stronger than $(UE)$, it implies $(DUE)$ and $(S_\beta)$ by Theorem 5.2. Next, $(UE_{loc}) \Rightarrow$("locality") by Theorem 2.14 above. The opposite implication

$$(DUE) + (\text{"locality"}) + (S_\beta) \Rightarrow (UE_{loc})$$

was stated in Theorem 3.8.

In order to state some consequence of Theorem 5.2, we need the following Proposition.

Define first the following condition:

($\mathbf{J}_{\geq}$) (*Lower bound of jump density*) There exist constants $C, \alpha, \beta > 0$ such that, for $\mu$ -almost all $x \neq y$,

$$J(x, y) \geq C^{-1} d(x, y)^{-(\alpha+\beta)}.$$

**Proposition 5.5** *Let $(M, d, \mu)$ be a metric measure space, and let $(\mathcal{E}, \mathcal{F})$ be a regular Dirichlet form in $L^2(M, \mu)$ with jump density $J$. Then*

$$(V_\alpha) + (J_\geq) \Rightarrow (DUE). \tag{5.9}$$

*Proof* As was proved in [HuK06, Theorem 3.1], under $(V_\alpha)$ the following inequality holds for all non-zero functions $u \in L^1 \cap L^2$:

$$\int_M \int_M \frac{(u(x) - u(y))^2}{d(x, y)^{\alpha+\beta}} d\mu(x) \, d\mu(y) \geq c \|u\|_2^{2(1+\beta/\alpha)} \|u\|_1^{-2\beta/\alpha},$$

where $c$ is a positive constant. Using (5.1), (5.3) and $(J_\geq)$ we obtain

$$\begin{aligned} \mathcal{E}(u) &= \mathcal{E}^{(L)}(u) + \mathcal{E}^{(J)}(u) \\ &\geq C \int_M \int_M \frac{(u(x) - u(y))^2}{d(x, y)^{\alpha+\beta}} d\mu(x) \, d\mu(y) \\ &\geq c \|u\|_2^{2(1+\beta/\alpha)} \|u\|_1^{-2\beta/\alpha} \end{aligned}$$

for all $u \in \mathcal{F} \cap L^1$. Hence, $(DUE)$ follows by Theorem 3.6. ■

We obtain the following consequence of Theorem 5.2.

**Theorem 5.6** (Grigor'yan et al. [GHL00]) *Let $(M, d, \mu)$ be a metric measure space with precompact balls, and let $(\mathcal{E}, \mathcal{F})$ be a regular conservative Dirichlet form in $L^2(M, \mu)$ with jump density $J$. If $(V_\alpha)$ holds and $J(x, y) \asymp d(x, y)^{-(\alpha+\beta)}$, then*

$$(UE) \Leftrightarrow (S_\beta). \tag{5.10}$$

*Proof* Let us show that $(S_\beta) \Rightarrow (UE)$. Indeed, $(DUE)$ holds by Proposition 5.5. Hence, $(UE)$ is satisfied by Theorem 5.2. The opposite implication $(UE) \Rightarrow (S_\beta)$ holds also by Theorem 5.2. ■

Therefore, if $(V_\alpha)$ holds and $J(x, y) \asymp d(x, y)^{-(\alpha+\beta)}$, then in order to obtain off-diagonal upper bounds of heat kernels, one needs only to verify the survival condition $(S_\beta)$. In the sequel, we will show that the survival condition $(S_\beta)$ holds for a class of measure spaces with effective resistance metrics.

## 5.2 Upper Bounds Using Effective Resistance

We will show how Theorem 5.2 can be applied for a certain class of metric measure spaces with effective resistance.

Let $(\mathcal{E}, \mathcal{F})$ be a regular Dirichlet form in $L^2(M, \mu)$ as before. Recall that the *effective resistance* $R(A, B)$ between two disjoint non-empty closed subsets $A$ and $B$ of $M$ is defined by

$$R(A, B)^{-1} = \inf \{\mathcal{E}(u) : u \in \mathcal{F} \cap C_0, \ u|_A = 1 \text{ and } u|_B = 0\}. \tag{5.1}$$

It follows from (5.1) that, for any fixed $A$, $R(A, B)$ is a non-increasing function of $B$. Denote by

$$R(x, B) := R(\{x\}, B) \text{ and } R(x, y) := R(\{x\}, \{y\}).$$

In general, it may happen that $R(x, y) = \infty$ for some points $x, y \in M$. Below we will exclude this case.

Fix a parameter $\gamma > 0$, and introduce conditions $(R_1)$ and $(R_2)$.

$(\mathbf{R}_1)$ : For all $u \in \mathcal{F} \cap C_0(M)$ and all $x, y \in M$, the following inequality holds:

$$|u(x) - u(y)|^2 \leq C d(x, y)^{\gamma} \mathcal{E}(u).$$

$(\mathbf{R}_2)$ : For all $x \in M$ and $r > 0$,

$$R(x, B(x, r)^c) \geq C^{-1} r^{\gamma}.$$

**Theorem 5.7** (Grigor'yan et al. [GHL00]) *Let $(\mathcal{E}, \mathcal{F})$ be a regular Dirichlet form in $L^2(M, \mu)$. Then*

$$(V_\alpha) + (R_1) + (R_2) \Rightarrow \left(S_\beta\right) + (DUE),$$

*where $\beta = \alpha + \gamma$. Consequently, under the standing conditions $(V_\alpha) + (R_1) + (R_2)$, we have that*

$$(UE) \Leftrightarrow (J_{\leq}). \tag{5.2}$$

**Sketch of proof for Theorem 5.7** The proof consists of the following five steps.

- *Step* 1. For any ball $B := B(x_0, r)$, using conditions $(R_1)$ and $(R_2)$, we can obtain the two-sided estimate of the Green functions $g_B(x, y)$ :

$$\sup_{x,y \in B} g_B(x, y) \leq C r^{\gamma}, \tag{5.3}$$

$$\inf_{y \in B(x_0, \eta r)} g_B(x_0, y) \geq C^{-1} r^{\gamma}, \tag{5.4}$$

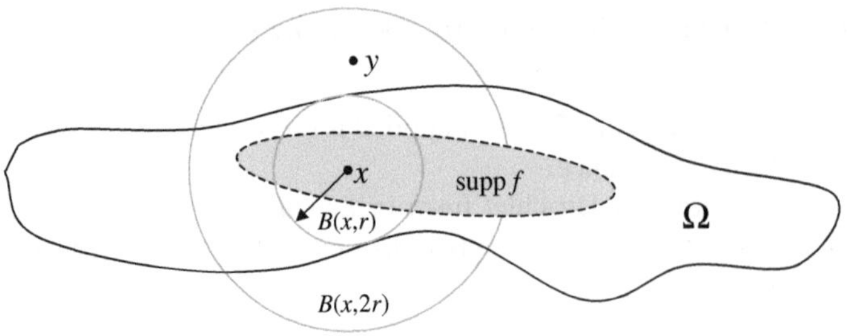


**Fig. 9** Points $x, y$

where $C > 0$ and $\eta \in (0, 1)$.

- *Step* 2. Therefore, under condition $(V_\alpha)$, it follows from (5.3), (5.4) that condition $\left(E'_\beta\right)$ holds:

$$\operatorname*{esup}_B E^B \le C r^{\alpha+\gamma}, \tag{5.5}$$

$$\operatorname*{einf}_{\delta_1 B} E^B \ge C^{-1} r^{\alpha+\gamma}, \tag{5.6}$$

where $E^B$ is the weak solution of the Poisson-type equation (4.7) as before, and $C > 0$ and $\delta_1 \in (0, 1/2)$.

- *Step* 3. To show condition $\left(S_\beta\right)$, observe that, for all $t > 0$ and $\mu$-almost all $x \in B$,

$$P_t^B 1_B(x) \ge \frac{E^B(x) - t}{\left\| E^B \right\|_\infty}, \tag{5.7}$$

which follows by using the parabolic maximum principle, nothing else. Hence, using (5.5), (5.6),

$$\begin{aligned} P_t^B 1_B(x) &\ge \frac{E^B(x) - t}{\left\| E^B \right\|_\infty} \\ &\ge c - c_1 t r^{-\beta} \\ &\ge \frac{c}{2}, \end{aligned}$$

for all $t > 0$ and $\mu$-almost all $x \in B(x_0, \delta_1 r)$, provided that $t r^{-\beta}$ is small enough, thus proving $\left(S_\beta\right)$.

- *Step* 4. We show that $(R_1) \Rightarrow$ (*DUE*). Consider a function $f \in \mathcal{F} \cap C_0(\Omega)$ normalized so that $\sup |f| = 1$, and let $x \in \Omega$ be a point such that $|f(x)| = 1$. Let $r$ be the largest radius such that $B(x, r) \subset \Omega$. Then the ball $B(x, 2r)$ is not covered by $\Omega$ so that there exists a point $y \in B(x, 2r) \setminus \Omega$ (note that $M$ is unbounded by condition $(V_\alpha)$). In particular, $y \notin \operatorname{supp} f$ (see Fig. 9). Noting that

$\mathcal{E}^{(J)}(f) \leq \mathcal{E}(f)$ and by the $\alpha$-regularity of $\mu$

$$r \leq C\,[\mu(B(x,r))]^{1/\alpha} \leq C\,[\mu(\Omega)]^{1/\alpha},$$

we obtain from $(R_1)$ that

$$\begin{aligned} 1 &= |f(y) - f(x)|^2 \\ &\leq Cd(y,x)^{\beta-\alpha}\mathcal{E}^{(J)}(f) \\ &\leq C\,(2r)^{\beta-\alpha}\,\mathcal{E}(f) \leq C2^{\beta-\alpha}\,[\mu(\Omega)]^{\beta/\alpha-1}\,\mathcal{E}(f). \end{aligned}$$

Since $\|f\|_2^2 \leq \mu(\Omega)$, it follows that

$$\frac{\mathcal{E}(f)}{\|f\|_2^2} \geq c\,[\mu(\Omega)]^{-\beta/\alpha},$$

for some $c > 0$, thus proving the Faber-Krahn inequality. Hence, condition *(DUE)* follows by using Theorem 3.6.

- *Step* 5. Finally, with a certain amount of effort [GHL00, Proposition 6.5, Lemma 6.4], one can show that

$$(R_1) + (R_2) \Rightarrow \text{conservativeness of } (\mathcal{E}, \mathcal{F}).$$

Therefore, the equivalence (5.2) follows directly by using Theorem 5.2.

## References


[AB] Andres, S., Barlow, M.: Energy inequalities for cutoff fucntions and some applications. J. Reine Angew. Math. (preprint)

[Aro67] Aronson, D.: Bounds for the fundamental solution of a parabolic equation. Bull. Amer. Math. Soc. **73**, 890–896 (1967)

[Aro68] Aronson, D.: Non-negative solutions of linear parabolic equations. Ann. Scuola Norm. Sup. Pisa. Cl. Sci. **22**, 607–694 (1968) (Addendum **25**:221–228, 1971)

[Bar98] Barlow, M.: Diffusions on fractals. Lecture Notes of Mathematics, vol. 1690, pp. 1–121. Springer, Berlin (1998)

[Bar04] Barlow, M.: Which values of the volume growth and escape time exponents are possible for graphs? Rev. Mat. Iberoamericana **20**, 1–31 (2004)

[Bar05] Barlow, M.: Some remarks on the elliptic harnack inequality. Bull. London Math. Soc. **37**, 200–208 (2005)

[BBCK09] Barlow, M., Bass, R., Chen, Z.-Q., Kassmann, M.: Non-local Dirichlet forms and symmetric jump processes. Trans. Amer. Math. Soc. **361**, 1963–1999 (2009)

[BBK06] Barlow, M., Bass, R., Kumagai, T.: Stability of parabolic Harnack inequalities on metric measure spaces. J. Math. Soc. Japan **58**, 485–519 (2006)

[BCK05] Barlow, M., Coulhon, T., Kumagai, T.: Characterization of sub-Gaussian heat kernel estimates on strongly recurrent graphs. Comm. Pure Appl. Math. **58**, 1642–1677 (2005)

[BGK09] Barlow, M., Grigor'yan, A., Kumagai, T.: Heat kernel upper bounds for jump processes and the first exit time. J. Reine Angew. Math. **626**, 135–157 (2009)

[BGK12] Barlow, M., Grigor'yan, A., Kumagai, T.: On the equivalence of parabolic harnack inequalities and heat kernel estimates. J. Math. Soc. Japan **64**, 1091–1146 (2012)

[BP88] Barlow, M., Perkins, E.: Brownian motion on the Sierpínski gasket. Probab. Theory Relat. Fields **79**, 543–623 (1988)

[BBW08] Barthelemy, P., Bertolotti, J., Wiersma, D.: A Lévy flight for light. Nature **453**, 495–498 (2008)

[BL02] Bass, R., Levin, D.: Transition probabilities for symmetric jump processes. Trans. Amer. Math. Soc. **354**, 2933–2953 (2002)

[CKS87] Carlen, E., Kusuoka, S., Stroock, D.: Upper bounds for symmetric Markov transition functions. Ann. Inst. H. Poincaré-Probab. Statist. **23**, 245–287 (1987)

[Car96] Carron, G.: Inégalités isopérimétriques de Faber-Krahn et conséquences. In: Actes de la table ronde de géométrie différentielle (Luminy, 1992) Collection SMF Séminaires et Congrès, **1** (1996), pp. 205–232

[CK03] Chen, Z., Kumagai, T.: Heat kernel estimates for stable-like processes on $d$-sets. Stoch. Process. Appl. **108**, 27–62 (2003)

[CK08] Chen, Z., Kumagai, T.: Heat kernel estimates for jump processes of mixed types on metric measure spaces. Probab. Theory Relat. Fields **140**, 277–317 (2008)

[CG98] Coulhon, T., Grigor'yan, A.: Random walks on graphs with regular volume growth. Geom. Funct. Anal. **8**, 656–701 (1998)

[Dav92] Davies, E.: Heat kernel bounds, conservation of probability and the Feller property. J. d'Analyse Math. **58**, 99–119 (1992)

[FOT11] Fukushima, M., Oshima, Y. Takeda, M.: Dirichlet forms and symmetric Markov processes. In: de Gruyter, W. (ed.) Studies in Mathematics, vol. 19, 2nd edn. Springer, Berlin (2011)

[Gol87] Goldstein, S.: Random walks and diffusion on fractals. In: Kesten, H. (ed.) Percolation Theory and Ergodic Theory of Infinite Particle Systems, vol. 8, pp. 121–129, IMA Math. Appl. Springer, New York (1987)

[Gri92] Grigor'yan, A.: The heat equation on noncompact riemannian manifolds. Mat. Sb. **182**(1), 55–87 (1991) (translation in Math. USSR-Sb. **72**(1) (1992), 47–77)

[Gri94] Grigor'yan, A.: Heat kernel upper bounds on a complete non-compact manifold. Rev. Mat. Iberoamericana **10**, 395–452 (1994)

[GH08] Grigor'yan, A., Hu, J.: Off-diagonal upper estimates for the heat kernel of the dirichlet forms on metric spaces. Invent. Math. **174**, 81–126 (2008)

[GH10] Grigor'yan, A., Hu, J.: Upper bounds of heat kernels on doubling spaces. Moscow Math. J. **14**, 505–563 (2014)

[GH00] Grigor'yan, A., Hu, J.: Heat kernels and green functions on metric measure spaces. Canad. J. Math. **66**, 641–699 (2014)

[GHL03] Grigor'yan, A., Hu, J., Lau, K.-S.: Heat kernels on metric-measure spaces and an application to semilinear elliptic equations. Trans. Amer. Math. Soc. **355**, 2065–2095 (2003)

[GHL08] Grigor'yan, A., Hu, J., Lau, K.-S.: Obtaining upper bounds of heat kernels from lower bounds. Comm. Pure Appl. Math. **61**, 639–660 (2008)

[GHL09] Grigor'yan, A., Hu, J., Lau, K.-S.: Heat kernels on metric spaces with doubling measure. In: Bandt, C., Mörters, P., Zähle, M. (eds.) Fractal Geometry and Stochastics IV, Progress in Probability, vol. 61, pp. 3–44. Birkhäuser, Basel (2009)

[GHL10] Grigor'yan, A., Hu, J., Lau, K.-S.: Comparison inequalities for heat semigroups and heat kernels on metric measure spaces. J. Funct. Anal. **259**, 2613–2641 (2010)

[GHL00] Grigor'yan, A., Hu, J., Lau, K.-S.: Estimates of heat kernels for non-local regular dirichlet forms. Trans. AMS (to appear)

[GK08] Grigor'yan, A., Kumagai, T.: On the dichotomy in the heat kernel two sided estimates. In: P. E. et al. (ed.) Analysis on Graphs and its Applications, Procedings of the Symposia in Pure Mathematics, American Mathematical Society, vol. 77, pp. 199–210 (2008)

[GT02] Grigor'yan, A., Telcs, A.: Harnack inequalities and sub-Gaussian estimates for random walks. Math. Ann. **324**, 521–556 (2002)

[GT12] Grigor'yan, A., Telcs, A.: Two-sided estimates of heat kernels on metric measure spaces. Ann. Probab. **40**, 1212–1284 (2012)

[HK99] Hambly, B., Kumagai, T.: Transition density estimates for diffusion processes on post critically finite self-similar fractals. Proc. London Math. Soc. **79**, 431–458 (1999)

[HuK06] Hu, J., Kumagai, T.: Nash-type inequalities and heat kernels for non-local Dirichlet forms. Kyushu J. Math. **60**, 245–265 (2006)

[Jon96] Jonsson, A.: Brownian motion on fractals and function spaces. Math. Zeit. **222**, 495–504 (1996)

[Kig01] Kigami, J.: Analysis on fractals. Cambridge University Press, Cambridge (2001)

[KS05] Kumagai, T., Sturm, K.: Construction of diffusion processes on fractals, $d$-sets, and general metric measure spaces. J. Math. Kyoto Univ. **45**, 307–327 (2005)

[Kus87] Kusuoka, S.: A diffusion process on a fractal. In: Symposium on Probabilistic Methods in Mathematical Physics, pp. 251–274. Katata, Academic Press, Amsterdam, Taniguchi (1987)

[Nas58] Nash, J.: Continuity of solutions of parabolic and elliptic equations. Amer. J. Math. **80**, 931–954 (1958)

[Pie00] Pietruska-PaŁuba, K.: On function spaces related to fractional diffusion on $d$-sets. Stoch. Stoch. Rep. **70**, 153–164 (2000)

[Sto00] Stós, A.: Symmetric $\alpha$-stable processes on $d$-sets. Bull. Pol. Acad. Sci. Math. **48**, 237–245 (2000)

# Stochastic Completeness of Jump Processes on Metric Measure Spaces

**Alexander Grigor'yan and Xueping Huang**

**Abstract** We give criteria for stochastic completeness of jump processes on metric measure spaces and on graphs in terms of volume growth.

## 1 Stochastic Completeness of a Diffusion

Let $\{X_t\}_{t\geq 0}$ be a reversible Markov process on a state space $M$. This process is called *stochastically complete* if its lifetime is almost surely $\infty$, that is

$$\mathbb{P}_x\left(X_t \in M\right) = 1.$$

If the process has no interior killing (which will be assumed) then the only way the stochastic incompleteness can occur is if the process leaves the state space in finite time. For example, diffusion in a bounded domain with the Dirichlet boundary condition is stochastically incomplete (Fig. 1).

A by far less trivial example was discovered by Azencott [Aze74] in 1974: he showed that Brownian motion on a *geodesically complete* non-compact manifold can be stochastically incomplete. In his example the manifold has negative sectional curvature that grows to $-\infty$ very fast with the distance to an origin. The stochastic incompleteness occurs because negative curvature plays the role of a drift towards infinity, and a very high negative curvature produces an extremely fast drift that sweeps the Brownian particle away to infinity in a finite time.

A. Grigor'yan (✉)
Department of Mathematics, University of Bielefeld, Bielefeld, Germany
e-mail: alexander.grigorian@googlemail.com

X. Huang
Department of Mathematics and Computer Science, Friedrich Schiller University of Jena, Jena, Germany



D.-J. Feng and K.-S. Lau (eds.), *Geometry and Analysis of Fractals*,
Springer Proceedings in Mathematics & Statistics 88,
DOI 10.1007/978-3-662-43920-3_7

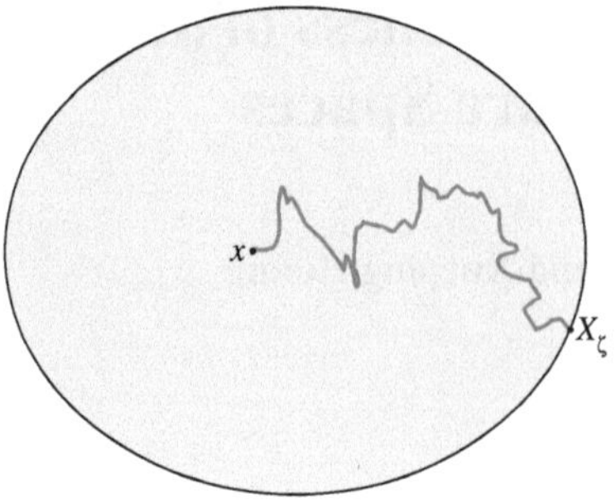


**Fig. 1** Diffusion in a bounded domain

Various sufficient conditions in terms of curvature bounds were obtained by Yau [Yau78], Hsu [Hsu89], etc. It is somewhat surprising that one can obtain a sufficient condition for stochastic completeness in terms of the volume growth. Let $V(x,r)$ be the volume of the geodesic ball of radius $r$ centered at some fixed $x$. Then

$$V(x,r) \leq \exp\left(Cr^2\right) \Rightarrow \text{stochastic completeness.}$$

Moreover,

$$\int^{\infty} \frac{rdr}{\ln V(x,r)} = \infty \Rightarrow \text{stochastic completeness.} \tag{1}$$

Let us sketch the construction of Brownian motion on a Riemannian manifold $M$ and approach to the proof of the volume test for stochastic completeness (cf. [Gri09] for more details). Let $M$ be a Riemannian manifold, $\mu$ be the Riemannian measure on $M$ and $\Delta$ be the Laplace-Beltrami operator on $M$. By the Green formula, $\Delta$ is a symmetric operator on $C_0^\infty(M)$ with respect to $\mu$, which allows to extend $\Delta$ to a a self-adjoint operator in $L^2(M,\mu)$. Assuming that $M$ is geodesically complete, it is possible to prove that this extension is unique. Hence, $\Delta$ can be regarded as a (non-positive definite) self-adjoint operator in $L^2$.

By functional calculus, the operator $P_t := e^{t\Delta}$ is a bounded self-adjoint operator for any $t \geq 0$. The family $\{P_t\}_{t\geq 0}$ is called the *heat semigroup* of $\Delta$. It can be used to solve the Cauchy problem in $\mathbb{R}_+ \times M$:

$$\begin{cases} \frac{\partial u}{\partial t} = \Delta u, \\ u|_{t=0} = f, \end{cases}$$

since $u(t,\cdot) = P_t f$ is solution for any $f \in L^2$.

Local regularity theory implies that $P_t$ is an integral operator, whose kernel $p_t(x,y)$ is a positive smooth function of $(t,x,y)$. In fact, $p_t(x,y)$ is the minimal positive fundamental solution to the heat equation.

The heat kernel can be used to construct a diffusion process $\{X_t\}$ on $M$ with transition density $p_t(x,y)$. For example, in $\mathbb{R}^n$ one has

$$p_t(x,y) = \frac{1}{(4\pi t)^{n/2}} \exp\left(-\frac{|x-y|^2}{4t}\right),$$

and the process $\{X_t\}$ with this transition density is Brownian motion.

In terms of the heat kernel the stochastic completeness of diffusion $\{X_t\}$ is equivalent to the following identity:

$$\int_M p_t(x,y)\,d\mu(y) = 1,$$

for all $t > 0$ and $x \in M$.

Another useful criterion for stochastic completeness is as follows: $M$ is stochastically complete if the homogeneous Cauchy problem

$$\begin{cases} \frac{\partial u}{\partial t} = \Delta u \\ u|_{t=0} = 0 \end{cases} \tag{2}$$

has a unique solution $u \equiv 0$ in the class of bounded functions (Khas'minskii [Kha60]).

By classical results, in $\mathbb{R}^n$ the uniqueness for (2) holds even in the class

$$|u(t,x)| \le \exp\left(C|x|^2\right)$$

(Tikhonov class [Tic35]), but not in

$$|u(t,x)| \le \exp\left(C|x|^{2+\varepsilon}\right).$$

More generally, uniqueness holds in the class

$$|u(t,x)| \le \exp(f|x|)$$

provided the positive increasing function $f$ satisfies

$$\int^{\infty} \frac{rdr}{f(r)} = \infty$$

(Täcklind class [Täc36]).

The following result can be regarded as an analogue of the latter uniqueness class.

**Theorem 1** (AG [Gri87]) *Let M be a complete connected Riemannian manifold, and let $u(x,t)$ be a solution to the Cauchy problem (2). Assume that, for some $x \in M$ and for some $T > 0$ and all $r > 0$,*

$$\int_0^T \int_{B(x,r)} u^2(y,t)\, d\mu(y) dt \leq \exp(f(r)), \tag{3}$$

*where $f(r)$ is a positive increasing function on $(0, +\infty)$ such that*

$$\int^{\infty} \frac{rdr}{f(r)} = \infty.$$

*Then $u \equiv 0$ in $(0,T) \times M$.*

If $u$ is a bounded solution, then replacing in (3) $u$ by const we obtain that if

$$V(x,r) \leq \exp(f(r))$$

then $u \equiv 0$, that is, $M$ is stochastically complete. Setting

$$f(r) = \ln V(x,r)$$

we obtain the volume test for stochastic completeness:

$$\int^{\infty} \frac{rdr}{\ln V(x,r)} = \infty.$$

The latter condition cannot be further improved: if $W(r)$ is an increasing function such that

$$\int^{\infty} \frac{rdr}{\ln W(r)} < \infty$$

then there exists a geodesically complete but stochastically incomplete manifold with $V(x,r) \leq W(r)$.

One may wonder why the geodesic balls can be used to determine the stochastic completeness, as the latter condition does not depend on the distance function at all. The reason is that the geodesic distance $d$ is by definition related to the gradient $\nabla$ (and, hence, to the Laplacian) by $|\nabla d| \leq 1$. An analogue of this condition will appear later also in jump processes.

## 2 Jump Processes

Let $(M, d)$ be a metric space such that all closed metric balls

$$B(x, r) = \{y \in M : d(x, y) \leq r\}$$

are compact. In particular, $(M, d)$ is locally compact and separable. Let $\mu$ be a Radon measure on $M$ with a full support.

Recall that a *Dirichlet form* $(\mathcal{E}, \mathcal{F})$ in $L^2(M, \mu)$ is a symmetric, non-negative definite, bilinear form $\mathcal{E} : \mathcal{F} \times \mathcal{F} \to \mathbb{R}$ defined on a dense subspace $\mathcal{F}$ of $L^2(M, \mu)$, that satisfies in addition the following properties:

- Closedness: $\mathcal{F}$ is a Hilbert space with respect to the following inner product:

$$\mathcal{E}_1(f, g) := \mathcal{E}(f, g) + (f, g).$$

- The Markov property: if $f \in \mathcal{F}$ then also $\widetilde{f} := (f \wedge 1)_+$ belongs to $\mathcal{F}$ and $\mathcal{E}(\widetilde{f}) \leq \mathcal{E}(f)$, where $\mathcal{E}(f) := \mathcal{E}(f, f)$.

For example, the classical Dirichlet form in $\mathbb{R}^n$ is

$$\mathcal{E}(f, g) = \int_{\mathbb{R}^n} \nabla f \cdot \nabla g \, dx$$

in $\mathcal{F} = W^{1,2}(\mathbb{R}^n)$.

A general Dirichlet form $(\mathcal{E}, \mathcal{F})$ has the *generator* $\mathcal{L}$ that is a non-positive definite, self-adjoint operator on $L^2(M, \mu)$ with domain $\mathcal{D} \subset \mathcal{F}$ such that

$$\mathcal{E}(f, g) = (-\mathcal{L}f, g)$$

for all $f \in \mathcal{D}$ and $g \in \mathcal{F}$. The generator $\mathcal{L}$ determines the *heat semigroup* $\{P_t\}_{t \geq 0}$ by $P_t = e^{t\mathcal{L}}$ in the sense of functional calculus of self-adjoint operators. It is known that $\{P_t\}_{t \geq 0}$ is a strongly continuous, contractive, symmetric semigroup in $L^2$, and is *Markovian*, that is, $0 \leq P_t f \leq 1$ for any $t > 0$ if $0 \leq f \leq 1$.

The Markovian property of the heat semigroup implies that the operator $P_t$ preserves the inequalities between functions, which allows to use monotone limits to extend $P_t$ from $L^2$ to $L^\infty$. In particular, $P_t 1$ is defined.

**Definition** The form $(\mathcal{E}, \mathcal{F})$ is called *conservative* or *stochastically complete* if $P_t 1 = 1$ for every $t > 0$.

Assume in addition that $(\mathcal{E}, \mathcal{F})$ is *regular*, that is, the set $\mathcal{F} \cap C_0(M)$ is dense both in $\mathcal{F}$ with respect to the norm $\mathcal{E}_1$ and in $C_0(M)$ "(compactly supported continuous functions)" with respect to the sup-norm. By a theory of Fukushima et al. [FOT11], for any regular Dirichlet form there exists a Hunt process $\{X_t\}_{t\geq 0}$ such that, for all bounded Borel functions $f$ on $M$,

$$\mathbb{E}_x f(X_t) = P_t f(x) \tag{4}$$

for all $t > 0$ and almost all $x \in M$, where $\mathbb{E}_x$ is expectation associated with the law of $\{X_t\}$ started at $x$.

Using the identity (4), one can show that the lifetime of $X_t$ is almost surely $\infty$ if and only if $P_t 1 = 1$ for all $t > 0$, which motivates the term "stochastic completeness" in the above definition.

One distinguishes local and non-local Dirichlet forms. The Dirichlet form $(\mathcal{E}, \mathcal{F})$ is called *local* if $\mathcal{E}(f, g) = 0$ for all functions $f, g \in \mathcal{F}$ with disjoint compact support. It is called *strongly local* if the same is true under a milder assumption that $f = \text{const}$ on a neighborhood of $\operatorname{supp} g$.

For example, the following Dirichlet form on a Riemannian manifold

$$\mathcal{E}(f, g) = \int_M \nabla f \cdot \nabla g d\mu$$

is strongly local. The generator of this form the self-adjoint Laplace-Beltrami operator $\Delta$, and the Hunt process is Brownian motion on $M$.

A well-studied non-local Dirichlet form in $\mathbb{R}^n$ is given by

$$\mathcal{E}(f, g) = \int_{\mathbb{R}^n \times \mathbb{R}^n} \frac{(f(x) - f(y))(g(x) - g(y))}{|x - y|^{n+\alpha}} dxdy \tag{5}$$

where $0 < \alpha < 2$. The domain of this form is the Besov space $B_{2,2}^{\alpha/2}$, the generator is (up to a constant multiple) the operator $-(-\Delta)^{\alpha/2}$, where $\Delta$ is the Laplace operator in $\mathbb{R}^n$, and the Hunt process is the symmetric stable process of index $\alpha$.

By a theorem of Beurling and Deny (cf. [FOT11]), any regular Dirichlet form can be represented in the form

$$\mathcal{E} = \mathcal{E}^{(c)} + \mathcal{E}^{(j)} + \mathcal{E}^{(k)},$$

where $\mathcal{E}^{(c)}$ is a strongly local part that has the form (assuming absolute continuity of energy measure for simplicity)

$$\mathcal{E}^{(c)}(f, g) = \int_M \Gamma(f, g) d\mu,$$

where $\Gamma(f, g)$ is the so called *energy density* (generalizing $\nabla f \cdot \nabla g$ on manifolds); $\mathcal{E}^{(j)}$ is a jump part that has the form

$$\mathcal{E}^{(j)}(f,g)=\frac{1}{2}\iint\limits_{X\times X}(f(x)-f(y))(g(x)-g(y))\,dJ(x,y)$$

with some measure $J$ on $X \times X$ that is called a *jump measure*; and $\mathcal{E}^{(k)}$ is a killing part that has the form

$$\mathcal{E}^{(k)}(f,g)=\int\limits_X fg\,dk$$

where $k$ is a measure on $X$ that is called a *killing measure*.

In terms of the associated process this means that $X_t$ is in some sense a mixture of diffusion and jump processes with a killing condition.

The ln-volume test of stochastic completeness of manifolds can be extended to strongly local Dirichlet forms as follows. Set as before $V(x,r)=\mu(B(x,r))$.

**Theorem 2** (Sturm [Stu94]) *Let $(\mathcal{E},\mathcal{F})$ be a regular strongly local Dirichlet form. Assume that the distance function $\rho(x)=d(x,x_0)$ on $M$ satisfies the condition*

$$\Gamma(\rho,\rho)\le C,$$

*for some constant $C$. If, for some $x\in M$,*

$$\int^{\infty}\frac{rdr}{\ln V(x,r)}=\infty$$

*then the Dirichlet form $(\mathcal{E},\mathcal{F})$ is stochastically complete.*

The method of proof is basically the same as for manifolds because for strongly local forms the same chain rule and product rules are available. The condition $\Gamma(\rho,\rho)\le C$ is analogous to $|\nabla\rho|\le 1$ that is automatically satisfied for the geodesic distance on any manifold.

Now let us turn to jump processes. For simplicity let us assume that the jump measure $J$ has a density $j(x,y)$. Namely, let $j(x,y)$ be a non-negative Borel function on $M\times M$ that satisfies the following two conditions:

(a) $j(x,y)$ is symmetric: $j(x,y)=j(y,x)$;
(b) there is a positive constant $C$ such that

$$\int\limits_M (1\wedge d(x,y)^2)j(x,y)d\mu(y)\le C\ \text{ for all } x\in M.$$

**Definition** We say that a distance function $d$ is *adapted* to a kernel $j(x,y)$ (or $j$ is adapted to $d$) if $(b)$ is satisfied.

The condition $(b)$ relates the distance function to the Dirichlet form and plays the same role as $\Gamma(\rho,\rho)\le C$ does for diffusion.

Consider the following bilinear functional

$$\mathcal{E}(f,g)=\frac{1}{2}\int\int_{X\times X}(f(x)-f(y))(g(x)-g(y))j(x,y)d\mu(x)d\mu(y)$$

that is defined on Borel functions $f$ and $g$ whenever the integral makes sense. Define the maximal domain of $\mathcal{E}$ by

$$\mathcal{F}_{\text{max}}=\left\{f\in L^2:\mathcal{E}(f,f)<\infty\right\},$$

where $L^2=L^2(M,\mu)$. By the polarization identity, $\mathcal{E}(f,g)$ is finite for all $f,g\in\mathcal{F}_{\text{max}}$. Moreover, $\mathcal{F}_{\text{max}}$ is a Hilbert space with the norm $\mathcal{E}_1$.

Denote by $\text{Lip}_0(M)$ the class of Lipschitz functions on $M$ with compact support. It follows from $(b)$ that

$$\text{Lip}_0(M)\subset\mathcal{F}_{\text{max}}.$$

Define the space $\mathcal{F}$ as the closure of $\text{Lip}_0(M)$ in $(\mathcal{F}_{\text{max}},\|\cdot\|_{\mathcal{E}_1})$. Under the above hypothesis, $(\mathcal{E},\mathcal{F})$ is a regular Dirichlet form in $L^2(M,\mu)$. The associated Hunt process $\{X_t\}$ is a pure jump process with the jump density $j(x,y)$.

Many examples of jump processes in $\mathbb{R}$ are provided by Lévy-Khintchine theorem where the Lévy measure $W(dy)$ corresponds to $j(x,y)d\mu(y)$. The condition $(b)$ appears also in Lévy-Khintchine theorem in the form

$$\int_{\mathbb{R}\setminus\{0\}}\left(1\wedge|y|^2\right)W(dy)<\infty.$$

Hence, the Euclidean distance in $\mathbb{R}$ is adapted to any Lévy process.

An explicit example of a jump density in $\mathbb{R}^n$ is

$$j(x,y)=\frac{\text{const}}{|x-y|^{n+\alpha}},$$

where $\alpha\in(0,2)$, which defines the Dirichlet form (5).

The next theorem is the main result.

**Theorem 3** *Assume that $j$ satisfies $(a)$ and $(b)$ and let $(\mathcal{E},\mathcal{F})$ be the jump form defined as above. If, for some $x\in M$, $c>0$ and for all large enough $r$,*

$$V(x,r)\leq\exp(cr\ln r), \tag{6}$$

*then the Dirichlet form $(\mathcal{E},\mathcal{F})$ is stochastically complete.*

This theorem was proved by Grigor'yan et al. [GHM12] for $c < \frac{1}{2}$, improving the work of Masamune and Uemura [MU11] for the sub-exponential volume growth case. Then it was observed ([MUW12]) that a minor modification of the proof of [GHM12] works for all $c$.

For the proof of Theorem 3 we split the jump kernel $j(x, y)$ into the sum of two parts:

$$j'(x, y) = j(x, y)\mathbf{1}_{\{d(x,y)\leq\varepsilon\}} \text{ and } j''(x, y) = j(x, y)\mathbf{1}_{\{d(x,y)>\varepsilon\}} \tag{7}$$

and show first the stochastic completeness of the Dirichlet form $(\mathcal{E}', \mathcal{F})$ associated with $j'$. For that we adapt the methods used for stochastic completeness for the local form.

The bounded range of $j'$ allows to treat the Dirichlet form $(\mathcal{E}', \mathcal{F})$ as "almost" local: if $f, g$ are two functions from $\mathcal{F}$ such that $d(\operatorname{supp} f, \operatorname{supp} g) > \varepsilon$ then $\mathcal{E}(f, g) = 0$. The condition $(b)$ plays in the proof the same role as the condition $|\nabla d| \leq 1$ in the local case. However, the lack of locality brings up in the estimates various additional terms that have to be compensated by a stronger hypothesis of the volume growth (6).

The tail $j''$ can be regarded as a small perturbation of $j'$ in the following sense: $(\mathcal{E}, \mathcal{F})$ is stochastically complete if and only if $(\mathcal{E}', \mathcal{F})$ is so. The proof is based on the fact that the integral operator with the kernel $j''$ is a bounded operator in $L^2(M, \mu)$, because by $(b)$

$$\int_M j''(x, y)\, d\mu(y) \leq C.$$

It is not yet clear if the volume growth condition (6) in Theorem 3 is sharp.

In contrast to the manifold case, we can not expect a corresponding uniqueness class result. Let us briefly mention a result about uniqueness class for the heat equation associated with the jump Dirichlet form on graphs satisfying $(a)$ and $(b)$.

Namely, Huang [Hua12] proved in 2011 that, for any $b < \frac{1}{2}$ the following inequality determines a uniqueness class

$$\int_0^T \int_{B(x,r)} u^2(t, x)\, d\mu(x)\, dt \leq \exp(br \ln r). \tag{8}$$

What is more surprising, that for $b > 2\sqrt{2}$ this statement fails even on the graph $\mathbb{Z}$.

The optimal value of $b$ in (8) is unknown. If $b < \frac{1}{2}$ then Huang's result can be used to obtain Theorem 3 on graphs provided the constant $c$ in (6) is smaller than $\frac{1}{2}$. However, in general the stochastic completeness test (6) does not follow from the uniqueness class (8), as can be seen from the range of constants. Indeed, even better results for stochastic completeness are known in the graph case, which we will discuss in the next section.

## 3 Random Walks on Graphs

Let us now turn to random walks on graphs. Let $(X, E)$ be a locally finite, infinite, connected graph, where $X$ is the set of vertices and $E$ is the set of edges. We assume that the graph is undirected, simple, without loops. Let $\mu$ be the counting measure on $X$. Define the jump kernel by $j(x, y) = 1_{\{x \sim y\}}$, where $x \sim y$ means that $x, y$ are neighbors, that is, $(x, y) \in E$. The corresponding Dirichlet form is

$$\mathcal{E}(f) = \frac{1}{2} \sum_{x,y:\, x \sim y} (f(x) - f(y))^2,$$

and its generator is

$$\Delta f(x) = \sum_{y, y \sim x} (f(y) - f(x)).$$

The operator $\Delta$ is called *unnormalized* (or *physical)* Laplace operator on $(X, E)$. This is to distinguish from the *normalized* or *combinatorial* Laplace operator

$$\hat{\Delta} f(x) = \frac{1}{\deg(x)} \sum_{y, y \sim x} (f(y) - f(x)),$$

where $\deg(x)$ is the number of neighbors of $x$. The normalized Laplacian $\hat{\Delta}$ is the generator of the same Dirichlet form but with respect to the degree measure $\deg(x)$.

Both $\Delta$ and $\hat{\Delta}$ generate the heat semigroups $e^{t\Delta}$ and $e^{t\hat{\Delta}}$ and, hence, associated continuous time random walks on $X$. It is easy to prove that $\hat{\Delta}$ is a bounded operator in $L^2(X, \deg)$, which then implies that the associated random walk is always stochastically complete. On the contrary, the random walk associated with the unnormalized Laplace operator can be stochastically incomplete.

We say that the graph $(X, E)$ is stochastically complete if the heat semigroup $e^{t\Delta}$ is stochastically complete.

Denote by $\rho(x, y)$ the graph distance on $X$, that is the minimal number of edges in an edge chain connecting $x$ and $y$. Let $B_\rho(x, r)$ be closed metric balls with respect to this distance $\rho$ and set $V_\rho(x, r) = \left|B_\rho(x, r)\right|$ where $|\cdot| := \mu(\cdot)$ denotes the number of vertices in a given set.

**Theorem 4** *If there is a point $x_0 \in X$ and a constant $c > 0$ such that*

$$V_\rho(x_0, r) \le c r^3 \ln r \tag{9}$$

*for all large enough $r$, then the graph $(X, E)$ is stochastically complete.*

Note that the function $r^3 \ln r$ is sharp here in the sense that it cannot be replaced by $r^3 \ln^{1+\varepsilon} r$. For any non-negative integer $r$, set

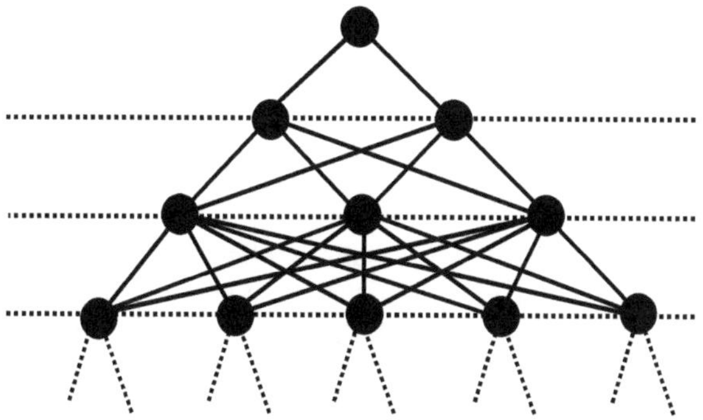

**Fig. 2** Anti-tree of Wojciechowski

$$S_r = \{x \in X : \rho(x_0, x) = r\}.$$

Wojciechowski [Woj11] considered the graph where every vertex on $S_r$ is connected to all vertices on $S_{r-1}$ and $S_r$ (see Fig. 2).

He proved that for such graphs the stochastic incompleteness is equivalent to the following condition:

$$\sum_{r=1}^{\infty} \frac{V_\rho(x_0, r)}{|S_{r+1}|\,|S_r|} < \infty. \tag{10}$$

Taking $|S_r| \simeq r^2 \ln^{1+\varepsilon} r$ we obtain $V_\rho(x_0, r) \simeq r^3 \ln^{1+\varepsilon}$ so that the condition (10) is satisfied and, hence, the graph is stochastically incomplete.

The proof of Theorem 4 is based on the following ideas. Observe first that the graph distance $\rho$ is in general not adapted. Indeed, the integral in $(b)$ is equal to

$$\sum_y \left(1 \wedge \rho^2(x, y)\right) j(x, y) = \sum_y j(x, y) = \deg(x)$$

so that $(b)$ holds if and only if the graph has uniformly bounded degree, which is not interesting as all graphs with bounded degree are automatically stochastically complete.

Let us construct an adapted distance as follows. For any edge $x \sim y$ define first its length $\sigma(x, y)$ by

$$\sigma(x, y) = \frac{1}{\sqrt{\deg(x)}} \wedge \frac{1}{\sqrt{\deg(y)}}.$$

Then, for all $x, y \in X$ define $d(x, y)$ as the smallest total length of all edges in an edge chain connecting $x$ and $y$. It is easy to verify that $d$ satisfies $(b)$:

$$\sum_y \left(1 \wedge d^2(x,y)\right) j(x,y) \le \sum_y \left(\frac{1}{\deg(x)} \wedge \frac{1}{\deg(y)}\right) j(x,y)$$
$$\le \sum_{y \sim x} \frac{1}{\deg(x)} = 1.$$

Then we will show that (9) for $\rho$-balls implies that the $d$-balls have at most quadratic exponential volume growth, so that the stochastic completeness will follow by the following result of Folz (stated in the current specific setting).

**Theorem 5** (Folz [Fol00]) *Let $(X, E)$ be a graph as above, with an adapted distance $d$. If the volume growth $V_d(x_0, r) = \mu(B_d(x_0, r))$ with respect to $d$ satisfies:*

$$\int^{\infty} \frac{r dr}{\ln V_d(x_0, r)} = \infty, \tag{11}$$

*for some reference point $x_0 \in X$, then the graph $(X, E)$ is stochastically complete.*

Roughly speaking, for a graph $(X, E)$ with an adapted distance $d$, Folz constructed a corresponding metric graph $Y$, which is enriched from $X$ by attaching intervals to the edges. The length and measure of intervals, which are used to define a strongly local Dirichlet form on $Y$, are determined by the adapted distance. Folz proved two significant relations between the metric graph $Y$ with the original graph $X$. First, the volume growth of $Y$ is controlled by that of $X$. More importantly, $X$ is stochastically complete if so is the diffusion on $Y$. Theorem 5 is then obtained as a consequence of Theorem 2. The second relation is the key to overcome the difficulty coming from lack of chain rule. It was first proven by Folz using probabilistic arguments. Two analytic proofs of this comparison result are obtained by Huang [Hua13]. We briefly describe one of them as it is rather concise.

By a well-known result in [FOT11], the stochastic completeness of a Dirichlet form $(\mathcal{E}, \mathcal{F})$ on a measure space $(M, \mu)$ is equivalent to the existence of a sequence of functions $\{v_n\} \subset \mathcal{F}$ such that

$$0 \le v_n \le 1, \lim_{n\to\infty} v_n = 1 \quad \mu\text{-a.e.}$$

and such that

$$\lim_{n\to\infty} \mathcal{E}(v_n, w) = 0$$

holds for any $w \in \mathcal{F} \cap L^1(M, \mu)$. Thus comparison of stochastic completeness boils down to comparing the existence of certain functions. There are natural ways to transfer back and forth between a function space on a graph and that on the corresponding metric graph. Assume for a graph $X$ that the corresponding metric graph $Y$ is stochastically complete, with a sequence $\{v_n\}$ as above. The sequence $\{\tilde{v}_n\}$ on $X$, as restrictions of $\{v_n\}$, is naturally expected to satisfy the conditions above. The condition

$$\lim_{n\to\infty} \mathcal{E}(\tilde{v}_n, \tilde{w}) = 0$$

for $\tilde{w}$ on $X$, can be checked by extending $\tilde{w}$ to $w$ on $Y$ through linear interpolation. The rest are simple calculations to make sure that $\tilde{v}_n$ and $w$ are in the correct function spaces.

Now we deduce Theorem 4 from Theorem 5. Without loss of generality, we assume that

$$V_\rho(x_0, r) \le c(r+1)^3 \ln(r+3), \tag{12}$$

for all $r \ge 0$. Observe that

$$V_\rho(x_0, n) = \sum_{r=0}^{n} \mu\left(S_\rho(r)\right).$$

Put $\varepsilon = \frac{1}{5}$ and $\alpha = 200c$ where $c$ is the constant in (12). It follows from (12) that, for any $n \ge 1$,

$$\left|\{r \in [n-1, 2n+1] : \mu(S_r) > \alpha(n+1)^2 \ln(n+3)\}\right| \le \frac{c(2n+2)^3 \ln(2n+4)}{\alpha(n+1)^2 \ln(n+3)} \le \varepsilon n.$$

Therefore,

$$\left|\{r \in [n+1, 2n] : \max_{i=-2,-1,0,1} \mu(S_{r+i}) > \alpha(n+1)^2 \ln(n+3)\}\right| \le 4\varepsilon n$$

and, hence,

$$\left|\{r \in [n+1, 2n] : \max_{i=-2,-1,0,1} \mu(S_{r+i}) \le \alpha(n+1)^2 \ln(n+3)\}\right| \ge (1-4\varepsilon)n. \tag{13}$$

For any point $x \in S_r$ we have

$$\deg x \le \mu\left(S_{r-1}\right) + \mu\left(S_r\right) + \mu\left(S_{r+1}\right) \tag{14}$$

(see Fig. 3).

It follows from (13) and (14) that

$$\left|\{r \in [n+1, 2n] : \max_{x \in S_{r-1} \cup S_r} \deg x \le 3\alpha(n+1)^2 \ln(n+3)\}\right| \ge (1-4\varepsilon)n. \tag{15}$$

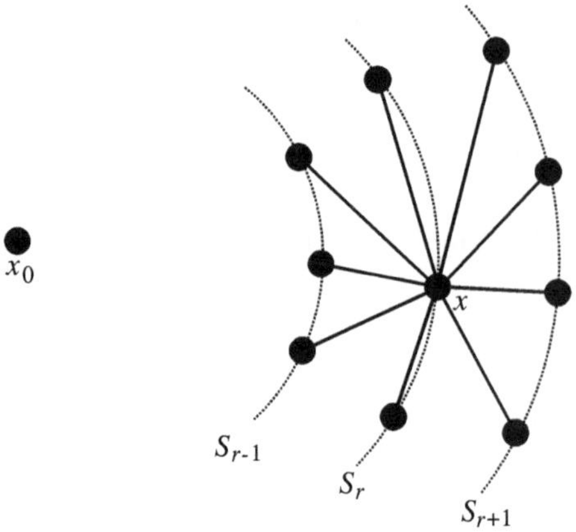


**Fig. 3** Neighbors of a vertex $x$ on $S_r$

It follows that, for $r$ as in (15), any pair of $x \sim y$ with $x \in S_{r-1}$, $y \in S_r$ necessarily satisfies

$$\sigma(x, y) \geq \frac{1}{\sqrt{3\alpha}(n+1)\sqrt{\ln(n+3)}}. \tag{16}$$

Fix a positive integer $n$ and two vertices $x \in S_n$ and $y \in S_{2n}$. Consider a chain of vertices connecting $x$ and $y \in S_{2n}$, and let us estimate from below the length $L$ of this chain. For any $r \in [n+1, 2n]$ there is an edge $x_r \sim y_r$ from this chain such $x_r \in S_{r-1}$ and $y_r \in S_r$. Clearly, we have

$$L \geq \sum_{r=n+1}^{2n} \sigma(x_r, y_r).$$

Restricting the summation to those $r$ that satisfy (15) and noticing that for any such $r$,

$$\sigma(x_r, y_r) \geq \frac{1}{\sqrt{3\alpha}(n+1)\sqrt{\ln(n+3)}},$$

we obtain

$$L \geq \frac{1}{\sqrt{3\alpha}(n+1)\sqrt{\ln(n+3)}}(1-4\varepsilon)\, n \geq \frac{\delta}{\sqrt{\ln(n+3)}} \geq \frac{\delta}{\sqrt{2+\ln n}}, \tag{17}$$

where $\delta = \frac{1-4\varepsilon.}{2\sqrt{3\alpha}}$.

Now we can estimate $d(x_0, x)$ for any vertex $x \notin B_\rho(x_0, R)$, where $R > 4$. Choose a positive integer $k$ so that

$$2^k \leq R < 2^{k+1}.$$

Any chain connecting $x_0$ and $x$ contains a subsequence $\{x_i\}_{i=1}^k$ of vertices such that $x_i \in S_{2^i}$. By (17) the length of the chain between $x_i$ and $x_{i+1}$ is bounded below by $\frac{\delta}{\sqrt{i+2}}$, for any $i = 1, ..., k-1$. It follows that the length of the whole chain is bounded below by

$$\delta \sum_{i=1}^{k-1} \frac{1}{\sqrt{i+2}},$$

whence

$$d(x_0, x) \geq \delta' \sqrt{k+1} \geq \delta' \sqrt{\ln R},$$

for some constant $\delta' > 0$. It follows that

$$B_d(x_0, \delta' \sqrt{\ln R}) \subset B_\rho(x_0, R).$$

Given a large enough $r$, define $R$ from the identity $r = \delta' \sqrt{\ln R}$, that is, $R = \exp(r^2/\delta'^2)$. Then we obtain

$$\mu(B_d(x_0, r)) \leq \mu\left(B_\rho(x_0, R)\right) \leq c(R+1)^3 \ln(R+3) \leq C \exp\left(br^2\right),$$

for some constants $C$ and $b$, which finishes the proof.

## References


[Aze74] Azencott, R.: Behavior of diffusion semi-groups at infinity. Bull. Soc. Math. (France) **102**, 193–240 (1974)

[Fol00] Folz, M.: Volume growth and stochastic completeness of graphs. Trans. Amer. Math. Soc. **366**(4), 2089–2119. ISSN: 0002-9947 (2014)

[FOT11] Fukushima, M., Oshima, Y., Takeda, M.: Dirichlet forms and symmetric Markov processes. In: de Gruyter (ed.) Studies in Mathematics, vol. 19. Walter de Gruyter and Co., Berlin (2011) [MR 2778606 (2011k:60249)]

[Gri87] Grigor'yan, A.: On stochastically complete manifolds. DAN SSSR **290** 534–537 (1986) (in Russian) [English translation: Soviet Math. Dokl. **34**(2) 310–313 (1987)]

[Gri09] Grigor'yan, A.: Heat kernel and analysis on manifolds. In: AMS-IP Studies in Advanced Mathematics, vol. 47. American Mathematical Society, Providence (2009)

[GHM12] Grigor'yan, A., Huang, X., Masamune, J.: On stochastic completeness of jump processes. Math. Z. **271**(3–4), 1211–1239 (2012)

[Hsu89] Hsu, E.P.: Heat semigroup on a complete Riemannian manifold. Ann. Probab. **17**, 1248–1254 (1989)

[Hua12] Huang, X.: On uniqueness class for a heat equation on graphs. J. Math. Anal. Appl. **393**(2), 377–388 (2012)

[Hua13] Huang, X.: A note on the volume growth criterion for stochastic completeness of weighted graphs. Potential Anal. **40**(2), 117–142 (2014)

[Kha60] Khas'minskii, R.Z.: Ergodic properties of recurrent diffusionprocesses and stabilization of solutions to the Cauchy problem forparabolic equations. Theor. Prob. Appl. **5**, 179–195 (1960)

[MU11] Masamune, J., Uemura, T.: Conservation property of symmetric jump processes. Ann. Inst. Henri. Poincaré Probab. Statist. **47**(3), 650–662 (2011)

[MUW12] Masamune, J., Uemura, T., Wang, J.: On the conservativeness and the recurrence of symmetric jump-diffusions, J. Funct. Anal. **263**(12) 3984–4008 (2012)

[Stu94] Sturm, K.T.: Analysis on local Dirichlet spaces. I. Recurrence, conservativeness and $L^p$-Liouville properties. J. Reine Angew. Math. **456**, 173–196 (1994)

[Täc36] Täcklind, S.: Sur les classes quasianalytiques des solutions deséquations aux dérivées partielles du type parabolique. Nova Acta Soc. Sci. Upsal., IV. Ser. **10**(3) 1–57 (1936)

[Tic35] Tichonov, A.N.: Uniqueness theorems for the equation of heat conduction, (in Russian) Matem. Sbornik **42**, 199–215 (1935)

[Woj11] Wojciechowski, R.K.: Stochastically incomplete manifolds and graphs. Prog. Probab. **64**, 163–179 (2011)

[Yau78] Yau, S.T.: On the heat kernel of a complete Riemannian manifold. J. Math. Pures Appl. **57**, 191–201 (1978)

# Self Similar Sets, Entropy and Additive Combinatorics

**Michael Hochman**

**Abstract** This article is an exposition of the main result of [Hoc12], that self-similar sets whose dimension is smaller than the trivial upper bound have "almost overlaps" between cylinders. We give a heuristic derivation of the theorem using elementary arguments about covering numbers. We also give a short introduction to additive combinatorics, focusing on inverse theorems, which play a pivotal role in the proof. Our elementary approach avoids many of the technicalities in [Hoc12], but also falls short of a complete proof; in the last section we discuss how the heuristic argument is turned into a rigorous one.

## 1 Introduction

### *1.1 Self-similar Sets*

Self-similar sets in the line are compact sets that are composed of finitely many scaled copies of themselves. These are the simplest fractal sets, the prototypical example being the famous middle-$\frac{1}{3}$ Cantor set $X \subseteq [0, 1]$, which satisfies the "geometric recursion"[1] relation $X = \frac{1}{3}X \cup (\frac{1}{3}X + \frac{2}{3})$, using the obvious notation for scaling and translation of a set. In general, a self-similar set is defined by a finite family $\Phi = \{f_i\}_{i\in\Lambda}$ of maps of the form $f_i(x) = r_i x + a_i$, where $0 < |r_i| < 1$ and $a_i \in \mathbb{R}$. The family $\Phi$ is called an *iterated function system* (or *IFS*),[2,3] and the self-similar set they define is unique compact set $X \neq \emptyset$ satisfying

[1] The mddle-1/3 Cantor set can also be described in other ways, e.g. by a recursive construction, or symbolically as the points in [0, 1] that can be written in base 3 without the digit 1. General self-similar sets also have representations of this kind, but in this paper we shall not use them.
[2] Iterated function systems consisting of non-affine maps and on other metric spaces than $\mathbb{R}$ are also of interest, but we do not discuss them here.
[3] Supported by ERC grant 306494.

M. Hochman (✉)
Department of Mathematics,The Hebrew University of Jerusalem, Jerusalem, Israel
e-mail: mhochman@math.huji.ac.il



D.-J. Feng and K.-S. Lau (eds.), *Geometry and Analysis of Fractals*,
Springer Proceedings in Mathematics & Statistics 88,
DOI 10.1007/978-3-662-43920-3_8

$$X = \bigcup_{i \in \Lambda} f_i X. \tag{1}$$

(existence and uniqueness are due to Hutchinson [Hut81]).

Throughout this paper we make a few simplifying assumptions. To avoid trivialities, we always assume that $\Phi$ contains at least two distinct maps, otherwise $X$ is just the common fixed point of the maps. We assume that $\Phi$ has *uniform contraction*, i.e. all the contraction ratios $r_i$ are equal to the same value $r$. Finally, we assume that $r > 0$, so the maps preserve orientation. These assumptions are not necessary but they simplify the statements and arguments considerably.

## 1.2 Dimension of Self-similar Sets

Despite the apparent simplicity of the definition, and of some of the better known examples, there are still large gaps in our understanding of the geometry of self-similar sets. In general, we do not even know how to compute their dimension. Usually one should be careful to specify the notion of dimension that one means, but it is a classical fact that, for self-similar sets, all the major notions of dimension coincide, and in particular the Hausdorff and box (Minkowski) dimensions agree (e.g. [Fal89, Theorem 4 and Example 2]). Thus we are free to choose either one of these, and we shall choose the latter, whose definition we now recall. For a subset $Y \subseteq \mathbb{R}$ denote its covering number at scale $\varepsilon$ by

$$N_\varepsilon(Y) = \min\{k \ : \ Y \text{ can be covered by } k \text{ sets of diameter } \leq \varepsilon\}$$

The *box dimension* of $Y$, if it exists, is the exponential growth rate of $N_\varepsilon(Y)$:

$$\dim_B Y = \lim_{\varepsilon \to 0} \frac{\log N_\varepsilon(Y)}{\log(1/\varepsilon)}$$

Thus $\dim_B Y = \alpha$ means that $N_\varepsilon(Y) = \varepsilon^{-\alpha + o(1)}$ as $\varepsilon \to 0$. It is again well known that the limit exists when $Y$ is self-similar, we shall see a short proof in Sect. 3.1.

It is easy to give upper bounds for the dimension of a self-similar set. Taking $X$ as in (1) and iterating the relation we get

$$X = \bigcup_{i \in \Lambda} f_i(\bigcup_{j \in \Lambda} f_j X) = \bigcup_{i,j \in \Lambda} f_i \circ f_j(X)$$

Writing $f_{i_1 \ldots i_n} = f_{i_1} \circ \ldots \circ f_{i_n}$ for $\underline{i} = i_1 \ldots i_n \in \Lambda^n$ and iterating $n$ times, we have

$$X = \bigcup_{\underline{i} \in \Lambda^n} f_{\underline{i}} X \tag{2}$$

This union consists of $|\Lambda|^n$ sets of diameter $r^n|X|$, so by definition,

$$N_{r^n \mathrm{diam}(X)}(X) \leq |\Lambda|^n$$

Hence

$$\dim_{\mathrm{B}}(X) = \lim_{n\to\infty} \frac{\log N_{r^n \mathrm{diam}(X)}(X)}{\log(1/r^n \mathrm{diam}(X))} \leq \frac{\log|\Lambda|}{\log(1/r)} \tag{3}$$

The right hand side of (4) is called the *similarity dimension* of $X$ and is denoted $\mathrm{sdim} X$.[4]

Is this upper bound an equality? Note that the bound is purely combinatorial and does not take into account the parameters $a_i$ at all. Equality is known to hold under some assumptions on the separation of the "pieces" $f_i X$, $i \in \Lambda$, for instance assuming *strong separation* (that the union (1) is disjoint), or the open set condition (that there exists open set $\emptyset \neq U \subseteq \mathbb{R}$ such that $f_i U \subseteq U$ and $f_i U \cap f_j U = \emptyset$ for $i \neq j$).

Without separation conditions, however, the inequality (3) can be strict. There are two trivial ways this can occur. First, there could be too many maps: if $|\Lambda| > 1/r$ then the right hand side of (3) is greater then 1, whereas $\dim_{\mathrm{B}} X \leq 1$ due to the trivial bound $N(X,\varepsilon) \leq \lceil \mathrm{diam}(X)/\varepsilon \rceil$. Thus we should adjust (3) to read

$$\dim_{\mathrm{B}}(X) \leq \min\{1, \frac{\log|\Lambda|}{\log(1/r)}\} \tag{4}$$

Second, the combinatorial bound may be over-counting if some of the sets in the union (2) coincide, that is, for some $n$ we have $f_{\underline{i}} = f_{\underline{j}}$ for some distinct $\underline{i}, \underline{j} \in \Lambda^n$. This situation is known as *exact overlaps*. If such $\underline{i}, \underline{j}$ exist then we can re-write (2) as $X = \bigcup_{\underline{u}\in\Lambda^n\setminus\{\underline{i}\}} f_{\underline{u}} X$, which presents $X$ as the attractor of the IFS $\Phi' = \{f_{\underline{u}}\}_{\underline{u}\in\Lambda'}$ for $\Lambda' = \Lambda^n \setminus \{\underline{i}\}$. This IFS consists of $|\Lambda'| = |\Lambda|^n - 1$ maps that contract by $r^n$, so, applying the trivial bound (3) to this IFS, we have $\dim_{\mathrm{B}} X \leq \log(|\Lambda|^n - 1)/\log(1/r^n)$, which is better than the previous bound of $\log|\Lambda|/\log(1/r)$. To take an extreme example, if all the maps $f_i$ coincide then the attractor $X$ is just the unique fixed point of the map, and its dimension is 0.

Are there other situations where a strict inequality occurs in (4)? A-priori, one does not need *exact* coincidences between sets in (2) to make the combinatorial bound very inefficient. It could happen, for example, that many of the sets $f_{\underline{i}} X$, $\underline{i} \in \Lambda^n$, align almost exactly, in which case one may need significantly fewer than $|\Lambda|^n$ $\varepsilon$-intervals to cover them. Nevertheless, although such a situation can easily be arranged for a fixed $n$, to get a drop in dimension one would need this to happen at all sufficiently small scales. No such examples are known, and the main subject of this paper is the conjecture that this cannot happen:

---

[4] It would be better to write $\mathrm{sdim}\Phi$, since this quantity depends on the presentation of $X$ and not on $X$ itself, but generally there is only one IFS given and no confusion should arise.

*Conjecture 1.1* A strict inequality in (4) can occur only in the presence of exact overlaps.

This conjecture appears in [PS00, Question 2.6], though special cases of it have received attention for decades, in particular Furstenberg's projection problem for the 1-dimensional Sierpinski gasket (see e.g. [Ken97]), the 0,1,3-problem (see e.g. [PS95]) and, for self-similar measures instead of sets, the Bernoulli convolutions problem, (e.g. [PSS00]).

One may also draw an analogy between this conjecture and rigidity statements in ergodic theory. Rigidity is the phenomenon that, for certain group actions of algebraic origin, the orbit of the point is as large as it can be (dense or possibly even equidistributed for the volume measure) unless there is an algebraic obstruction to this happening. To see the connection with the conjecture above, note that $X$ is just the orbit closure of (any) $x \in \mathbb{R}$ under the semigroup $\{f_{\underline{i}} : \underline{i} \in \bigcup_{n=1}^{\infty} \Lambda^N\}$ of affine maps, and that exact overlaps occur if and only if this semigroup is not generated freely by $\{f_i\}_{i \in \Lambda}$. Thus the conjecture predicts that the orbit closure of any point is as large as it can be unless there are algebraic obstructions.

## 1.3 Progress Towards the Conjecture

Our main subject here is a weakened form of the conjecture which proves the full conjecture in some important examples and special classes of IFSs. In order to state it we must first quantify the degree to which the sets $f_{\underline{i}}(X)$ are separated from each other. Since all of the maps in $\Phi$ contract by the same ratio, any two of the sets $f_{\underline{i}}(X)$ and $f_{\underline{j}}(X)$ for $\underline{i}.\underline{j} \in \Lambda^n$ are translates of each other. We define the distance between them as the magnitude of this translation, which is given by $f_{\underline{i}}(x) - f_{\underline{j}}(x)$ for any $x \in \mathbb{R}$; we shall choose $x = 0$ for concreteness. Thus a measure of the degree of concentration of cylinders $f_{\underline{i}}(X)$, $\underline{i} \in \Lambda^n$, is provided by

$$\Delta_n = \min\{|f_{\underline{i}}(0) - f_{\underline{j}}(0)| \ : \ \underline{i}, \underline{j} \in \Lambda^n \ , \ \underline{i} \neq \underline{j}\}$$

Evidently, exact overlaps occur if and only if there exists an $n$ such that $\Delta_n = 0$. Fixing $x \in X$, the points $f_{\underline{i}}(x)$ , $\underline{i} \in \Lambda^n$, all lie in $X$, and so there must be a distinct pair $\underline{i}, \underline{j} \in \Lambda^n$ with $|f_{\underline{i}}(x) - f_{\underline{j}}(x)| \leq \text{diam}(X)/|\Lambda|^n$; hence $\Delta_n \to 0$ at least exponentially. In general there may be an exponential lower bound on $\Delta_n$ as well, i.e. $\Delta_n \geq cr^n$ for some $c, r > 0$. This is always the case when the IFS satisfies strong separation or the open set condition, but there are examples where it holds even when these conditions fail (see Garsia [Gar62]). Therefore the following theorem from [Hoc12] gives nontrivial information and should be understood as a weak form of Conjecture 1.1.

**Theorem 1.2** *If* $X \subseteq \mathbb{R}$ *is a self-similar set and* $\dim X < \min\{1, sdimX\}$, *then* $\Delta_n \to 0$ *super-exponentially, that is,* $-\frac{1}{n}\log \Delta_n \to \infty$.

In practice, one applies the theorem after establishing an exponential lower bound on $\Delta_n$ to deduce that $\dim X = \min\{1, \text{sdim}X\}$. For example,

**Proposition 1.3** *Let $\mathcal{R}$ denote the set of rational IFSs, i.e. such that $r, a_i \in \mathbb{Q}$. Then Conjecture 1.1 holds in $\mathcal{R}$.*

*Proof* First, a useful identity: For $\underline{i} \in \Lambda^n$, a direct calculation shows that

$$f_{\underline{i}}(x) = r^n x + \sum_{k=1}^{n} a_{i_k} r^{k-1} \tag{5}$$

$$= r^n x + f_{\underline{i}}(0) \tag{6}$$

Now let that $f_i(x) = rx + a_i$ where $r = p/q$ and $a_i = p_i/q_i$ for $p, p_i, q, q_i$ integers and write $Q = \prod_{i \in \Lambda} q_i$. Then $f_{\underline{i}}(0) = \sum_{k=1}^{n} a_{i_k} r^{n-k}$ is a rational number with denominator $Qq^n$. Suppose that no overlaps occur, so that $\Delta_n > 0$ for all $n$. Given $n$, by definition there exist distinct $\underline{i}, \underline{j} \in \Lambda^n$ such that $\Delta_n = f_{\underline{i}}(0) - f_{\underline{j}}(0)$. Therefore $\Delta_n$ is a non-zero rational number with denominator $Qq^n$ so we must have $\Delta_n \geq 1/Qq^n$. By Theorem 1.2 we conclude that $\dim X = \min\{1, \text{sdim}X\}$. □

The same argument works in the class of IFSs with algebraic coefficients, using a similar lower bound on polynomial expressions in a given set of algebraic numbers. See [Hoc12, Theorem 1.5]. A simple (but non-trivial) calculation, due to B. Solomyak and P. Shmerkin, also allows one to deal with the case that one of the translation parameters $a_i$ is irrational, resolving Furstenberg's question about linear projections of the one-dimensional Sierpinski gasket [Hoc12, Theorem 1.6]. Theorem 1.2 leads to strong results about parametric families of self-similar sets [Hoc12, Theorem 1.8], and there is a version for measures which has also led to substantial progress on the Bernoulli convolutions problem, see [Hoc12, Theorem 1.9] and the recent advance by Shmerkin [Shm13]. Another interesting application is given in [Orp13].

The rest of this paper is an exposition of the proof of the theorem. Our goal is to present the ideas as transparently as possible, and to this end we frame the argument in terms of covering numbers (rather than entropy as in [Hoc12]). This leads to simpler statements and to an argument that is conceptually correct but, unfortunately, incomplete; some crucial steps of this simplified argument are flawed. In spite of this deficiency we believe that such an exposition will be useful as a guide to the more technical proof in [Hoc12]. To avoid any possible misunderstandings, we have indicated the false statements in quotation marks ("Lemma", "Proof", etc.).

As we shall see, the main idea is to reduce (the negation of) the theorem to a statement about sums of self-similar sets with other sets. Problems about sums of sets fall under the general title of additive combinatorics, and in the next section we give a brief introduction to the parts of this theory that are relevant to us. In Sect. 3 we explain the reduction to a statement about sumsets, and show how an appropriate inverse theorem essentially settles the matter. Finally, in Sect. 4, we discuss how the heuristic argument can be made rigorous.

# 2 A Birds-Eye View of Additive Combinatorics

## 2.1 *Sumsets and Inverse Theorems*

The sum (or sumset) of non-empty sets $A, B \subseteq \mathbb{R}^d$ is

$$A + B = \{a + b \, : \, a \in A \, , \, b \in B\}$$

Additive combinatorics, or at least an important chapter of it, is devoted to the study of sumsets and the relation between the structure of $A$, $B$ and $A + B$. We focus here on so-called inverse problems, that is the problem of describing the structure of sets $A, B$ such that $A + B$ is "small" relative to the sizes of the original sets. The general flavor of results of this kind is that, if the sumset is small, there must be an algebraic or geometric reason for it. It will become evident in later sections that this question comes up naturally in the study of self-similar sets.

To better interpret what "small" means, first consider the trivial bounds. Assume that $A, B$ are finite and non-empty. Then $|A + B| \geq \max\{|A|, |B|\}$, with equality if and only if at least one of the sets is a singleton. In the other direction, $|A + B| \leq |A||B|$, and equality can occur (consider $A = \{0, 10, 20, 30, \ldots, 10n\}$ and $B = \{0, 1, \ldots, 9\}$). For "generic" pairs of sets the upper bound is close to the truth. For example, when $A, B \subseteq \{1, \ldots, n\}$ are chosen randomly by including each $1 \leq i \leq n$ in $A$ with probability $p$ and similarly for $B$, with all choices independent, there is high probability that $|A + B| \geq c|A||B|$. The question becomes, what can be said between these two extremes.

## 2.2 *Minimal Growth*

One of the earliest inverse theorems is the Brunn-Minkowski inequality of the late 19th century. The setting is $\mathbb{R}^d$ with the volume measure, and it states that if $A, B \subseteq \mathbb{R}^d$ are convex sets then, given the volumes of $A, B$, the volume of $A+B$ is minimized when $A, B$ are balls with respect to some common norm. Since the volume of a ball scales like the $d$-th power of the radius, this means that $\mathrm{vol}(A + B) \geq (\mathrm{vol}(A)^{1/d} + \mathrm{vol}(B)^{1/d})^d$, and equality occurs if and only if, up to a nullset, $A, B$ are dilates of the same convex set. The inequality was later extended to arbitrary Borel sets (note that $A+B$ may not be a Borel set but it is an analytic set and hence Lebesgue measurable). For a survey of this topic see Gardner [Gar02].

Similar tight statements hold in the discrete setting. The analog of a convex body is an *arithmetic progression (AP)*, namely a set of the form $P = \{a, a + d, a + 2d, \ldots, a + (1 - k)d\}$, where $d$ is called the gap (we assume $d \neq 0$) and $k$ is called the length of $P$. Then for finite sets $A, B \subseteq \mathbb{Z}^d$ with $|A|, |B| \geq 2$ we always have $|A + B| \geq |A| + |B| - 1$, with equality if and only if $A, B$ are APs of the same gap [TV06, Proposition 5.8].

## 2.3 Linear Growth: Small Doubling and Freiman's Theorem

Now suppose that $A = B \subseteq \mathbb{Z}^d$ but weaken the hypothesis, assuming only that

$$|A + A| \leq C|A| \tag{7}$$

where we think of $A$ as large and $C$ as constant. Such sets are said to have *small doubling.*

The simplest example of small doubling in $\mathbb{Z}^d$ is when $A = \{1, \ldots, n\}^d$, in which case $|A + A| \leq 2^d|A|$. This example can be pushed down to any lower dimension as follows. For $i = 1, \ldots, k$, take intervals of integers $I_i = \{1, 2, \ldots, n_i\}$, and let $T : \mathbb{Z}^k \to \mathbb{Z}^d$ be an affine map given by integer parameters. Suppose that $T$ is injective on $I = I_1 \times \ldots \times I_k$. Then $A = T(I) \subseteq \mathbb{Z}^d$ has the property that

$$|A + A| = |T(I) + T(I)| = |T(I + I)| \leq |I + I| \leq 2^k|I| = 2^k|A|$$

A set $A$ as above is called a *(proper) generalized arithmetic progression (GAP) of rank $k$.*

GAPs are still extremely algebraic objects but one can get away from this a little using another cheap trick: Begin with a set $A$ satisfying $|A+A| \leq C|A|$ (e.g. a GAP) and choose any $A' \subseteq A$ with cardinality $|A'| \geq D^{-1}|A|$ for some $D > 1$. Then

$$|A' + A'| \leq |A + A| \leq C|A| \leq CD|A'|$$

One of the central results of additive combinatorics is Freiman's theorem, which says that, remarkably, these are the only ways to get small doubling.

**Theorem 2.1** (Freiman) *If $A \subseteq \mathbb{Z}^d$ and $|A+A| \leq C|A|$, then $A \subseteq P$ for a GAP $P$ of rank $C'$ and satisfying $|P| \leq C''|A|$. The constants satisfy $C' = O(C(1+\log C))$ and $C'' = C^{O(1)}$.*

For more information see [TV06, Theorem 5.32 and Theorem 5.33].

Combined with some standard arguments (e.g. the Plünnecke-Rusza inequality), the symmetric version leads to an asymmetric versions: assuming $A, B \subseteq \mathbb{Z}^d$ and $C^{-1} \leq |A|/|B| \leq C$, if $|A + B| \leq C|A|$ then $A, B$ are contained in a GAP $P$ of rank and $\leq C'$ and size $|P| \leq C'|A|$, with similar bounds on the constants.

## 2.4 Power Growth, the "Fractal" Regime

Now relax the growth condition even more and consider finite sets $A \subseteq \mathbb{Z}$ (or $A \subseteq \mathbb{R}$) such that

$$|A + A| \leq |A|^{1+\delta} \tag{8}$$

This is the discrete analog of the condition

$$\dim_B(X+X) \leq (1+\delta)\dim_B X \tag{9}$$

for $X \subseteq \mathbb{R}$. Indeed, given $X \subseteq \mathbb{R}$ and $n \in \mathbb{N}$ let $X_n$ denote the set obtained by replacing each $x \in X$ with the closest point $k/2^n, k \in \mathbb{Z}$. Then $|X_n| \sim 2^{n(\dim_B X+o(1))}$ and $|X_n+X_n| \sim 2^{n(\dim_B(X+X)+o(1))}$ for large $n$, so (9) is equivalent to $|X_n+X_n| \lesssim |X_n|^{1+o(1)}$. Thus, the difference between (7) and (8) is roughly the difference between using Lebesgue measure or dimension to quantify the size of a set $X \subseteq \mathbb{R}$.

Here is a typical example of a set satisfying (8). Write $P_n = \{0,\ldots,n-1\}$ and let

$$\begin{aligned} A_n &= \sum_{i=1}^{n} \frac{1}{2^{i^2}} P_{2^i} \\ &= \{\sum_{i=1}^{n} a_i 2^{-i^2} \; : \; 1 \leq a_i \leq 2^i\} \end{aligned}$$

(again, one can think of this either as a subset of $\mathbb{R}$, or of $\frac{1}{4^{n^2}}\mathbb{Z}$). It is easy to verify that the distance between distinct points $x, x' \in A_n$ is at least $1/4^{n^2}$, and that such $x$ has a unique representation as a sum $\sum_{i=1}^{n} a_i 4^{-i^2} \; : \; 1 \leq a_i \leq 2^i$. Indeed, each term in the sum $\sum_{i=1}^{n} a_i 2^{-i^2}$ determines a distinct block of binary digits. Thus $A_n$ is a GAP, being the image of $P_2 \times P_4 \times \ldots \times P_{2^n}$ by the map $(x_1,\ldots,x_n) \mapsto \sum \frac{1}{2^{i^2}} x_i$. The rank is $n$, and so, as we saw in the previous section,

$$|A_n + A_n| \leq 2^n |A_n|$$

Since

$$|A_n| = \prod_{i=1}^{n} |P_n| = 2^{\sum_{i=1}^{n} i} = 2^{n(n+1)/2}$$

we have

$$|A_n + A_n| = |A_n|^{1+o(1)} \qquad \text{as } n \to \infty$$

The reader may recognize the example above as the discrete analog of a Cantor set construction, where at stage $n$ we have a collection of intervals $2^{n(n+1)/2}$ of length $2^{-n^2}$, and from each of these intervals we keep $2^{n+1}$ sub-intervals of length $2^{-(n+1)^2}$, separated by gaps of length $2^{-n^2-(n+1)}$. For the resulting Cantor set $X$ it is a standard exercise to see that $\dim X = \dim_B X = 1/2$, and the calculation above shows that $\dim X + X = 1/2$ as well. Such constructions appear in the work of Erdős-Volkmann [EV66], and also in the papers of Schmeling-Shmerkin [SS10] and

Körner [Kör08], who showed that for any sequence $\alpha_1 \leq \alpha_2 \leq \ldots$ there is a set $X$ with $\dim \sum_{i=1}^{n} X = \alpha_n$.

Do all examples of (9) look essentially like this one? In principle one can apply Freiman's theorem, since the hypothesis (9) can be written as $|A + A| \leq C|A|$ for $C = |A|^{\delta}$. What one gets, however, is that $A$ is a $|A|^{O(\delta)}$-fraction of a GAP or rank $|A|^{O(\delta)}$, and this gives rather coarse information about $A$ (note that, trivially, every set is a GAP of rank $|A|$).

Instead, it is possible to apply a multi-scale analysis, showing that at some scales the set looks quite "dense" and at others quite "sparse". The best way to explain this is in the language of trees, which we introduce next.

## 2.5 Trees and Tree-Measures

Denote the length of a finite sequence $\sigma = \sigma_1 \ldots \sigma_n$ by $|\sigma| = n$ and write $\emptyset$ for the empty word, which by definition has $|\emptyset| = 0$. Denote the concatenation of words $\sigma$ and $\tau$ by $\sigma\tau$, in which case we say that $\sigma$ is a prefix of $\tau$, and that $\sigma\tau$ extends $\sigma$.

The *full binary tree* of height $h$ is the set $\{0,1\}^{\leq h} = \bigcup_{k=0}^{h} \{0,1\}^k$ of 0, 1-valued sequences of length $\leq h$, where our convention is that $\{0,1\}^0 = \{\emptyset\}$, so the empty word is included. We define a *tree of height h* is a subset $T \subseteq \bigcup_{i=0}^{h} \{0,1\}^i$ satisfying

(T1) $\emptyset \in T$.
(T2) If $\sigma \in T$ and $\eta$ is an initial segment of $\sigma$ then $\eta \in T$.
(T3) If $\sigma \in T$ then there is an $\eta \in T$ which extends $\sigma$ and $|\eta| = h$.

One may think of $T$ as a set of vertices and introduce edges between every pair $\sigma_1 \ldots \sigma_i, \sigma_1 \ldots \sigma_i \sigma_{i+1} \in T$. Then $T$ is a tree if $\emptyset \in T$ and in the associated graph there is a path from $\emptyset$ to every node, and all maximal paths are of length $h$.

The *level (or depth)* of $\sigma \in T$ is its length (the graph-distance from $\emptyset$ to $\sigma$). The *leaves* of a tree $T$ of height $h$ are the elements of the lowest (deepest) level, namely $h$:

$$\partial T = T \cap \{0,1\}^h$$

The *descendants* of $\sigma \in T$ are the nodes $\eta \in T$ that extend $\sigma$. The nodes $m$ generations below $\sigma$ in $T$ are the nodes of the form $\eta = \sigma\sigma' \in T$ for $\sigma' \in \{0,1\}^m$.

We also shall need to work with measures "on trees", or, rather, measures on their leaves. For notational purposes it is better to introduce the notion of a *tree-measure*[5] on the full tree $\{0,1\}^{\leq h}$, namely, a function $\mu : \{0,1\}^{\leq h} \to [0,1]$ satisfying

(M1) $\mu(\emptyset) = 1$.
(M2) $\mu(\sigma) = \sum_{i \in \{0,1\}} \mu(\sigma i)$

It is easily to derive from (M1) and (M2) that $\sum_{\sigma \in \{0,1\}^k} \mu(\sigma) = 1$ for every $1 \leq k \leq h$, so a tree-measure induces genuine probability measures on every level

[5] This notion is identical to a flow on the tree in the sense of network theory.

of the full tree, and in particular on $\partial T$. Conversely, if we have a genuine probability measure $\mu$ on the set of leaves $\{0,1\}^h$ of the full tree of height $h$ then it induces a tree-measure by $\mu(\sigma) = \sum_{\eta:\sigma\eta\in\partial T}\mu(\{\sigma\eta\})$. Given a tree-measure, the set $T = \{\sigma : \mu(\sigma) > 0\}$ is a tree which might be called the support of $\mu$.

Every tree-measure $\mu$ on $\{0,1\}^{\leq h}$ defines a distribution on the nodes of the tree as follows: first choose a level $0 \leq i \leq h$ uniformly, and then choose a node $\sigma \in \{0,1\}^i$ in level $i$ with the probability given by $\mu$, i.e. $\mu(\sigma)$ (we have already noted that at each level the masses sum to 1). Thus the probability of $A \subseteq T$ is

$$\mathbb{P}_\mu(A) = \frac{1}{h+1}\sum_{\sigma\in A}\mu(\sigma)$$

and the expectation of $f : \{0,1\}^{\leq h} \to \mathbb{R}$ is

$$\mathbb{E}_\mu(f) = \frac{1}{h+1}\sum_{k=0}^{n}\sum_{\sigma\in\{0,1\}^k}\mu(\sigma)f(\sigma)$$

Sometimes we write $\mathbb{P}_{\sigma\sim\mu}$ or $\mathbb{E}_{\sigma\sim\mu}$ to define $\sigma$ as a random node, as in the expression

$$\mathbb{P}_{\sigma\sim\mu}(\sigma \in T \text{ and } \sigma \text{ has two children in } T) = \frac{1}{h+1}\sum_{\sigma\in T}\mu(\sigma)1_{\{\sigma 0,\sigma 1\in T\}}$$

Given the tree $T$ of height $h$, it is natural to consider the uniform probability measure $\mu_{\partial T}$ on $\partial T$ and, as described above, extend it to a tree-measure, which we denote $\mu_T$. In this case we abbreviate the probability and expectation operators above by $\mathbb{P}_T$ and $\mathbb{E}_T$, etc. It is important to note that *choosing a node according to* $\mu_T$ *is not the same as choosing a node uniformly from* $T$. The latter procedure is usually heavily biased towards sampling from the leaves, since these generally constitute a large fraction of the nodes (in the full binary tree, sampling this way gives a leaf with probability $> 1/2$). In contrast, $\mu_T$ samples uniformly from the levels, and within each level we sample according to the relative number of leaves descended from each node.

Trees and tree-measures are naturally related to sets and measures on $[0,1)$ using binary coding. Given a set $X \subseteq [0,1)$ and $h \in \mathbb{N}$, we lift $X$ to a tree $T$ of height $h$ by taking all the initial sequences of length $\leq h$ of binary expansions of points in $X$, with the convention that the expansion terminates in 1s if there is an ambiguity. We remark that for $k \leq h$,

$$N_{1/2^k}(X) \leq \left|T \cap \{0,1\}^k\right| \leq 2N_{1/2^k}(X)$$

Similarly, a probability measure $\mu$ on $[0,1)$ can be lifted to a tree-measure $\widetilde{\mu}$ on $\{0,1\}^{\leq h}$ by defining $\widetilde{\mu}(\sigma)$ equal to the mass of the interval of numbers whose binary expansion begins with $\sigma$.

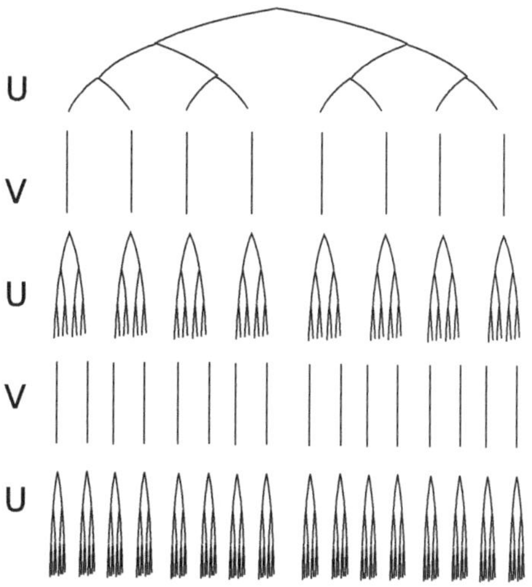


**Fig. 1** A tree with alternating levels having full branching at some levels, full concentration at others, and a few levels omitted. Schematically this is what the tree associated to $A_n$ from Sect. 2.4 looks like, as well as the conclusion of Theorem 2.2 (with $W$ indicated by the small space between levels)

## *2.6 Inverse Theorems in the Power-Growth Regime*

We need some terminology for describing the local structure of trees. We say that $T$ has *full branching for m generations at* $\sigma$ if $\sigma$ has all $2^m$ possible descendants $m$ generations below it, that is, $\sigma\eta \in T$ for all $\eta \in \{0,1\}^m$. At the other extreme, we say that $T$ is *fully concentrated for m generations at* $\sigma$ if $\sigma$ has a single descendant $m$ generations down, that is, there is a unique $\eta \in \{0,1\}^m$ with $\sigma\eta \in T$.

Let us return to the example $A_n$ from Sect. 2.4 and examine the associated tree $T_n$ of height $n^2$. For every $i < n$, every node at level $i^2$ has full branching for $i$ generations; and every node at level $i^2+i$ is fully concentrated for $i+1$ generations. Consequently, for every $j \in [i^2, i^2+i)$ every node of level $j$ has full branching for one generation; for $j \in [i^2+i, (i+1)^2)$, every node at level $j$ is fully concentrated for one generation. We also have the following statement: For every $m$ we can partition the levels $0, 1, \ldots, n^2$ into three sets $U, V, W$, such that (a) For every $i \in U$, every level-$i$ node has full branching for $m$ generations; (b) For every $j \in V$, every level-$j$ node is fully concentrated for $m$ generations; and (c) $W$ is a negligible fraction of the levels, specifically $|W|/n^2 = o(1)$ as $n \to \infty$ (with $m$ fixed). Of course, $U = \bigcup_{i>m}[i^2, i^2+i-m)$, $V = \bigcup_{i>m}[i^2+i, (i+1)^2-m)$, and $W$ is the set of remaining levels. This is pictured schematically in Fig. 1.

Does this picture hold in general when $|A+A| \leq |A|^{1+\delta}$? Certainly not exactly, since we can always pass to a subset $A' \subseteq A$ of size $|A'| \geq |A|^{1-\delta}$ and get a set with similar doubling behavior (for a constant loss in $\delta$), but much less structure. One can also perturb it in other ways. However, in a looser sense, the picture above is quite general. One approach is to pass to a subtree of reasonably large relative size. Such an approach was taken by Bourgain in [Bou03, Bou10]. The approach taken in [Hoc12] is more statistical, and in a sense it gives a description of the entire tree, but requires us to weaken the notion of concentration. Given $\delta > 0$, we say that $T$ is

*$\delta$-concentrated* for $m$ generations at $\sigma \in T$ if there exists $\eta \in \{0,1\}^m$ such that

$$\mu_T(\sigma\eta) \geq (1-\delta)\mu_T(\sigma)$$

where $\mu_T$ is the tree-measure associated to $T$. In other words, $T$ is $\delta$-concentrated at $\sigma$ if it is possible to remove an $\delta$-fraction of the leaves descended from $\sigma$ in such a way that the resulting tree becomes fully concentrated for $m$ generations at $\sigma$. Note that this definition is not purely local, since it depends not only on the depth-$m$ subtree of $T$ rooted at $\sigma$, but on the entire subtree rooted at $\sigma$, since the weights on $S = \{\sigma\eta \,:\, \eta \in \{0,1\}^m\}$ are determined by the number of leaves of $T$, not by $S$ itself.

**Theorem 2.2** *For every $\varepsilon > 0$ and $m > 1$, there is a $\delta > 0$ such that for all sufficiently small $\rho > 0$ the following holds. Let $X \subseteq [0,1]$ be a finite set such that*

$$N_\rho(X+X) \leq N_\rho(X)^{1+\delta}$$

*and let $T$ be the associated tree of height $h = \lceil \log(1/\rho) \rceil$. Then the levels $0, 1, \ldots, h$ can be partitioned into sets $U, V, W$ such that*

1. *For every $i \in U$,*

$$\mathbb{P}_{\sigma\sim T}(T \text{ has full branching at } \sigma \text{ for } m \text{ generations} \mid \sigma \text{ is in level } i) > 1-\varepsilon.$$

2. *For every $j \in V$,*

$$\mathbb{P}_{\sigma\sim T}(T \text{ is } \varepsilon\text{-concentrated at } \sigma \text{ for } m \text{ generations} \mid \sigma \text{ is in level } j) > 1-\varepsilon.$$

3. *$|W| < \varepsilon h$.*

Note that if $X$ is $\rho$-separated, the hypothesis is essentially the same as $|X+X| \leq |X|^{1+\varepsilon}$.

Our analysis of self-similar sets requires the following asymmetric variant, which is easily seen to imply the symmetric one above. To motivate it, note that $|A+B| \leq C|A|$ can occur for two trivial reasons: One is that $A = \{1,\ldots,n\}$ for some $n$ and $B \subseteq \{1,\ldots,n\}$ is arbitrary. The second is that $B = \{b\}$, a singleton, and $A$ is arbitrary. The following theorem says that when $|A+B| \leq |A|^{1+\delta}$ then there are essentially two kinds of scales: those where, locally, the sets $A, B$ look like in the first trivial case, and those where, locally, $A, B$ look like the second trivial case. See Fig. 2.

**Theorem 2.3** *For every $\varepsilon > 0$ and $m > 1$, there is a $\delta > 0$ such that for all sufficiently small $\rho > 0$ the following holds. Let $X, Y \subseteq [0,1]$ be finite sets such that*

$$N_\rho(X+Y) \leq N_\rho(X)^{1+\delta}$$

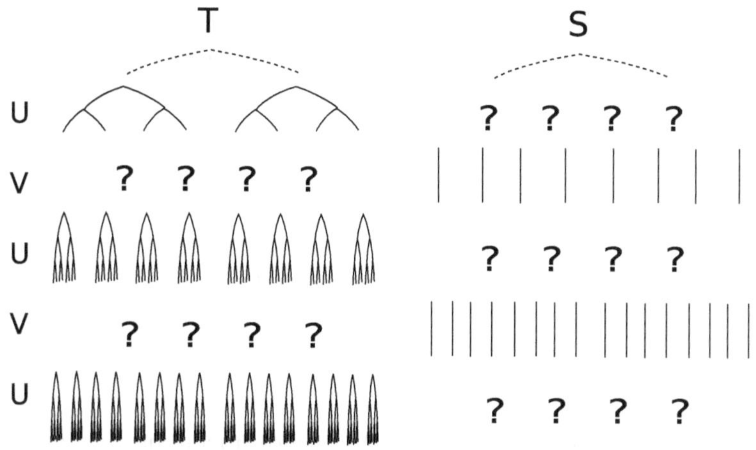


**Fig. 2** Schematic representation of the conclusion of Theorem 2.3 (with $W$ indicated by the small space between levels)

*Let $T, S$ be the associated trees of height $h = \lceil \log(1/\rho) \rceil$, respectively. Then the levels $0, 1, \ldots, h$ can be partitioned into sets $U, V, W$ such that*

1. *For every $i \in U$,*

$$\mathbb{P}_{\sigma \sim T}(T \textit{ has full branching at } \sigma \textit{ for } m \textit{ generations} \mid \sigma \textit{ is in level } i) > 1 - \varepsilon$$

*(but we know nothing about $S$ at level $i$).*

2. *For every $j \in V$,*

$$\mathbb{P}_{\sigma \sim S}(S \textit{ is } \varepsilon\textit{-concentrated at } \sigma \textit{ for } m \textit{ generations} \mid \sigma \textit{ is in level } j) > 1 - \varepsilon$$

*(but we know nothing about $T$ at level $j$).*

3. *$|W| < \varepsilon h$.*

The theorems above follow from [Hoc12, Theorems 2.7 and 2.9], using the fact that high enough entropy at a given scale implies full branching, and small enough entropy at a given scale implies $\delta$-concentration.

# 3 A Conceptual Proof of Theorem 1.2

In this section we give a heuristic proof of Theorem 1.2. We begin with some general observations about self-similar sets. Then we explain how the theorem is reduced to a statement about sumset growth. Finally, we demonstrate how the inverse theorems of the previous section are applied.

From now on let $\Phi = \{f_i\}_{i \in \Lambda}$ be an IFS with attractor $X$, as in the introduction. We assume that $0 \in X \subseteq [0, 1)$; this can always be achieved by a change of coordinates, which does not affect the statement of Theorem 1.2.

## *3.1 Sumset Structure of Self-similar Sets*

Our analysis will focus on finite approximations of $X$. Define the $n$-th approximations by

$$X_n = \{f_{\underline{i}}(0) \,:\, \underline{i} \in \Lambda^n\}$$

Clearly $X_n \subseteq X$. Also note that $|X_n| \leq |\Lambda|^n$, with a strict inequality for some $n$ if and only if exact overlaps occur. Self similarity enters our argument via the following lemma.

**Lemma 3.1** *For any $m, n \in \mathbb{N}$,*

$$X = X_m + r^m X \tag{10}$$

$$X_{m+n} = X_m + r^m X_n \tag{11}$$

*Proof* By (2) and (6),

$$\begin{aligned} X &= \bigcup_{\underline{i} \in \Lambda^n} f_{\underline{i}}(X) \\ &= \bigcup_{\underline{i} \in \Lambda^n} \{f_{\underline{i}}(0) + r^m x \,:\, x \in X\} \\ &= X_m + r^m X \end{aligned}$$

which is the first identity. To prove the second, for $\underline{i} \in \Lambda^m$ and $\underline{j} \in \Lambda^n$ denote by $\underline{i}\,\underline{j}$ their concatenation. By (5),

$$\begin{aligned} f_{\underline{i}\,\underline{j}}(0) &= \sum_{k=1}^{m} a_{i_k} r^{k-1} + r^m \sum_{k=1}^{n} a_{j_k} r^{k-1} \\ &= f_{\underline{i}}(0) + r^m f_{\underline{j}}(0) \end{aligned}$$

hence

$$\begin{aligned} X_{m+n} &= \{f_{\underline{i}\,\underline{j}}(0) \,:\, \underline{i}\underline{j} \in \Lambda^{m+n}\} \\ &= \{f_{\underline{i}}(0) + r^m f_{\underline{j}}(0) \,:\, \underline{i} \in \Lambda^m, \underline{j} \in \Lambda^n\} \\ &= X_m + r^m X_n \end{aligned}$$

□

Let us demonstrate the usefulness of this lemma by showing that $\dim_B(X)$ exists. First, since $r^m X$ is of diameter $\leq r^m$, it is easy to deduce from (10) that $N_{r^n}(X_n)$, $N_{r^n}(X)$ differ by at most a factor of 2. Thus the existence of $\dim_B X$ is equivalent to existence of the limit $\frac{1}{m} \log N_{r^m}(X_m)$ as $n \to \infty$. Next, we have a combinatorial lemma.

**Lemma 3.2** *Let $A, B \subseteq \mathbb{R}$ with $B \subseteq [0, \varepsilon)$. Then for any $\gamma < \varepsilon$,*

$$N_\gamma(A + B) \geq \frac{1}{3} \cdot N_\varepsilon(A) \cdot N_\gamma(B)$$

*Proof* Let $\mathcal{I} = \{I_i\}_{i=1}^{N_\varepsilon(A)}$ be an optimal cover of $A$ by disjoint intervals of length $\varepsilon$. Let $\mathcal{J} = \{J_j\}_{j=1}^{N_\gamma(A+B)}$ be an optimal cover of $A + B$ by intervals of length $\gamma$. For each $I_i \in \mathcal{I}$ fix a point $a_i \in A \cap I_i$ and note that $a_i + B \subseteq A + B$ is covered by $\mathcal{J}$, so $a_i + B$ intersects at least $N_\gamma(B)$ intervals in $\mathcal{J}$. If each interval $J_j$ intersects a unique translate $a_i + B$, we would conclude that $N_\gamma(A+B) \geq N_\varepsilon(A)N_\gamma(B)$. While $a_i$ may not be unique, we can argue as follows: Since $B \subseteq [0, \varepsilon)$, if $J_j = [u, u + \varepsilon]$ and intersects $a + B$ for some $a \in A$, then $a \in [u - \varepsilon, u + 2\varepsilon)$. Since the intervals $I_i$ are disjoint and of length $\varepsilon$, there are most 3 intervals $I_i \in \mathcal{I}$ that $a$ could belong to. The claim follows. □

Since $X \subseteq [0, 1)$ we have $r^m X_n \subseteq [0, r^m)$, so by the lemma,

$$\begin{aligned} N_{r^{m+n}}(X_{m+n}) &= N_{r^{m+n}}(X_m + r^m X_n) \\ &\geq \frac{1}{3} \cdot N_{r^m}(X_m) \cdot N_{r^{m+n}}(r^m X_n) \\ &= \frac{1}{3} \cdot N_{r^m}(X_m) \cdot N_{r^n}(X_n) \end{aligned}$$

where in the last equality we used the identity $N_{t\varepsilon}(tZ) = N_\varepsilon(Z)$. Taking logarithms, this shows that the sequence $s_n = \log N_{r^m}(X_m)$ is approximately super-additive in the sense that $s_{m+n} \geq s_m + s_n - C$ for a constant $C$. The existence of the limit of $\frac{1}{n}s_n$ as $n \to \infty$ is then well known (perhaps it is better known when $C = 0$ and $s_n$ is (really) super-additive. The proof for $C = 0$ works also in the $C > 0$ case; alternatively, note that $s'_n = s_n - \log n$ becomes super-additive after excluding finitely many terms, so $\lim \frac{1}{n}s'_n$ exists, and $\frac{1}{n}s'_n - \frac{1}{n}s_n \to 0$).

## *3.2 From Theorem 1.2 to Additive Combinatorics*

Let us return to our main objective, Theorem 1.2. Continuing with the previous notation, write

$$\begin{aligned} \alpha &= \dim_B X \\ \beta &= \min\{1, \mathrm{sdim} X\} \end{aligned}$$

and suppose, by way of contradiction, that $\alpha < \beta$ and that for some $k \in \mathbb{N}$ we have $\Delta_n \geq 2^{-kn}$ for all $n$ (in particular, there are no exact overlaps). We make a number of observations. The first is rather trivial: that "too small" dimension means that there are intervals of length $r^m$ containing exponentially many points from $X_m$. Precisely,

**Proposition 3.3** *Let $\sigma = \frac{1}{2}(\beta - \alpha) > 0$. Then for every large enough $m$, there is an interval $I_m$ of length $r^m$ such that $|X_m \cap I_m| > r^{-\sigma m}$.*

*Proof* As we have already noted, $\frac{1}{m\log(1/r)} \log N_{r^m}(X_m) \to \alpha$ as $m \to \infty$. Thus for large enough $m$,

$$N_{r^m}(X_m) < r^{-(\beta-\sigma)m}$$

On the other hand, since there are no exact overlaps,

$$|X_m| = |\Lambda|^m = r^{-m\mathrm{sdim}(X)} \geq r^{-m\beta}$$

Thus in an optimal cover of $X_m$ by $r^m$-intervals, at least one must contain $|X_m|/N_{r^m}(X_m) \geq (1/r)^{\sigma m}$ points. □

We now wish to extract more information from the sumset identity $X_{m+n} = X_m + r^m X_n$. In itself it provides limited information about the covering number $N_{m+n}(X_n)$, since the summands live at different scales. This is what was used earlier in proving super-additivity of $s_n = \log N_{r^m}(X_m)$. The next step is to localize the sumset relation.

**Proposition 3.4** *For all $\delta > 0$, for all large $m$ there exists an interval $J_m$ of length $r^m$ such that $X_m \cap J_m \neq \emptyset$ and, writing $n = km$,*

$$N_{r^{m+n}}((X_m \cap J_m) + r^m X_n) < r^{-(1+\delta)\alpha n} \tag{12}$$

*Proof* Fix $m$, set $n = km$, and let $\mathcal{J}$ denote the partition of $\mathbb{R}$ into intervals $[ur^m, (u+1)r^m)$, $u \in \mathbb{Z}$, whose lengths are $r^m$. Since $X_m = \bigcup_{J \in \mathcal{J}}(X_m \cap J)$, we can re-write (11) as

$$\begin{aligned} X_{m+n} &= X_m + r^m X_n \\ &= \bigcup_{J \in \mathcal{J}} ((X_m \cap J) + r^m X_n) \end{aligned} \tag{13}$$

Since $X_m \cap J \subseteq [ur^m, (u+1)r^m)$ for some $u$ and $r^m X_n \subseteq [0, r^m)$, we have $(X_m \cap J) + r^m X_n \subseteq [ur^m, (u+2)r^m)$ and in particular each set in the union (13) is of diameter $\leq 2r^m$. On the other hand, no interval of length $r^{m+n}$ intersects more than three of the sets $[ur^m, (u+2)r^m)$. Therefore, arguing as in the proof of Lemma (3.2),

$$N_{r^{m+n}}(X_{m+n}) \geq \frac{1}{3} \cdot N_{r^m}(X_m) \cdot \min_{J \in \mathcal{J} : X_m \cap J \neq \emptyset} N_{r^{m+n}}((X_{m+n} \cap J) + r^m X_n)$$

so

$$\begin{aligned}\min_{J\in\mathcal{J}\,:\,X_m\cap J\neq\emptyset} N_{r^{m+n}}((X_{m+n}\cap J)+r^m X_n) &\le 3\cdot\frac{N_{r^{m+n}}(X_{m+n})}{N_{r^m}(X_m)}\\ &\le 3\cdot\frac{r^{-(\alpha+o(1))(m+n)}}{r^{-(\alpha+o(1))m}}\\ &= r^{-(\alpha+o(1))n}\qquad \text{as } m\to\infty\end{aligned}$$

The proposition follows. □

Now suppose that it so happens that, for large $m$, the propositions above produce the *same* interval: $I_m = J_m$. Then we would have the following:

**Proposition 3.5** *Suppose that* $\dim X < \min\{1, sdimX\}$ *and* $\Delta_n \ge 2^{-kn}$ *for all* $n$. *Then there is a constant* $\tau > 0$ *such that, for every* $\delta > 0$ *and all suitably large* $n$, *there is a subset* $Y_n \subseteq [0,1]$ *with*

$$N_{r^n}(Y_n) \ge 2^{\tau n} \tag{14}$$

$$N_{r^n}(X_n + Y_n) \le N_{r^n}(X_n)^{1+\delta} \tag{15}$$

*Proof* Let $\sigma$ be as in Proposition 3.3 and take $\tau = \sigma/(k\log(1/r))$. As before write $n = (k+1)m$, and assume that the intervals $I_m, J_m$ provided by the two previous propositions coincide for arbitrarily large $m$: $I_m = J_m = [a_m, b_m)$. Let

$$Y_m = r^{-m}(X_m \cap I_m - a_m)$$

and note that $Y_m \subseteq [0,1)$. Now, by choice of $I_m$ we know that $|X_m \cap I_m| \ge r^{-\sigma m}$, and since $\Delta_m \ge 2^{-km} = 2^{-n}$, we know that every two points in $X_m \cap I_m$ are separated by at least $2^{-n}$. Therefore,

$$\begin{aligned}N_{r^n}(Y_m) &= N_{r^{m+n}}(X_m \cap I_m)\\ &\ge r^{-\sigma m}\end{aligned}$$

Using the identity $N_{t\varepsilon}(tZ) = N_\varepsilon(Z)$ with $t = r^m$ and $Z = Y_m$, we conclude that

$$N_{r^{-n}}(Y_m) \ge r^{-\sigma m} = r^{\tau n}$$

Similarly, since $X_n + Y_m = r^{-m}((X_m \cap I_m) + r^m X_n)$, from the definition of $J_m$ and the identity $N_{t\varepsilon}(tZ) = N_\varepsilon(Z)$ again, we find that for large enough $n$ (equivalently, $m$),

$$\begin{aligned}N_{r^n}(X_n + Y_n) &= N_{r^{m+n}}((X_m \cap I_m) + r^m X_n)\\ &\le r^{-(1+\delta)\alpha n}\\ &\le N_{r^n}(X_n)^{(1+2\delta)}\end{aligned}$$

where in the last inequality we again used the fact that $N_{r^m}(X_m) \sim r^{-n\alpha}$. □

The task of showing that the conclusion of the "Proposition" is impossible falls within the scope of additive combinatorics. Heuristically, it cannot happen because, being a fractal, $X_n$ has very little "additive structure". This intuition is correct, as we discuss in the next section.

But can one really ensure that $I_m$,$J_m$ coincide? A natural attempt would be to show that, for a fixed optimal $r^m$-cover of $X_m$, "most" intervals of length $r^m$ can play each of the roles, and hence a positive fraction can play both. In fact, for every $\eta > 0$, for large $m$ at least a $(1-\eta)$-proportion of these intervals will be a good choice for $J_m$. Unfortunately, although the number of candidates for $I_m$ can be shown to be exponential in $m$, it could still be exponentially small compared to $N_{r^m}(X_m)$, and so we cannot conclude that the two families of "good" intervals have members in common. It is possible that more sophisticated counting can make this work, but the approach that is currently simplest is to replace covering numbers by the entropy, at an appropriate scale, of the uniform measure on $X_m$. We return to this in Sect. 4.

## 3.3 Getting a Contradiction

Our goal now is to demonstrate that the conclusion of "Proposition" 3.5 is impossible. The argument we give again falls short of this goal, but it gives the essential ideas of the proof. Thus, we ask the reader to suspend his disbelief a little longer.

Let $\tau > 0$ be as given in "Proposition" 3.5. Choose a very small parameter $\varepsilon > 0$ which we shall later assume is small compared to $\tau$. Choose $m$ large enough that

$$N_{r^m}(X_m) \geq r^{-m(1-\varepsilon)\alpha}$$

Apply the inverse Theorem 2.3 with parameters $\varepsilon, m$ and obtain the promised $\delta > 0$. From "Proposition" 3.5 obtain the corresponding $Y_n \subseteq [0, 1)$ satisfying (14) and (15).

Write $T^n$ for the tree of height $h_n = [1/r^n]$ associated to $X_n$ and $S^n$ for the tree of the same height associated to $Y_n$. From our choice of $\delta$ and (15), by the inverse theorem there is a partition $U_n \cup V_n \cup W_n$ of $\{1, \ldots, h_n\}$ such that

(I) At scales $i \in U_n$, a $1-\varepsilon$ fraction of nodes of $T^n$ at level $i$ have full branching for $m$-generations.
(II) At scales $j \in V_n$, a $1-\varepsilon$ fraction of nodes of $S^n$ at level $j$ are $\varepsilon$-concentrated for $m$ generations.
(III) $|W_n| \leq \varepsilon h_n$.

Our first task is to show that $V_n$ is not too large. It is quite clear (or at least believable) that if a tree has few nodes with more than one child, then it can have only an exponentially small number of leaves. The same is true if we only assume, for a small $\lambda > 0$, that most nodes are $\lambda$-concentrated. More precisely,

**Lemma 3.6** *Let $S$ be a tree of height $h$, let $\lambda > 0$ and $\ell \geq 1$. Suppose that*

$$\mathbb{P}_S(\sigma \in S \, : \, S \text{ is } \lambda\text{-concentrated at } \sigma \text{ for } \ell \text{ generations}\} > 1 - \lambda$$

*Then $|\partial S| \leq 2^{\lambda' \cdot h}$ where $\lambda' \to 0$ as $\lambda \to 0$ and $h/\ell \to \infty$.*

We leave the proof to the motivated reader. We note that this lemma is superseded by Proposition 4.5, which gives a stronger statement and has a simpler proof.

We apply the lemma to $S = S^n$ with $\ell = m$. Choose $\lambda$ small enough that $\lambda' < \tau$ for large $n$ (hence large $h_n$). Thus $\lambda$ depends only on $\tau$ and we may assume that at the start we chose $\varepsilon < \frac{1}{2}\lambda$. Suppose that we had $|V_n| > (1 - \lambda/2)h_n$. Since in each level $j \in V_n$ a $(1 - \varepsilon)$-fraction of the nodes (with respect to the tree measure $\mu_{S^n}$) is $\varepsilon$-concentrated, at least the same fraction is $\lambda$-concentrated, and we would conclude

$$\begin{aligned}
\mathbb{P}_{S^n}(\sigma \in S^n \, : \, S^n \text{ is } \lambda\text{-concentrated at } \sigma \text{ for } m \text{ generations}) &\geq \frac{1}{h_n}|V_n| \cdot (1 - \varepsilon) \\
&> (1 - \frac{\lambda}{2})(1 - \varepsilon) \\
&> 1 - \lambda
\end{aligned}$$

From the lemma we would have $N_{r^n}(Y_n) \leq |\partial S^n| < 2^{\lambda' h_n} < 2^{\tau h_n}$, contradicting (14). Thus, we conclude that

$$|V_n| < (1 - \frac{\lambda}{2})h_n$$

Consequently, assuming as we may that $\varepsilon < \lambda/6$,

$$|U_n| = h_n - |V_n| - |W_n| \geq (\frac{\lambda}{2} - \varepsilon)h_n > \frac{\lambda}{3}h_n \tag{16}$$

So far we have seen that $U_n$ consists of a positive fraction of the levels of $T^n$, and hence a positive fraction of nodes in $T^n$ have full branching for $m$ generations. Our next task will be to show that most of the remaining nodes have roughly $r^{-\alpha m}$ descendants $m$ generations down. This is where we use self-similarity again in an essential way.

**Proposition 3.7** *If $m$ is large enough, then for all large enough $n$,*

$$\mathbb{P}_{\sigma \sim \mu_{T^n}} \begin{pmatrix} \sigma \text{ has } \geq 2^{(1-\varepsilon)\alpha m} \text{ descendants} \\ m \text{ generations down in } T \end{pmatrix} > 1 - \varepsilon$$

*Proof (sketch.)* A node $\sigma \in T^n$ of level $\ell$ corresponds to an interval $I = [\frac{u}{2^\ell}, \frac{u+1}{2^\ell})$. We call such intervals level-$\ell$ intervals, and recall that the probability induced from $\mu_{T^n}$ on level-$\ell$ intervals is just proportional to $|I \cap X_n|$. The claim is then that if we choose $0 \leq \ell \leq h_n$ uniformly and then choose a level-$\ell$ interval $I$ at random, then with probability at least $1-\varepsilon$, we will have $N_{2^{-\ell-m}}(I \cap X_n) \geq 2^{-(1-\varepsilon)\alpha m}$. In order to prove this, it is enough to show that for all levels $0 \leq \ell \leq (1-\frac{\varepsilon}{2})h_n$, if we choose a level-$\ell$ interval $I$ at random, then with probability at least $1-\frac{\varepsilon}{2}$ we have $N_{2^{-\ell-m}}(I \cap X_n) \geq 2^{-(1-\varepsilon)\alpha m}$.

Fix a parameter $m_0$ depending on $\varepsilon$ and assume $m, n$ large with respect to it. Observe that $X_n$ decomposes into a union of copies of $X_{n'}$ scaled by approximately $2^{-\ell-m_0}$. More precisely, choosing $u \in \mathbb{N}$ such that $r^u \approx 2^{-\ell-m_0}$, by (11) we have

$$X = X_u + r^u X_{n-u} = \bigcup_{x \in X_u} (x + r^u X_{n-u})$$

The idea is now the following. The translates $x + r^u X_{n-u}$ in the union are of diameter $r^u \approx 2^{-\ell}/2^{m_0}$, which is much smaller than $2^{-\ell}$, and hence with probability at least $1-\frac{\varepsilon}{2}$ a level-$\ell$ interval $I$ will contain an entire translate $x + r^u X_{n-u}$ from the union above. The details of the proof are somewhat tedious and we omit them. The point is that, if $x + r^u X_{n-u} \subseteq I$, and assuming that $m$ is large enough relative to $\varepsilon, m_0$, we have

$$\begin{aligned} N_{2^{-\ell-m}}(I \cap X_n) &\geq N_{2^{-\ell-m}}(x + r^u X_{n-u}) \\ &= N_{2^{-\ell-m} r^{-u}}(X_{n-u}) \\ &\approx N_{2^{-(m-m_0)}}(X_{n-u}) \\ &> 2^{(1-\varepsilon)\alpha m} \end{aligned}$$

which is what we wanted to prove. □

Now that we know that most nodes in $T^n$ have many descendants, and a positive fraction have the maximal possible number of descendants, $m$ generations down, the last ingredient we need is a way to use this information to get a lower bound on the number of leaves in $T^n$. Heuristically, this is the analog of the upper bound we had in Lemma 3.6.

**Proposition 3.8** *Let $T$ be a tree of height $h$, let $m \geq 0$, and suppose that the nodes of $T$ can be partitioned into disjoint sets $A_1, \ldots, A_\ell$ such that each node $\sigma \in A_i$ has $2^{c_i m}$ descendants $m$ generations down. Write $p_i = \mathbb{P}_{\mu_T}(A_i)$. Then*

$$|\partial T| \geq \prod_{i=1}^{\ell} 2^{c_i \cdot p_i h}$$

This "Proposition" is, unfortunately, incorrect, and the reader may find it instructive to look for a counterexample. The statement could be fixed if we made stronger assumptions than just bounding the branching in each of the sets $A_i$, but the resulting argument would almost certainly be more complicated than the proof in [Hoc12], and we do not pursue it. The correct statement is given in Proposition 4.5 below.

We can now put the pieces together. By the defining property (I) of $U_n$ and equation (16), the set $A_1^n \subseteq T^n$ of nodes with full branching for $m$-generations satisfies

$$\begin{aligned}\mathbb{P}_{T^n}(A_1^n) &\geq \frac{1}{h_n}|U_n| \cdot (1-\varepsilon) \\ &\geq \frac{\lambda}{3}(1-\varepsilon) \\ &\geq \frac{\lambda}{4}\end{aligned}$$

assuming again $\varepsilon$ small compared to $\lambda$ (equivalently $\tau$). Let $A_2^n$ denote the set of nodes of $T^n \setminus A_1^n$ which do *not* have at least $2^{m(1-2\varepsilon)\dim X}$ descendants $m$ generations down; by Proposition 3.7,

$$\mathbb{P}_{T^n}(A_2^n) < \varepsilon$$

Therefore if we define $A_3^n = T^n \setminus \{A_1^n \cup A_2^n \backslash\}$ then all nodes in $A_3^n$ have at least $2^{m(1-2\varepsilon)\dim X}$ descendants $m$ generations down and

$$\mathbb{P}_{T^n}(A_3^n) = 1 - \mathbb{P}_{T^n}(A_1^n) - \mathbb{P}_{T^n}(A_2^n)$$

In the terminology of the "Proposition", we have $p_1 \geq \lambda/4$ and $p_2 < \varepsilon$, hence $p_3 \geq 1 - p_1 - \varepsilon$. Also $c_1 = 1$, $c_3 = (1-2\varepsilon)\dim X$ and by default $c_2 \geq 0$. From the "Proposition" we find that

$$\begin{aligned}|\partial T^n| &\geq 2^{p_1 h_n} \cdot 2^{p_2 \cdot c_2 h_n} \cdot 2^{p_3 (1-2\varepsilon)\dim X \cdot h_n} \\ &\geq 2^{p_1 h_n + (1-2\varepsilon)\dim X \cdot (1-p_1-\varepsilon) h_n} \\ &\geq 2^{(\dim X + \varepsilon) h_n}\end{aligned}$$

where in the last inequality we assumed that $\varepsilon$ is small compared to $p_1$ and $\dim X$. Since $N_{r^n}(X) = |\partial T^n|^{1+o(1)}$ as $n \to \infty$, this contradicts the definition of $\dim X$.

## *3.4 Sums with Self-similar Sets*

What we "proved" above is the following statement which is of independent interest, and is, moreover, true (a proof follows easily from the methods of [Hoc12]).

**Theorem 3.9** *For every any self-similar set $X$ with $\dim X < 1$ and every $\tau > 0$ there is a $\delta > 0$ such that for all small enough $\rho > 0$ and any set $Y \subseteq \mathbb{R}$,*

$$N_\rho(Y) > (1/\rho)^\tau \qquad \Longrightarrow \qquad N_\rho(X+Y) > N_\rho(X)^{1+\delta}$$

There is also a fractal version for Hausdorff dimension:

**Theorem 3.10** *For every any self-similar set $X$ with* $\dim X < 1$ *and every $\tau > 0$ there is a $\delta > 0$ such that for any set $Y \subseteq \mathbb{R}$,*

$$\dim Y > \tau \qquad \Longrightarrow \qquad \dim(X+Y) > \dim X + \delta$$

For box dimension (lower or upper) the analogous statement follows directly from the previous theorem. The version for Hausdorff dimension requires slightly more effort and will appear in [Hoc13] along with the analog for measures.

# 4 Entropy

In this final section we discuss how to turn the outline above into a valid proof. The main change is to replace sets by measures and covering numbers by entropy. Each of the three parts of the argument (inverse theorem, reduction to a statement about sumsets, and the analysis of the sums) has an entropy analog which we indicate below, along with a reference to the relevant part of [Hoc12].

The reader should note that the outline given below is designed to match as closely as possible the argument from the previous section, rather than the proof from [Hoc12]. Although the ideas and many of the details are the same, the original proof is direct, whereas the one here is by contradiction. For this reason not all of the statements below have exact analogs in [Hoc12].

## *4.1 Entropy*

We assume that the reader is familiar with the basic properties of Shannon entropy, see for example [CT06]. Let $\mathcal{I}_\varepsilon = \{[k\varepsilon, (k+1)\varepsilon)\}_{k\in\mathbb{Z}}$, which is a partition of $\mathbb{R}$ into intervals of length $\varepsilon$. The entropy $H(\mu, \mathcal{I}_\varepsilon)$ of $\mu$ at scale $\varepsilon$ is the natural measure-analog of the covering number $N_\varepsilon(X)$, albeit in a logarithmic scale. For a measure $\nu$ supported on a set $X$, the two quantities are related by the basic inequality

$$0 \le H(\nu, \mathcal{I}_\varepsilon) \le \log \#\{I \in \mathcal{I}_\varepsilon \,:\, X \cap I \neq \emptyset\} \le \log N_\varepsilon(X) + O(1)$$

(the $O(1)$ error is because we are choosing a sub-cover of $X$ from a fixed cover of $\mathbb{R}$ rather than allowing arbitrary $\varepsilon$-intervals). We introduce the normalized $\varepsilon$-scale entropy:

$$H_\varepsilon(\nu) = \frac{1}{\log(1/\varepsilon)} H(\nu, \mathcal{I}_\varepsilon)$$

Thus, for $\nu$ supported on a set $X$ with well-defined box dimension, the previous inequality implies

$$\limsup_{\varepsilon\to 0} H_\varepsilon(\nu) \leq \dim_B X \tag{17}$$

## 4.2 Inverse Theorems for Entropy

The measure-analog of the sumset operation is convolution, which for discrete probability measures $\mu = \sum p_i \delta_{x_i}$ and $\nu = \sum q_j \delta_{y_j}$ is[6]

$$\mu * \nu = \sum_{i,j} p_i q_j \delta_{x_i + y_j}$$

The entropy-analog of the small doubling condition $|A + A| \leq C|A|$ is the inequality $H(\mu * \mu) \leq H(\mu) + C'$, where $H(\mu)$ is the entropy of a measure with respect to the partition into points (remember that entropy is like cardinality, but in logarithmic scale). Alternatively we could discretize at scale $\varepsilon$, giving $H_\rho(\mu*\mu) \leq H_\rho(\mu)+O(1/\log(1/\varepsilon))$. Tao [Tao10] has shown that such inequalities have implications similar to Freiman's theorem. Related results were also obtained by Madiman [Mad08], see also [MMT12].

The regime that interests us is, as before, the analog of $|A + B| \leq |A|^{1+\delta}$, which by formal analogy takes the form $H_\rho(\mu * \nu) \leq (1+\delta)H_\rho(\mu)$. When $\mu$ is supported on $[0, 1]$ we have $H_\rho(\mu) \leq 1 + o(1)$ as $\rho \to 0$, and this inequality is implied (and in the cases that interest us essentially equivalent to)

$$H_\rho(\mu * \nu) \leq H_\rho(\mu) + \delta \tag{18}$$

Before stating the inverse theorem for entropy we need a few more definitions. Consider the lift of $\mu$ to a tree-measure $\widetilde{\mu}$ on the full binary tree of height $h$ (see Sect. 2.5). Given a node $\sigma = \sigma_1 \ldots \sigma_k$ and $m \in \mathbb{N}$, write $\sigma\{0, 1\}^m$ for the set of descendants of $\sigma$ $m$-generations down. Let $\widetilde{\mu}_{\sigma,m}$ denote the probability measure on $\sigma\{0, 1\}^m$ that assigns to each node its normalized weight according to $\widetilde{\mu}$. Since $\sum_{\eta\in\sigma\{0,1\}^m} \widetilde{\mu}(\eta) = \widetilde{\mu}(\sigma)$, this measure is given by $\widetilde{\mu}_{\sigma,m}(\eta) = \widetilde{\mu}(\eta)/\widetilde{\mu}(\sigma)$.

We say that $\widetilde{\mu}$ is $\delta$-concentrated at $\sigma$ for $m$ generations if $H(\widetilde{\mu}_{\sigma,m}) < \delta$, that is, if

$$-\frac{1}{m} \sum_{\eta\in\sigma\{0,1\}^m} \frac{\widetilde{\mu}(\eta)}{\widetilde{\mu}(\sigma)} \log \frac{\widetilde{\mu}(\eta)}{\widetilde{\mu}(\sigma)} < \delta.$$

---

[6] In general there is a similar formula: $\mu*\nu = \int\int \delta_{x+y} d\mu(x) d\nu(y)$, where the integral is interpreted as a measure by integrating against Borel functions.

For a tree measure $\mu_T$ associated to a tree $T$ and for fixed $m$, this is equivalent to $T$ being $\delta'$-concentrated for $m$ generations at $\sigma$ for an appropriate $\delta'$ which tends to 0 together with $\delta$. We say that $\widetilde{\mu}$ is $\delta$-uniform at $\sigma$ for $m$ generations if $H(\widetilde{\mu}_{\sigma,m}) > \log m - \delta$. Note that for $m$ fixed, when $\delta$ is small enough this implies that $\widetilde{\mu}(\eta) > 0$ for all $\eta \in \sigma\{0,1\}^m$, so this indeed generalizes full branching. We can now state the inverse theorem:

**Theorem 4.1** (Theorem 2.7 of [Hoc12]) *For every $\varepsilon > 0$ and $m \geq 1$, there is a $\delta > 0$ such that for sufficiently small $\rho > 0$ the following holds. Let $\mu, \nu$ be probability measures on $[0,1]$ and suppose that*

$$H_\rho(\mu * \nu) \leq H_\rho(\mu) + \delta$$

*Let $\widetilde{\mu}, \widetilde{\nu}$ denote the lifts of $\mu, \nu$ to the full binary trees of height $h = \lceil \log_2(1/\rho) \rceil$. Then there is a partition of the levels $\{0, \ldots, h\}$ into three sets $U \cup V \cup W$ such that*

1. *For $i \in U$,*

$$\mathbb{P}_{\sigma\sim\widetilde{\mu}}(\widetilde{\mu} \text{ is } \varepsilon\text{-uniform at } \sigma \text{ for } m \text{ generations} \mid \sigma \text{ is in level } i) > 1 - \varepsilon$$

2. *For $j \in V$,*

$$\mathbb{P}_{\sigma\sim\widetilde{\mu}}(\widetilde{\nu} \text{ is } \varepsilon\text{-concentrated at } \sigma \text{ for } m \text{ generations} \mid \sigma \text{ is in level } i) > 1 - \varepsilon$$

3. *$|W| < \delta n$.*

## 4.3 Reduction of Theorem 1.2 to a Convolution Inequality

We return to our IFS $\Phi$ with attractor $0 \in X \subseteq [0,1]$, as in Sect. 3. Define measures $\mu^{(n)}$ analogous to $X_n$ by

$$\mu^{(n)} = \frac{1}{|\Lambda|^n} \sum_{\underline{i}\in\Lambda^n} \delta_{f_{\underline{i}}(0)}$$

Write $S_t\mu(A) = \mu(t^{-1}A)$ (this is the usual push-forward of $\mu$ by $S_t$). Then the analog of the sumset relation $X_{m+n} = X_m + r^m X_n$ is

$$\mu^{(m+n)} = \mu^{(m)} * S_{r^m}\mu^{(n)}$$

The derivation is elementary, using the definition of convolution, equation (6) and the identity $S_t\delta_y = \delta_{ty}$. Next, as in Sect. 3.1, if we define $s_m = H(\mu^{(m)}, \mathcal{I}_{r^m})$ then the sequence $s_n$ is almost super-additive in the sense that $s_{m+n} \geq s_m + s_n - O(1)$. This is proved by a similar argument to the covering number case but in the language of entropy. It follows that the limit

$$\alpha = \lim_{m\to\infty} H_{r^m}(\mu^{(m)})$$

exists. Since $\mu^{(m)}$ is supported on $X$, by (17) we have $\alpha \leq \dim X$.

Turning to Theorem 1.2, write

$$\beta = \min\{1, \mathrm{sdim}\, X\}$$

and assume for the sake of contradiction that $\dim X < \beta$ and $\Delta_n \geq 2^{-kn}$ for some $k$. Since $\alpha \leq \dim X$, we can choose $\varepsilon > 0$ so that $\alpha < \beta - \varepsilon$. Arguing analogously to Proposition 3.3 one obtains the analogous result:

**Proposition 4.2** *There is a constant $c$ (depending on $\beta, \varepsilon$) such that for large enough $m$,*

$$\mu^{(m)}\left(\bigcup\left\{I \in \mathcal{I}_{r^m} \,:\, H_{r^{m+n}}(\mu_I^{(m)}) > cm\right\}\right) > c$$

This lemma does not appear explicitly in [Hoc12], since that is a direct proof. Ours is a proof by contradiction, and the contradiction can be interpreted as showing that the lemma above is false. This falsehood is demonstrated directly in the last displayed equation of Sect. 5.3 of [Hoc12].

Next, for a probability measure $\nu$ and set $E$ with $\nu(E) > 0$, write $\nu_E$ for the conditional measure on $E$, that is, $\nu_E(A) = \frac{1}{\nu(E)}\nu(E \cap A)$. The analog of Proposition 3.4 then holds, again with an analogous proof:

**Proposition 4.3** (See Equation (40) of [Hoc12]) *For every $\delta > 0$, as $m \to \infty$*

$$\mu^{(m)}\left(\bigcup\left\{I \in \mathcal{I}_{r^m} \,:\, \frac{1}{n}H_{r^{m+n}}(\mu_I^{(m)} * S_{r^m}\mu^{(n)}) \leq \alpha + \delta\right\}\right) \geq 1 - o(1)$$

From the last two propositions one sees that for given $\delta > 0$ and large enough $m$, there are intervals $I = I_m \in \mathcal{I}_{r^m}$ that appear in the unions in the conclusions of both propositions. Taking $\nu_m$ to be the re-scaling of $\mu_I^{(m)}$ by $r^{-m}$ (translated back to $[0, 1)$), we have the rigorous analog of "Proposition" 3.5:

**Proposition 4.4** *There is a $\tau > 0$ such that for every $\delta > 0$, for all sufficiently large $m$, there is a measure $\nu_m$ supported on $[0, 1)$ with*

$$\frac{1}{m}H_{r^m}(\nu_m) > \tau \tag{19}$$

$$\frac{1}{m}H_{r^m}(\mu^{(m)} * \nu_m) < \frac{1}{m}H_{r^m}(\mu^{(m)}) + \delta \tag{20}$$

## 4.4 Getting a Contradiction

The missing ingredient in Sect. 3.3 was the ability to estimate the number of leaves of a tree from the average amount of branching of its nodes. This is where entropy really comes in handy, because of the following (easy!) lemma. Recall that given a tree-measure $\theta$, we write $\theta_{\sigma,m}$ for the normalized weights on the nodes $m$ generations down from $\sigma$.

**Lemma 4.5** (Lemma 3.4 of [Hoc12]) *Let $\widetilde{\theta}$ be a tree-measure on the full binary tree $T$ of height $h$. Write $\partial\widetilde{\theta}$ for the measure induced by $\widetilde{\theta}$ on the leaves of the tree. Then for any $m$,*

$$\frac{1}{h}H(\partial\widetilde{\theta}) = \mathbb{E}_{\sigma\sim\widetilde{\theta}}(\frac{1}{m}H(\widetilde{\theta}_{\sigma,m})) + O(\frac{m}{h})$$

From here the argument proceeds exactly as in Sect. 3.3. Let $\tau > 0$ be the constant provided by Proposition 4.4. Choose a small parameter $\varepsilon > 0$. Choose $m$ large enough that

$$H_{r^m}(\mu^{(m)}) \geq (1-\varepsilon)\alpha$$

Apply the inverse theorem 4.1 with parameters $\varepsilon, m$ and let $\delta > 0$ be the resulting number. Applying Proposition 4.4 with this $\delta$, there exist probability measures $\nu_n$ on $[0, 1]$ satisfying (19) and (20). Write $\widetilde{\mu}^{(n)}, \widetilde{\nu}_n$ for the lift of $\mu^{(n)}, \nu_n$, respectively, to the binary tree of height $h_n = \lceil 1/\log(r^n)\rceil$. By the inverse theorem there is a partition $U_n \cup V_n \cup W_n$ of the levels $\{1, \ldots, h_n\}$ such that

(I) At scales $i \in U_n$, the $\widetilde{\mu}^{(n)}$-mass of nodes at level $i$ that are $\varepsilon$-uniform for $m$ generations is at least $1-\varepsilon$.
(II) At scales $j \in V_n$, the $\widetilde{\nu}_n$-mass of nodes at level $i$ that are $\varepsilon$-concentrated for $m$ generations is at least $1-\varepsilon$
(III) $|W_n| \leq \varepsilon h_n$.

If $|V_n| > (1-\tau/2)h_n$ and $\varepsilon$ is small enough compared to $\tau$, then sufficiently many nodes (with respect to $\widetilde{\nu}_n$) would have $H(\widetilde{\nu}^n_{\sigma,m}) < \varepsilon$ that we could invoke Lemma 4.5 and conclude that the entropy $H_{r^n}(\nu_n) \approx \frac{1}{n\log(1/r)}H(\widetilde{\nu}^n) < \tau$, contradicting (19). Therefore $|V_n| \leq (1-\tau/2)h_n$. In particular, assuming $\varepsilon$ is small enough compared to $\tau$,

$$|U_n| \geq h_n - |V_n| - |W_n| \geq \frac{\tau}{3}h_n$$

Next, suppose that $m$ is large enough so that $H_{r^n}(\mu^{(n)}) > (1-\varepsilon)\alpha$. Using self-similarity of $X$ and an argument analogous to the one outlined in Proposition 3.7, we get the analogous result:

**Lemma 4.6** (Lemma 5.4 of [Hoc12]) *For all large enough $n$,*

$$\frac{1}{h_n+1}\sum_{\sigma}\left\{\widetilde{\mu}^{(n)}(\sigma)\ :\ \frac{1}{m}H(\widetilde{\mu}^{(n)}_{\sigma,m})>(1-2\varepsilon)\alpha\right\}>1-\varepsilon$$

Now, from the definition of $U_n$ and our bound $|U_n|\geq\frac{\tau}{3}h_n$, we know that at least a $(1-\varepsilon)\tau/3$-fraction of the nodes of $\widetilde{\mu}^{(n)}$ satisfy $\frac{1}{m}H(\widetilde{\mu}^{(n)}_{\sigma,m})>(1-\varepsilon)m$. Of the remaining nodes, by the last lemma all but a $\varepsilon$-fraction satisfy $\frac{1}{m}H(\widetilde{\mu}^{(n)}_{\sigma,m})\geq(1-2\varepsilon)\alpha$. Therefore by Lemma 4.5 again, for all large enough $n$,

$$H_{r^n}(\mu^{(n)})\approx\frac{1}{h_n}H(\widetilde{\mu}^{(n)})>(1-\varepsilon)^2\frac{\tau}{3}+(1-(1-\varepsilon)\frac{\tau}{3}-\varepsilon)\alpha>\alpha+\varepsilon$$

assuming $\varepsilon$ is small compared to $\tau$. This contradicts the definition of $\alpha$.

**Acknowledgments** Many thanks to Boris Solomyak for his coments on the paper.

## References


[Bou03] Bourgain, J.: On the Erdős-Volkmann and Katz-Tao ring conjectures. Geom. Funct. Anal. **13**(2), 334–365 (2003)

[Bou10] Bourgain, J.: The discretized sum-product and projection theorems. J. Anal. Math. **112**, 193–236 (2010)

[CT06] Thomas, M.C., Joy A.T.: Elements of Information Theory. 2nd edn. Wiley, Hoboken (2006)

[EV66] Erdős, P., Volkmann, B.: Additive Gruppen mit vorgegebener Hausdorffscher Dimension. J. Reine Angew. Math. **221**, 203–208 (1966)

[Fal89] Falconer, K.J.: Dimensions and measures of quasi self-similar sets. Proc. Amer. Math. Soc. **106**(2), 543–554 (1989)

[Gar02] Gardner, R.J.: The Brunn-Minkowski inequality. Bull. Amer. Math. Soc. (N.S.), **39**(3), 355–405 (2002)

[Gar62] Garsia, A.M.: Arithmetic properties of Bernoulli convolutions. Trans. Amer. Math. Soc. **102**, 409–432 (1962)

[Hoc12] Michael H.: Self-similar sets with overlaps and sumset phenomena for entropy. preprint (2012)

[Hoc13] Michael H.: Hausdorff dimension of sums and convolutions. in preparation (2013)

[Hut81] Hutchinson, J.E.: Fractals and self-similarity. Indiana Univ. Math. J. **30**(5), 713–747 (1981)

[Ken97] Kenyon, R.: Projecting the one-dimensional Sierpinski gasket. Israel J. Math. **97**, 221–238 (1997)

[Kör08] Körner, T.W.: Hausdorff dimension of sums of sets with themselves. Studia Math. **188**(3), 287–295 (2008)

[Mad08] Madiman, M.: On the entropy of sums. In: Information Theory Workshop, 2008. ITW '08. IEEE, pp. 303–307 (2008)

[MMT12] Madiman, M., Marcus, A.W., Tetali, P.: Entropy and set cardinality inequalities for partition-determined functions. Random Struct. Algorithms **40**(4), 399–424 (2012)

[Orp13] Tuomas O.: On the packing measure of self-similar sets. preprint (2013). http://arxiv.org/abs/1212.4813

[PSS00] Yuval, P., Wilhelm, S., Boris, S.: Sixty years of Bernoulli convolutions. In: Fractal Geometry and Stochastics, II (Greifswald/Koserow, 1998), Progr. Probab., vol. 46, pp. 39–65. Birkhäuser, Basel (2000)

[PS00] Yuval, P., Boris, S.: Problems on self-similar sets and self-affine sets: an update. In: Fractal Geometry and stochastics, II (Greifswald/Koserow, 1998), Progr. Probab., vol. 46, pp. 95–106. Birkhäuser, Basel (2000)

[PS95] Pollicott, M., Simon, K.: The Hausdorff dimension of $\lambda$-expansions with deleted digits. Trans. Amer. Math. Soc. **347**(3), 967–983 (1995)

[SS10] Jörg, S., Pablo, S.: On the dimension of iterated sumsets. In: Recent Developments in Fractals and Related Fields, Appl. Numer. Harmon. Anal., pp. 55–72. Birkhäuser Boston Inc., Boston (2010)

[Shm13] Pablo, S.: On the exceptional set for absolute continuity of bernoulli convolutions. preprint (2013). http://arxiv.org/abs/1303.3992

[Tao10] Tao, T.: Sumset and inverse sumset theory for Shannon entropy. Combin. Probab. Comput. **19**(4), 603–639 (2010)

[TV06] Terence, T., Van V.: Additive Combinatorics, Cambridge Studies in Advanced Mathematics. vol. 105, Cambridge University Press, Cambridge (2006)

# Quasisymmetric Modification of Metrics on Self-Similar Sets

**Jun Kigami**

**Abstract** Using the notions of scales and their gauge functions associated with self-similar sets, we give a necessary and sufficient condition for two metrics on a self-similar set being quasisymmetric to each other. As an application, we construct metrics on the Sierpinski carpet which is quasisymmetric with respect to the Euclidean metrics and obtain an upper estimate of the conformal dimension of the Sierpinski carpet.

## 1 Introduction

The main purpose of this paper is to give a characterization of quasisymmetry for self-similar sets in terms of scales and related notions introduced in [Kig09]. As an application, we will construct a series of metrics on the Sierpinski carpet which are quasisymmetric to the restriction of the Euclidean metric and give an upper estimate of the quasiconformal dimension of the Sierpinski carpet (Fig. 1).

Quasisymmetric maps have been introduced by Tukia and Väisälä in [TV80] as a generalization of quasiconformal mappings in the complex plane.

**Definition 1.1** (*Quasisymmetry*).

(1) Let $(X, d)$ and $(X, \rho)$ be metric spaces. $\rho$ is said to be quasisymmetric, or QS for short, with respect to $d$ if and only if there exists a homeomorphism $h$ from $[0, +\infty)$ to itself such that $h(0) = 0$ and, for any $t > 0$, $\rho(x, z) < h(t)\rho(x, y)$ whenever $d(x, z) < td(x, y)$. We write $\rho \underset{\text{QS}}{\sim} d$ if $\rho$ is quasisymmetric with respect to $d$.

(2) Let $(X, d)$ be a metric space. A homeomorphism $f : X \to X$ is called quasisymmetric if and only if $d \underset{\text{QS}}{\sim} d_f$, where $d_f(x, y)$ is defined by $d_f(x, y) = d(f(x), f(y))$.

The above definition immediately implies the following facts.

J. Kigami (✉)
Graduate School of Informatics, Kyoto University, Kyoto 606-8501, Japan
e-mail: kigami@i.kyoto-u.ac.jp



D.-J. Feng and K.-S. Lau (eds.), *Geometry and Analysis of Fractals*, Springer Proceedings in Mathematics & Statistics 88,
DOI 10.1007/978-3-662-43920-3_9

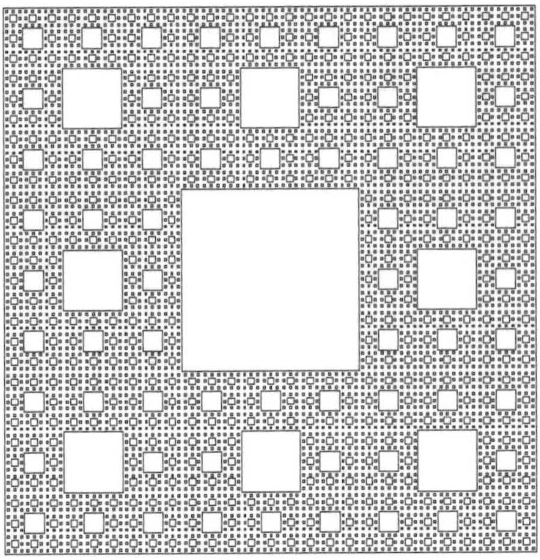

**Fig. 1** The Sierpinski carpet

**Proposition 1.2** *Let* $(X, d)$ *and* $(X, \rho)$ *be metric spaces.*

(1) *If* $\rho \underset{\mathrm{QS}}{\sim} d$*, then the identity may of* $X$ *is a homeomorphism from* $(X, d)$ *to* $(X, \rho)$.

(2) *The relation* $\underset{\mathrm{QS}}{\sim}$ *is an equivalence relation among metrics on* $X$*. In particular,* $\rho \underset{\mathrm{QS}}{\sim} d$ *if and only if* $d \underset{\mathrm{QS}}{\sim} \rho$.

Associated with the notion of quasisymmetry, the quasiconformal dimension of a metric space has been introduced by Pansu in [Pan89] as an invariant under quasisymmetric modification of a metric.

**Definition 1.3** (*Quasiconformal dimension*) Let $(X, d)$ be a metric space. We define the conformal dimension of $(X, d)$, $\dim_C(X, d)$, by

$$\dim_C(X, d) = \inf\{\dim_H(X, \rho) | \rho \text{ is a metric on } X \text{and } d \underset{\mathrm{QS}}{\sim} \rho\},$$

where $\dim_H(X, \rho)$ is the Hausdorff dimension of $(X, \rho)$.

Quasisymmetric maps on self-similar sets have been paid much attentions in recent years as well as their conformal dimensions. For example, Bonk and Merenkov have shown that any quasisymmetric homeomorphism from the Sierpinski carpet to itself is a composition of rotations and reflections in [BM00]. About the conformal dimensions, Tyson and Wu have proven that the conformal dimension of the Sierpinski gasket is one in [TW06]. For the Sierpinski carpet, it is known that

$$1 + \frac{\log 3}{\log 2} \leq \dim_C(\mathrm{SC}, d_E) < \dim_H(\mathrm{SC}, d_E) = \frac{\log 8}{\log 3}, \tag{1.1}$$

where SC is the Sierpinski carpet and $d_E$ is the restriction of the Euclidean metric. The strict inequality between the Hausdorff and the quasiconformal dimensions in (1.1) has shown by Keith and Laakso [KL04]. See [MT10] for details.

The first problem we are going to study is to obtain a verifiable characterization of quasisymmetric metrics. It will turn out that scales and related notions introduced in [Kig09] are useful in dealing with such a problem. Let $K$ be a connected self-similar set associated with the family of contractions $\{F_1, \ldots, F_N\}$, i. e. $K = F_1(K) \cup \ldots \cup F_N(K)$. Define $F_{w_1 \ldots w_m} = F_{w_1} \circ \ldots \circ F_{w_m}$ and $K_{w_1 \ldots w_m} = F_{w_1 \ldots w_m}(K)$ for any $w_1, \ldots, w_m \in \{1, \ldots, N\}$. The notion of scales has been introduced in order to study how to find a metric under which the contraction mappings $\{F_1, \ldots, F_N\}$ have prescribed values of contraction ratios. A scale essentially gives "diameters" of $K_{w_1 \ldots w_m}$'s and induces a family of assumed "balls" $U_s(x)$ around $x \in K$ with radius $s > 0$. See Sect. 2 for precise definitions. In the language of scales, we are going to present an equivalent condition in Theorem 3.4 for metrics being quasisymmetric to each other which is easy to verify for concrete examples, in particular, in the case of "self-similar" metrics.

As an application, we will present a systematic way of constructing a self-similar metric on the Sierpinski carpet which is quasisymmetric to $d_E$ and Ahlfors regular. The main idea is to find an "invisible" set introduced in Sect. 4. Roughly speaking, an invisible set is a collection of places where the shortest paths between two separated boundary points will not visit. (We define the "boundary" of the Sierpinski carpet by the union of four line segments, namely, the most upper, lower, right and left line segments of the square which is the convex hull of the Sierpinski carpet.) Putting an arbitrary weight on an invisible set, we will obtain a self-similar metric having the desired properties mentioned above with an explicit formula for its Hausdorff dimension in Theorem 5.3. Constructing series of invisible sets and taking advantage of the associated metrics, we will show that

$$\dim_C(\mathrm{SC}, d_E) \leq \frac{\log\left(\frac{9+\sqrt{41}}{2}\right)}{\log 3} = 1.858183... < \frac{\log 8}{\log 3} = 1.892789....$$

in Sect. 6.[1]

Note that the conformal dimension in the above inequality can be replaced by the Ahlfors regular conformal dimension since our metrics are Ahlfors regular. See [MT10] for the definition of the Ahlfors regular conformal dimension.

The following is a convention in notations in this paper.

Let $f$ and $g$ be functions with variables $x_1, \ldots, x_n$. We use "$f \asymp g$ for any $(x_1, \ldots, x_n) \in A$" if and only if there exist positive constants $c_1$ and $c_2$ such that

[1] After completion of the preliminary version of this paper, B. Kleiner informed me that he had obtained better upper bound of $\dim_C(\mathrm{SC}, d_E)$ around 1999 by a different method in [Kle00]. His upper bound is about 1.856685 .... The author would like to express his gratitude to Professor Bruce Kleiner for his detailed comments.

$$c_1 f(x_1, \ldots, x_n) \le g(x_1, \ldots, x_n) \le c_2 f(x_1, \ldots, x_n)$$

for any $(x_1, \ldots, x_n) \in A$.

## 2 Basic Notions

This section is devoted to introducing fundamental notions and results regarding scales and self-similar sets and scales.

The following is the standard definitions on (finite and infinite) sequences of finite symbols.

**Definition 2.1** Let $S$ be a finite set. For $m \ge 0$, define $W_m(S) = S^m = \{w|w = w_1 \ldots w_m, w_i \in S\}$, where $W_0(S) = \{\emptyset\}$. Define $W_* = \cup_{m \ge 0} W_m$. Also $\Sigma(S) = S^{\mathbb{N}} = \{\omega|\omega = \omega_1\omega_2 \ldots, \omega_i \in S\}$. For $w = w_1 \ldots w_m \in W_*(S)$, the length $|w|$ of $w$ is defined by $|w| = m$. For $w = w_1 \ldots w_m$ and $v = v_1 \ldots v_n \in W_*(S)$, we define $w \cdot v$ (or $wv$ for short) by $w{\cdot}v = w_1 \ldots w_m v_1 \ldots v_n$. For a subseteq $A, B \in W_*(S)$, $A \cdot B$ (or $AB$ for short) is defined by $A{\cdot}B = \{wv|w \in A, v \in B\}$.

*Remark* The notion of "gauge function" given in the above definition is not related to the notion of "conformal gauge" which is commonly used in literatures concerning the conformal dimension, for example, [MT10].

With the product topology, $\Sigma(S)$ is compact, perfect and totally disconnected. In other words, $\Sigma(S)$ is a Cantor set. A scale is defined by a gauge function which assign a "diameter" to every $w \in W_*(S)$.

**Definition 2.2** (*Scale*). Let $S$ be a finite set.

(1) A function $g : W_*(S) \to (0, 1]$ is called a gauge function if and only if $g(\emptyset) = 1$, $g(w_1 \ldots w_m) \le g(w_1 \ldots w_{m-1})$ and $\max_{w \in W_m(S)} g(w) \to 0$ as $m \to 0$. A gauge function $g$ is said to be elliptic if and only if there exists $c \in (0, 1)$ and $n$ such that $g_{wi} \ge cg(w)$ for any $i \in S$ and any $w \in W_*(S)$ and $g_{wv} \le cg(w)$ for any $w \in W_*(S)$ and $v \in W_n$.

(2) Let $g$ be a gauge function. Define

$$\Lambda_s^g = \{w = w_1 \ldots w_m | g(w_1 \ldots w_{m-1}) \ge s > g(w_1 \ldots w_m)\}$$

We call $\mathcal{S}^g = \{\Lambda_s^g\}_{s \in (0,1]}$ a scale on $\Sigma$ associated with the gauge function $g$.

If no confusion may occur, we omit $S$ in $W_m(S)$, $W_*(S)$ and $\Sigma(S)$ and simply write $W_m$, $W_*$ and $\Sigma$ respectively.

The notion of self-similar structure describes topological feature of self-similar sets.

**Definition 2.3** $(K, S, \{F_i\}_{i \in S})$ is called a self-similar structure if the following four conditions (S1), (S2), (S3) and (S4) are satisfied:

(S1) $K$ is a compact metrizable set.
(S2) $S$ is a finite set.
(S3) $F_s : K \to K$ is continuous for any $s \in S$.
(S4) There exists a continuous surjection $\pi : \Sigma(S) \to K$ such that $F_s \circ \pi = \pi \circ \sigma_s$ for any $s \in K$, where $\sigma_s : \Sigma(S) \to \Sigma(S)$ is defined by $\sigma_s(\omega_1\omega_2\dots) = s\omega_1\omega_2\dots$.

Hereafter in this paper, $(K, S, \{F_s\}_{s\in S})$ is always a self-similar structure.

**Notation** Define $F_{w_1\dots w_m} = F_{w_1} \circ \cdots \circ F_{w_m}$ and $K_w = F_w(K)$. Moreover, define $K(A) = \cup_{w\in A} K_w$ for a subset $A \subseteq W_*$.

A scale $\mathcal{S}$ on $\Sigma(S)$ induces a family of "balls" $U^{(n)}(x, s)$ around $x \in X$ with "radius" $s$. One of the main concerns is the existence of a metric under which those "balls" are really balls, in other words, the existence of adapted metric according to the following definition.

**Definition 2.4** Let $\mathcal{S} = \{\Lambda_s\}_{s\in(0,1]}$ be a scale on $\Sigma$ associated with a gauge function $g$.

(1) For $x \in K$, define $(\Lambda_s)_x^{(n)}$ and $U_{\mathcal{S}}^{(n)}(x, s)$ inductively by

$$(\Lambda_s)_x^{(0)} = \{w | w \in \Lambda_s, x \in K_w\}$$
$$U_{\mathcal{S}}^{(n)}(x, s) = K\left((\Lambda_s)_x^{(n)}\right)$$
$$(\Lambda_s)_x^{(n)} = \{w | w \in \Lambda_s, K_w \cap U_{\mathcal{S}}^{(n-1)}(x, s) \neq \emptyset\}$$

(2) A metric $d$ on $K$ is said to be adapted to the scale $\mathcal{S}$ if and only if there exist $\alpha, \beta > 0$ and $n \geq 1$ such that

$$B_d(x, \alpha s) \subseteq U_{\mathcal{S}}^{(n)}(x, s) \subseteq B_d(x, \beta s)$$

for any $x \in K$ and any $s$.

The notion of gentleness between scales is introduced in [Kig09] as a part of the equivalence condition for a measure being volume doubling with respect to a scale. Roughly, if two scales are gentle with respect to each other, then the transition to one scale to the other is "smooth".

**Definition 2.5** Let $\mathcal{S}^g$ and $\mathcal{S}^l$ be scales on $\Sigma$ associated with gauge functions $g$ and $l$ respectively. We say $\mathcal{S}^l$ is gentle with respect to $\mathcal{S}^g$ if and only if there exists $c > 0$ such that $l(w) \leq cl(v)$ whenever $w, v \in \Lambda_s$ for some $s > 0$ and $K_w \cap K_v \neq \emptyset$. We write $\mathcal{S}^g \underset{\text{GE}}{\sim} \mathcal{S}^l$ if $\mathcal{S}^l$ is gentle with respect to $\mathcal{S}^g$.

**Proposition 2.6** *Among elliptic scales, i.e. scales whose gauge functions are elliptic, $\underset{\text{GE}}{\sim}$ is an equivalent relation. In particular, if $g$ and $l$ are elliptic, then $\mathcal{S}^g \underset{\text{GE}}{\sim} \mathcal{S}^l$ implies $\mathcal{S}^l \underset{\text{GE}}{\sim} \mathcal{S}^g$.*

There exists a natural "pseudo"metric associated with a scale which is defined by the infimum of the "length" of paths between two points.

**Definition 2.7** (1) A sequence $(w(1), \ldots, w(n))$ is called a path in $K$ if and only if $w(1), \ldots, w(n) \in W_*$, $K_{w(i)} \cap K_{w(i+1)} \neq \emptyset$ for any $i = 1, \ldots, n-1$. The collection of all the paths is denoted by $\mathcal{CH}$. For $U, V \subseteq K$, a path $(w(1), \ldots, w(n))$ is called a path between $U$ and $V$ if and only if $K_{w(1)} \cap U \neq \emptyset$ and $K_{w(n)} \cap V \neq \emptyset$. We use $\mathcal{CH}(U, V)$ to denote the collection of paths between $U$ and $V$. For two paths $\mathbf{p}_1 = (w(1), \ldots, w(n))$ and $\mathbf{p}_2 = (v(1), \ldots, v(m))$, if $K_{w(n)} \cap K_{v(1)} \neq \emptyset$, we define $\mathbf{p}_1 \vee \mathbf{p}_2 \in \mathcal{CH}$ by $\mathbf{p}_1 \vee \mathbf{p}_2 = (w(1), \ldots, w(n), v(1), \ldots, v(m))$.
(2) Let $\mathcal{S}$ be a scale on $\Sigma$ associated with a gauge function $g$. For any $x, y \in K$, we define

$$D_{\mathcal{S}}(x, y) = \inf\{\sum_{i=1}^{n} g(w(i)) | (w(1), \ldots, w(n)) \in \mathcal{CH}(x, y)\}.$$

*Remark* We identify a point $x \in X$ and a set $\{x\}$ if no confusion may occur.

*Remark* We often use $D_g$ instead of $D_{\mathcal{S}}$ if $\mathcal{S}$ is the scale associated with a gauge function $g$.

**Proposition 2.8** *$D_{\mathcal{S}}$ is a pseudometric, i.e. $D_{\mathcal{S}}(x, y) = D_{\mathcal{S}}(y, x)$, $D_{\mathcal{S}}(x, y) \geq 0$, $D_{\mathcal{S}}(x, x) = 0$ and $D_{\mathcal{S}}(x, y) \leq D_{\mathcal{S}}(x, z) + D_{\mathcal{S}}(z, y)$.*

By [Kig09, Lemma 2.3.10], we have the following theorem, which says that a metric adapted to a scale $\mathcal{S}$, if such a metric exists at all, is essentially $D_{\mathcal{S}}$.

**Theorem 2.9** *Let $\mathcal{S}$ be a scale. There exists a metric $d$ on $K$ such that $d$ is adapted to $\mathcal{S}$ if and only if $D_{\mathcal{S}}$ is a metric on $K$ which is adapted to $\mathcal{S}$.*

## 3 Quasisymmetric Metrics and Scales

In this section, we give an equivalent condition for two metrics on a self-similar set being quasisymmetric in terms of scales and related notions introduced in Sect. 2.

Let $(K, S, \{F_i\}_{i \in S})$ be a self-similar structure. Assume that $K \neq \overline{V_0}$. Hereafter in this section, every metric on $K$ is assumed to satisfy the following two properties:

1. It produces the same topology as the original topology of $K$.
2. The diameter of $K$ under it equals one.

The next lemma can be verified immediately by the definitions in the previous section.

**Lemma 3.1** *Let $\mathcal{S}_1 = \{\Lambda_s^1\}$ and $\mathcal{S}_2 = \{\Lambda_s^2\}$ be scales. If $\mathcal{S}_1 \underset{\mathrm{GE}}{\sim} \mathcal{S}_2$, then for any $n \geq 1$, there exists $c_n \in (0, 1)$ such that*

$$U_1^{(n)}(x, c_n t) \subseteq U_2^{(n)}(x, s) \subseteq U_1^{(n)}(x, t/c_n)$$

*for any* $x \in K$, *any* $(s, t)$ *with* $\Lambda^1_{t,x} \cap \Lambda^2_{s,y} \neq \emptyset$, *where* $U_i^{(n)}(x, s) = U_{\mathcal{S}_i}^{(n)}(x, s)$ *and* $\Lambda^i_{t,x} = (\Lambda^i_s)_x^{(0)}$ *for* $i = 1, 2$.

First we define a scale associated with a metric.

**Definition 3.2** Let $d$ be a metric on $K$ with $\mathrm{diam}(K, d) = 1$. Define $\mathcal{S}^d = \{\Lambda^d_s\}$ be the scale with the gauge function $d_w = \mathrm{diam}(K_w, d)$.

**Lemma 3.3** *Let* $\mathcal{S} = \{\Lambda_s\}$ *be an elliptic scale and let* $d$ *be a metric on* $K$ *which is adapted to* $\mathcal{S}$. *Let* $l(w)$ *be the gauge function of* $\mathcal{S}$. *Then*

(1) $d_w \asymp l(w)$ *for any* $w \in W_*$.
(2) *The pseudometric* $D_{\mathcal{S}}$ *associated with* $\mathcal{S}$ *is a metric and* $\mathcal{D}_{\mathcal{S}}(x, y) \asymp d(x, y)$ *for any* $x, y \in K$.
(3) $\mathcal{S}^d$ *is elliptic and* $d$ *is adapted to* $\mathcal{S}^d$.

*Proof* Write $U^{(n)}(x, r) = U_{\mathcal{S}}^{(n)}(x, r)$. Since $d$ is adapted to $\mathcal{S}$, we have

$$U^{(n)}(x, \beta s) \subseteq B_d(x, s) \subseteq U^{(n)}(x, \alpha r) \tag{3.2}$$

(1) For $w \in W_*$, $U^{(n)}(x, l(w)) \subseteq B_d(x, \alpha l(w))$. Hence $d_w \leq \alpha l(w)$. Now by [Kig09, Lemma 1.3.12], there exists $y \in K_w$ and $\gamma \in (0, 1)$ such that $U^{(n)}(y, \gamma l(w)) \subseteq K_w$. Hence $B_d(x, \beta\gamma l(w)) \subseteq K_w$. Since $K$ is connected, we have $\beta\gamma l(w) \leq d_w$.
(2) This is immediate from Theorem 2.9.
(3) These claims are immediate from (1) and Lemma 3.1. □

Now we present one of the main results of this paper. The following theorem gives an equivalent condition for certain metrics on a self-similar set being quasisymmetic. It plays a crucial role in the proof of Theorem 5.3.

**Theorem 3.4** *Let* $d$ *be a metric on* $K$ *and let* $\mathcal{S} = \{\Lambda_s\}$ *be an elliptic scale. Assume that* $d$ *is adapted to* $\mathcal{S}$. *Let* $\rho$ *be a metric on* $K$. *Then* $d \underset{\mathrm{QS}}{\sim} \rho$ *if and only if* $\mathcal{S}^\rho$ *is elliptic,* $\mathcal{S} \underset{\mathrm{GE}}{\sim} \mathcal{S}^\rho$ *and* $\rho$ *is adapted to* $\mathcal{S}^\rho$.

The rest of this section is devoted to the proof of Theorem 3.4.

*Proof* First we show $\Rightarrow$. Assume $d \underset{\mathrm{QS}}{\sim} \rho$. By Lemma 3.3, we may regard the gauge function of $\mathcal{S}$ is $d_w$ and hence $\mathcal{S} = \mathcal{S}^d$.

By the results in [Kig00, Part 2], $d \underset{\mathrm{QS}}{\sim} \rho$ is equivalent to the facts that there exists $\delta \in (0, 1)$ such that

$$\begin{aligned} B_d(x,r) &\supseteq B_\rho\left(x, \delta\overline{\rho}_d(x,s)\right) \\ B_\rho(x,r) &\supseteq B_d\left(x, \delta\overline{d}_\rho(x,r)\right) \end{aligned} \tag{3.3}$$

and

$$\begin{aligned} \overline{\rho}_d(x,r/2) &\geq \delta\overline{\rho}_d(x,r) \\ \overline{d}_\rho(x,r/2) &\geq \delta\overline{d}_\rho(x,r), \end{aligned} \tag{3.4}$$

where $\overline{\rho}_d(x,r) = \sup_{y\in B_d(x,r)} \rho(x,y)$ and $\overline{d}_d(x,r) = \sup_{y\in B_\rho(x,r)} d(x,y)$.

First we show the following claim.

**Claim 1** *Let $w \in \Lambda_s^d$. Then there exists $z \in K_w$ such that $\rho_w \geq c\overline{\rho}_d(z,s)$, where $c$ is a constant which is independent of $w$ and $s$.*

*Proof of Claim 1* By [Kig09, Lemma 1.3.12] and (3.3), we may find $z \in K_w$ such that

$$K_w \supseteq U_d^{(n)}(z,\gamma s) \supseteq B_d(z,\gamma s/\alpha) \supseteq B_\rho\left(z, \delta\overline{\rho}_d(z,\gamma s/\alpha)\right)$$

Hence by (3.4)

$$\rho_w \geq \delta\overline{\rho}_d(z,\gamma s/\alpha) \geq c\overline{\rho}_d(z,s).$$

□

Step 1: $\mathcal{S}^\rho$ is elliptic.

Proof of "$\rho_{wi} \geq c\rho_w$ for any $w \in W_*$ and any $i \in S$:

By Claim 1, it follows that

$$\rho_{wi} \geq c'\overline{\rho}_d(z,d_{wi}) \geq c'\overline{\rho}_d(z,2d_{wi}). \tag{3.5}$$

for some $z \in K_{wi}$". On the other hand,

$$K_w \subseteq B_d(z,2d_w) \subseteq B_\rho(z,\rho_d(x,2d_w)).$$

Hence

$$\rho_w \leq \overline{\rho}_d(z,2d_w).$$

This with (3.5) suffices.

Proof of "there exists $c \in (0,1)$ and $m$ such that $\rho_{wv} \leq c\rho_w$ for any $w \in W_*$ and any $v \in W_m$".

Since $K_{wv} \subseteq B_\rho(x,\overline{\rho}_d(x,2d_{wv}))$, we have

$$\rho_{wv} \leq \overline{\rho}_d(x,2d_{wv}) \leq \delta\overline{\rho}_d(x,d_{wv}), \tag{3.6}$$

where $x \in K_{wv}$. On the other hand, by [Kig09, Lemma 1.3.12], there exists $z \in K_w$ such that

$$K_w \supseteq U_d^{(n)}(x, \gamma d_w) \supseteq B_\rho\left(x, \delta\overline{\rho}_d(x, \gamma d_w)\right).$$

Hence

$$\rho_w \geq \delta\overline{\rho}_d(x, \gamma d_w) \geq \delta'\overline{\rho}_d(x, d_w) \tag{3.7}$$

Now, since $\mathcal{S}^d$ is elliptic, there exists $a \in (0, 1)$ such that

$$d_{wv} \leq ca^{|v|}d_w$$

for any $w$ and $v$. Hence by (3.6) and (3.7), the uniform decay of $\rho$ with respect to $d$, (See [Kig00, Proposition 10.7]),

$$\rho_{wv} \leq \delta\overline{\rho}_d(x, d_{wv}) \leq \delta\overline{\rho}_d(x, ca^{|v|}d_w) \leq cb^{|v|}\overline{\rho}_d(x, d_w) \leq c'b^{|v|}\rho_w,$$

where $b \in (0, 1)$. Hence choosing sufficiently large $m = |v|$, we obtain the desired inequality.

Thus we have shown that $\mathcal{S}^\rho$ is elliptic.

Step 2: $\mathcal{S} \underset{\text{GE}}{\sim} \mathcal{S}^\rho$

Let $w, v \in \Lambda_s^d$ with $K_w \cap K_v \neq \emptyset$. Choose $x \in K_w$ and $y \in K_v$. Then $d(x, y) \leq 2s$ and hence $B_d(x, 3s) \supseteq B_d(y, s)$. This implies $\overline{\rho}_d(x, 3s) \geq \overline{\rho}_d(y, s)$. By (3.4),

$$\overline{\rho}_d(x, s) \asymp \overline{\rho}_d(y, s).$$

By Claim 1, choosing $y \in K_v$ properly, we see that $\rho_v \geq c\overline{\rho}_d(y, s)$. Since $\overline{\rho}_d(x, 2s) \geq \rho_w$, (3.4) shows that $\mathcal{S}^d \underset{\text{GE}}{\sim} \mathcal{S}$.

Step 3: $\rho$ is adapted to $\mathcal{S}^\rho$.

Let $x \in K$ and let $w \in \Lambda_{r,x}^d \cap \Lambda_{s,x}^\rho$. Then by Lemma 3.1, (3.3) and (3.4),

$$\begin{aligned} U_\rho^{(n)}(x, cs) \supseteq U_d^{(n)}(x, r) \supseteq B_d(x, r/\alpha) \supseteq B_\rho\left(x, \delta\overline{\rho}_d(x, r/\alpha)\right) \\ \supseteq B_\rho\left(x, \delta'\overline{\rho}_d(x, 2r)\right) \supseteq B_\rho(x, \delta'\rho_w) \supseteq B_\rho(x, \delta''s). \end{aligned}$$

On the other hand, let $x \in K$ and let $w \in \Lambda_s^\rho \cap \Lambda_t^d$. Then

$$B_\rho(x, s) \supseteq B_d\left(x, \delta\overline{d}_\rho(x, s)\right) \supseteq U_d^{(n)}\left(x, \beta\delta\overline{d}_\rho(x, s)\right) \supseteq U_\rho^{(n)}(x, c'r), \tag{3.8}$$

where $wv \in \Lambda_{\beta\delta\overline{d}_\rho(x,s),x}^d \cap \Lambda_{r,x}^\rho$. Since $B_\rho(x, 2s) \supseteq K_w$, we see that $\overline{d}_\rho(x, 2s) \geq d_w$. Hence $\overline{d}_\rho(x, s) \geq c_1 d_w$. Consequently, $d_{wv} \geq c_2 d_w$, where $c_2$ is independent of $w$ and $v$. This implies that $|v|$ is uniformly bounded. Since $\mathcal{S}^\rho$ is elliptic, $\rho_{wv} \geq c_3\rho_w$. This implies $U_\rho^{(n)}(x, c'r) \supseteq U_\rho^{(n)}(x, c_4 s)$. By (3.8), it follows that $B_\rho(x, s) \supseteq U_\rho^{(n)}(x, c_5 s)$. Thus we have shown that $\rho$ is adapted to $\mathcal{S}^\rho$.

This concludes the proof of $\Rightarrow$. □

To show the converse direction of Theorem 3.4, we need the following lemma.

**Lemma 3.5** *Assume that* $d$ *is adapted to* $\mathcal{S}^d$. *Then, for any* $n$ *and* $k$, *there exists* $\lambda \in (0, 1)$ *such that*

$$U_d^{(n)}(x, r) \supseteq U_d^{(n+k)}(x, \lambda r)$$

*for any* $x \in K$ *and any* $r$.

*Proof* Since $d$ is adapted to $\mathcal{S}^d$, there exists $c > 0$ such that $U_d^{(n)}(x, r) \supseteq B_d(x, cr)$. Then $B_d(x, cr) \supseteq U_d^{(n+k)}(x, cr/(n + k + 2))$. □

*Proof* (of $\Leftarrow$ of Theorem 3.4) Since $d$ and $\rho$ are adapted to $\mathcal{S}^d$ and $\mathcal{S}^\rho$ respectively,

$$U_d^{(n)}(x, \beta_1 r) \subseteq B_d(x, r) \subseteq U_d^{(n)}(x, \alpha_1 r)$$
$$U_\rho^{(m)}(x, \beta_2 r) \subseteq B_\rho(x, r) \subseteq U_\rho^{(m)}(x, \alpha_2 r).$$

First we show (3.3). By Lemma 3.1,

$$B_d(x, r) \subseteq U_d^{(n)}(x, \alpha_1 r) \subseteq U_\rho^{(n)}(x, c\rho_w), \tag{3.9}$$

where $w \in \Lambda^d_{\alpha_1 r, x}$. Using Lemma 3.5 if necessary, we obtain

$$B_d(x, r) \subseteq U_\rho^{(m)}(x, c_1\rho_w) \subseteq B_\rho(x, c_2\rho_w).$$

Hence $\overline{\rho}_d(x, r) \leq c_2\rho_w$. Now by Lemma 3.1,

$$B_d(x, r) \supseteq U_d^{(n)}(x, \beta_1 r) \supseteq U_\rho^{(n)}(x, c'\rho_{wv}), \tag{3.10}$$

where $wv \in \Lambda^d_{\beta_1 r, x}$. By making use of Lemma 3.5 if necessary, we have

$$B_d(x, r) \supseteq U_\rho^{(m)}(x, c''\rho_{wv}) \supseteq B_\rho(x, c''\beta_2\rho_{wv}).$$

Since $\mathcal{S}^d$ is elliptic, the fact that $w \in \Lambda^d_{\alpha_1 r, x}$ and $wv \in \Lambda^d_{\beta_1 r, x}$ implies that $|v|$ is uniformly bounded with respect to $x$ and $r$. Since $\mathcal{S}^\rho$ is also elliptic, we see that $\rho_w v \geq c_3\rho_w \geq c_4\overline{\rho}_d(x, r)$. Hence (3.3) holds. (By exchanging $\rho$ and $d$, we also obtain the other one.)

Next we show (3.4). By (3.9),

$$\overline{\rho}_d(x, r) \leq c(n + 1)\rho_w,$$

where $w \in \Lambda^d_{\alpha_1 r}$. Replacing $r$ by $\lambda r$ for $\lambda \in (0, 1)$ in (3.10), we have

$$B_d(x, \lambda r) \supseteq U^{(n)}(x, c'\rho_{wv}),$$

where $wv \in \Lambda^d_{\lambda\beta_1 r,x}$. This implies $\overline{\rho}_d(x, \lambda r) \geq c'\rho_{wv}$. The same arguments as above show that $|v|$ is uniformly bounded and $\rho_{wv} \geq c\rho_w$. Combining all these, we obtain

$$\overline{\rho}_d(x, \lambda r) \geq c'\rho_{wv} \geq c''\rho_w \geq c'''\overline{\rho}_d(x, r).$$

Again the other one is obtained by exchanging $d$ and $\rho$. Thus we have obtained (3.4). □

# 4 Sierpinski Carpet and Its Invisible Sets

In this and the following sections, we are going to apply Theorem 3.4 to the Sierpinski carpet. First we give the definition of the Sierpinski carpet.

**Definition 4.1** Let $S = \{\swarrow, \downarrow, \searrow, \rightarrow, \nearrow, \uparrow, \nwarrow, \leftarrow\}$. Define $p_{\swarrow} = -1 - \sqrt{-1}$, $p_{\downarrow} = -\sqrt{-1}$, $p_{\searrow} = 1 - \sqrt{-1}$, $p_{\rightarrow} = 1$, $p_{\nearrow} = 1 + \sqrt{-1}$, $p_{\uparrow} = \sqrt{-1}$, $p_{\nwarrow} = -1 + \sqrt{-1}$ and $p_{\leftarrow} = -1$. Moreover, define $F_s : \mathbb{C} \to \mathbb{C}$ for $s \in S$ by

$$F_s(z) = \frac{(z - p_s)}{3} + p_s.$$

The Sierpinski carpet $K$ is the unique non-empty compact set which satisfies

$$K = \bigcup_{s \in S} F_s(K).$$

Let $d_E$ be the restriction of the Euclidean metric on the Sierpinski carpet $K$.

We consider $d_E$ as the standard metric on $K$ and are going to construct metrics which is quasisymmetric with respect to $d_E$. Obviously, the scale $\mathcal{S}_{d_E}$ associated with $d_E$ is elliptic and $d_E$ is adapted to the scale $\mathcal{S}_{d_E}$. In fact, the gauge function associated with $d_E$ is given by $3^{-|w|}$ for any $w \in W_*$.

Next we introduce notions and notations regarding the boundary of the Sierpinski carpet (Fig. 2).

**Definition 4.2**

(1) Define $L = K \cap \{z|\mathrm{Re}\, z = -1\}$, $R = K \cap \{z|\mathrm{Re}\, z = 1\}$, $T = K \cap \{z|\mathrm{Im}\, z = 1\}$ and $B = K \cap \{z|\mathrm{Im}\, z = -1\}$. Let $H_w = F_w(H)$ for any $w \in W_*$ and any $H \in \{L, R, T, B\}$. Moreover define $\partial_m = \{L_w, R_w, T_w, B_w | w \in W_m\}$.

(2) Define $L^m = \{\swarrow, \leftarrow, \nwarrow\}^m$, $R^m = \{\searrow, \rightarrow, \nearrow\}^m$, $T^m = \{\nwarrow, \uparrow, \nearrow\}^m$, $B^m = \{\swarrow, \downarrow, \searrow\}^m$ and $\delta_m = L^m \cup R^m \cup T^m \cup B^m$.

*Remark* Recall that $K(A) = \cup_{w \in A} K_w$ for a subset $A \subseteq W_*$. The map $A \to K(A)$ can be regarded as a map from the subsets of $W_*$ to the subsets of $K$. In the case

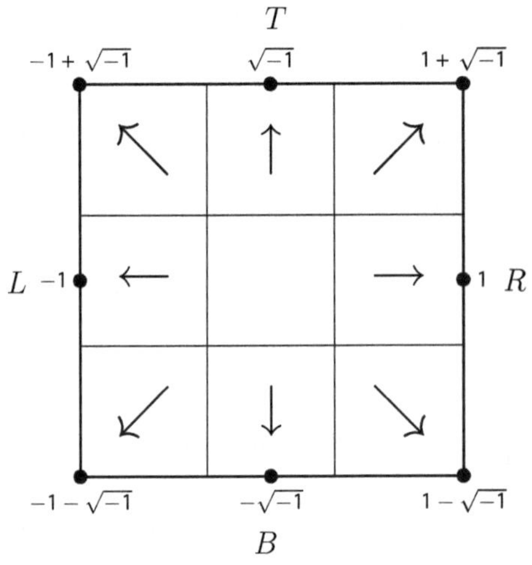


**Fig. 2** Generation of the Sierpinski carpet

of the Sierpinski carpet, this map is injective, i.e. if $A \neq B$, then $K(A) \neq K(B)$. Therefore, if no confusion may occur, we identify $A \subseteq W_*$ with $K(A) \subseteq K$.

Note that $D_{d_E}(x, y) \geq 1$ for any $(x, y) \in (L \times R) \cup (T \times B)$. This fact may remain true even if you put 0 as weights (length) of some pieces of $w$'s. Such a collection of $w$'s is called an invisible set, whose precise definition is given below.

**Definition 4.3**

(1) Let

$$\mathcal{CH}_m = \{(w(1), \ldots, w(n)) | (w(1), \ldots, w(n)) \in \mathcal{CH}, w(i) \in W_m\}$$

and let $\mathcal{CH}_m(U, V) = \mathcal{CH}(U, V) \cap \mathcal{CH}_m$ for $U, V \subseteq K$.

(2) Let $A \subseteq W_m$. For $\mathbf{p} = (w(1), \ldots, w(n)) \in \mathcal{CH}_m$, define

$$\ell_A(\mathbf{p}) = \frac{\#\{i | i = 1, \ldots, n, w(i) \notin A\}}{3^m}$$

(3) Let $A \subseteq W_m$. $A$ is said to be an invisible set if and only if

$$\inf_{\mathbf{p} \in \mathcal{CH}_m(L,R) \cup \mathcal{CH}_m(T,B)} \ell_A(\mathbf{p}) \geq 1$$

(4) Let $A \subseteq W_m$. $A$ is said to be $+$-invariant if and only if $K(A)$ is symmetric with respect to both the real and imaginary axes.

Since $L^m$, $R^m$, $T^m$ and $B_m$ are the shortest paths, we have the following proposition.

**Proposition 4.4** *Let $A \subseteq W_m$. If $A$ is invisible, then $A \cap \delta_m = \emptyset$.*

The next theorem is one of the fundamental property of an invisible set. It will play a key role in constructing a metric associated with an invisible set in the next section.

**Theorem 4.5** *Let $A \subseteq W_m$ be an invisible set and let $X \subseteq W_n$ be an invisible and $+$-invariant set. Then $AW_n \cup W_m X$ is an invisible set.*

The rest of this section is devoted to the proof of Theorem 4.5.

**Definition 4.6**

(1) Let $A \subseteq W_m$. Define $\partial_m A = \{F | F \in \partial_m, F \subseteq K(A) \cap \overline{K \backslash K(A)}\}$.
(2) Define $f_{m,\rightarrow}(z) = z + 3^{-m}$, $f_{m,\leftarrow}(z) = z - 3^{-m}$, $f_{m,\uparrow}(z) = z + 3^{-m}\sqrt{-1}$ and $f_{m,\downarrow}(z) = z - 3^{-m}\sqrt{-1}$. Moreover, let $f_{m,\swarrow} = f_{m,\downarrow} \circ f_{m,\leftarrow}$, $f_{m,\searrow} = f_{m,\downarrow} \circ f_{m,\rightarrow}$, $f_{m,\nwarrow} = f_{m,\uparrow} \circ f_{m,\leftarrow}$ and $f_{m,\nearrow} = f_{m,\uparrow} \circ f_{m,\rightarrow}$.
(3) Let $w \in W_m$. For $s \in S$, if there exists $w' \in W_m$ such that $f_{m,s}(K_w) = K_{w'}$, then define $(w)_s = w'$. Otherwise define $(w)_s = \%$, where % is used as the symbol which represents non-existence (Fig. 3).

**Lemma 4.7** *Let $F \in \partial_m$ and let $G \in \partial_m(W_m(F))$. If $X \subseteq W_n$ is invisible and $+$-invariant, then $\ell_{W_m X}(\mathbf{p}) \geq 3^{-m}$ for any $\mathbf{p} \in \mathcal{CH}_{m+n}(F, G)$.*

*Proof* Note that $\#(W_m(F)) \leq 6$. Up to parallel translations, the reflections in the real and the imaginary axes and the $\pi/2$-rotation, we may assume that $F = B_w$ for some $w \in W_m$. Then $W_m(F) \subseteq \{w, (w)_\leftarrow, (w)_\swarrow, (w)_\downarrow, (w)_\searrow, (w)_\rightarrow\}$, where some of them may be %. In fact there are 7 cases. (See Fig. 4.)

Case 1 $\#(W_m(F)) = 6$.
Case 2 $\#(W_m(F)) = 5$ and $(w)_\searrow = \%$.
Case 3 $\#(W_m(F)) = 5$ and $(w)_\downarrow = \%$.
Case 4 $\#(W_m(F)) = 4$ and $(w)_\downarrow = (w)_\searrow = \%$.
Case 5 $\#(W_m(F)) = 3$ and $(w)_\downarrow = (w)_\searrow = (w)_\swarrow = \%$.
Case 6 $\#(W_m(F)) = 3$ and $(w)_\leftarrow = (w)_\swarrow = (w)_\searrow = \%$.
Case 7 $\#(W_m(F)) = 2$ and $(w)_\downarrow = (w)_\swarrow = (w)_\searrow = (w)_\leftarrow = \%$.

We consider the first case. The other cases can be treated by the similar discussion. If $D = \cup_{U \in \partial_m(W_m(F))} U$, then $D = \partial K(W_m(F))$. Let $\mathbf{p} = (w(1), \ldots, w(k)) \in \mathcal{CH}_{m+n}(F, G)$. The reflection in the line containing $F$ induces a natural bijection from $W_m(F) \cdot W_n$ to itself, which is denoted by $\eta$. Define $\theta : W_m(F) \cdot W_n \rightarrow \{(w)_\leftarrow, w, (w)_\rightarrow\} \cdot W_n$ by

$$\theta(uv) = \begin{cases} uv & \text{if } u \in \{(w)_\leftarrow, w, (w)_\rightarrow\} \text{ and } v \in W_n, \\ \eta(uv) & \text{if } u \in \{(w)_\swarrow, (w)_\downarrow, (w)_\searrow\} \text{ and } v \in W_n. \end{cases}$$

Define $v(i) = \theta(w(i))$ and $\widetilde{\mathbf{p}} = (v(1), \ldots, v(k))$. Then the $+$-invariant property of $X$ implies that $\widetilde{\mathbf{p}} \in \mathcal{CH}_{m+n}(F, D_1)$, where $D_1 = L_{(w)_\leftarrow} \cup T_{(w)_\leftarrow} \cup T_w \cup T_{(w)_\rightarrow} \cup R_{(w)_\rightarrow}$, and

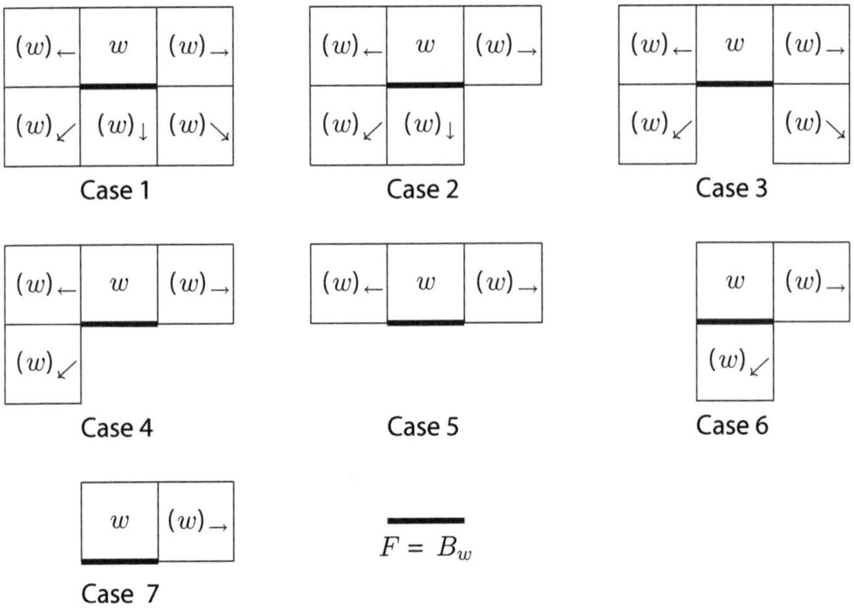


**Fig. 3** Structures of $W_m(F)$

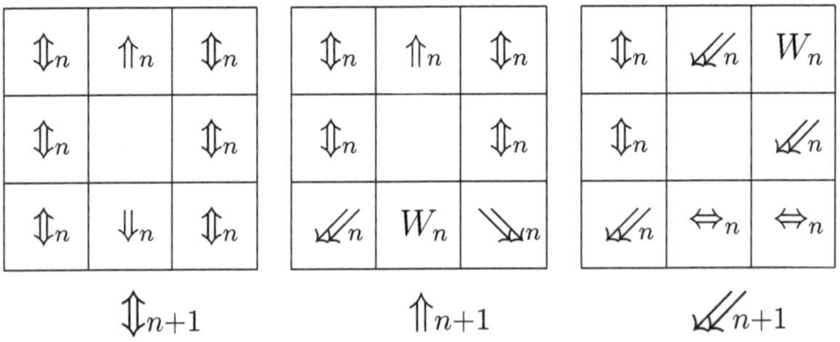

**Fig. 4** Construction of $\Updownarrow_n$, $\Uparrow_n$ and $\swarrow\!\!\swarrow_n$

$$\ell_{W_m X}(\mathbf{p}) = \ell_{W_m X}(\widetilde{\mathbf{p}}),$$

If $v(k) \cap L_{(w)_\leftarrow} \neq \emptyset$, then there exists $j$ such that $(v(j), v(j+1), \ldots, v(k)) \in \mathcal{CH}_m(R_{(w)_\leftarrow}, L_{(w)_\leftarrow})$ and $K_{v(i)} \subseteq (w)_\leftarrow \cdot W_n$ for any $i \in \{j, j+1, \ldots, k\}$. Since $X$ is invisible, it follows that

$$\ell_{W_m X}(\mathbf{p}) \geq \ell_{W_m X}((v(j), \ldots, v(k))) \geq 3^{-m}.$$

The same discussion shows that $\ell_{W_m X}(\mathbf{p}) \geq 3^{-m}$ if $K_{v(k)} \cap R_{(w)_\rightarrow} \neq \emptyset$.

Next suppose $v(K) \cap \big(T_{(w)_{\leftarrow}} \cup T_w \cup T_{(w)_{\rightarrow}}\big) \neq \emptyset$. Then using the reflections in the lines containing $L_w$ and $R_w$, we may construct $(u(1), \ldots, u(k)) \in \mathcal{CH}_{m+n}(B_w, T_w)$ which satisfies $u(i) \in w \cdots W_n$ for any $i$ and $\ell_{W_m X}(\mathbf{p}) = \ell_{W_m X}((u(1), \ldots, u(k)))$. Since $X$ is invisible, it follows that $\ell_{W_m X}(\mathbf{p}) \geq 3^{-m}$. □

**Lemma 4.8** *Let $F, G \in \partial_m$ with $F \cap G = \emptyset$ and let $\mathbf{p} = (w(1), \ldots, w(k)) \in \mathcal{CH}_{m+n}(F, G)$. If $\{w(i)\}_{i=1}^{k} \cap AW_n = \emptyset$, then there exists $\mathbf{p}_* \in \mathcal{CH}_m(F, G)$ such that $\ell_A(\mathbf{p}_*) \leq \ell_{AW_n \cup W_m X}(\mathbf{p})$.*

*Proof* Let $k_1 = \max\{j | \{w(i)\}_{i=1}^{j} \subseteq W_{m+n}(F)\}$. Define $v(1) = [w(k_1)]_m$. Note that $v(1) \notin A$. There exists a unique $F_1 \in \{L_{v(1)}, R_{v(1)}, T_{v(1)}, B_{v(1)}\} \cap \partial_m(W_m(F))$ such that $K_{w(k_1)} \subseteq F_1$. By Lemma 4.7,

$$\ell_{AW_n \cup W_m X}((w(1), \ldots, w(k_1))) \geq 3^{-m} = \ell_A((v(1))).$$

Now, if $F_1 \cap G \neq \emptyset$, $(v(1)) \in \mathcal{CH}_m(F, G)$ and $\ell_{AW_n \cup W_m X}(\mathbf{p}) \geq \ell_A((v(1))$. Hence we have constructed $\mathbf{p}_* = (v(1))$. Otherwise, replacing $(w(1), \ldots, w(k))$ and $F$ by $(w(k_1), \ldots, w(k))$ and $F_1$ respectively, we repeat the same procedure as above and obtain $k_2$, $v(2)$ and $F_2$. Inductively, we have $\mathbf{p}_* = (v(1), \ldots, v(l))$ with the desired properties. □

**Lemma 4.9** *Let $F, G \in \partial_m$ with $F \cap G = \emptyset$. Then for any $\mathbf{p} \in \mathcal{CH}_{m+n}(F, G)$, there exists $\mathbf{p}_* \in \mathcal{CH}_m(F, G)$ such that $\ell_A(\mathbf{p}_*) \leq \ell_{AW_n \cup W_m X}(\mathbf{p})$.*

*Proof* Let $\mathbf{p} = (w(1), \ldots, w(k))$. If $w(i) \notin AW_n$ for any $i$, then Lemma 4.8 suffices. Hence we assume that there exists $i$ such that $w(i) \in AW_n$.

**Claim 1** *Without loss of generality, we may assume that there exists $p_1 \geq 1$ and $G_1 \in \partial_m$ such that $w(1), \ldots, w(p_1) \in W_{m+n} \backslash AW_n, w(p_1+1) \in AW_n, G_1 \cap F = \emptyset$, $G_1 \subseteq K_{[w(p_1+1)]_m}$ and $(w(1), \ldots, w(p_1)) \in \mathcal{CH}_{m+n}(F, G_1)$.*

*Proof of Claim 1* Case 1: $F \cap K(A) = \emptyset$

In this case, define

$$p_1 = \min\{i | w(i) \in AW_n\} - 1$$

and choose $G_1 \in \partial_m$ so that $G_1 \cap K_{w(p_1)} \cap K_{w(p_1+1)} \neq \emptyset$ and $G_1 \subseteq K_{[w(p_1)]_m}$.

Case 2: $F \cap K(A) \neq \emptyset$

In this case, $F$ intersects at most two connected components of $K(A)$. Let $C_1$ and $C_2$ be those connected components of $K(A)$ (It is possible that $C_1 = C_2$).

Case 2.1: $\{i | K_{w(i)} \subseteq C_1 \cup C_2\} = \emptyset$.

Define $p_1$ and choose $G_1$ as in Case 1. Then $p_1$ and $G_1$ satisfies the desired property.

Case 2.2: $\{i | K_{w(i)} \subseteq C_1 \cup C_2\} \neq \emptyset$.

Define

$$q = \max\{i | K_{w(i)} \in C_1 \cup C_2\}.$$

We may choose $F_0 \in \partial_m$ so that $F_0 \cap K_{w(q)} \cap K_{w(q+1)} \neq \emptyset$ and $F_0 \subseteq K_{[w(q)]_m}$. Moreover, we may choose $\mathbf{p}^0 = (v(1), \ldots, v(k_0)) \in \mathcal{CH}_m(F, F_0)$ so that $v(i) \in AW_n$ for any $i = 1, \ldots, k_0$ and $v(k_0) = [w(q)]_m$. Note that $\ell_A(\mathbf{p}^0) = 0$. If $F_0 \cap G \neq \emptyset$, then $K_{v(k_0)} \cap G \neq \emptyset$ and $\mathbf{p}^0 \in \mathcal{CH}_m(F, G)$. Hence letting $\mathbf{p}_* = \mathbf{p}^0$, we have constructed $\mathbf{p}_*$ which satisfies all the conditions. Assume that $F_0 \cap G = \emptyset$. Since $(w(1), \ldots, w(q)) \in \mathcal{CH}_{m+n}(F, F_0)$ corresponds $\mathbf{p}^0 \in \mathcal{CH}_m(F, F_0)$, it is enough to show the statement of the lemma in the case where $F$ and $\mathbf{p}$ are replaced by $F_0$ and $(w(q+1), \ldots, w(k))$ respectively. In this situation, the counterpart of Case 2.1 holds and so does Claim 1 (End of Proof of Claim 1). □

**Claim 2** *Without loss of generality, we may assume that there exists $k_*$ and $F_* \in \partial_m$ such that $w(k_*), \ldots, w(k) \in W_{m+n} \backslash AW_n$, $w(k_* - 1) \in AW_n$, $F_* \cap G = \emptyset$, $F_* \subseteq K_{w(k_*)}$ and $(w(k_*), \ldots, w(k)) \in \mathcal{CH}_{m+n}(F_*, G)$.*

*Proof of Claim 2* By considering the chain $(w(k), w(k-1), \ldots, w(1)) \in \mathcal{CH}_{m+n}(G, F)$, the same argument as in the proof of Claim 1 yields this claim (End of Proof of Claim 2).

Now under Claim 1 and Claim 2, we may choose $p_1, \ldots, p_{j+1}$ and $q_0, q_1, \ldots, q_j$ which satisfy the following conditions (A), (B), (C) and (D):

(A) $q_0 = 0$, $p_{j+1} = k$, $q_i < p_{i+1} < q_{i+1}$ for any $i$.
(B) $\{(w(q_{i-1}+1), \ldots, w(p_i)\} \cap AW_n = \emptyset$ for any $i = 1, \ldots, j+1$
(C) $K_{w(p_i+1)}$ and $K_{w(q_i)}$ belong to the same connected component of $K(A)$ for any $i = 1, \ldots, j$.
(D) $K_{w(q_i)}$ and $K_{w(p_{i+1}+1)}$ belong to the different connected components of $K(A)$ for any $i = 1, 2, \ldots, j-1$

Let $\mathbf{p}_i = (w(q_{i-1}+1), \ldots, w(p_i))$ for $i = 1, \ldots, j+1$. Define $F_1 = F$. For $i \geq 2$, we may choose $F_i \in \partial_m$ so that $F_i \cap K_{w(q_{i-1})} \cap K_{w(q_{i-1}+1)} \neq \emptyset$ and $F \subseteq K_{[w(q_{i-1})]_m}$. Moreover, for $i = 1, \ldots, j$, we may choose $G_i \in \partial_m$ so that $G_i \cap K_{w(p_i)} \cap K_{w(p_i+1)} \neq \emptyset$ and $G_i \subseteq K_{[w(p_i+1)]_m}$. Also let $F_{j+1} = G$. By the condition (D), $F_i \cap G_i = \emptyset$ for any $i = 1, \ldots, j+1$. Hence letting $F = F_i$, $G = G_i$ and $\mathbf{p} = \mathbf{p}_i$ and applying Lemma 4.8, we obtain $\mathbf{p}_{*,i} = (v(i,1), \ldots, v(i,k_i)) \in \mathcal{CH}_m(F_i, G_i)$ which satisfies $\ell_A(\mathbf{p}_{*,i}) \leq \ell_{AW_n \cup W_m X}(\mathbf{p}_i)$.

Note that $G_i$ and $F_i$ belong to the same connected component of $K(A)$ by the condition (C). Hence there exists $\mathbf{p}_i^1 = (u(i,1), \ldots, u(i,l_i)) \in \mathcal{CH}_m(G_i, F_i)$ such that $u(i,1), \ldots, u(i,l_i) \in A$.

Finally let $\mathbf{p}_* = (\mathbf{p}_{*,1}, \mathbf{p}_1^1, \mathbf{p}_{*,2}, \ldots, \mathbf{p}_j^1, \mathbf{p}_{*,j+1})$. Then $\mathbf{p}_* \in \mathcal{CH}_m(F, G)$ and $\ell_A(\mathbf{p}_*) \leq \ell_{AW_n \cup W_m X}(\mathbf{p})$. □

*Proof of Theorem 4.5* Let $\mathbf{p} = (w(1), \ldots, w(k)) \in \mathcal{CH}_{m+n}(L, R)$. Set $F = L_{[w(1)]_m}$ and $G = R_{[w(k)]_m}$. By Lemma 4.9, there exists $\mathbf{p}_* \in \mathcal{CH}_m(F, G)$ such that $\ell_A(\mathbf{p}_*) \leq \ell_{AW_n \cup W_m X}(\mathbf{p})$. Since $A$ is invisible, we have $\ell_A(\mathbf{p}_*) \geq 1$. Hence $\ell_{AW_n \cup W_m X}(\mathbf{p}) \geq 1$. In the same way, if $\mathbf{p}' \in \mathcal{CH}_{m+n}(T, B)$, it follows that $\ell_{AW_n \cup W_m X}(\mathbf{p}') \geq 1$. Thus $AW_n \cup W_m X$ is invisible. □

## 5 Metric Associated with Invisible Set

In this section, we construct a metric associated with a $+$-invariant invisible set and characterize the Hausdorff dimension and the Hausdorff measure with respect to the metric.

Throughout this section, we fix a $+$-invariant invisible set $A \subseteq W_m$.

**Notation** We write $W_{m,n} = (W_m)^n = W_{mn}$, $W_{m,*} = \cup_{n \geq 0} W_{m,n}$ and $\Sigma^{(m)} = (W_m)^{\mathbb{N}}$.

Naturally $W_{m,*}$ is regarded as a subset of $W_*$ and $\Sigma^{(m)}$ is identified with $\Sigma$.

**Definition 5.1**

(1) Let $\epsilon > 0$. Define $D_\epsilon^A(w)$ for $w \in W_m$ by

$$D_\epsilon^A(w) = \begin{cases} 3^{-m} & \text{if } w \notin A, \\ \epsilon & \text{if } w \in A. \end{cases}$$

and $D_\epsilon^A(\emptyset) = 1$ for $\emptyset \in W_0$. For any $w = w^{(1)} \cdots w^{(n)} \in W_{m,n}$, where $w^{(i)} \in W_m$, define

$$D_\epsilon^A(w) = D_\epsilon^A(w^{(1)}) D_\epsilon^A(w^{(2)}) \cdots D_\epsilon^A(w^{(n)}).$$

(2) Define

$$\mathcal{CH}^{(m)} = \{(w(1), \ldots, w(k)) | \\ (w(1), \ldots, w(k)) \in \mathcal{CH}, w(i) \in W_{m,*} \text{ for any } i = 1, \ldots, k\}.$$

and $\mathcal{CH}^{(m)}(U, V) = \mathcal{CH}(U, V) \cap \mathcal{CH}^{(m)}$ for $U, V \subseteq K$. Moreover, define $\ell^{A,\epsilon}(\mathbf{p}) = \sum_{i=1}^{k} D_\epsilon^A(w(i))$ for any $\mathbf{p} = (w(1), \ldots, w(k)) \in \mathcal{CH}^{(m)}$ and, for $x, y \in K$,

$$d_\epsilon^A(x, y) = \inf\{\ell^{A,\epsilon}(\mathbf{p}) | \mathbf{p} \in \mathcal{CH}^{(m)}(x, y)\}.$$

$D_\epsilon^A(\cdot)$ is a gauge function on $\Sigma^{(m)}$ and $d_\epsilon^A$ is the associated pseudometric.

The next fact is obvious from the definition.

**Proposition 5.2** *$d_0^A(x, y) \leq d_\epsilon^A(x, y)$ for any $x, y \in K$ and any $\epsilon > 0$.*

The next theorem shows that $d_\epsilon^a$ is really a metric and $d_\epsilon^A \underset{\text{QS}}{\sim} d_E$.

**Theorem 5.3** *For any $\epsilon > 0$, $d_\epsilon^A$ is a metric on $K$ which is quasisymmetric with respect to $d_E$. The Hausdorff dimension of $K$ with respect to the metric $d_\epsilon^A$, $\dim_H(K, d_\epsilon^A)$ is given by the unique $\alpha$ which satisfies*

$$(8^m - \#(A)) 3^{-m\alpha} + \#(A) \epsilon^\alpha = 1. \tag{5.1}$$

*Furthermore, let* $\mathcal{H}^\alpha$ *be the* $\alpha$*-dimensional Hausdorff measure on* $(X, d_\epsilon^A)$*. Then the metric measure space* $(X, d_\epsilon^A, \mathcal{H}^\alpha)$ *is Ahlfors* $\alpha$*-regular, i.e.*

$$\mathcal{H}^\alpha(B_d(x,r)) \asymp r^\alpha$$

*for any* $x \in K$ *and* $r \in [0, \mathrm{diam}(X, d_\epsilon^A))$.

Letting $\epsilon \downarrow 0$ in (5.1), we obtain the following corollary.

**Corollary 5.4**

$$\dim_C(K, d_E) \le \frac{\log 8}{\log 3} + \frac{1}{m \log 3} \log\left(1 - \frac{\#(A)}{8^m}\right).$$

In the rest of this section, we are going to prove the above theorem. Hereafter, we omit $A$ in the notations $D_\epsilon^A(w)$, $\ell^{A,\epsilon}(\mathbf{p})$ and $d_\epsilon^A(x,y)$ and write $D_\epsilon(w)$, $\ell^\epsilon(\mathbf{p})$ and $d_\epsilon(x,y)$ respectively.

**Lemma 5.5** *Define* $A_n \subseteq W_{mn}$ *inductively by* $A_1 = A$ *and*

$$A_{n+1} = A W_{mn} \cup W_m A_n.$$

*Then* $A_n$ *is* $+$*-invariant and invisible.*

*Proof* Letting $X = A_n$ and applying Theorem 4.5, we see inductively that $A_{n+1}$ is $+$-invariant and invisible. □

**Lemma 5.6** $d_0^A(x,y) \ge 1$ *for any* $(x,y) \in (L \times R) \cup (T \times B)$.

*Proof* Define $I(\mathbf{p}) = \max_{i=1,\ldots,k} |w(i)|/m$ for any $\mathbf{p} = (w(1), \ldots, w(k)) \in \mathcal{CH}^{(m)}(L,R)$. We are going to show that $\ell^0(\mathbf{p}) \ge 1$ by an induction in $I(\mathbf{p})$. If $I(\mathbf{p}) = 0$, then $\mathbf{p} = (\emptyset)$ and $\ell^0(\mathbf{p}) = D_0(\emptyset) = 1$. Let $J = \{i | i = 1, \ldots, k, |w(i)| = I(\mathbf{p})m\}$. Then there exists $k_1, \ldots, k_l$ and $j_1, \ldots, j_l$ such that $k_i \le j_i < k_{i+1}$ and $J = \cup_{i=1,\ldots,l}\{j | k_i \le j \le j_i\}$. Let $\mathbf{p}^i = (w(k_i), \ldots, w(j_i))$. Since $|w(k_i - 1)| \le (I(\mathbf{p}) - 1)m$ and $|w(j_i + 1) \le (I(\mathbf{p}) - 1)m$, there exist $F, G \in \partial_M$, where $M = (I(\mathbf{p}) - 1)m$, such that $F \subseteq K_{w(k_i - 1)}$, $F \cap K_{w(k_i)} \ne \emptyset$, $G \subseteq K_{w(j_i+1)}$ and $G \cap K_{w(j_i)} \ne \emptyset$. If $F \cap G = \emptyset$, then $K_{w(k_i-1)} \cap K_{w(j_i+1)} \ne \emptyset$. Hence if $\mathbf{p}' = (w(1), \ldots, w(k_i - 1), w(j_i + 1), \ldots, w(k)) \in \mathcal{CH}^{(m)}(L,R)$, then we define $\mathbf{p}_*^i$ as the empty sequence. Note that $\ell^0(\mathbf{p}) \ge \ell^0(\mathbf{p}')$. Now assume that $F \cap G = \emptyset$. Set $X = A_M$. Lemma 5.5 shows that $X$ is $+$-invariant and invisible. Then by Lemma 4.9, there exists $\mathbf{p}_*^i = (v(1), \ldots, v(l)) \in \mathcal{CH}_M(F,G)$ such that $\ell_{A_M}(\mathbf{p}_*^i) \le \ell_{A_M W_m \cup W_M A}(\mathbf{p}^i)$. Note that $A_M W_m \cup W_M A = A_{I(\mathbf{p})m}$, that $\ell_{A_M}(\mathbf{p}_*^i) = \ell^0(\mathbf{p}_*^i)$ and that $\ell_{A_M W_m \cup W_M A}(\mathbf{p}^i) = \ell^0(\mathbf{p}^i)$. Let $\mathbf{p}_*$ be the chain where $\mathbf{p}^i$ is replaced by $\mathbf{p}_*^i$ for all $i$. Then $\mathbf{p}_* \in \mathcal{CH}^{(m)}(L,R)$, $I(\mathbf{p}_*) < I(\mathbf{p})$ and $\ell^0(\mathbf{p}) \ge \ell^0(\mathbf{p}_*)$. Now we have $\ell^0(\mathbf{p}) \ge \ell^0(\mathbf{p}_*) \ge 1$ by induction hypothesis.

Now, $d_0^A(x,y) \ge \inf\{\ell^0(\mathbf{p}) | \mathbf{p} \in \mathcal{CH}^{(m)}(x,y)\} \ge 1$ for any $x \in L$ and any $y \in R$. In the same manner, it follows that $d_0^A(x,y) \ge 1$ for any $x \in T$ and any $y \in B$ as well. □

**Lemma 5.7** *$d_\epsilon^A(\cdot,\cdot)$ is a metric on $K$ for any $\epsilon > 0$.*

*Proof* Let $x, y \in K$ with $x \neq y$. Then $\mathrm{Re}\, x \neq \mathrm{Re}\, y$ or $\mathrm{Im}\, x \neq \mathrm{Im}\, y$. Suppose $\mathrm{Re}\, x < \mathrm{Re}\, y$. Then there exist $n$ and $i \in \{0, 1, \ldots, 3^{mn} - 1\}$ such that $\mathrm{Re}\, x \leq (2i - 3^{mn})3^{-mn} < (2i + 2 - 3^{mn})3^{-mn} \leq \mathrm{Re}\, y$.

**Claim** $d_\epsilon^A(x, y) \geq \min\{D_\epsilon^A(w) | w \in W_{m,k}, k = 0, 1, \ldots, n\}$.

*Proof of Claim* Define $W_{m,n}^i = \{w | w \in W_{m,n}, K_w \subseteq \{z | (2i - 3^{mn})3^{-mn} \leq \mathrm{Re}\, z \leq (2i + 2 - 3^{mn})3^{-mn}\}$. Let $D_{mn,i} = \min\{D_\epsilon(w) | w \in W_{m,n}^i\}$. We also define $L_{mn,i} = \bigcup_{w \in W_{m,n}^i} L_w$ and $R_{mn,i} = \bigcup_{w \in W_{m,n}^i} R_w$. Let $\mathbf{p} = (w(1), \ldots, w(k)) \in \mathcal{CH}^{(m)}(x, y)$. If $|w(i)| \leq mn$ for some $i$, then the claim is trivial. Hence assume that $|w(i)| < mn$ for any $i = 1, \ldots, k$. Then $\mathbf{p}$ contains $(w(p), w(p+1), \ldots, w(q)) \in \mathcal{CH}(L_{mn,i}, R_{mn,i})$ which satisfies $w(i) \in \bigcup_{w \in W_{m,n}^i} w W_{m,*}$. Let $w(i) = u(i)v(i)$ for $i = p, \ldots, q$, where $u(i) \in W_{m,n}^i$ and $v(i) \in W_{m,*}$. It follows that

$$\ell^{A,\epsilon}(\mathbf{p}) \geq \ell^{A,\epsilon}((w(p), w(p+1), \ldots, w(q))) \geq D_{mn,i} \sum_{i=p}^{q} D_\epsilon(v(i)). \qquad (5.2)$$

Now the reflection $\psi$ in the real axis induces a natural bijection $\varphi_\leftrightarrow : W_* \to W_*$ defined by $\psi(K_w) = K_{\varphi_\leftrightarrow(w)}$ which satisfies $\varphi_\leftrightarrow(\varphi_\leftrightarrow(w)) = w$. Hereafter in this section, we write $\varphi = \varphi_\leftrightarrow$. There exist $p_1, p_2, \ldots p_l$ such that $p_1 = p$, $p_l = q + 1$, $p_i < p_{i+1}$, $u(p_i) = u(p_i + 1) = \ldots = u(p_{i+1} - 1)$ and $u(p_i) \neq u(p_{i+1})$ for any $i$. Let $\bar{v}(j) = \varphi^i(v(j))$ for $j = p_i, p_i + 1, \ldots, p_{i+1} - 1$, where $\varphi^j$ is the $j$-th iteration of $\varphi$. Then $(\bar{v}(p), \bar{v}(p + 1), \ldots, \bar{v}(q)) \in \mathcal{CH}^{(m)}(L, R)$. Since $A$ is $+$-invariant, $\sum_{i=p}^q D_\epsilon(v(i)) = \sum_{i=p}^q D_\epsilon(\bar{v}(i))$. Hence Lemma 5.6 implies that

$$\sum_{i=p}^{q} D_\epsilon(v(i)) = \sum_{i=p}^{q} D_\epsilon(\bar{v}(i)) \geq \sum_{i=p}^{q} D_0(\bar{v}(i)) \geq 1.$$

Combining this with (5.2), we have $\ell^{A,\epsilon}(\mathbf{p}) \geq D_{mn,i}$. Hence the claim holds (End of Proof of Claim).

The claim shows that $d_\epsilon^A(x, y) > 0$ if $\mathrm{Re}\, x \neq \mathrm{Re}\, y$. Similar discussion implies $d_\epsilon^A(x, y) > 0$ if $\mathrm{Im}\, x \neq \mathrm{Im}\, y$. □

*Proof of Theorem 5.3* Let $\mathcal{S}^{(m)}(A, \epsilon) = \{\Lambda_s^{(m)}(A, \epsilon)\}_{s \in (0,1]}$ be the scale on $\Sigma^{(m)}$ whose gauge function is $D_\epsilon^A$ and let $\mathcal{S}^{(m)}$ by the scale on $\Sigma^{(m)}$ whose gauge function $g$ is given by $g(w(1) \ldots w(k)) = 3^{-mk}$ for $w(1) \ldots w(k) \in W_{m,*}$ with $w(1), \ldots, w(k) \in W_m$. Obviously $\Sigma^{(m)}$ is adapted to the Euclidean metric on $K$. Also since $\mathcal{S}^{(m)}(A, \epsilon)$ and $\mathcal{S}^{(m)}$ are self-similar, they are elliptic.

Note that $(K, W_m, \{F_w\}_{w \in W_m})$ is a rationally ramified self-similar structure. (See [Kig09, Sect. 1.5] for the definition of rationally ramified self-similar structures.) In fact, define $h : L^1 \to R^1$ by $h(\nwarrow) = \nearrow, h(\leftarrow) = \rightarrow, h(\swarrow) = \searrow$ and $g : T^1 \to B^1$ by $g(\nwarrow) = \swarrow, g(\uparrow) = \downarrow, g(\nearrow) = \searrow$. Then define $h_m : L^m \to R^m$

by $h_m(w_1 \ldots w_m) = h(w_1) \ldots h(w_m)$ for $w_1 \ldots w_m \in L^m$ and $g_m : T^m \to B^m$ by $g_m(w_1 \ldots w_m) = g(w_1) \ldots g(w_m)$ for $w_1 \ldots w_m \in T^m$. Then a relation set $\mathcal{R}_m$ of $(K, W_m, \{F_w\}_{w \in W_m})$ is given by

$$\mathcal{R}_m = \{(L^m, R^m, h_m, w, v) | w, v \in W_m, L_w = R_v\} \cup \{(T^m, B^m, g_m, w, v) | w, v \in W_m, T_w = B_v\}$$

By Proposition 4.4, $D_\epsilon^A(w) = 3^{-m}$ for any $w \in L^m \cup R^m \cup T^m \cup B^m$. Using [Kig09, Theorem 1.6.6], we see that $\mathcal{S}^{(m)}(A, \epsilon) \underset{\text{GE}}{\sim} \mathcal{S}^{(m)}$.

Theorems 1.6.1 and 2.2.7 in [Kig09] imply that $\mathcal{S}^{(m)}(A, \epsilon)$ is intersection type finite. Since $d_\epsilon^A$ is a metric on $K$ by Lemma 5.7, we may apply [Kig09, Theorem 2.3.16] and show that $d_\epsilon^A$ is adapted to the scale $\mathcal{S}^{(m)}(A, \epsilon)$. Thus we have obtained all the conditions in Theorem 3.4 and hence shown that $d_\epsilon^A$ is quasisymmetric with respect to the Euclidean metric.

The Hausdorff dimension and Ahlfors regularity of the Hausdorff measure of $(K, d_\epsilon^A)$ are immediately obtained by [Kig01, Theorem 1.5.7]. □

## 6 Construction of Invisible Sets

Under the existence of an invisible set, we have constructed a corresponding metric which is quasisymmetric with respect to $d_E$ and characterized the associated Hausdorff dimension in the previous two sections. In this section, it is shown that invisible sets do exist. In fact, we construct a series of invisible sets inductively.

**Definition 6.1** Let $\psi_\updownarrow$ and $\psi_\leftrightarrow$ be the reflections in the real and complex axes respectively. Then $\psi_\updownarrow$ induces a natural bijection $\varphi_\updownarrow$ from $W_*$ to itself defined by $\psi_\updownarrow(K_w) = K_{\varphi_\updownarrow(w)}$. In the same way, we define a bijection $\varphi_\leftrightarrow$ from $W_*$ to itself by $\psi_\leftrightarrow(K_w) = K_{\varphi_\leftrightarrow(w)}$. Moreover, let $R$ be the $\pi/2$-rotation around the origin 0 and let $\rho : W_* \to W_*$ be the bijection defined by $R(K_w) = K_{\rho(w)}$.

The idea to have invisible sets is to divide the notion of a invisible set into a vertically invisible set and a horizontally invisible set. The final existence of invisible sets are established by taking intersections of vertically invisible set and horizontally invisible set in Theorem 6.4.

**Definition 6.2** Define $\Updownarrow_n$, $\Uparrow_n$ and $\Swarrow_n$ as subsets of $W_n$ inductively by

$$\Updownarrow_{n+1} = \{\nwarrow, \leftarrow, \swarrow, \nearrow, \rightarrow, \searrow\} \cdot \Updownarrow_n \cup \uparrow \cdot \Uparrow_n \cup \downarrow \cdot \Downarrow_n \tag{6.1}$$

$$\Uparrow_{n+1} = \{\nearrow, \leftarrow, \nwarrow, \rightarrow\} \cdot \Updownarrow_n \cup \uparrow \cdot \Uparrow_n \cup \swarrow \cdot \Swarrow_n \cup \downarrow \cdot W_n \cup \searrow \cdot \Searrow_n \tag{6.2}$$

$$\Swarrow_{n+1} = \{\nwarrow, \leftarrow\} \cdot \Updownarrow_n \cup \{\downarrow, \searrow\} \cdot \Leftrightarrow_n \cup \{\uparrow, \rightarrow, \swarrow\} \cdot \Swarrow_n \cup \nearrow \cdot W_n \tag{6.3}$$

and $\Updownarrow_0 = \Uparrow_0 = \Swarrow_0 = \emptyset$, where $\Downarrow_n = \varphi_\updownarrow(\Uparrow_n)$, $\Searrow_n = \varphi_\leftrightarrow(\Swarrow_n)$ and $\Leftrightarrow_n = \rho(\Updownarrow_n)$.

Lemma 6.13 will show that $\Updownarrow_n$ and $\Leftrightarrow_n$ are vertically and horizontally invisible sets respectively.

**Lemma 6.3**

$$\#(\Updownarrow_n) = 8^n - \frac{7+\sqrt{41}}{2\sqrt{41}}\left(\frac{9+\sqrt{41}}{2}\right)^n + \frac{7-\sqrt{41}}{2\sqrt{41}}\left(\frac{9-\sqrt{41}}{2}\right)^n$$

*Proof* Write $a_n = \#(\Updownarrow_n)$, $b_n = \#(\Uparrow_n)$ and $c_n = \#(\Swarrow_n)$. By (6.1), (6.2) and (6.3), it follows that

$$\begin{aligned} a_{n+1} &= 6a_n + 2b_n \\ b_{n+1} &= 4a_n + b_n + 2c_n + 8^n \\ c_{n+1} &= 4a_n + 3c_n + 8^n. \end{aligned}$$

Solving these with $a_0 = b_0 = c_0 = 0$, we obtain $a_n$ as in the statement of the lemma. □

Now we have the main theorem of this section.

**Theorem 6.4** *Let $A_n = \Updownarrow_n \cap \Leftrightarrow_n$. Then $A_n$ is a $+$-invariant invisible set and*

$$\alpha_n \leq 8^n - \#(A_n) \leq 2\alpha_n,$$

*where*

$$\alpha_n = \frac{7+\sqrt{41}}{2\sqrt{41}}\left(\frac{9+\sqrt{41}}{2}\right)^n - \frac{7-\sqrt{41}}{2\sqrt{41}}\left(\frac{9-\sqrt{41}}{2}\right)^n.$$

*Example 6.5* $A_0 = A_1 = A_2 = A_3 = \emptyset$.

$$A_4 = \{\uparrow\downarrow\leftarrow\rightarrow, \uparrow\downarrow\rightarrow\leftarrow, \downarrow\uparrow\leftarrow\rightarrow, \downarrow\uparrow\leftarrow\rightarrow, \leftarrow\rightarrow\uparrow\downarrow, \leftarrow\rightarrow\downarrow\uparrow, \rightarrow\leftarrow\uparrow\downarrow, \rightarrow\leftarrow\downarrow\uparrow\}.$$

Applying Corollary 5.4 and letting $n \to \infty$, we obtain the following upper estimate of the conformal dimension of the Sierpinski carpet.

**Corollary 6.6**

$$\dim_C(K, d_E) \leq \frac{\log\left(\frac{9+\sqrt{41}}{2}\right)}{\log 3} = 1.858183... < \frac{\log 8}{\log 3} = 1.892789....$$

*Remark* The known lower bound of $\dim_C(K, d_E)$ given in (1.1) is $\frac{\log 6}{\log 3} = 1.630929....$

The rest of this section is devoted to proving Theorem 6.4.

**Lemma 6.7** (1) $\varphi_{\leftrightarrow}(\Updownarrow_n) = \Updownarrow_n$ *and* $\varphi_{\updownarrow}(\Updownarrow_n) = \Updownarrow_n$.
(2) $\varphi_{\leftrightarrow}(\Uparrow_n) = \Uparrow_n$.
(3) $\varphi_{\leftrightarrow} \circ \rho(\Swarrow_n) = \Swarrow_n$.

**Definition 6.8** Define the vertical index $I^n_{\updownarrow} : W_n \to \{1, \ldots, 3^n\}$ by

$$I^n_{\updownarrow}(w) = \frac{3^n(\mathrm{Im}\,(F_w(\sqrt{-1})) + 1)}{2}$$

For $H \in \{L, R\}$, define $w^n_H(i)$ for $i = 1, \ldots, 3^n$ as the unique $w \in H^n$ which satisfies $I^n_{\updownarrow}(w) = i$. Moreover, for $w, v \in W_n$, define $\mathbf{p}^n_H(w, v) \in \mathcal{CH}_n$ by

$$\mathbf{p}^n_H(w, v) = \begin{cases} (w^n_H\left(I^n_{\updownarrow}(w)\right), w^n_H\left(I^n_{\updownarrow}(w) + 1\right), \ldots, w^n_H\left(I^n_{\updownarrow}(v)\right) & \text{if } I^n_{\updownarrow}(w) \le I^n_{\updownarrow}(v), \\ (w^n_H\left(I^n_{\updownarrow}(w)\right), w^n_H\left(I^n_{\updownarrow}(w) - 1\right), \ldots, w^n_H\left(I^n_{\updownarrow}(v)\right) & \text{if } I^n_{\updownarrow}(v) \le I^n_{\updownarrow}(w). \end{cases}$$

In the same way, we define the horizontal index $I^n_{\leftrightarrow} : W_n \to \{1, \ldots, 3^n\}$, $w^n_T(i)$, $w^n_B(i)$, $\mathbf{p}^n_T(w, v)$ and $\mathbf{p}^n_B(w, v)$.

**Lemma 6.9** *Let* $A \subseteq W_n$. *Assume that*

$$\inf\{\ell_A(\mathbf{p}_*)|\mathbf{p}_* \in \mathcal{CH}_n(T, p_{\swarrow})\} \ge 1 \tag{6.4}$$

*Let* $\mathbf{p} = (w(1), \ldots, w(k)) \in \mathcal{CH}_n$. *If* $(w(1), w(k)) \in (T^n \cup L^n) \times L^n$, *then*

$$\ell_A(\mathbf{p}) \ge \frac{|I^n_{\updownarrow}(w(1)) - I^n_{\updownarrow}(w(k))| + 1}{3^n} = \ell_A(\mathbf{p}^n_L(w(1), w(k))). \tag{6.5}$$

*Remark* Using the symmetries, we may exchange $(T, L, p_{\swarrow})$ in the statement of Lemma 6.9 by $(T, R, p_{\searrow})$, $(B, L, p_{\nwarrow})$ and $(B, R, p_{\nearrow})$.

*Proof* Since $\ell_A((w^n_L(i))_{i=1,\ldots,3^n}) \ge 1$, it follows that $\{w^n_L(i)|i = 1, \ldots, 3^n\} \cap A = \emptyset$. Let $\mathbf{p} = (w(1), \ldots, w(k)) \in \mathcal{CH}_n$.

Suppose that $(w(1), w(k)) \in T^n \times L^n$. Note that $w(k) = w^n_L(i)$ for some $i$. Then $\mathbf{p}^n_L(w(1), w(k)) = (w^n_L(3^n), w^n_L(3^n - 1), \ldots, w^n_L(i))$ and $\ell_A(\mathbf{p}^n_L(w(1), w(k))) = 1 - (i-1)/3^n$. Since $\mathbf{p} \vee \mathbf{p}^n_L(w^n_L(i-1), w^n_L(1)) \in \mathcal{CH}_n(T, p_{\swarrow})$, (6.4) implies

$$\ell_A(\mathbf{p}) + \ell_A(\mathbf{p}^n_L(w^n_L(i-1), w^n_L(1))) = \ell_A(\mathbf{p} \vee \mathbf{p}^n_L(w^n_L(i-1), w^n_L(1))) \ge 1.$$

This shows (6.5) in this case.

Suppose that $(w(1), w(k)) \in L^n \times L^n$. Set $w(1) = w^n_L(j)$ and $w(k) = w^n_L(i)$. If $j < i$, then we consider $(w(k), \ldots, w(1))$ in place of $(w(1), \ldots, w(k))$. In this way, we may assume that $j \ge i$ without loss of generality. Let $\widetilde{\mathbf{p}} = \mathbf{p}^n_L\left(w^n_L(3^n), w^n_L(j+1)\right) \vee \mathbf{p} \vee \mathbf{p}^n_L\left(w^n_L(j-1), w^n_L(1)\right)$. Since $\widetilde{\mathbf{p}} \in \mathcal{CH}_n(T, p_{\swarrow})$, we have

$$\frac{3^n - j}{3^n} + \ell_A(\mathbf{p}) + \frac{i-1}{3^n} = \ell_A(\widetilde{\mathbf{p}}) \geq 1$$

This immediately implies (6.5) in this case. □

**Lemma 6.10** *Let $X, Y \subseteq W_n$. Assume that*

$$\inf\{\ell_X(\mathbf{p}) | \mathbf{p} \in \mathcal{CH}_n(T, B)\} \geq 1$$

*and that*

$$\inf\{\ell_Y(\mathbf{p}) | \mathbf{p} \in \mathcal{CH}_n(T, p_{\swarrow})\} \geq 1.$$

*Define $A = \nwarrow \cdot X \cup \uparrow \cdot Y$. If $\mathbf{p} = (w(1), \ldots, w(k)) \in \mathcal{CH}_{n+1}(T, B_{\nwarrow})$ and $\{w(1), \ldots, w(k)\} \subseteq \{\nwarrow, \uparrow\} \cdot W_n$, then*

$$\ell_A(\mathbf{p}) \geq \frac{1}{3}.$$

*Proof* Let $w(i) = s(i)v(i)$, where $s(i) \in \{\nwarrow, \uparrow\}$ and $v(i) \in W_n$.

First assume that $s(1) = \nwarrow$. Then there exist $j_1, j_2, \ldots, j_{2p+2}$ which satisfies the following three conditions (C1), (C2) and (C3):

(C1) $j_1 = 1$, $j_{2p+2} = k + 1$ and $j_1 < j_2 < \ldots < j_{2p+2}$
(C2) $s(i) = \nwarrow$ for $i = j_{2q-1}, \ldots, j_{2q} - 1$ and $q = 1, \ldots, p + 1$
(C3) $s(i) = \uparrow$ for $i = j_{2q}, \ldots, j_{2q+1} - 1$ and $q = 1, \ldots, p$.

Set $\mathbf{p}_{1,q} = \big(w(j_{2q-1}), \ldots, w(j_{2q} - 1)\big)$ and $\widetilde{\mathbf{p}}_{1,q} = \big(v(j_{2q-1}), \ldots, v(j_{2q} - 1)\big)$. Since $\big(v(j_{2q-1}), v(j_{2q} - 1)\big) \in (T^n \times R^n) \cup (R^n \times R^n) \cup (R^n \times B^n)$, Lemma 6.9 and its variants explained in the remark imply

$$\ell_A\left(\mathbf{p}_{1,q}\right) = \frac{1}{3}\ell_X\left(\widetilde{\mathbf{p}}_{1,q}\right) \geq \frac{1}{3}\ell_X\left(\mathbf{p}_R^n\left(v\left(j_{2q-1}\right), v\left(j_{2q} - 1\right)\right)\right). \tag{6.6}$$

Set $\mathbf{p}_{2,q} = \left(w\left(j_{2q}\right), \ldots, w\left(j_{2q+1} - 1\right)\right)$ and $\widetilde{\mathbf{p}}_{2,q} = \left(v\left(j_{2q}\right), \ldots, v\left(j_{2q+1} - 1\right)\right)$. Since $\left(v\left(j_{2q}\right), v\left(j_{2q+1} - 1\right)\right) \in L^n \times L^n$, Lemma 6.9 shows that

$$\ell_A\left(\mathbf{p}_{2,q}\right) = \frac{1}{3}\ell_Y\left(\widetilde{\mathbf{p}}_{2,q}\right) \geq \frac{1}{3}\ell_Y\left(\mathbf{p}_L^n\left(v\left(j_{2q}\right), v\left(j_{2q+1} - 1\right)\right)\right). \tag{6.7}$$

Note that for any $i = 1, \ldots, 3^n$, there exists $l = 1, 2, \ldots, 2q + 1$ such that $I_{\updownarrow}^n(v(j_l)) \leq i \leq I_{\updownarrow}^n(v(j_{l+1} - 1))$ or $I_{\updownarrow}^n(v(j_l)) \geq i \geq I_{\updownarrow}^n(v(j_{l+1} - 1))$. Hence

$$\sum_{q=1}^{p+1} \ell_X\left(\mathbf{p}_R^n\left(v\left(j_{2q-1}\right), v\left(j_{2q} - 1\right)\right)\right) + \sum_{q=1}^{p} \ell_Y\left(\mathbf{p}_L^n\left(v\left(j_{2q}\right), v\left(j_{2q+1} - 1\right)\right)\right) \geq 1.$$

Combining this with (6.6) and (6.7), we obtain

$$\ell_A(\mathbf{p}) = \sum_{q=1}^{p+1} \ell_A\left(\mathbf{p}_{1,q}\right) + \sum_{q=1}^{p} \ell_A\left(\mathbf{p}_{2,q}\right) \geq \frac{1}{3}.$$

Thus we have shown the desired statement in the case when $s(1) = \nwarrow$.

If $s(1) = \uparrow$, slight modification of the above arguments yields the lemma as well. □

**Definition 6.11** Define $\pi : W_* \to W_*$ by

$$\pi(w) = \begin{cases} w & \text{if } \mathrm{Re}\, F_w(0) \leq 0, \\ \varphi_{\leftrightarrow}(w) & \text{if } \mathrm{Re}\, F_w(0) > 0. \end{cases}$$

For $\mathbf{p} = (w(1), \ldots, w(k)) \in \mathcal{CH}_n$, we define $\pi_n(\mathbf{p}) = (\pi(w(1)), \ldots, \pi(w(k))$. Also define $\xi : W_* \to W_*$ by

$$\xi(w) = \begin{cases} w & \text{if } \mathrm{Re}\, F_w(0) \leq \mathrm{Im}\, F_w(0), \\ \varphi_{\leftrightarrow}(\rho(w)) & \text{if } \mathrm{Re}\, F_w(0) > \mathrm{Im}\, F_w(0). \end{cases}$$

For $\mathbf{p} = (w(1), \ldots, w(k)) \in \mathcal{CH}_n$, we define $\xi_n(\mathbf{p}) = (\xi(w(1)), \ldots, \xi(w(k))$.

By the symmetry of $\Updownarrow_n$, $\Uparrow_n$ and $\Swarrow_n$ given in Lemma 6.7, we have the following lemma.

**Lemma 6.12** (1) $\pi_n : \mathcal{CH}_n \to \mathcal{CH}_n$, $\ell_{\Updownarrow_n}(\pi_n(\mathbf{p})) = \ell_{\Updownarrow_n}(\mathbf{p})$ *and* $\ell_{\Uparrow_n}(\pi_n(\mathbf{p})) = \ell_{\Uparrow_n}(\mathbf{p})$.
(2) $\xi_n : \mathcal{CH}_n \to \mathcal{CH}_n$ *and* $\ell_{\Swarrow_n}(\xi_n(\mathbf{p})) = \ell_{\Swarrow_n}(\mathbf{p})$.

**Lemma 6.13** *Suppose that*

$$\inf\{\ell_{\Updownarrow_n}(\mathbf{p}) | \mathbf{p} \in \mathcal{CH}_n(T, B)\} \geq 1 \tag{6.8}$$

*and*

$$\inf\{\ell_{\Uparrow_n}(\mathbf{p}) | \mathbf{p} \in \mathcal{CH}_n(T, \{p_{\swarrow}, p_{\searrow}\}) \geq 1. \tag{6.9}$$

*If* $\mathbf{p} = (w(1), \ldots, w(k)) \in \mathcal{CH}_{n+1}(T, B_{\nwarrow} \cup B_{\nearrow})$ *and* $\{w(i)\}_{i=1}^k \subseteq \{\nwarrow, \uparrow, \nearrow\} \cdot W_n$, *then* $\ell_{\Updownarrow_{n+1}}(\mathbf{p}) \geq 1/3$.

*Proof* Replacing $\mathbf{p}$ by $\pi_{n+1}(\mathbf{p})$, we may assume that $w(1), \ldots, w(k) \in \{\nwarrow, \uparrow\} \cdot W_n$ and $w(k) \in \nwarrow \cdot B^n$ without loss of generality. Set $X = \Updownarrow_n$ and $Y = \Uparrow_n$. Then the assumptions (6.8) and (6.9) of Lemma 6.10 follows. Hence $\ell_{\Updownarrow_{n+1}}(\mathbf{p}) \geq 1/3$. □

**Lemma 6.14** *Suppose that* (6.8) *holds and that*

$$\inf\{\ell_{\Swarrow_n}(\mathbf{p})|\mathbf{p} \in \mathcal{CH}_n(T \cup R, p_{\swarrow}) \geq 1. \tag{6.10}$$

*Let* $\mathbf{p} = (w(1), \ldots, w(k)) \in \mathcal{CH}_{n+1}(T, B_{\swarrow})$. *If* $\{w(1), \ldots, w(k)\} \subseteq \nwarrow, \uparrow, \nearrow \cdot W_n$, *then* $\ell_{\Swarrow_{n+1}}(\mathbf{p}) \geq 1/3$.

*Proof* First assume that $\{w(1), \ldots, w(k)\} \subseteq \{\nwarrow, \uparrow\}\cdot W_n$. By (6.3), applying Lemma 6.10 with $X = \Updownarrow_n$ and $Y = \Swarrow_n$, we have $\ell_{\Swarrow_{n+1}}(\mathbf{p}) \geq 1/3$.

Next, suppose that $w(i) \in \nearrow \cdot W_n$ for some $i$. Let

$$i_* = \max\{i|w(i) \in \nearrow\cdot W_n\} + 1.$$

and let

$$j_* = \min\{j|w(j) \in \nwarrow\cdot W_n, j \geq i_*\} - 1.$$

Then, for $i = \{i_*, \ldots, j_*\}$, there exists $v(i) \in W_n$ such that $w(i) = \uparrow \cdot v(i)$. Define

$$\mathbf{p}_* = \left(\uparrow \cdot\xi\left(v\left(i_*\right)\right), \uparrow \cdot\xi\left(v\left(i_* + 1\right)\right), \ldots, \uparrow \cdot\xi\left(v\left(j_*\right)\right), w\left(j_* + 1\right), \ldots, w\left(k\right)\right).$$

By (6.3) and Lemma 6.12,

$$\begin{aligned}\ell_{\Swarrow_{n+1}}\left(w\left(i_*\right), \ldots, w\left(j_*\right)\right) &= \frac{1}{3}\ell_{\Swarrow_n}\left(v\left(i_*\right), \ldots, v\left(j_*\right)\right) \\ &= \frac{1}{3}\ell_{\Swarrow_n}\left(\xi\left(v\left(i_*\right)\right), \ldots, \xi\left(v\left(j_*\right)\right)\right) \\ &= \ell_{\Swarrow_{n+1}}\left(\uparrow \cdot\xi\left(v\left(i_*\right)\right), \ldots, \uparrow \cdot\xi\left(v\left(j_*\right)\right)\right).\end{aligned}$$

Hence $\ell_{\Swarrow_{n+1}}(\mathbf{p}) \geq \ell_{\Swarrow_{n+1}}(\mathbf{p}_*)$. Let $\mathbf{p}_* = (w_*(1), w_*(2), \ldots, w_*(l))$. Then $w^*(i) \in \{\nwarrow, \uparrow\}\cdot W_n$ for any $i = 1, \ldots, l$. Now replacing $\mathbf{p}$ by $\mathbf{p}_*$, we are exactly in the first case and hence the desired inequality is satisfied. □

**Lemma 6.15** (6.8), (6.9) *and* (6.10) *hold for any* $n \geq 0$.

*Proof* We use induction on $n$. Obviously (6.8), (6.9) and (6.10) holds for $n = 0$ since $\Updownarrow_n$, $\Uparrow_n$ and $\Swarrow_n$ are the empty sets. Assume that (6.8), (6.9) and (6.10) are true for $n = m$.

First we show (6.8) holds for $n = m + 1$. Let $\mathbf{p} = (w(1), \ldots, w(k)) \in \mathcal{CH}_{m+1}(T, B)$. Note that by Lemma 6.7-(1), $\pi_{n+1}(\mathbf{p}) \in \mathcal{CH}_{m+1}(T, B)$ and $\ell_{\Updownarrow_{m+1}}(\mathbf{p}) = \ell_{\Updownarrow_{m+1}}(\pi_{n+1}(\mathbf{p}))$. Hence replacing $\mathbf{p}$ by $\pi_{n+1}(\mathbf{p})$, we may assume that $w(i) \in \{\nwarrow, \uparrow, \leftarrow, \swarrow, \downarrow\}\cdot W_m$ for any $i = 1, \ldots, k$ without loss of generality. Set $w(i) = s(i)v(i)$, where $s(i) \in \{\nwarrow, \uparrow, \leftarrow, \swarrow, \downarrow\}$ and $v(i) \in W_m$. We may choose $i_1, i_2, i_3$ and $i_4$ which satisfies $i_1 < i_2 < i_3 < i_4$ and the following tree conditions (a1), (b1) and (c1):

(a1) $s(1), \ldots, s(i_1) \in \{\nwarrow, \uparrow\}$, $(v(1), \ldots, v(i_1)) \in \mathcal{CH}_m(T, B_{\swarrow})$,

(b1) $s(i) = \leftarrow$ for $i = i_2, \ldots, i_3$, $(v(i_2), \ldots, v(i_3)) \in \mathcal{CH}_m(T, B)$,
(c1) $s(i_4), \ldots, s(k) \in \{\swarrow, \downarrow\}$, $w_*(i_4) \in \swarrow\cdot T^m$.

Let $\mathbf{p}_1 = (w(1), \ldots, w(i_1))$. Then by the induction hypothesis, we may apply Lemma 6.13 and see that $\ell_{\Updownarrow_{m+1}}(\psi_1) \geq 1/3$.

Let $\mathbf{p}_2 = (w(i_2), \ldots, w(i_3))$. Since $(v(i_2), \ldots, v(i_3)) \in \mathcal{CH}_m(T, B)$, the induction hypothesis implies

$$\ell_{\Updownarrow_{m+1}}(\mathbf{p}_2) = \frac{1}{3}\ell_{\Updownarrow_m}(v(i_2), \ldots, v(i_3)) \geq \frac{1}{3}.$$

Set $\mathbf{p}_3 = (w(i_3), \ldots, w(k))$ and $\widetilde{\mathbf{p}}_3 = \big(\varphi_\updownarrow(w(k)), \varphi_\updownarrow(w(k-1)), \ldots, \varphi_\updownarrow(w(i_3))\big)$. Then $\ell_{\Updownarrow_{m+1}}(\mathbf{p}_3) = \ell_{\Updownarrow_{m+1}}(\widetilde{\mathbf{p}}_3)$. As $\mathbf{p}_1$, we may apply Lemma 6.13 to $\widetilde{\mathbf{p}}_3$ and obtain $\ell_{\Updownarrow_{m+1}}(\widetilde{\mathbf{p}}_3) \geq 1/3$. Combining all the estimates on $\ell_{\Updownarrow_{m+1}}(\mathbf{p}_i)$ for $i = 1, 2, 3$, we have $\ell_{\Updownarrow_{m+1}}(\mathbf{p}) \geq 1$.

Secondly, we show that (6.9) holds for $n = m + 1$. Let $\mathbf{p} = (w(1), \ldots, w(k)) \in \mathcal{CH}_{m+1}(T, \{p_\swarrow, p_\searrow\})$. As in the first case, we may assume that $w(i) \in \{\nwarrow, \uparrow, \leftarrow, \swarrow, \downarrow\}$ for any $i = 1, \ldots, k$ without loss of generality. Set $w(i) = s(i)v(i)$, where $s(i) \in \{\nwarrow, \uparrow, \leftarrow, \swarrow, \downarrow\}$ and $v(i) \in W_m$. We may choose $i_1, i_2, i_3$ and $i_4$ which satisfies $i_1 < i_2 < i_3 < i_4$ and the following tree conditions (a2), (b2) and (c2):

(a2) $s(1), \ldots, s(i_1) \in \{\nwarrow, \uparrow\}$, $(v(1), \ldots, v(i_1)) \in \mathcal{CH}_m(T, B_\swarrow)$,
(b2) $s(i) = \leftarrow$ for $i = i_2, \ldots, i_3$, $(v(i_2), \ldots, v(i_3)) \in \mathcal{CH}_m(T, B)$,
(c2) $s(i) = \swarrow$ for $i = i_4, \ldots, k$. $(v(i_4), \ldots, v(k)) \in \mathcal{CH}_m(T \cup R, p_\swarrow)$.

Define $\mathbf{p}_1$, $\mathbf{p}_2$ and $\mathbf{p}_3$ as in the first case. Then using the same discussion as in the first case, we obtain $\ell_{\Uparrow_{m+1}}(\mathbf{p}_j) \geq 1/3$ for $j = 1, 2$. Since $(v(i_4), \ldots, v(k)) \in \mathcal{CH}_m(T \cup R, p_\swarrow)$, The induction hypothesis and Lemma 6.14 yield that $\ell_{\Swarrow_m}((v(i_4), \ldots, v(k))) \geq 1$. By (6.2), it follows that

$$\ell_{\Uparrow_{m+1}}(\mathbf{p}_3) = \frac{1}{3}\ell_{\Swarrow_m}((v(i_4), \ldots, v(k))) \geq 1/3.$$

Thus, we have shown that $\ell_{\Uparrow_{m+1}}(\mathbf{p}) \geq 1$.

Finally we show that (6.10) holds for $n = m + 1$. Let $\mathbf{p} = (w(1), \ldots, w(k)) \in \mathcal{CH}_{m+1}(T \cup R, p_\swarrow)$. Note that $\xi_{m+1}(\mathbf{p}) \in \mathcal{CH}_{m+1}(T, p_\swarrow)$ and $\ell_{\Swarrow_{m+1}}(\mathbf{p}) = \ell_{\Swarrow_{m+1}}(\xi_{m+1}(\mathbf{p}))$. Hence replacing $\mathbf{p}$ by $\xi_{m+1}(\mathbf{p})$, we may assume that $w(i) \in \{\nwarrow, \uparrow, \nearrow, \leftarrow, \swarrow\}\cdot W_m$ for any $i = 1, \ldots, k$ without loss of generality (Fig. 5). Set $w(i) = s(i)v(i)$, where $s(i) \in \{\nwarrow, \uparrow, \nearrow, \leftarrow, \swarrow\}$ and $v(i) \in W_m$. We may choose $i_1, i_2, i_3$ and $i_4$ which satisfies $i_1 < i_2 < i_3 < i_4$ and the following tree conditions (a3), (b3) and (c3):

(a3) $s(1), \ldots, s(i_1) \in \{\nwarrow, \uparrow\}$, $(v(1), \ldots, v(i_1)) \in \mathcal{CH}_m(T, B_\swarrow)$,
(b3) $s(i) = \leftarrow$ for $i = i_2, \ldots, i_3$, $(v(i_2), \ldots, v(i_3)) \in \mathcal{CH}_m(T, B)$,
(c3) $s(i) = \swarrow$ for $i = i_4, \ldots, k$. $(v(i_4), \ldots, v(k)) \in \mathcal{CH}_m(T \cup R, p_\swarrow)$.

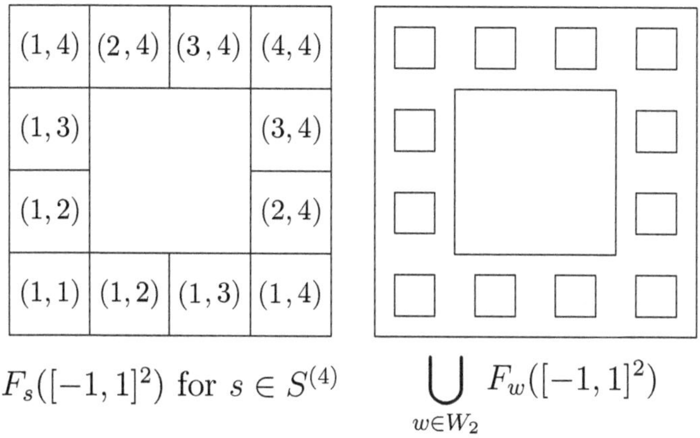


**Fig. 5** Construction of $K^{(4)}$

Define $\mathbf{p}_1$, $\mathbf{p}_2$ and $\mathbf{p}_3$ as in the above two cases. Then by the induction hypothesis and (6.3), it follows that $\ell_{\Swarrow_{m+1}}(\mathbf{p}_j) \geq 1/3$ for $j = 2, 3$. Furthermore, Lemma 6.14 implies $\ell_{\Swarrow_{m+1}}(\mathbf{p}_1) \geq 1/3$. Hence we have $\ell_{\Swarrow_{m+1}}(\mathbf{p}) \geq 1$.

Thus we have obtained (6.8), (6.9) and (6.10) for $n = m + 1$. □

*Proof of Theorem 6.4* Since $A_n \subseteq \Updownarrow_n$, $\ell_{A_n}(\mathbf{p}) \geq 1$ for any $\mathbf{p} \in \mathcal{CH}_n(T, B)$. By the fact that $\Leftrightarrow_n = \rho(\Updownarrow_n)$, it follows that $\ell_{\Leftrightarrow_n}(\mathbf{p}) \geq 1$ for any $\mathbf{p} \in \mathcal{CH}_n(L, R)$. Hence $\ell_{A_n}(\mathbf{p}) \geq 1$ for any $\mathbf{p} \in \mathcal{CH}_n(L, R)$. Thus $A_n$ is invisible. By Lemma 6.7-(1), it follows that $A_n$ is $+$-invariant.

Lemma 6.3 shows that $8^n - \#(\Updownarrow_n) = \#(W_n \backslash \Updownarrow_n) = \alpha_n$. Since $W_n \backslash \Updownarrow_n \subseteq W_n \backslash A_n \subseteq (W_n \backslash \Updownarrow_n) \cup (W_n \backslash \Leftrightarrow_n)$, we have $\alpha_n \leq 8^n - \#(A_n) \leq 2\alpha_n$. □

## 7 Generalized Sierpinski Carpet

The idea of invisible sets can be exploited for the generalized Sierpinski carpets. We will present results for a special class of the generalized Sierpinski carpet in this section. We fix $N \geq 3$. The complex plane $\mathbb{C}$ is identified with $\mathbb{R}^2$ in the usual manner.

**Definition 7.1**

(1) For any $(i, j) \in \{1, \ldots, N\}^2$, we define $J_{(i,j)} = [-1+2(i-1)/N, -1+2i/N] \times [-1 + 2(j-1)/N, -1 + 2j/N]$ and $F_{(i,j)} : \mathbb{R}^2 \to \mathbb{R}^2$ by $F_{(i,j)}(x, y) = (x/N + a_{(i,j)}, y/N + b_{(i,j)})$, where $a_{(i,j)} = -1 + (2i - 1)/N$ and $b_{(i,j)} = -1 + (2j - 1)/N$.
(2) Define $S^{(N)} = \{(i, j) | (i, j) \in \{1, \ldots, N\}^2, i \in \{1, N\}$ or $j \in \{1, N\}\}$. Let $K^{(N)}$ be the unique compact set which satisfies

$$K^{(N)} = \bigcup_{(i,j)\in S^{(N)}} F_{(i,j)}\left(K^{(N)}\right).$$

When $N = 3$, $K^{(3)}$ is the Sierpinski carpet.

**Proposition 7.2** *$\#(S^{(N)}) = 4N-4$ and $\dim_H\left(K^{(N)}, d_E\right) = \dfrac{\log(4N-4)}{\log N}$, where $d_E$ is the restriction of the Euclidean metric.*

In the following, we occasionally omit $N$ in $S^{(N)}$ and $K^{(N)}$ and write them $S$ and $K$ respectively. Also we use $W_m$, $W_*$ and $\Sigma$ in place of $W_m\left(S^{(N)}\right)$, $W_*\left(S^{(N)}\right)$ and $\Sigma\left(S^{(N)}\right)$.

**Definition 7.3** Let $A \subseteq W_m$.

(1) Let $A \subseteq W_m$. For $\mathbf{p} = (w(1), \ldots, w(k)) \in \mathcal{CH}_m$, define

$$\ell_A(\mathbf{p}) = \frac{\#(\{i \mid i = 1, \ldots, k,\ w(i) \notin A\})}{N^m}.$$

(2) $A$ is called an invisible set if and only if

$$\inf_{\mathbf{p}\in\mathcal{CH}_m(T,B)\cup\mathcal{CH}_m(L,R)} \ell_A(\mathbf{p}) \geq 1,$$

where $T$, $B$, $L$ and $R$ are the same as in the last tree sections.

We also define the notion of $+$-invariance exactly same as in the previous sections. Then the analogous results as Theorems 4.5 and 5.3 hold. As a consequence we have the following statement.

**Theorem 7.4** *Let $A \subset W_m$ be a $+$-invariant invisible set. Then*

$$\dim_{\mathcal{C}}(K^{(N)}, d_E) \leq \frac{\log\left((4N-4)^m - \#(A)\right)}{m \log N}.$$

A procedure which is similar to that in Sect. 6 produces a sequence of invisible sets. We assume $N \geq 4$ hereafter. The maps $\varphi_{\leftrightarrow}$, $\varphi_{\updownarrow}$ and $\rho$ from $W_*$ to itself associated with symmetries can be defined in the same way as in the last section.

**Definition 7.5** Define $\Updownarrow_n \subseteq W_n$ and $\Searrow_n \subseteq W_n$ inductively by

$$\begin{aligned}\Updownarrow_{n+1} = \{(i,j) \mid (i,j) \in S. i \in \{1, N\}\}\cdot \Updownarrow_n \\ \cup\, (2,1)\cdot \Searrow_n \cup (2,N)\cdot \Nearrow_n \cup (N-1,1)\cdot \Swarrow_n \cup (N-1,N)\cdot \Nwarrow_n \\ \cup \{(i,j) \mid (i,j) \in S,\ j \in \{1, N\},\ i \notin \{1, 2, N-1, N\}\}\cdot W_n,\end{aligned}$$

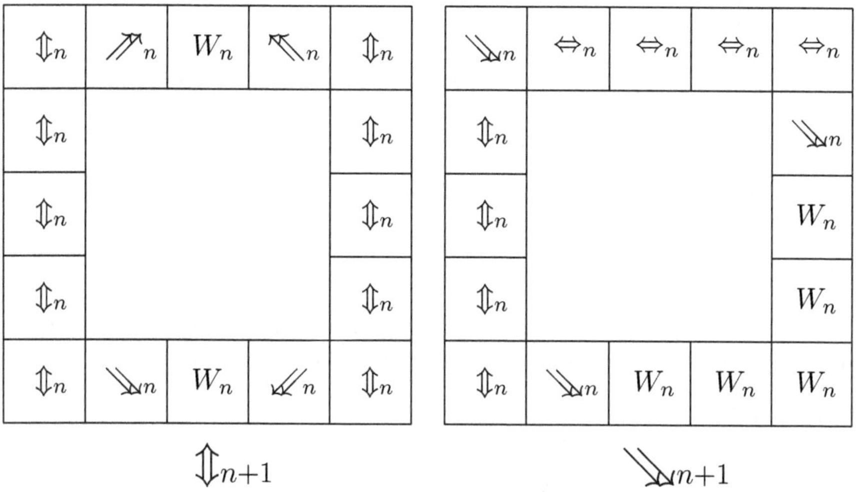

**Fig. 6** Construction of $\Updownarrow_n$ and $\Searrow_n$ for $N = 5$

$$\begin{aligned}\Searrow_{n+1} = \{(1,N),(2,1),(N-1,N)\}\cdot \Searrow_n \\ \cup\{(1,j)|j=1,\ldots,N-1\}\cdot \Updownarrow_n \cup\{(i,N)|i=2,\ldots,N\}\cdot \Leftrightarrow_n \\ \cup\{(1,j)|j=3,\ldots,N\}\cdot W_n \cup\{(i,N)|i=1,\ldots,N-2\}\cdot W_n,\end{aligned}$$

$\Updownarrow_0 = \emptyset$ and $\Searrow_0 = \emptyset$, where $\Leftrightarrow_n = \rho(\Updownarrow_n)$, $\Swarrow_n = \varphi_{\leftrightarrow}(\Searrow_n)$, $\Nearrow_n = \varphi_{\updownarrow}(\Searrow_n)$ and $\Nwarrow_n = \varphi_{\leftrightarrow}(\Nearrow_n)$ (Fig. 6).

By the above definition, it follows that

$$\begin{aligned} x_{n+1} &= 2Nx_n + 4y_n + 2(N-4)(4N-4)^n \\ y_{n+1} &= 2(N-1)x_n + 3y_n + (2N-5)(4N-4)^n, \end{aligned}$$

where $x_n = \#(\Updownarrow_n)$ and $y_n = \#(\Searrow_n)$. Define

$$\tau_N = \sqrt{4N^2 + 20N - 23}.$$

Then we have

$$\begin{aligned} x_n = (4N-4)^n - \Big(\frac{2N+5}{2\tau_N} + \frac{1}{2}\Big)\Big(\frac{2N+3+\tau_N}{2}\Big)^n \\ + \Big(\frac{2N+5}{2\tau_N} - \frac{1}{2}\Big)\Big(\frac{2N+3-\tau_N}{2}\Big)^n \end{aligned}$$

The same discussion as in the last section shows

$$\inf_{\mathbf{p} \in \mathcal{CH}_n(T,B)} \ell_{\Updownarrow_n}(\mathbf{p}) \geq 1.$$

Hence we obtain the counterpart of Theorem 6.4.

**Theorem 7.6** *Let $A_n = \Updownarrow_n \cap \Leftrightarrow_n$. Then $A_n$ is $+$-invariant invisible set and there exist $c_1, c_2 > 0$ such that*

$$c_1\left(\frac{2N+3+\tau_N}{2}\right)^n \leq (4N-4)^n - \#(A_n) \leq c_2\left(\frac{2N+3+\tau_N}{2}\right)^n$$

*for sufficiently large $n$.*

As an corollary, we obtain the following estimate of the conformal dimension of $K^{(N)}$. The lower estimate is shown by applying [MT10, Example 4.1.9].

**Corollary 7.7**

$$\begin{aligned}\frac{\log(2N)}{\log N} \leq \dim_C\left(K^{(N)}, d_E\right) &\leq \frac{\log\frac{2N+3+\tau_N}{2}}{\log N}\\ &< \frac{\log(4N-4)}{\log N} = \dim_H\left(K^{(N)}, d_E\right).\end{aligned}$$

*Remark*

$$2N+3 \leq \frac{2N+3+\tau_n}{2} < 2N+4.$$

## References


[BM00] Bonk, M., Merenkov, S.: Rigidity of square Sierpinski carpets, preprint.

[KL04] Keith, S., Laakso, T.: Conformal Assouad dimension and modulus. Geom. Funct. Anal. **14**, 1278–1321 (2004)

[Kig00] Kigami, J.: Resistance Forms, Quasisymmetric Maps and Heat Kernel Estimates. Memoirs of the American Mathematical Society. American Mathematical Society, Providence, RI (2012)

[Kle00] Kleiner, B.: Personal communication

[Kig01] Kigami, J.: Analysis on Fractals, Cambridge Tracts in Mathematics. Cambridge University Press, Cambridge (2001)

[Kig09] Kigami, J.: Volume doubling measures and heat kernel estimates on self-similar sets. Mem. Am. Math. Soc. **199**(932), XXX (2009)

[MT10] Mackay, J.M., Tyson, J.T.: Conformal Dimension, Theory and Application. University Lecture Series. American Mathematical Society, Providence, RI (2010)

[Pan89] Pansu, P.: Dimension conforme et sphère á l'infini des variétés à courbure nègative. Ann. Acad. Sci. Fenn. Ser. A I Math. **14**, 177–212 (1989)

[TV80] Tukia, P., Väisälä, J.: Quasisymmetric embeddings of metric spaces. Ann. Acad. Sci. Fenn. Ser. A I Math. **5**, 97–114 (1980)

[TW06] Tyson, J.T., Wu, J.-M.: On quasiconformal dimensions of self-similar fractals. Rev. Mat. Iberoamericana **22**, 205–258 (2006)

# Recent Progress on Dimensions of Projections

**Pertti Mattila**

**Abstract** This is a survey on recent progress on the question: how do projections effect dimensions generically? I shall also discuss briefly dimensions of plane sections.

**Keywords** Hausdorff dimension · Projections · Heisenberg group

**2000 Mathematics subject Classification** Primary 28A75

## 1 Introduction

I give a survey on the question how projection-type transformations change dimensions of sets. I shall mainly discuss Hausdorff dimension but packing and Minkowski dimensions will also be briefly looked at. First I review classical Marstrand's projection theorem and give Kaufman's proof for it. Then I present recent partial analogues of Marstrand's projection theorem in Heisenberg groups due to Balogh, Durand-Cartagena, Fässler, Tyson and myself. After that I discuss generalized projections of Peres and Schlag. In Heisenberg groups and other situations one encounters small, restricted, families of transformations. I review recent results of E. Järvenpää, M. Järvenpää, Ledrappier, Leikas and Keleti and of Fässler and Orponen on them. Then I mention briefly older results of Falconer and Howroyd and recent results of Fässler and Orponen on packing and Minkowski dimensions. For them one has generally only inequalities, but Falconer and Howroyd proved also a constancy theorem. I present this and some recent constancy theorem of Fässler and Orponen on Hausdorff dimension for a particular restricted family of projections. Finally we shall have

The author was supported by the Academy of Finland.

P. Mattila (✉)
Department of Mathematics and Statistics, University of Helsinki, P.O. Box 68, 00014 Helsinki, Finland
e-mail: pertti.mattila@helsinki.fi



D.-J. Feng and K.-S. Lau (eds.), *Geometry and Analysis of Fractals*,
Springer Proceedings in Mathematics & Statistics 88,
DOI 10.1007/978-3-662-43920-3_10

a look at Marstrand's classical result on Hausdorff dimension of plane sections and recent analogues of it in Heisenberg groups.

Background on this topic can be found in the books [Fa85, Map2]. Recent related surveys are [Map3, Map4].

I would like to thank Katrin Fässler and Tuomas Orponen for several useful comments.

## 2 Marstrand's Projection Theorem

Marstrand proved in 1954 the following theorem in [Ma54]:

**Theorem 2.1** *Suppose* $A \subset \mathbb{R}^2$ *is a Borel set and denote by* $P_\theta, \theta \in [0, \pi)$, *the orthogonal projection onto the line* $L_\theta = \{t(\cos\theta, \sin\theta) : t \in \mathbb{R}\}$*:* $P_\theta(x, y) = (\cos\theta)x + (\sin\theta)y$.

(1) *If* $\dim A \leq 1$, *then* $\dim P_\theta(A) = \dim A$ *for almost all* $\theta \in [0, \pi)$.

(2) *If* $\dim A > 1$, *then* $\mathcal{L}^1(P_\theta(A)) > 0$ *for almost all* $\theta \in [0, \pi)$.

Here dim means Hausdorff dimension and $\mathcal{L}^1$ is the one-dimensional Lebesgue measure.

Marstrand's original proof was based on the definition and basic properties of Hausdorff measures. Kaufman used in [Ka68] potential theoretic and Fourier analytic methods to give a different proof. To present Kaufman's proof let us first look at the required preliminaries.

The Hausdorff dimension of a Borel set $A \subset \mathbb{R}^n$ can be determined by looking at the behaviour of Borel measures $\mu$ with compact support $\text{spt}\mu \subset A$. Denote the family of such measures $\mu$ with $0 < \mu(A) < \infty$ by $\mathcal{M}(A)$. By the well-known Frostman's lemma $\dim A$ is the supremum of the numbers $s$ such that there exists $\mu \in \mathcal{M}(A)$ for which

$$\mu(B(x, r)) \leq r^s \quad \text{for } x \in \mathbb{R}^n. \tag{2.1}$$

This is easily transformed into an integral condition. Let

$$I_s(\mu) = \iint |x - y|^{-s} d\mu x d\mu y$$

be the $s$-energy of $\mu$. Then $\dim A$ is the supremum of the numbers $s$ such that there exists $\mu \in \mathcal{M}(A)$ for which

$$I_s(\mu) < \infty. \tag{2.2}$$

For a fixed $\mu$ (2.1) and (2.2) may not be equivalent, but they are closely related: (2.2) implies that the restriction of $\mu$ to a suitable set with positive $\mu$ measure satisfies (2.1), and (2.1) implies that $\mu$ satisfies (2.2) for any $s' < s$. Defining the Riesz kernel $k_s$, $k_s(x) = |x|^{-s}$, the $s$-energy of $\mu$ can written as

$$I_s(\mu) = \int k_s * \mu d\mu.$$

For $0 < s < n$ the Fourier transform of $k_s$ is (in the sense of distributions) $\widehat{k_s} = c(s,n)k_{n-s}$. Thus we have by Plancherel's theorem

$$I_s(\mu) = \int \widehat{k_s}|\widehat{\mu}|^2 = c(s,n)\int |x|^{s-n}|\widehat{\mu}(x)|^2 dx.$$

Consequently, $\dim A$ is the supremum of the numbers $s \le n$ such that there exists $\mu \in \mathcal{M}(A)$ for which

$$\int |x|^{s-n}|\widehat{\mu}(x)|^2 dx < \infty. \tag{2.3}$$

To prove (1) of Theorem 2.1 let $0 < s < \dim A$ and choose by (2.2) a measure $\mu \in \mathcal{M}(A)$ such that $I_s(\mu) < \infty$. Let $\mu_\theta \in \mathcal{M}(P_\theta(A))$ be the push-forward of $\mu$ under $P_\theta$: $\mu_\theta(B) = \mu(P_\theta^{-1}(B))$. Then

$$\begin{aligned}\int_0^\pi I_s(\mu_\theta)d\theta &= \int_0^\pi \iint |P_\theta(x-y)|^{-s} d\mu x d\mu y d\theta \\ &= \iint \int_0^\pi |P_\theta(\tfrac{x-y}{|x-y|})|^{-s} d\theta |x-y|^{-s} d\mu x d\mu y \\ &= c(s)I_s(\mu) < \infty,\end{aligned}$$

where for $v \in S^1$, $c(s) = \int_0^\pi |P_\theta(v)|^{-s} d\theta < \infty$ as $s < 1$. Referring again to (2.2) we see that $\dim P_\theta(A) \ge s$ for almost all $\theta \in [0,\pi)$. By the arbitrariness of $s, 0 < s < \dim A$, we obtain $\dim P_\theta(A) \ge \dim A$ for almost all $\theta \in [0,\pi)$. The opposite inequality follows from the fact that the projections $P_\theta$ are Lipschitz.

To prove (2) choose by (2.3) a measure $\mu \in \mathcal{M}(A)$ such that $\int |x|^{-1}|\widehat{\mu}(x)|^2 dx < \infty$. Directly from the definition of the Fourier transform we see that $\widehat{\mu_\theta}(t) = \widehat{\mu}(t(\cos\theta, \sin\theta))$ for $t \in \mathbb{R}$, $\theta \in [0,\pi)$. Integrating in polar coordinates we obtain

$$\int_0^\pi \int_{-\infty}^\infty |\widehat{\mu_\theta}(t)|^2 dt d\theta = 2 \int_0^\pi \int_0^\infty |\widehat{\mu}(t(\cos\theta, \sin\theta))|^2 dt d\theta$$
$$= \int |x|^{-1} |\widehat{\mu}(x)|^2 dx < \infty.$$

Thus for almost all $\theta \in [0, \pi)$, $\widehat{\mu_\theta} \in L^2(\mathbb{R})$ which means that $\mu_\theta$ is absolutely continuous with $L^2$-density and hence $\mathcal{L}^1(P_\theta(A)) > 0$.

It is not difficult to prove (2) without Fourier transform: application of Fubini's theorem and some simple estimates yield

$$\int_0^\pi \int \liminf_{\delta \to 0} \delta^{-1} \mu_\theta(x - \delta, x + \delta) d\mu_\theta x d\theta \leq C I_1(\mu), \tag{2.4}$$

from which (2) follows by standard results on differentiation of measures.

Theorem 2.1 has the following generalization:

**Theorem 2.2** *Suppose $A \subset \mathbb{R}^2$ is a Borel set.*

(1) *If* $0 \leq t \leq \dim A \leq 1$, *then*

$$\dim\{\theta \in [0, \pi) : \dim P_\theta(A) < t\} \leq t.$$

(2) *If* $\dim A > 1$, *then*

$$\dim\{\theta \in [0, \pi) : \mathcal{L}^1(P_\theta(A)) = 0\} \leq 2 - \dim A.$$

Part (1) was proved by Kaufman with a similar method as above; one uses Frostman's lemma also for the exceptional set of directions. Part (2) was proved by Falconer with a Fourier-analytic method.

To formulate the higher dimensional version of Theorem 2.2, denote by $G(n, m)$ the Grassmannian manifold of linear $m$-dimensional subspaces of $\mathbb{R}^n$. For $V \in G(n, m)$, let

$$P_V : \mathbb{R}^n \to V$$

be the orthogonal projection. As above, we shall often write also $P_V : \mathbb{R}^n \to \mathbb{R}^m$ in a natural way. Identifying $V$ with $P_V$, $G(n, m)$ becomes a smooth submanifold of dimension $m(n - m)$ of $\mathbb{R}^{n^2}$.

**Theorem 2.3** *Suppose $A \subset \mathbb{R}^n$ is a Borel set.*

(1) *If* $\dim A \leq m$, *then*

$$\dim\{V \in G(n, m) : \dim P_V(A) < t\} \leq m(n - m) - m + t.$$

(2) *If* $\dim A > m$, *then*

$$\dim\{V \in G(n,m) : \mathcal{L}^1(P_V(A)) = 0\} \le m(n-m) + m - \dim A.$$

Part (1) was proved in [Map5] and (2) in [Fa82]. The bound in (1) is sharp when $t = \dim A$. This was shown by Kaufman and myself in [KM75] with examples based on Jarnik's results on dimension and diophantine approximation, see also [Fa85], Sect. 8.5. Similar examples work also for (2). As far as I know the sharp bound in (1) for $t < \dim A$ is unkown. Anyway, the one given in Theorem 2.2 is not always sharp due to the following result of Bourgain in [Bo10] and Oberlin in [Ob12]:

**Theorem 2.4** *Suppose* $A \subset \mathbb{R}^2$ *is a Borel set. Then*

$$\dim\{\theta \in [0,\pi) : \dim P_\theta(A) < \dim A/2\} = 0.$$

The construction in [KM75] can be used to get for any $0 < t \le s < 2$ a compact set $A \subset \mathbb{R}^2$ with $\dim A = s$ such that

$$\dim\{\theta \in [0,\pi) : \dim P_\theta(A) \ge t\} \ge 2t - s.$$

Could $2t - s$ be the sharp upper bound in the range $s/2 \le t \le \min\{1, s\}$? In any case this shows that to get dimension 0 for the exceptional set, the bound $\dim A/2$ is the best possible.

Bourgain's estimate is somewhat stronger than the above. He obtained his result as part of deep investigations in additive combinatorics, whereas Oberlin's proof is much simpler and more direct. Oberlin also had another exceptional set estimate in [Ob13].

Some improvements on part (2) of Theorem 2.3 will be given soon in Sect. 4 in a more general setting.

## 3 Projection Theorems in Heisenberg Groups

Heisenberg group $\mathbb{H}^n$ is $\mathbb{R}^{2n+1}$ equipped with a non-abelian group structure, with a left invariant metric and with natural dilations. The first Heisenberg group $\mathbb{H}^1$ is the simplest of these. We can write $\mathbb{H}^1 = \mathbb{C} \times \mathbb{R}$, where the points are written as $p = (w,s), q = (z,t) \in \mathbb{H}^1$. Then the product of $p$ and $q$ is

$$p \cdot q = (w + z, s + t + 2Im(w\bar{z})).$$

To define the distance between $p$ and $q$, set first

$$\|p\| = (|z|^4 + t^2)^{1/4},$$

and then

$$d(p,q) = \|p^{-1} \cdot q\| = (|z-w|^4 + |t-s-2Im(w\bar{z})|^2)^{1/4}.$$

It is easy to check that $d$ really is a metric and that it is left invariant. We have the dilations

$$\delta_r(p) = (rz, r^2t)$$

for which

$$d(\delta_r(p), \delta_r(q)) = rd(p,q).$$

When the distance is restricted to the $t$-axis $\{0\} \times \mathbb{R}$ it is just the square root distance. Essentially because of this the Heisenberg Hausdorff dimension of $\mathbb{H}^1$ is

$$\dim_H \mathbb{H}^1 = 4.$$

Here $\dim_H$ refers to the Hausdorff dimension with respect to the Heisenberg metric. Always dim will refer to the Hausdorff dimension with respect to the Euclidean metric. It is easy to check that

$$\dim(A) \leq \dim_H(A) \leq 2\dim(A),\ A \subset \mathbb{H}^1.$$

These inequalities are sharp. For example, if $A$ is a subset of the $x$-axis, $\dim_H(A) = \dim(A)$, and if $A$ is a subset of the $t$-axis, $\dim_H(A) = 2\dim(A)$. However, one can improve them for sets $A$ with $\dim(A) > 1$. Very precise inequalities were obtained by Balogh et al. [BT09].

We define the projections in $\mathbb{H}^1$ in the group sense. Good subgroups of $\mathbb{H}^1$ for this purpose are those which are invariant under the dilations and have a complementary subgroup in the sense described below. They are precisely the horizontal lines

$$V_\theta = \{te_\theta : t \in \mathbb{R}\},\ e_\theta = (\cos\theta, \sin\theta, 0),\ 0 \leq \theta < \pi,$$

and the vertical planes

$$W_\theta = V_\theta^\perp.$$

The horizontal lines $V_\theta$ are Euclidean, the distance restricted to them is the Euclidean distance, whereas the vertical planes $W_\theta$ are non-Euclidean; for them $\dim_H W_\theta = \dim W_\theta + 1$. We have the splitting

$$\mathbb{H}^1 = W_\theta \cdot V_\theta,$$

that is, for $p \in \mathbb{H}^1$ we have the unique factorization

$$p = Q_\theta(p) \cdot P_\theta(p),\ \ P_\theta(p) \in V_\theta,\ Q_\theta(p) \in W_\theta.$$

Thus we get the group projections

$$P_\theta : \mathbb{H}^1 \to V_\theta,\ Q_\theta : \mathbb{H}^1 \to W_\theta,\ 0 \le \theta < \pi.$$

Writing $p = (z, t) = (x + iy, t) \in \mathbb{H}^1$ we have the explicit formulas

$$P_\theta(p) = ((x \cos\theta + y \sin\theta)e_\theta, 0),$$

$$Q_\theta(p) = ((y \cos\theta - x \sin\theta)e_\theta^\perp, t - 2\cos(2\theta)xy + \sin(2\theta)(x^2 - y^2)),$$

where $e_\theta^\perp = (-\sin\theta, \cos\theta)$. So $P_\theta$ is the standard linear projection, essentially the one we considered above in $\mathbb{R}^2$, but $Q_\theta$ is a non-linear projection. $P_\theta$ is nice, it is Lipschitz and group homomorphism, but $Q_\theta$ is neither of those, it is only Hölder continuous with exponent $1/2$.

Now we have the following analogue for horizontal projections of Marstrand's projection theorem from [BD13]:

**Theorem 3.1** *Let $A \subset \mathbb{H}^1$ be a Borel set. Then for almost all $\theta \in [0, \pi)$,*

$$\dim_H P_\theta(A) \ge \dim_H A - 2 \text{ if } \dim_H A \le 3,$$

$$\mathcal{H}^1(P_\theta(A)) > 0 \text{ if } \dim_H A > 3.$$

This is sharp: consider $A = \{(x, 0, t) : x \in C, t \in [0, 1]\}, C \subset \mathbb{R}$. Then $\dim_H A = \dim C + 2$ and

$$\dim_H P_\theta(A) = \dim P_\theta(A) = \dim P_\theta(C) = \dim C$$

for all but one $\theta$.

Theorem 3.1 follows easily applying Marstrand's projection theorem to the projection of $A$ on $\mathbb{C} \times \{0\}$.

For the vertical projections we have:

**Theorem 3.2** *Let $A \subset \mathbb{H}^1$ be a Borel set. If $\dim_H A \le 1$, then for almost all $\theta \in [0, \pi)$,*

$$\dim_H A \le \dim_H Q_\theta(A) \le 2 \dim_H A.$$

For $A$ with $\dim_H A \le 1$ this is sharp:
if $A \subset t$-axis, $\dim_H Q_\theta(A) = \dim_H A$ for all $\theta$,
if $A \subset x$-axis, $\dim_H Q_\theta(A) = 2 \dim_H A$ for all but one $\theta$.

The upper bound $2 \dim_H A$ follows from the Hölder continuity of $Q_\theta$. For the lower bound we use again the energy integrals. Let

$$p = (z, t), q = (\zeta, \tau) \in \mathbb{H}^1$$

and denote

$$\varphi_1 = \arg(z - \zeta), \varphi_2 = \arg(z + \zeta).$$

Then one can check that

$$d(p,q)^4 = |z - \zeta|^4 + (t - \tau + |z^2 - \zeta^2| \sin(\varphi_1 - \varphi_2))^2$$

and

$$d(Q_\theta(p), Q_\theta(q))^4 = |z - \zeta|^4 \sin^4(\varphi_1 - \theta)$$
$$+ (t - \tau - |z^2 - \zeta^2| \sin(\varphi_2 + \varphi_1 - 2\theta))^2$$

To get $\int_0^\pi d(Q_\theta(p), Q_\theta(q))^{-s} d\theta \lesssim d(p,q)^{-s}$, one needs for $a \in \mathbb{R}$,

$$\int_0^\pi \frac{d\theta}{|a + \sin\theta|^{s/2}} \lesssim 1, \tag{3.1}$$

which is easy to check when $s < 1$.

If $\dim_H A > 1$, we have some estimates which quite likely are not sharp. For example, we do not know if $\dim_H A > 3$ implies $\mathcal{H}^2(Q_\theta(A)) > 0$ for almost all $\theta \in [0, \pi)$. Here $\mathcal{H}^2$ is the Euclidean two-dimensional Hausdorff measure. When restricted to a vertical plane it agrees with the three-dimensional Heisenberg Hausdorff measure, both give the Haar measure for this subgroup.

A related Euclidean question is: does $\dim A > 2$ imply $\mathcal{H}^2(Q_\theta(A)) > 0$ for almost all $\theta \in [0, \pi)$?

Let us now consider higher dimensions, these were treated in [BF12]. Then the basic notions and facts are

- $\mathbb{H}^n = \mathbb{C}^n \times \mathbb{R}$, $p = (w, s), q = (z, t) \in \mathbb{H}^n$,
- $\omega(w, z) = 2Im(w \cdot z) = 2\sum_{j=1}^n (v_j x_j - u_j y_j)$, $w = (u_j + iv_j), z = (x_j + iy_j)$,
- $p \cdot q = (w + z, s + t + \omega(w, z))$,
- $||p|| = (|z|^4 + t^2)^{1/4}$,
- $d(p, q) = ||p^{-1} \cdot q|| = (|z - w|^4 + |t - s - \omega(w, z)|^2)^{1/4}$,
- $\delta_r(p) = (rz, r^2 t)$,
- $d(\delta_r(p), \delta_r(q)) = rd(p, q)$,
- $d(p \cdot q_1, p \cdot q_2) = d(q_1, q_2)$,
- $\dim_H \mathbb{H}^n = 2n + 2$.

The subgroups invariant under dilations split again to horizontal and vertical subgroups. The horizontal ones are those $m$-dimensional linear subspaces of $\mathbb{R}^{2n}$, $0 < m \le n$, on which the bilinear form $\omega$ vanishes. That is, the elements of

$$G_h(n, m) = \{V \in G(2n, m) : \omega(w, z) = 0\ \forall w, z \in V\}.$$

They are called isotropic subspaces. The unitary group $U(n) \subset O(2n)$ acts transitively on $G_h(n,m)$; by definition $g \in U(n)$ if $\omega(g(w), g(z)) = \omega(w, z)$ for all $w, z \in \mathbb{C}^n$. The vertical subgroups are all linear subspaces of $\mathbb{R}^{2n+1}$ which contain the $t$-axis. The horizontal subgroups are again Euclidean and for the vertical subgroups $W$ we have $\dim_H W = \dim W + 1$. Then

$$\mathbb{H}^n = V^\perp \cdot V, V^\perp \subset \mathbb{R}^{2n+1}, V \in G_h(n,m),$$

$$p = Q_V(p) \cdot P_V(p), \ P_V(p) \in V, Q_V(p) \in V^\perp, \text{ for } p \in \mathbb{H}^n.$$

Again $P_V : \mathbb{H}^n \to V$ is the standard linear projection, but $Q_V : \mathbb{H}^n \to V^\perp$,

$$Q_V(z,t) = (P_{V^\perp}(z), t - \omega((P_{V^\perp}(z), P_V(z))),$$

is a non-linear projection.

Notice that in the above splitting the linear dimension of $V$ is always at most $n$. The vertical subgroups $W$ of linear dimension $1 \le \dim W \le n$ have no complementary subgroups in the above sense.

We have the following horizontal projection theorem in $\mathbb{H}^n$:

**Theorem 3.3** *Let $A \subset \mathbb{H}^n$ be a Borel set. If $\dim_H A \le m+2$, then*

$$\dim P_V(A) \ge \dim_H A - 2$$

*for $\mu_{n,m}$ almost all $V \in G_h(n,m)$. Furthermore, if $\dim_H A > m+2$, then*

$$\mathcal{H}^m(P_V(A)) > 0 \textit{ for } \mu_{n,m} \textit{ almost } V \in G_h(n,m).$$

This is again sharp. Above $\mu_{n,m}$ is the unique $U(n)$-invariant Borel probability measure on $G_h(n,m)$.

For the vertical projections we have

**Theorem 3.4** *Let $A \subset \mathbb{H}^n$ be a Borel subset with $\dim_H A \le 1$. Then for $\mu_{n,m}$ almost $V \in G_h(n,m)$,*

$$\dim_H A \le \dim_H Q_V A \le 2\dim_H A.$$

This is sharp when $\dim_H A \le 1$. Some, probably rather imprecise, partial results are known when $\dim_H A > 1$. One might expect that the methods would yield this theorem for $\dim_H A \le m$, but there are some serious obstacles. Let us see what they are. We can now write

$$d_H(p,q) = \sqrt[4]{|z-w|^4 + (t-s-2\omega(\zeta,z))^2},$$

and

$$d_H(Q_V(p), Q_V(q))^4 = |P_{V^\perp}(z-w)|^4 + (t - s - \omega(P_{V^\perp}(z), P_V(z)) \\ + \omega(P_{V^\perp}(w), P_V(w)) - \omega(P_{V^\perp}(w), P_{V^\perp}(z)))^2.$$

The key estimate in the proof is

$$\int_{G_h(n,m)} |a - \omega(v, P_V(w))|^{-s/2} d\mu_{n,m} V \lesssim 1$$

for all $0 < s < 1$, $a \in \mathbb{R}$ and $v, w \in S^{2n-1}$. In local coordinates for $V$ the expression $a - 2\omega(v, P_V(w))$ is a second degree polynomial which can vanish to second order. Because of this the above estimate is false for $s \geq 1$ and it seems to be difficult to find anything to replace it.

There are various other results in the papers [BD13, BF12]. In particular, quite precise information is obtained on inequalities that hold for all projections.

## 4 Generalized Projections

Studying Kaufman's proof of Marstrand's projection theorem one notices quickly that it applies to much more general families of mappings than orthogonal projections onto lines and planes. Peres and Schlag developed this idea in [PS00] much farther. The following is still a special case of their general setting:

Let $(\Omega, d)$ be a compact metric space, $Q \subset \mathbb{R}^k$ an open connected set. We have mappings

$$\pi_\lambda : \Omega \to \mathbb{R}^m, \quad \lambda \in Q,$$

such that the mapping $\lambda \mapsto \pi_\lambda(x)$ is in $C^\infty(Q)$ for every fixed $x \in \Omega$, and to every compact $K \subset Q$ and any multi-index $\eta = (\eta_1, \dots, \eta_n) \in \mathbb{N}^n$ there corresponds a finite constant $C_{\eta,K} > 0$ such that

$$|\partial_\lambda^\eta \pi_\lambda(x)| \leq C_{\eta,K}, \qquad \lambda \in K. \tag{4.1}$$

**Definition 4.1** Define

$$\Phi_\lambda(x, y) = \frac{\pi_\lambda(x) - \pi_\lambda(y)}{d(x, y)}.$$

The family $\{\pi_\lambda, \lambda \in Q\}$ is said to be *transversal*, if there exists a finite constant $C_0 > 0$ such that

$$|\Phi_\lambda(x, y)| \leq C_0 \implies \det(D_\lambda \Phi_\lambda(x, y)(D_\lambda \Phi_\lambda(x, y)^t)) \geq C_0 \tag{4.2}$$

for $\lambda \in Q$ and $x, y \in \Omega, x \neq y$. The family $\{\pi_\lambda, \lambda \in Q\}$, is said to be *regular*, if to every multi-index $\eta = (\eta_1, \ldots, \eta_n) \in \mathbb{N}^n$ there correspond a finite constant $C_\eta > 0$ such that

$$|\Phi_\lambda(x, y)| \leq C_0 \implies \left|\partial_\lambda^\eta \Phi_\lambda(x, y)\right| \leq C_\eta \tag{4.3}$$

for $\lambda \in Q$ and $x, y \in \Omega, x \neq y$.

Orthogonal projections when restricted to some compact set are easily seen to form transversal and regular families of mappings. When considering projections from $\mathbb{R}^n$ onto $m$-planes, we can take $k = \dim G(n, m) = m(n - m)$.

Since we are looking for lower bounds for dimensions of projections, the bad pairs of points are such $x$ and $y$ which are mapped close to each other. The point in transversality is that if $(x, y)$ is a pair of bad points for some $\lambda$, then it becomes quickly better when $\lambda$ moves a bit. For real-valued maps ($m = 1$), such as projections onto lines, the transversality means

$$|\Phi_\lambda(x, y)| \leq C_0 \implies |\nabla_\lambda \Phi_\lambda(x, y)| \geq C_0.$$

Here is a special case of a theorem of Peres and Schlag:

**Theorem 4.2** *Suppose the above transversality and regularity conditions hold. Let $A \subset \Omega$ be a Borel set and $s = \dim A$.*

(a) *If $s \leq m$ and $t \in (0, s)$, then*

$$\dim\{\lambda \in Q : \dim \pi_\lambda(A) < t\} \leq k - m + t.$$

(b) *If $s > m$, then*

$$\dim\{\lambda \in Q : \dim \pi_\lambda(A) < t\} \leq k - s + t$$

*and*

$$\dim\{\lambda \in Q : \mathcal{L}^m(\pi_\lambda(A)) = 0\} \leq k - s + m.$$

(c) *If $s > 2m$, then*

$$\dim\{\lambda \in Q : \textit{the interior of } \pi_\lambda(A) \textit{ is empty}\} \leq n - s + 2.$$

In addition to being applicable to many families of mappings, this theorem also improves Theorem 2.3 in the case of orthogonal projections. As Peres and Schlag showed it can be applied in many interesting situations, for example to Bernoulli convolutions, sum sets and pinned distance sets.

In $\mathbb{R}^{2n}$ the horizontal Grassmannian, the Grassmannian of isotropic subspaces, $G_h(n, m)$, discussed before in the case of Heisenberg groups, is a proper lower dimensional submanifold of the full Grassmannian $G(2n, m)$ when $1 < m \leq n$. Nevertheless Marstrand's projection theorem holds for this submanifold. We proved this in [BF12]. Hovila established it in [Ho18] by verifying that the family $P_V : \mathbb{R}^{2n} \to V, V \in G_h(n, m)$, is transversal. This has two further consequnces: exceptional set estimates and Besicovitch-Federer projection theorem. The first follows from the above results of Peres and Schlag, the second from Hovila's joint work with Järvenpää et al. in [HLJ2]. There they proved Besicovitch-Federer projection theorem for transversal families of generalized projections. The classical Besicovitch-Federer projection theorem says that an $\mathcal{H}^m$ measurable set $A \subset \mathbb{R}^n$ with $\mathcal{H}^m(A) < \infty$ p! rojects into zero $\mathcal{H}^m$ measure in almost all $m$-planes $V \in G(n, m)$ if an and only if it meets every $m$-dimensional $C^1$-surface in a set of zero $\mathcal{H}^m$ measure, see [Fe69] or [Map2].

Neither the vertical nor the horizontal projections in Heisenberg groups satisfy transversality; these families are too small for that.

# 5 Restricted Families of Projections

The reason that it is not possible to get precise almost everywhere equalities for dimensions of projections in Heisenberg groups is that we have too few projections. It is of interest to search projection theorems for such restricted families of projections also in Euclidean spaces. That is, one considers a proper lower dimensional submanifold $G$ of the Grassmannian $G(n, m)$ and the projections $P_V, V \in G$. This splits into two cases: $G$ is general allowing flat submanifolds or $G$ is required to possess some curvature properties. What these mean becomes clearer below. In the first case less can be said and it is completely solved by E. Järvenpää, M. Järvenpää and Keleti as we shall see soon. The second case is extremely difficult and some partial results have been obtained by Fässler and Orponen.

One motivation for studying restricted families of projection-type transformations comes from the work of E. Järvenpää, M. Järvenpää, Ledrappier and their co-workers on measures invariant under geodesic flows on manifolds, see [HJL2] and the references given there.

A simple restricted family of projections in $\mathbb{R}^3$ is given by the horizontal projections, or the projections onto the lines $L_\theta = \{t(\cos\theta, \sin\theta, 0) : t \in \mathbb{R}\}$,

$$P_\theta : \mathbb{R}^3 \to \mathbb{R}, P_\theta(x, y, z) = x\cos\theta + y\sin\theta, 0 \le \theta < \pi. \tag{5.1}$$

Since $P_\theta(A) = P_\theta((\pi(A))$ where $\pi(x, y, z) = (x, y)$, and $\dim A \le \dim \pi(A) + 1$, it is easy to conclude using Marstand's projection theorem that for any Borel set $A \subset \mathbb{R}^3$, for almost all $\theta \in [0, \pi)$,

$$\dim P_\theta(A) \ge \dim A - 1 \text{ if } \dim A \le 2,$$
$$\dim P_\theta(A) = 1 \text{ if } \dim A \ge 2.$$

This is sharp by trivial examples; consider product sets $A = B \times C, B \subset \mathbb{R}^2$, $C \subset \mathbb{R}$.

A simple example of projections onto planes is given by

$$\Pi_\theta : \mathbb{R}^3 \to \mathbb{R}^2, \Pi_\theta(x, y, z) = (x \sin\theta - y \cos\theta, z), 0 < \theta < \pi. \tag{5.2}$$

These are essentially orthogonal projections onto the orthogonal complements of the lines $L_\theta$.

Also now it is easy to prove that for any Borel set $A \subset \mathbb{R}^3$, for almost all $\theta \in [0, \pi)$,

$$\begin{aligned}
&\dim \Pi_\theta(A) \geq \dim A \text{ if } \dim A \leq 1,\\
&\dim \Pi_\theta(A) \geq 1 \text{ if } 1 \leq \dim A \leq 2,\\
&\dim \Pi_\theta(A) \geq \dim A - 1 \text{ if } \dim A \geq 2.
\end{aligned}$$

Again by easy examples these inequalities are sharp.

Järvenpää et al. proved in [JJ05] that the above sets of inequalities remain in force for any smooth, in a suitable sense non-degenerate, one-dimensional families of orthogonal projections onto lines and planes in $\mathbb{R}^3$. In fact, they proved such inequalities in more general dimensions and in [JJ13] Järvenpää et al. found the complete solution in all dimensions; sharp inequalities for smooth non-degenerate families of orthogonal projections onto $m$-planes in $\mathbb{R}^n$.

Consider now a slightly modified family of one-dimensional projections; let $p_\theta, \theta \in [0, 2\pi)$, be the orthogonal projection onto the line $l_\theta$ spanned by $(\cos\theta, \sin\theta, 1)$. The previous lines $L_\theta$ spanned a plane, but the lines $l_\theta$ span a cone. The trivial counter-examples do not work anymore and in fact one can now improve the above estimates for $p_\theta$'s relatively easily by showing that if $A \subset \mathbb{R}^3$ is a Borel set with $\dim A \leq 1/2$, then

$$\dim p_\theta(A) \geq \dim A \text{ for almost all } \theta \in [0, 2\pi).$$

The restriction $1/2$ comes because using Kaufman's method one is now lead to estimate integrals of the type

$$\int_0^{2\pi} \frac{d\theta}{|a + \sin\theta|^s}$$

for $s < \dim A$, and they are bounded only if $s < 1/2$. So this is the best one get without new ideas. Introducing some geometric arguments Fässler and Orponen were able to prove in [FO50] the following theorem for the packing dimensions, $\dim_p$, of the projections:

**Theorem 5.1** *Let $U \subset \mathbb{R}$ be an open interval and $\gamma : U \to S^2$ be a $C^3$ curve such that for all $\theta \in U$ the vectors $\gamma(\theta)$, $\gamma'(\theta)$ and $\gamma''(\theta)$ span $\mathbb{R}^3$. Let*

$$p_\theta(x) = \gamma(\theta) \cdot x$$

*be the orthogonal projection onto the line $l_\theta$ spanned by $\gamma(\theta)$ and*

$$\pi_\theta(x) = x - (\gamma(\theta) \cdot x)\gamma(\theta)$$

*the orthogonal projection onto the orthogonal complement of $l_\theta$. Suppose $A \subset \mathbb{R}^3$ is a Borel set with* $\dim A = s$.

(1) *If $s > 1/2$, there exists a number $\sigma_1(s) > 1/2$ such that*

$$\dim_p p_\theta(A) \geq \sigma_1(s) \text{ for almost all } \theta \in U.$$

(2) *If $s > 1$, there exists a number $\sigma_1(s) > 1$ such that*

$$\dim_p \pi_\theta(A) \geq \sigma_2(s) \text{ for almost all } \theta \in U.$$

It is not known if here the packing dimension could be replaced by the Hausdorff dimension. In [Orp3] Orponen was able to do this for the special family of orthogonal projections onto the lines $l_\theta$ spanned by $(\cos\theta, \sin\theta, 1)$ and their orthogonal complements.

It would be very interesting to find similar results for some non-linear families of mappings, for example for the vertical projections $Q_\theta$ of the Heisenberg group $\mathbb{H}^1$ considered just as mappings in $\mathbb{R}^3$:

$$Q_\theta(x, y, t) = (y\cos\theta - x\sin\theta, t - 2\cos(2\theta)xy + \sin(2\theta)(x^2 - y^2)), \theta \in [0, \pi).$$

Although, as said before, for the corresponding linear projections $\Pi_\theta$ as in (5.2) nothing more can be said than what we get from Marstrand's theorem, the non-linear mappings might be better.

## 6 Minkowski and Packing Dimensions

The analogue of Marstrand's theorem fails for Minkowski and packing dimensions; the dimension of the set does not prescribe the dimensions of the typical projections. However, Falconer and Howroyd proved in [FH96] the following sharp inequalities:

**Theorem 6.1** *Let $A \subset \mathbb{R}^n$ be a Borel set. Then*

$$\dim_p P_V(A) \geq \frac{\dim_p A}{1 + (1/m - 1/n)\dim_p A} \text{ for almost all } V \in G(n, m).$$

The same inequality holds also for upper and lower Minkowski dimensions. Examples of Järvenpää in [Jm94] show that the lower bound is sharp. In these examples the Hausdorff dimension of $A$ is 0. Later on we proved with Falconer in [Fm96] a version of this result which gives a sharp lower bound for the packing dimension of the typical projections given both Hausdorff and packing dimension of $A$.

Finding good dimension estimates for exceptional sets in packing dimension projection theorems has turned out to be a very delicate question, Rams obtained some results in [Ram2] for self-conformal sets. Orponen proved in [Ort2] several such estimates and constructed many illustrative examples. He also established Baire category results.

# 7 Constancy Results for Projections

Although there is no dimension preservation for the packing and Minkowski dimensions under projections, Falconer and Howroyd proved in [FH97] that given the set $A$, the dimensions equal almost surely a constant called $\mathrm{Dim}_m A$. The number $\mathrm{Dim}_m A$ comes from certain potentials. More precisely, we first define it for measures $\mu \in \mathcal{M}(A)$:

$$\mathrm{Dim}_m \mu = \sup\{t \geq 0 : \liminf_{r \to 0} r^{-t} F_m^{\mu}(x, r) = 0 \text{ for } \mu \text{ almost all } x \in \mathbb{R}^n\},$$

where

$$F_m^{\mu}(x, r) = \int \min\{1, r^m |x - y|^{-m}\} d\mu y.$$

For sets we define

$$\mathrm{Dim}_m A = \sup\{\mathrm{Dim}_m \mu : \mu \in \mathcal{M}(A)\}.$$

The theorem of Falconer and Howroyd now reads

**Theorem 7.1** *Let $A \subset \mathbb{R}^n$ be a Borel set. Then*

$$\dim_p P_V(A) = \mathit{Dim}_m A \textit{ for almost all } V \in G(n, m).$$

The relation of the potentials $F_m^{\mu}(x, r)$ to projections comes from the following observation:

$$\begin{aligned} F_m^{\mu}(x, r) &\approx \int \gamma_{n,m}(\{V \in G(n, m) : |P_V(x - y)| \leq r\}) d\mu y \\ &= \int P_V \mu(B(P_V(x), r)) d\gamma_{n,m} V, \end{aligned}$$

where $P_V \mu$ is the push forward of $\mu$ under $P_V$.

Similar tools were also used in [FH96, Fm96].

Perhaps such constancy results hold also in Heisenberg groups. This is not known but Fässler and Orponen proved in [FO13] constancy results for some restricted families of projections in $\mathbb{R}^3$. They did it in general dimensions but for simplicity I state their result only in $\mathbb{R}^3$:

**Theorem 7.2** *Let $A \subset \mathbb{R}^3$ be a Borel set. Then for the projections $\Pi_\theta$ as in (5.2)*

$$\dim \Pi_\theta(A) = \sup\{\Pi_\theta(A) : \theta \in (0, \pi)\} \textit{ for almost all } \theta \in (0, \pi).$$

Notice that for the projections $P_\theta$ onto lines given in (5.1) the constancy is trivial by Marstrand's projection theorem: for almost all $\theta \in (0, \pi)$,

$$\dim P_\theta(A) = \dim P_\theta(\pi(A)) = \min\{\dim \pi(A), 1\},$$

where $\pi(x, y, z) = (x, y)$.

Fässler and Orponen proved analogous results also for packing and Minkowski dimensions.

## 8 Slicing Theorems

Marstrand's line intersection theorem says that if $A$ is a Borel subset of the plane with $\dim A > 1$, then the typical lines which intersect $A$ intersect it in dimension $\dim A - 1$. Here is a way to state it more precisely and in higher dimensions:

**Theorem 8.1** *Let $A \subset \mathbb{R}^n$ be a Borel set, $s > m$ and $0 < \mathcal{H}^s(A) < \infty$. Then for $\gamma_{n,m}$ almost $V \in G(n, m)$,*

$$\mathcal{H}^m(\{v \in V : \dim(A \cap (V^\perp + v)) = s - m\}) > 0.$$

Here $\gamma_{n,m}$ is the orthogonally invariant Borel probability measure on $G(n, m)$.

With Balogh et al. we proved in [BF12] the analogous result in Heisenberg groups:

**Theorem 8.2** *Let $A \subset \mathbb{H}^n$ be a Borel set, $s > m + 2$ and $0 < \mathcal{H}^s_H(A) < \infty$. Then for $\mu_{n,m}$ almost $V \in G_h(n, m)$,*

$$\mathcal{H}^m(\{v \in V : \dim_H(A \cap (V^\perp \cdot v)) = s - m\}) > 0.$$

Here for $v \in V$, $V^\perp \cdot v$ is the coset $\{p \cdot v : p \in V^\perp\}$. The assumption $\dim_H A > m + 2$ is necessary.

Another way to formulate such a result, actually the one Marstrand used, is

**Theorem 8.3** *Let $A \subset \mathbb{H}^n$ be a Borel set, $s > m + 2$ and $\mathcal{H}^s_H(A) < \infty$. Then for $\mathcal{H}^s_H$ almost all $p \in A$ we have*

$$\dim_H(A \cap (V^\perp \cdot p)) = s - m \text{ for } \mu_{n,m} \text{ almost all } V \in G_h(n,m).$$

Here $\mathcal{H}_H^s$ is the $s$-dimensional Hausdorff measure with respect to the Heisenberg metric.

Orponen studied in [Or12] the problem of the dimension of exceptional sets for line sections. A higher dimensional version of his results is

**Theorem 8.4** *Let $A \subset \mathbb{R}^n$ be a Borel set, $s > m$ and $0 < \mathcal{H}^s(A) < \infty$. Then there is a Borel set $E \subset G(n,m)$ such that*

$$\dim E \leq m(n-m) + m - s$$

*and*

$$\mathcal{H}^m(\{a \in V : \dim A \cap (V^\perp + a) = s - m\}) > 0 \text{ for all } V \in G(n,m) \setminus E.$$

The upper bound is again sharp. Observe that this strengthens part (2) of Theorem 2.3.

To get his result, Orponen proved the following inequality: if $s > m$, $\mu \in \mathcal{M}(\mathbb{R}^n)$ and $V \in G(n,m)$, then

$$\int_V I_{s-m}(\mu_{V,a})\, d\mathcal{H}^m a \lesssim \int_{\mathbb{R}^n} |P_{V^\perp}(x)|^{s-n} |\widehat{\mu}(x)|^2 dx.$$

Here the measures $\mu_{V,a}$ are natural slices of $\mu$ with planes $V+a$. They have supports in $\operatorname{spt}\mu \cap (V+a)$ and they disintegrate $\mu$ if the projection of $\mu$ on $V$ is absolutely continuous with respect to the $m$-dimensional Hausdorff measure on $V$.

Fraser et al. found in [FOS3] another interesting application for this inequality: they showed that any one-dimensional graph has Fourier dimension 1. More precisely, in general dimensions

**Theorem 8.5** *For any function $f : A \to \mathbb{R}^{n-m}$, $A \subset \mathbb{R}^m$, and for its graph $G_f = \{(x, f(x) : x \in A\}$, if $\mu \in \mathcal{M}(G_f)$, $s > 0$ and*

$$|\widehat{\mu}(x)| \leq |x|^{-s/2} \text{ for } x \in \mathbb{R}^n,$$

*then $s \leq m$.*

Notice that we make no assumptions on $f$, not even measurability. Still, before the work of Fraser, Orponen and Sahlsten this question was open even for Brownian graphs.

## 9 Final Comments

I have restricted here to general sets. Many of the above results are formulated, and are more natural and general, for measures and their dimensions. I have completely ignored the very interesting question on what can be said in various special cases, for example for self-similar and related sets and measures. For these one can often obtain results which hold for all directions or the exceptional directions can be specified. Outstanding work on self-similar and other dynamically generated sets has been recently done by Furstenberg in [Fu08], by Peres and Shmerkin in [PSh9], by Hochman and Shmerkin in [HS12], and by Hochman in [Ho16]. There have also been many results on dimensions of sections in various special cases, for example in [BFS12, LXZ7, MS03, WWXl, WX10].

## References


[BD13] Balogh, Z.M., Cartagena, D.E., Fässler, K., Mattila, P., Tyson, J.T.: The effect of projections on dimension in the Heisenberg group, Revista Mat. Iberoamericana **29**, 381–432 (2013)

[BF12] Balogh, Z.M., Fässler, K., Mattila, P., Tyson, J.T.: Projection and slicing theorems in Heisenberg groups. Adv. Math. **231**, 569–604 (2012)

[BT09] Balogh, Z.M., Tyson, J.T., Warhurst, B.: Sub-Riemannian versus Euclidean dimension comparision and fractal geometry on Carnot groups. Adv. Math. **220**, 560–619 (2009)

[BFS12] Barany, B., Ferguson, A., Simon, K.: Slicing the Sierpinski gasket. Nonlinearity **25**, 1753–1770 (2012)

[Bo10] Bourgain, J.: Discretized sum-product and projection theorems. J. Anal. Math. **112**, 193–236 (2010)

[Fa82] Falconer, K.J.: Hausdorff dimension and the exceptional set of projections. Mathematika **29**, 90–96 (1982)

[Fa85] Falconer, K.J.: Geometry of Fractal Sets. Cambridge University Press, Cambridge (1985)

[FH96] Falconer, K.J., Howroyd, J.D.: Projection theorems for box and packing dimensions. Math. Proc. Camb. Phil. Soc. **119**, 287–295 (1996)

[FH97] Falconer, K.J., Howroyd, J.D.: Packing dimension of projections and dimension profiles. Math. Proc. Camb. Phil. Soc. **121**, 269–286 (1997)

[Fm96] Falconer, K.J., Mattila, P.: Packing dimension of projections and sections of measures. Math. Proc. Camb. Phil. Soc. **119**, 695–713 (1996)

[FO13] Fässler, K., Orponen, T.: Constancy results for special families of projections. Math. Proc. Camb. Philos. Soc. **154**, 549–568 (2013). arXiv:1208.1876

[FO50] Fässler, K., Orponen, T.: On restricted families of projections in $\mathbb{R}^3$. arXiv:1302.6550

[Fe69] Federer, H.: Geometric Measure Theory. Springer, New York (1969)

[FOS3] Fraser, J., Orponen, T., Sahlsten, T.: On Fourier analytic properties of graphs. IMRN. arXiv:1212.4813 (to appear)

[Fu08] Furstenberg, H.: Ergodic fractal measures and dimension conservation. Ergodic Theo. Dynam. Syst. **28**, 405–422 (2008)

[Ho16] Hochman, M.: On self-similar sets with overlaps and inverse theorems for entropy. arXiv:1212.1873

[HS12] Hochman, M., Shmerkin, P.: Local entropy averages and projections of fractal measures. Ann. Math. **175**, 1001–1059 (2012)

[Ho18] Hovila, R.: Transversality of isotropic projections, unrectifiability and Heisenberg groups. Revista Mat. Iberoamericana (to appear)

[HLJ2] Hovila, R., Järvenpää, E., Järvenpää, M., Ledrappier, F.: Besicovitch-Federer projection theorem and geodesic flows on Riemann surfaces. Geom. Dedicata. **161**, 51–61 (2012)

[HJL2] Hovila, R., Järvenpää, E., Järvenpää, M., Ledrappier, F.: Singularity of projections of 2-dimensional measures invariant under the geodesic flow. Comm. Math. Phys. **312**, 127–136 (2012)

[JJ13] Järvenpää, E., Järvenpää, M., Keleti, T.: Hausdorff dimension and non-degenerate families of projections. J. Geom. Anal. (2013)

[JJ05] Järvenpää, E., Järvenpää, M., Ledrappier, F., Leikas, M.: (Non)regularity of projections of measures invariant under geodesic flow. Comm. Math. Phys. **254**, 695–717 (2005)

[JJ28] Järvenpää, E., Järvenpää, M., Leikas, M.: One-dimensional families of projections. Nonlinearity **21**, 453–463 (2008)

[Jm94] Järvenpää, M.: On the upper Minkowski dimension, packing dimension, and orthogonal projections. Ann. Acad. Sci. Fenn. Ser. A I Math. Dissertationes **99**, 34 (1994)

[Ka68] Kaufman, R.: On Hausdorff dimension of projections. Mathematika **15**, 153–155 (1968)

[KM75] Kaufman, R., Mattila, P.: Hausdorff dimension and exceptional sets of linear transformations. Ann. Acad. Sci. Fenn. Ser. A I Math. **1**, 387–392 (1975)

[LXZ7] Liu, Q.H., Xi, L.F., Zhao, Y.F.: Dimensions of intersections of the Sierpinski carpet with lines of rational slopes. Proc. Edinb. Math. Soc. **50**, 411–428 (2007)

[MS03] Manning, A., Simon, K.: Dimension of slices through the Sierpinski carpet. Trans. Amer. Math. Soc. **365**, 213–250 (2013)

[Ma54] Marstrand, J.M.: Some fundamental geometrical properties of plane sets of fractional dimensions. Proc. London Math. Soc. **3**(4), 257–302 (1954)

[Map5] Mattila, P.: Hausdorff dimension, orthogonal projections and intersections with planes. Ann. Acad. Sci. Fenn. Ser. A I Math. **1**, 227–244 (1975)

[Map2] Mattila, P.: Geometry of Sets and Measures in Euclidean Spaces. Cambridge University Press, Cambridge (1995)

[Map3] Mattila, P.: Hausdorff dimension, projections, and the Fourier transform. Publ. Mat. **46**, 3–48 (2004)

[Map4] Mattila, P.: Marstrand's theorems. In: Córdoba, A., Férnandez, J.L., Férnandez, P. (eds.) All That Math, Portraits of Mathematicians as Young Readers, pp. 235–245. Revista Mathematica Iberoamericana (2011)

[Ob12] Oberlin, D.M.: Restricted Radon transforms and projections of planar set. Canad. Math. Bull. **55**, 815–820 (2012)

[Ob13] Oberlin, D.M.: Exceptional sets of projections, unions of k-planes, and associated transforms. arXiv:1107.4913

[Or12] Orponen, T.: Slicing sets and measures, and the dimension of exceptional parameters. J. Geom. Anal. **24**, 47–80 (2012)

[Ort2] Orponen, T.: On the packing dimension and category of exceptional sets of orthogonal projections. arXiv:1204.2121

[Orp3] Orponen, T.: Hausdorff dimension estimates for some restricted families of projections in $\mathbb{R}^3$. arXiv:1304.4955

[PS00] Peres, Y., Schlag, W.: Smoothness of projections, Bernoulli convolutions, and the dimension of exceptions. Duke Math. J. **102**, 193–251 (2000)

[PSh9] Peres, Y., Shmerkin, P.: Resonance between Cantor sets. Ergodic Theor. Dynam. Syst. **29**, 201–221 (2009)

[Ram2] Rams, M.: Packing dimension estimation for exceptional parameters. Israel J. Math. **130**, 125–144 (2002)

[WWXl] Wen, Z., Wu, W., Xi, L.: Dimension of slices through a self-similar set with initial cubical pattern. Ann. Acad. Sci. Fenn. Ser. A I (to appear)

[WX10] Wen, Z.-Y., Xi, L.: On the dimension of sections for the graph-directed sets. Ann. Acad. Sci. Fenn. Ser. A **I**(35), 515–535 (2010)

# The Geometry of Fractal Percolation

**Michał Rams and Károly Simon**

**Abstract** A well studied family of random fractals called fractal percolation is discussed. We focus on the projections of fractal percolation on the plane. Our goal is to present stronger versions of the classical Marstrand theorem, valid for almost every realization of fractal percolation. The extensions go in three directions:

- the statements work for all directions, not almost all,
- the statements are true for more general projections, for example radial projections onto a circle,
- in the case $\dim_H > 1$, each projection has not only positive Lebesgue measure but also has nonempty interior.

**Keywords** Random fractals · Hausdorff dimension · Processes in random environment

**2000 Mathematics Subject Classification:** Primary 28A80 Secondary 60J80 · 60 J85

Rams was partially supported by the MNiSW grant N201 607640 (Poland). The research of Simon was supported by OTKA Foundation # K 104745.

M. Rams
Institute of Mathematics, Polish Academy of Sciences, ul. Śniadeckich 8, Warsaw 00-956, Poland
e-mail: rams@impan.gov.pl

K. Simon (✉)
Institute of Mathematics, Technical University of Budapest, Budapest H-1529 B.O.box 91, Hungary
e-mail: simonk@math.bme.hu



D.-J. Feng and K.-S. Lau (eds.), *Geometry and Analysis of Fractals*, Springer Proceedings in Mathematics & Statistics 88,
DOI 10.1007/978-3-662-43920-3_11

# 1 Introduction

To model turbulence, Mandelbrot [Ma74, Ma83] introduced a statistically self-similar family of random Cantor sets. Since that time this family has got at least three names in the literature: fractal percolation, Mandelbrot percolation and canonical curdling, among which we will use the first one.

In 1996 Lincoln Chayes [Ch96] published an excellent survey giving an account about the most important results known in that time. His survey focused on the percolation related properties while we place emphasis on the geometric measure theoretical properties (projections and slices) of fractal percolation sets.

About the projections of a general Borel set the celebrated Marstrand Theorem gives the following information:

**Theorem 1** ([Ma54]) *Let* $E \subset \mathbb{R}^2$ *be a Borel set.*

- *If* $\dim_H(E) < 1$ *then for Lebesgue almost all* $\theta$ $\dim_H(\text{proj}_\theta(E)) = \dim_H(E)$.
- *If* $\dim_H(E) > 1$ *then for Lebesgue almost all* $\theta$ *we have* $\mathcal{L}\text{eb}(\text{proj}_\theta(E)) > 0$.

*where* $\text{proj}_\theta$ *is the orthogonal projection in direction* $\theta$.

In this paper we review some recent results which give more precise information in the special case of the projections of fractal percolation Cantor sets.

# 2 The Construction and Its Immediate Consequences

The construction consists of the infinite iteration of two steps. We start from the unit cube in $\mathbb{R}^d$.

- All cubes we have after the $n$-th iteration of the process (they will be called level $n$ cubes) we subdivide into smaller cubes of equal size,
- Among them some are retained and some are discarded. Retaining or discarding of different cubes are independent random events. The cubes that were retained are the level $n + 1$ cubes.

Those points that have never been discarded form the fractal percolation set.

Please note that in literature the term fractal percolation is often used to denote object which we call homogeneous fractal percolation. That is, the fractal percolation for which all squares have equal probabilities of being retained.

## *2.1 An Informal Description of Fractal Percolation*

We fix integer $M \geq 2$. We partition the unit cube $Q \subset \mathbb{R}^d$ into $M^d$ congruent cubes of side length $M^{-1}$ and we assign a probability to each of the cubes in this partition (Fig. 1a). We retain each of the cubes of this partition with the corresponding

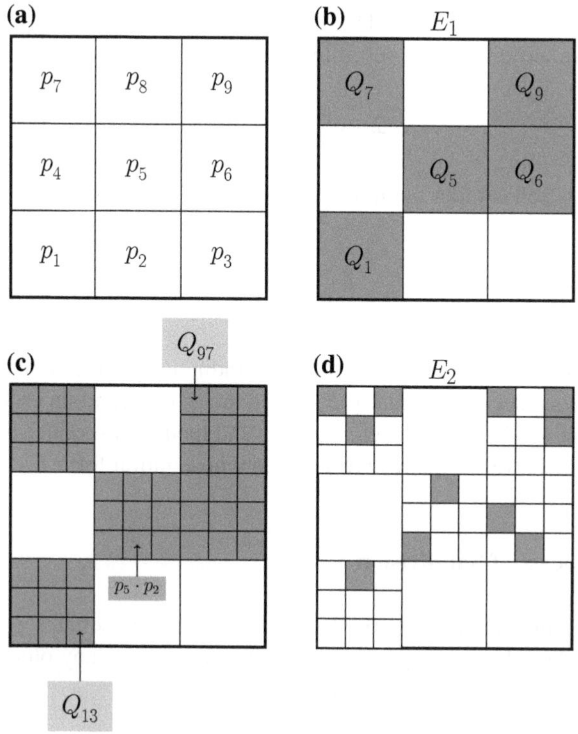


**Fig. 1** The first two steps of the construction. $\mathbb{P}(Q_{52} \text{ retained}) = p_5 \cdot p_2$. For this realization $\mathcal{E}_1 = \{1, 5, 6, 7, 9\}$, $\mathcal{E}_2 = \{17, 51, 58, 62, 64, 75, 77, 79, 96, 97, 99\}$

probability independently and discard it with one minus the corresponding probability. The union of the retained squares is the first approximation of the random set to be constructed ( Fig. 1b). We obtain the second approximation by repeating this process independently of everything in each of the retained squares ( Fig. 1c, d). We continue this process at infinitum.

The object of our investigation is the collection of those points which have not been discarded. It will be called **fractal percolation set** and denoted by $E = (d, M, \mathbf{p})$, where $\mathbf{p}$ is the chosen vector of the probabilities $\{p_i\}$. In the special case when all $p_i$ are equal we obtain the **homogeneous fractal percolation set** which is denoted by $E^h = E^h(d, M, p)$.

## 2.2 Fractal Percolation Set in More Details

For simplicity we give the construction on the plane but the definition works with obvious modifications in $\mathbb{R}^d$ for all $d \geq 1$. Besides the dimension of the ambient space the two other parameters of the construction are: the natural number $M \geq 2$ and a vector of probabilities $\mathbf{p} \in [0,1]^{M^2}$ (note: not a probabilistic vector). To shorten the notation we write $\mathcal{I}$ for the set of indices of $\mathbf{p}$:

$$\mathcal{I} := \left\{1, \ldots, M^2\right\}$$

The statistically self-similar random set which is the object of our study is defined as

$$E := \bigcap_{n=1}^{\infty} E_n, \tag{2.1}$$

where $E_n$ is the $n$-th approximation of $E$. The inductive definition of $E_n$ will occupy the rest of this subsection. Actually $E_n$ is the union of a random collection of level $n$ squares. First we define the level $n$ squares and then we introduce the random rule with which those level $n$ squares are selected whose union form $E_n$.

### 2.2.1 The Process of Subdivision

We divide the unit square $Q = [0,1]^2$ into $M^2$ congruent squares $Q_1, \ldots, Q_{M^2}$ of size $M^{-1}$ numbered according to lexicographical order (or any other order). These squares are the level one $M$-adic squares. Let

$$\mathcal{N}_1 := \{x_i\}_{i \in \mathcal{I}}$$

be the set of midpoints of the level one squares ( Fig. 2). For each midpoint $x_i$ we define the homothetic map $\varphi_i : Q \to Q_i$:

$$\varphi_i(y) := x_i + M^{-1} \cdot \left(y - \left(\frac{1}{2}, \frac{1}{2}\right)\right).$$

For every $\mathbf{i} \in \mathcal{I}^n$, $\mathbf{i} = (i_1, \ldots, i_n)$ we write

$$x_{\mathbf{i}} := \varphi_{\mathbf{i}}\left(\frac{1}{2}, \frac{1}{2}\right).$$

and we define the map

$$\varphi_{\mathbf{i}}(y) := x_{\mathbf{i}} + M^{-n} \cdot \left(y - \left(\frac{1}{2}, \frac{1}{2}\right)\right).$$

To simplify the notation, we will not distinguish the set of the centers of level $n$ squares

$$\mathcal{N}_n := \left\{\varphi_{\mathbf{i}}\left(\frac{1}{2}, \frac{1}{2}\right) : \mathbf{i} \in \mathcal{I}^n\right\}$$

and the family of level $n$-squares:

$$\left\{Q_{\mathbf{i}} := \varphi_{\mathbf{i}}(Q) : \mathbf{i} \in \mathcal{I}^n\right\}. \tag{2.2}$$

### 2.2.2 The Process of Retention

The square $Q = Q_\emptyset$ is retained. For any $\mathbf{i} \in \mathcal{I}^n$ for which the square $Q_{\mathbf{i}}$ is retained and for each $j \in \mathcal{I}$, the square $Q_{\mathbf{i}j}$ is retained with probability $p_j$. The events '$Q_{\mathbf{i}j}$ is retained' and '$Q_{\mathbf{i}'j'}$ is retained' are independent whenever $\mathbf{i} \neq \mathbf{i}'$ or $j \neq j'$.

We define $E_1$ as the union of retained squares $Q_i, i \in \mathcal{I}$. Similarly, $E_n$ is the union of retained squares $Q_{\mathbf{i}}, \mathbf{i} \in \mathcal{I}^n$. We write

$$\mathcal{E}_n := \left\{\mathbf{i} \in \mathcal{I}^n : Q_{\mathbf{i}} \text{ retained}\right\}.$$

## *2.3 The Corresponding Probability Space and Statistical Self-similarity*

The probability space corresponding to this random construction is best described by Dekking [De09]. For the convenience of the reader we repeat it here. Let $\mathcal{T}$ be the $M^d$ array tree that is

$$\mathcal{T} := \bigcup_{n=0}^{\infty} \mathcal{I}^n,$$

where $\mathcal{I}^0 := \emptyset$ is the root of three . Let $\Omega := \{0, 1\}^{\mathcal{T}}$ that is $\Omega$ is the set of labeled trees where we label every node of $\mathcal{T}$ by 0 or 1. The probability measure $\mathbb{P}_{\mathbf{p}}$ on $\Omega$ is define in such a way that the family of labels $X_{\mathbf{i}} \in \{0, 1\}$ of nodes $\mathbf{i} \in \mathcal{T}$ satisfy:

- $\mathbb{P}_{\mathbf{p}}(X_\emptyset = 1) = 1$
- $\mathbb{P}_{\mathbf{p}}(X_{i_1,\dots,i_n}) = p_{i_n}$
- $\{X_{\mathbf{i}}\}_{\mathbf{i}\in\mathcal{T}}$ are independent.

Following [De09] we define the survival set of level $n$ by

$$S_n := \left\{\mathbf{i} \in \mathcal{I}^n : X_{i_1\dots,i_k} = 1,\ \forall 1 \leq k \leq n\right\}.$$

Then

$$E_n = \bigcup_{\mathbf{i}\in S_n} Q_{\mathbf{i}}, \quad E = \bigcap_{n=1}^{\infty} E_n.$$

It follows from the construction that generalized fractal percolation set is statistically self-similar and the number of retained cubes form a branching process:

**Lemma 2** *(a)* $\{\#\mathcal{E}_n\}$ *is a branching process with average number of offsprings* $\sum_{i\in\mathcal{I}} p_i$. *In particular if* $p_i \equiv p$ *then the offspring distribution is* `Binomial` $(M^d, p)$.

*(b) For every* $n \geq 1$ *and* $\mathbf{i} \in \mathcal{E}_n$ *the rescaled copy* $\varphi_{\mathbf{i}}^{-1}(E \cap Q_{\mathbf{i}})$ *has the same distribution as E itself.*

*(c) The sets* $\{E \cap Q_{\mathbf{i}}\}_{\mathbf{i}\in\mathcal{E}_n}$ *are independent.*

Using this it is not hard to prove that

$$E \neq \emptyset \text{ implies that } \dim_{\mathrm{H}}(E) = \dim_{\mathrm{B}}(E) = \frac{\log \sum_{i\in\mathcal{I}} p_i}{\log M} \text{ a.s.} \tag{2.3}$$

This was proved by Kahane and Peyriere [KP76], Hawkes [Ha81], Falconer [Fa86], Mauldin and Williams [MW86] independently. A canonical example of the inhomogeneous fractal percolation set is:

*Example 3* [Random Sierpiński Carpet] Let $SC_p := E(2, 3, \mathbf{p})$, where using the notation of Fig. 1c:

$$p_5 = 0 \, and \; for \, i \in \{1, \ldots, 9\} \setminus \{5\} : \; p_i = p.$$

□

# 3 Percolation and Projection to Coordinate Axes

In this section we work on the plane so $Q = [0, 1]^2$. The connectivity properties of $E^h(2, M, p)$ for an arbitrary $M \geq 2$ was first investigated by Chayes, Chayes and Durrett [CCD88]. Dekking and Meester [DM90] gave a simpler proof and extended the scope of the theorem for some inhomogeneous fractal percolation sets like the random Sierpiński carpet $SC_p$. Here we summarize briefly some of the most interesting results of this area. For a much more detailed account see by Chayce [Ch96].

We say that $E$ **percolates** if $E$ contains a connected set which intersects both the left and the right sides of $Q$. If $E$ percolates then $E$ has a large connected component.

## 3.1 The Homogeneous Case

The following very important result was proved by Chayce,Chayce, Durrett.

**Theorem 4** ([CCD88]) *Fix an arbitrary* $M \geq 2$. *Then there is a critical probability* $\frac{1}{M} < p_c < 1$ *such that*

*(1) If* $p < p_c$ *then* $E^h(2, M, p)$ *is a random dust that is totally disconnected almost surely.*

(2) *If $p \geq p_c$ then $E^h(2, M, p)$ percolates with positive probability. This implies that $E^h(2, M, p)$ is not totally disconnected almost surely.*

This shows a remarkable difference in between the fractal percolation and the usual percolation: in the latter case, the probability of percolation at critical parameter $p = p_c$ is 0.

## 3.2 The Inhomogeneous Case

Using some earlier works of Dekking and Grimmett [DG88], the results above were extended by Dekking and Meester [DM90]. They proved that by changing the components of $\mathbf{p}$ the inhomogeneous fractal percolation set $E(2, M, \mathbf{p})$ can go through the six stages below. Here the projection to the $x$-axis is denoted by $\text{proj}_x$. That is $\text{proj}_x(a, b) = a$.

**The DM stages of $\mathbf{E(2, M, p)}$:**

I: $E = \emptyset$ almost surely.
II: $\mathbb{P}(E \neq \emptyset) > 0$ but $\dim_{\mathrm{H}}(\text{proj}_x E) = \dim_{\mathrm{H}}(E)$ almost surely.
III: $\dim_{\mathrm{H}}(\text{proj}_x E) < \dim_{\mathrm{H}}(E)$ if $E \neq \emptyset$ but $\mathcal{L}\text{eb}(\text{proj}_x E) = 0$ almost surely.
IV: $0 < \mathcal{L}\text{eb}(\text{proj}_x E) < 1$ almost surely.
V: $\mathcal{L}\text{eb}(\text{proj}_x E) = 1$ holds with positive probability but $E$ does not percolate almost surely.
VI: $E$ percolates with positive probability.

It was proved in [DM90] that the random Sierpiński Carpet $SC_p$ goes through all of these stages as we increase the value of $p$. The following theorem gives the precise answer when exactly a system appears in stages I, II, III.

**Theorem 5** ([DG88, Fa86]) *Let $m_r$ be the sum of the probabilities in the $r$-th column, that is the expected number of squares in column $r$. Then*

(1) $E = \emptyset$ *almost surely iff* $\sum_{i=1}^{M^2} p_i \leq 1$. *Except when $\exists i$ such that $p_i = 1$ and $p_j = 0$ for all $i \neq j$. In this case $E$ is a singleton.*

(2) $\dim_{\mathrm{H}}(\text{proj}_x(E)) = \dim_{\mathrm{H}}(E)$ *holds almost surely, iff* $\sum_{r=1}^{M} m_r \log m_r \leq 0$.

(3) $\mathcal{L}\text{eb}(\text{proj}_x E) = 0$ *holds almost surely iff* $\sum_{r=1}^{M} \log m_r \leq 0$.

This result was strengthened by Falconer and Grimmett:

**Theorem 6** ([FG92, FG94]) *Assume that $m := \min\{m_r\} > 1$. Then $\text{proj}_x(E)$ contains an interval almost surely, conditioned on non-extinction.*

We will present the proof in the fifth section.

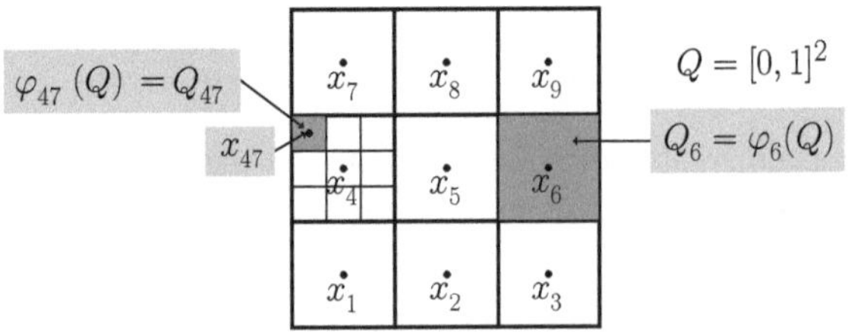


**Fig. 2** Definition of level $n$ squares

### 3.3 The DM Stages for The Homogeneous Case

For the homogeneous case $m_r = M \cdot p$ Hence we obtain that almost surely:

- If $0 < p \leq \frac{1}{M^2}$ then $E = \emptyset$.
- If $\frac{1}{M^2} < p \leq \frac{1}{M}$ then the system is in stage II.
- If $\frac{1}{M} < p < p_c$ then the system is in stage V.

Stages III and IV do not appear in the homogeneous case.

## 4 The Arithmetic Sum/Difference of Two Fractal Percolations

There is a very nice and more detailed survey of this field due to M. Dekking [De09]. In the previous section we studied the connectivity properties and the 90° projections of random Cantor sets. In this section we consider sets which are products of inhomogeneous fractal percolation sets and we take their 45°, (−45°) projections in order to study the arithmetic difference (arithmetic sum) respectively of independent copies of $E(1, M, \mathbf{p})$.

### 4.1 The Arithmetic Sum and Its Visualization

Let $A, B \subset \mathbb{R}$ be arbitrary. Then the arithmetic sum $A + B := \{a + b : a \in A, b \in B\}$ is the −45°-projection of $A \times B$ to the $x$-axis (this is the direction of the line $\ell_a$ on Fig. 3). Similarly, we can visualize the arithmetic difference by taking the projection of the product set with the line of +45° angle.

The motivation for studying the arithmetic difference (or sum) of random Cantor sets comes from a conjecture of Palis which states that typically (in a natural sense which depends on the actual setup), the arithmetic difference of two dynamically defined Cantors is either small in the sense that it has Lebesgue measure zero or big in the sense that it contains some intervals, but at least typically, it does not occur that the arithmetic difference set is a set of positive Lebesgue measure with empty interior. This conjecture does not hold for the algebraic difference of inhomogeneous

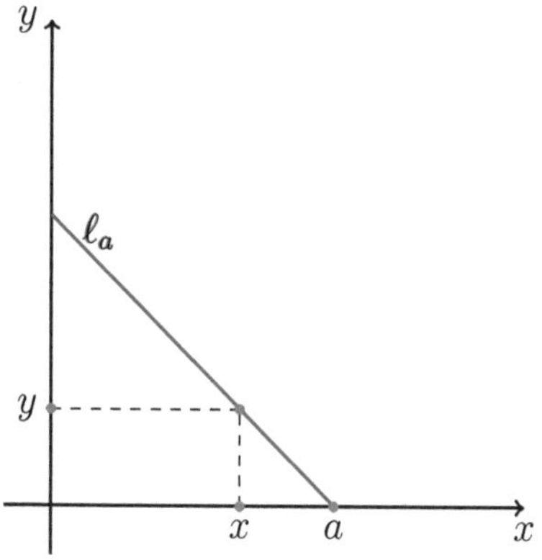


**Fig. 3** Algebraic sum as $-45°$ projection: $a = x + y = \text{proj}_{-45°}(x, y)$

fractal percolation sets, but it holds in the homogeneous case. The way to prove this is via the 45°-projections of $E(1, M, \mathbf{p}) \times E(1, M, \mathbf{p})$.

## 4.2 *The Product of Two One Dimensional Fractal Percolation Versus a Two Dimensional Fractal Percolation*

We explain this relation in the case when $M = 3$. Assume that we are given the inhomogeneous fractal percolations $E(1, 3, \mathbf{a})$, and $E(1, 3, \mathbf{b})$, where $\mathbf{a} = (a_1, a_2, a_3)$, $\mathbf{b} = (b_1, b_2, b_3)$ are the vectors of probabilities. We define the vector $\mathbf{p} \in [0, 1]^9$ as their product $\mathbf{p} = \mathbf{a} \bigotimes \mathbf{b}$ in the natural way which is suggested by looking at Fig. 1a. That is:

$$p_i := a_u \cdot b_v \text{ if } i - 1 = 3 * (v - 1) + (u - 1), \; 1 \leq u, v \leq 3.$$

The reason that $E(2, 3, \mathbf{p})$ and $E(1, 3, \mathbf{a}) \times E(1, 3, \mathbf{b})$ are similar is explained in **(a)** and the essential difference between them is pointed out in **(b)** below:

(a) Let $\mathbf{i} \in \{1, \ldots, 9\}^n$. Then the probability that $Q_\mathbf{i}$ is retained is the same during the construction of $E(2, 3, \mathbf{p})$ and the construction of $E(1, 3, \mathbf{a}) \times E(1, 3, \mathbf{b})$.
(b) Let $K$ and $L$ be level $n$ squares for some $n$. Assume that both $K$ and $L$ are retained during the construction of $E(2, 3, \mathbf{p})$ and $E(1, 3, \mathbf{a}) \times E(1, 3, \mathbf{b})$. Then

- In the construction of $E = E(2, 3, \mathbf{p})$ the sets $E \cap K$ and $E \cap L$ are independent.
- In the construction of $E(1, 3, \mathbf{a}) \times E(1, 3, \mathbf{b})$ the sets $E \cap K$ and $E \cap L$ are independent iff $\text{proj}_x K \neq \text{proj}_x L$ and $\text{proj}_y K \neq \text{proj}_y L$ hold.

In dimension $d \geq 2$ the analogy is the same: the probability of the retention of a level $n$ cube is the same for the $d$-dimensional percolation and for the $d$-fold product of the corresponding one dimensional percolations. On the other hand, the future of what ever happens in two distinct retained level $n$ cubes is:

- always independent in the $d$-dimensional percolation case,
- independent for the $d$-fold product of the corresponding one dimensional fractal percolations iff the two cubes do not share any common projections to coordinate axes.b

## 4.3 The Existence of An Interval in the Arithmetic Difference Set

Let $E_1 := E(1, M, \mathbf{p})$ and $E_2 := E(1, M, \mathbf{q})$. We define the cyclic cross correlation coefficients:

$$\gamma_k := \sum_{i=1}^{M} p_i q_{i-k(\text{mod } M)} \text{ for } k = 1, \ldots, M. \tag{4.1}$$

**Theorem 7** ([DS08]) *Assuming that $E_1, E_2 \neq \emptyset$, we have*

*(a) If $\forall i = 1, \ldots, M : \ \gamma_i > 1$ then almost surely*

$$E_2 - E_1 \; contains \; an \; interval.$$

*(b) If $\exists i \in \{1, \ldots, M\} : \gamma_i, \gamma_{i+1 \text{ mod } M} < 1$ then almost surely*

$$E_2 - E_1 \; does \; not \; contain \; any \; interval.$$

In the homogeneous case and in the case when $M = 3$ this gives complete characterization. Otherwise we can change to higher order Cantor sets (collapsing $n \geq 2$ steps of the construction into one) and we can apply the same theorem in that case. The fact that this can be done is not trivial because higher order fractal percolations are correlated. That is the way as the random set develops in one level $n$ square is dependent how it develops in some other squares. Nevertheless, M. Dekking and H. Don proved that this can be done by pointing out that the proof of the theorem above can be carried out for more general, correlated random sets than the inhomogeneous fractal percolations. This more general family includes the higher order fractal percolation sets.

## 4.4 The Lebesgue Measure of the Arithmetic Difference Set

Let $E_2, E_2$ be two independent realizations of $E(1, M, \mathbf{p})$. Then

$$\gamma_k := \sum_{i=1}^{M} p_i p_{i-k(\text{mod } M)} \text{ for } k = 1, \ldots, M.$$

Let $\Gamma := \gamma_1 \cdots \gamma_M$.

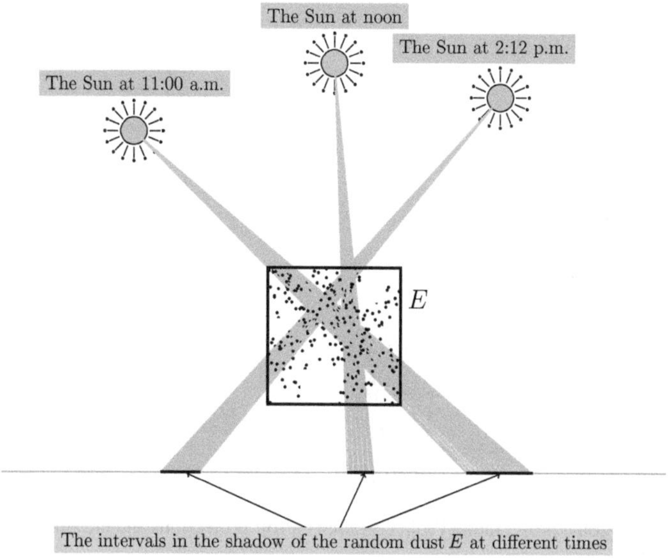


**Fig. 4** Projections of fractal percolation

**Theorem 8** ([MSS09]) *If* $\Gamma > 1$ *then*
$\mathcal{L}\mathrm{eb}(E_2 - E_1) > 0$.

Combined application of Theorems 7 and 8 yields that the Palis conjecture does not hold in the case when for $M = 3$ and $\mathbf{p} = (0.52, 0.5, 0.72)$. Namely, in this case $\gamma_1 = 1.0388$ and $\gamma_2 = \gamma_3 = 0.941$. Let $E_1, E_2$ be two independent copies of $E(1, 3, \mathbf{p})$. Then by Theorem 7 there is no interval in $E_1 - E_2$ (since there are two consecutive $\gamma$'s that are smaller than one) and by Theorem 8 we have $\mathcal{L}\mathrm{eb}\,(E_1 - E_2) > 0$ since $\gamma_1 \cdot \gamma_2 \cdot \gamma_3 = 1.0272 > 1$.

## 5 General Projections: The Opaque Case

In this and in the following sections we study the projections of fractal percolation sets in general directions. In this section we consider the case $\dim_{\mathrm{H}}(E) > 1$. Under some mild assumption, almost surely projections of $E$ have not only positive Lebesgue measure, as per Marstrand theorem, but also non-empty interior. Furthermore it holds for all and not only almost all directions. Moreover, this remains valid if we replace the orthogonal projection with a much more general family of projections.

One practical application of our result is shown above (Fig. 4). One does not need to rotate such a set to use it as an umbrella.

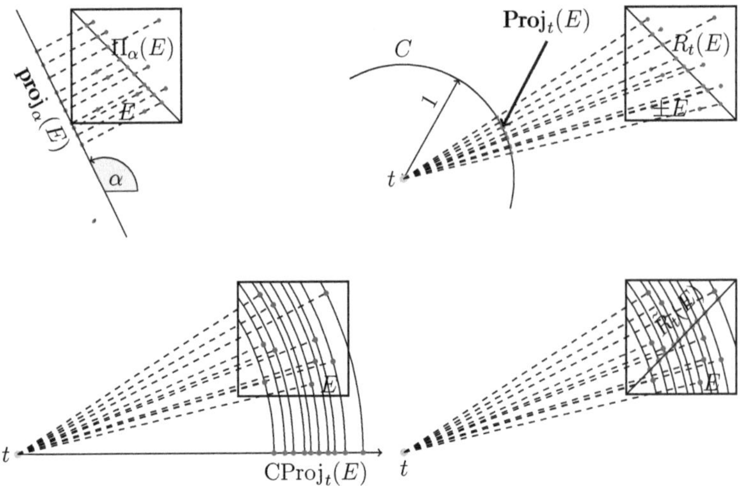


**Fig. 5** The orthogonal $\mathrm{proj}_\alpha$, radial $\mathrm{Proj}_t$, co-radial $\mathrm{CProj}_t$ projections and the auxiliary projections $\Pi_\alpha$, $R_t$, and $\tilde{R}_t$

We have already studied the horizontal and vertical projections. So we can restrict our attention to the directions $\alpha \in \mathcal{D} := (0, 90°)$ A condition $A(\alpha)$, $\alpha \in \mathcal{D}$ on the vector of probabilities $\mathbf{p}$ will be defined below.

**Theorem 9** ([RS00b]) *Let* $\alpha \in \mathcal{D}$. *If* $A(\alpha)$ *holds and* $E$ *is nonempty then almost surely* $\mathrm{proj}_\alpha(E)$ *contains an interval.*

**Theorem 10** ([RS00b]) *If* $A(\alpha)$ *holds for all* $\alpha \in \mathcal{D}$ *and* $E \neq \emptyset$ *then almost surely all projections* $\mathrm{proj}_\alpha(E)$ *contain an interval.*

*Remark 11* The assertions of Theorems 9 and 10 remain valid if we replace $\mathrm{proj}_\alpha$ with more general families of projections, see [RS00b, Sect. 6]. In particular, radial or co-radial projections (see Fig. 5) are included. □

*Example 12* If either

(1) Homogeneous case: $p_i = p > M^{-1}$ for all $i$, or
(2) Generalized random Sierpiński Carpet: $M = 3$, $p_5 = q$, $p_i = p$ for $i \neq 5$, and

$$p > \max\left(\frac{1}{3}, \frac{1-q}{2}\right)$$

then Condition A($\alpha$) is satisfied for all $\alpha \in \mathcal{D}$. Note that (1) is equivalent to $\dim_{\mathrm{H}}(E) > 1$ almost surely. □

## 5.1 Horizontal and Vertical Projections

Let us start by presenting the large deviation argument (by Falconer and Grimmett) working for horizontal and vertical projections. If $\dim_H \Lambda > 1$ then from the dimension formula for some $n$ one can find a level $n$ column with (exponentially) many squares. We prove inductively that in its every $N$-th level sub-column, $N > n$, we typically have exponentially many squares on each level (probability of existence of $N > n$ and an $N$-th level subcolumn which does not have exponentially many squares is super-exponentially small). When we move from level $n$ column to its level $n+1$ subcolumns, each square in the column gives birth to an expected number of $pM > 1$ number of level $n+1$ squares in each of the subcolumns. By large deviation theorem there is only a superexponentially small probability that the number of level $n+1$ squares in a subcolumn is smaller than a fixed $\alpha \in (1, pM)$ multiple of the level $n$ squares in the column. By induction, if this exceptional situation does not happen (or happens only finitely many times), for each $N > n$ the number of squares of level $N$ in each subcolumn will be at least of order $\alpha^{N-n}$.

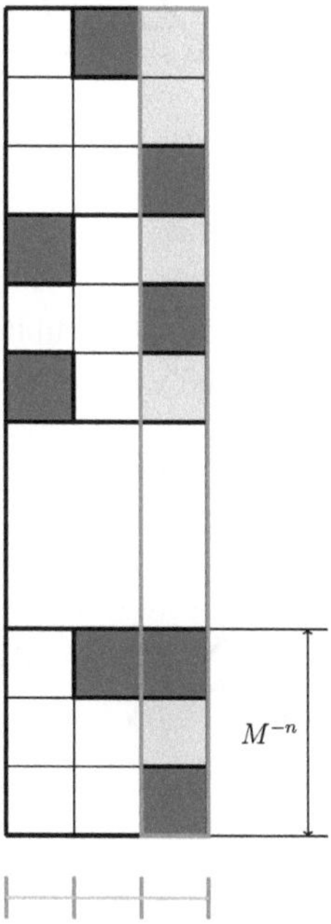


## 5.2 Condition A

Our goal in this subsection is to modify this argument to work in a more complicated situation of projections in general directions. Indeed, contrary to the horizontal/vertical projections case, here it is in general not true that if a line intersects a square of level $n$ then the expected number of squares of level $n+1$ it intersects is greater than 1. It is still true if the line intersects 'central' part of the square, but not if it hits it close to the corners.

Nevertheless, we are able to find a modified version of the argument. We fix $\alpha \in \mathcal{D}$. We are going to consider $\Pi_\alpha$ instead of $\text{proj}_\alpha$, i.e. we are projecting onto a diagonal $\Delta_\alpha$ of $Q$, see Fig. 5. For any $\mathbf{i} \in \mathcal{I}^n$ the map $\Pi_\alpha \circ \varphi_{\mathbf{i}} : \Delta_\alpha \to \Delta_\alpha$ is a linear contraction of ratio $M^{-n}$. We will use its inverse: a map $\psi_{\alpha,\mathbf{i}} : \Pi_\alpha(Q_{\mathbf{i}}) \to \Delta_\alpha$. It is a linear expanding map (of ratio $M^n$) and it is onto.

Consider the class of nonnegative real functions on $\Delta_\alpha$, vanishing on the endpoints. There is a natural random inverse Markov operator $G_\alpha$ defined as

$$G_\alpha f(x) = \sum_{i \in \mathcal{E}_1; x \in \Pi_\alpha(Q_i)} f \circ \psi_{\alpha,i}(x).$$

The corresponding operator on the $n$-th level is

$$G_\alpha^{(n)} f(x) = \sum_{\mathbf{i} \in \mathcal{E}_n; x \in \Pi_\alpha(Q_{\mathbf{i}})} f \circ \psi_{\alpha,\mathbf{i}}(x).$$

In particular for any $H \subset \Delta^\alpha$ we have

$$G_\alpha^{(n)} \mathbf{1}_H(x) = \#\{\mathbf{i} \in \mathcal{E}_n : x \in \Pi_\alpha(\varphi_{\mathbf{i}}(H))\}.$$

Although $G_\alpha^{(n)}$ should not be thought of as the $n$-th iterate of $G_\alpha$, the expected value of $G_\alpha^{(n)}$ is the $n$-th iterate of the expected value of $G_\alpha$. Namely, let

$$F_\alpha = \mathbb{E}[G_\alpha] \text{ and } F_\alpha^n = \mathbb{E}\left[G_\alpha^n\right]$$

We then have the formulas

$$F_\alpha f(x) = \sum_{i \in \mathcal{I}; x \in \Pi_\alpha(Q_i)} p_i \cdot f \circ \psi_{\alpha,i}(x)$$

and

$$F_\alpha^n f(x) = \sum_{\mathbf{i}; x \in \Pi_\alpha(Q_{\mathbf{i}})} p_{\mathbf{i}} \cdot f \circ \psi_{\alpha,\mathbf{i}}(x),$$

where

$$p_{\mathbf{i}} = \prod_{k=1}^{n} p_{i_k}.$$

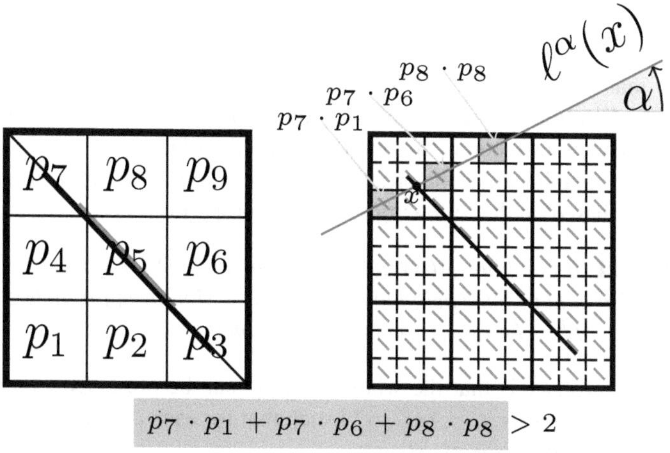


**Fig. 6** Condition $A(\alpha)$: $I_1^\alpha$ is the small red, $I_2^\alpha$ is the big blue interval on the left. $r_\alpha = 2$ and the small red intervals are the scaled copies of $I_1^\alpha$

**Definition 13** We say the percolation model satisfies **Condition A**($\alpha$) if there exist closed intervals $I_1^\alpha, I_2^\alpha \subset \Delta_\alpha$ and a positive integer $r_\alpha$ such that

(i) $I_1^\alpha \subset \text{int} I_2^\alpha$, $I_2^\alpha \subset \text{int}\Delta_\alpha$,
(ii) $F_\alpha^{r_\alpha} \mathbf{1}_{I_1^\alpha} \geq 2 \cdot \mathbf{1}_{I_2^\alpha}$.

It will be convenient to use additional notation. For $x \in \Delta_\alpha$, $\alpha \in \mathcal{D}$, and $I \subset \Delta_\alpha$ we denote

$$D_n(x, I, \alpha) = \{\mathbf{i} \in \mathcal{I}^n; x \in \Pi_\alpha \circ \varphi_{\mathbf{i}}(I)\}.$$

That is, if we write $\ell^\alpha(x)$ for the line segment through $x \in \Delta_\alpha$ in direction $\alpha$, $D_n(x, I, \alpha)$ is the set of $\mathbf{i}$ for which $\ell^\alpha(x)$ intersects $\varphi_{\mathbf{i}}(I)$.

The point ii) of Definition 13 can then be written as

$$\forall_{x \in I_2^\alpha} \sum_{\mathbf{i} \in D_{r_\alpha}(x, I_1^\alpha, \alpha)} p_{\mathbf{i}} \geq 2.$$

In other words, Condition $A(\alpha)$ is satisfied if for given $\alpha$ one can define 'small central' and 'large central' part of each square in such a way that for some $r \in \mathbb{N}$ if a line in direction $\alpha$ intersects the 'large central' part of some $n$-th level square then the expected number of 'small central' parts of its $n + r$-th level subsquares it intersects is uniformly greater than 1 (Fig. 6).

## 5.3 Consequences of Condition $A(\alpha)$

It is clear that if $A(\alpha)$ holds then one can apply the large deviation argument for projection in direction $\alpha$-modulo a minor technical problem that the random variables in the large deviations theorem are not identically distributed.

A bit more complicated is the proof that almost surely all the projections contain intervals. It is based on the following robustness properties:

**Proposition 14** *If condition $A(\alpha)$ holds for some $\alpha \in \mathcal{D}$ for some $I_1^\alpha, I_2^\alpha$ and $r_\alpha$ then it will also hold in some neighbourhood $J \ni \alpha$. Moreover, for all $\theta \in J$ we can choose $I_1^\theta = I_1', I_2^\theta = I_2, r_\theta = r_\alpha$ not depending on $\theta$.*

A natural corollary is that the whole range $\mathcal{D}$ can be presented as a countable union of closed intervals $J_i = [\alpha_i^-, \alpha_i^+]$ such that for each $i$ Condition A($\alpha$) holds for all $\alpha \in J_i$ with the same $I_1^i, I_2^i, r_i$.

**Proposition 15** *Let $I \subset B(I, \ell) \subset J \subset \Delta_\alpha$. If $\mathbf{i} \in D_n(x, I, \alpha)$ then $\mathbf{i} \in D_n(x, J, \beta)$ for all $\beta \in (\alpha - \ell M^{-n}, \alpha + \ell M^{-n})$.*

Hence, inside each $J_i$ one does not need to repeat the large deviation argument separately for each $\alpha$. At level $n$ it is enough to check it for approximately $M^n$ directions. As the number of directions one needs to check grows only exponentially fast with $n$, the proof goes through.

## 5.4 Checking Condition $A(\alpha)$

One last thing needed is an efficient way to check whether $A(\alpha)$ holds.

**Definition 16** We say that the fractal percolation model satisfies **Condition B**($\alpha$) if there exists a nonnegative continuous function $f : \Delta_\alpha \to \mathbb{R}$ such that $f$ is strictly positive except at the endpoints of $\Delta_\alpha$ and that

$$F_\alpha f \geq (1 + \varepsilon) f \tag{5.1}$$

for some $\varepsilon > 0$.

**Proposition 17** *$B(\alpha)$ implies $A(\alpha)$.*

In particular, for homogeneous case $p_i = p > M^{-1}$ for any $\alpha$ one can choose $f(x)$ as the length of the intersection of $Q$ with the line in direction $\alpha$ passing through $x$. It is easy to check that this function satisfies (5.1) for $\varepsilon = pM - 1$.

### 5.5 Application: Visibility

For a given set $E$, we define the visible subset (from direction $\alpha$) as the set of points $x \in E$ such that the half-line starting at $x$ and going in direction $\alpha$ does not meet any other point $y \in E$. Similarly, given $z \in \mathbb{R}^2$, the visible subset (from $z$) is the set of points $x \in E$ such that the interval $\overline{xz}$ does not meet any other point $y \in E$.

Let $E$ be a homogeneous fractal percolation with $p > M^{-1}$. By Theorem 9, $E$ is quite opaque: the orthogonal projection in any direction almost surely contain intervals. In particular, with large probability it contains large intervals. By stochastical self-similarity of $E$, the same is true for each $E \cap Q_{\mathbf{i}}$. Hence, not many points can be visible:

**Theorem 18** ([Ar12]) *If $E$ is nonempty, almost surely the visible set from direction $\alpha$ has finite one-dimensional Hausdorff measure for each $\alpha$ and the visible set from point $z$ has Hausdorff dimension 1 for each $z \in \mathbb{R}^2$.*

## 6 General Projections: The Transparent Case

In this section we present results analogous to the second part of the Marstrand theorem. For homogeneous fractal percolation with Hausdorff dimension smaller than 1 almost surely $\dim_{\mathrm{H}}(\mathrm{proj}_\alpha(E)) = \dim_{\mathrm{H}} E$ for all $\alpha$. Together with the results of the previous section, it implies

**Theorem 19** ([RS00]) *In the homogeneous case, that is $E = E^h(2, M, p)$ for almost all realizations of $E$*

$$\forall\alpha,\ \dim_{\mathrm{H}}(\mathrm{proj}_\alpha E) = \min\{1, \dim_{\mathrm{H}}(E)\}. \tag{6.1}$$

**Principal Assumption for this Section:** In this section we always work in the homogeneous case:

$$E = E^h(2, M, p),$$

where

$$M^{-2} < p \le M^{-1}. \tag{6.2}$$

That is $p$ is chosen to ensure that $E \neq \emptyset$ with positive probability and $\dim_{\mathrm{H}}(E) \le 1$ almost surely conditioned on non-extinction. To prove Theorem 19 one needs to analyze the structure of the slices of $E_n$:

**Informal description of the structure of slices of $\mathbf{E_n}$** (which was defined as the $n$-th approximation of $E$): Namely, for almost all realizations of $E$ and for **all** straight lines $\ell$: the number of level $n$ squares having nonempty intersection with $E$ is at most const $\cdot\, n$. On the other hand, almost surely for $n$ big enough, we can find some line of 45° angle which intersects const $\cdot\, n$ level $n$ squares.

Let $\mathcal{L}^{\varepsilon}$ be the set of lines on the plane whose angle is separated both from 0° and 90° at least by $\varepsilon$. Further for a line $\ell$ let $\mathcal{E}_n(\ell)$ be the set of retained level $n$ squares that intersect $\ell$. That is,

$$\mathcal{E}_n(\ell) := \{\mathbf{i} \in \mathcal{E}_n : Q_{\mathbf{i}} \cap \ell \neq \emptyset\} .$$

**Theorem 20** ([RS00]) *For almost all realizations of $E$ we have*

$$\forall \varepsilon \in \left(0, \frac{\pi}{2}\right), \ \exists N, \ \forall n \geq N, \ \forall \ell \in \mathfrak{L}^{\varepsilon}; \quad \#\mathcal{E}_n(\ell) \leq \text{const} \cdot n. \tag{6.3}$$

For simplicity, the proof in horizontal/vertical direction only (for general directions one needs to apply techniques presented in previous subsection). The proof is once again based on the large deviation argument, but working in the opposite direction. This time the expected number of squares in a subcolumn is smaller (by a constant bounded away from 1) than the number of squares in the column (and not greater, like in the opaque case). Hence, we can guarantee that if the column has sufficiently many squares for the large deviation theory to work, the number of squares in all subcolumns will shrink. This leads to an estimation on the possible rate of growth.

This estimation is sharp:

**Proposition 21** ([RS00]) *There exists a constant $0 < \lambda < 1$ such that for almost all realizations, conditioned on $E \neq \emptyset$, there exists an $N$ such that for all $n > N$ there exists a line $\ell$ with*

$$\#\mathcal{E}_n(\ell) > \lambda n. \tag{6.4}$$

Theorem 19 is an immediate consequence of Theorem 20.

# 7 The Arithmetic Sum of at Least Three Fractal Percolations

To study arithmetic sums of more than two fractal percolations we need to combine results of the previous three sections. Like in Sect. 4, we look at the projection $(x_1, \ldots, x_d) \to \sum x_i$ from the cartesian product of fractal percolations to the real line. The proof is based on the large deviation argument presented in Sect. 5. However, the main technical difficulty is the presence of dependencies. We will use the results from Sect. 6 to bound their impact.

Let

$$E^i := E^h(1, M, p_i), \ i = 1, 2, 3, \quad p := p_1 \cdot p_2 \cdot p_3 \text{ and } E := E^h(3, M, p).$$

Then

$$\dim_{\mathrm{H}}\left(E^1 \times E^2 \times E^3\right) = \dim_{\mathrm{H}}(E) = \frac{\log M^3 \cdot p}{\log M}.$$

Moreover, the probability that a level $n$ cube $C$ is contained in any of the two random Cantor sets above is equal to $p^n$.

Let $S_a$ be the plane $\{\sum x_i = a\}$. We can write

$$E^{\text{sum}} := E^1 + E^2 + E^3 = \left\{a : S_a \cap \left(E^1 \times E^2 \times E^3\right) \neq \emptyset\right\}.$$

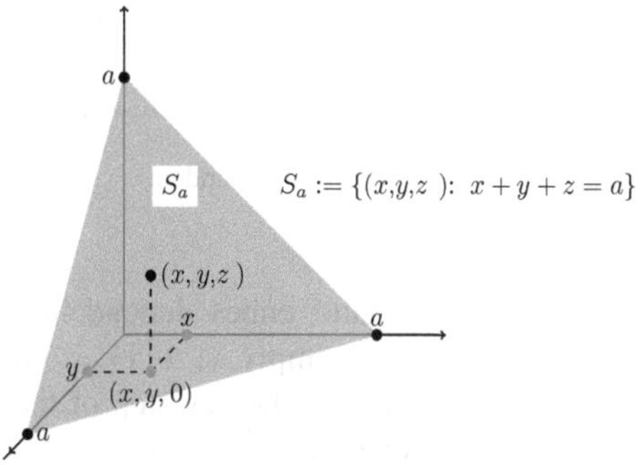


That is we can consider $E^{\text{sum}}$ as the projection of $E_1 \times E_2 \times E_3$ to the $x$-axis with planes orthogonal the vector $(1, 1, 1)$. So, $E^{\text{sum}}$ can contain an interval only if its dimension is greater than one, that is $p > M^{-2}$. It is a sufficient condition as well:

**Theorem 22** ([RS00]) *Let $d \geq 2$ and for $i = 1, \ldots, d$ let $E^i := E^h(1, M, p_i)$ satisfying*

$$p := \prod_{i=1}^{d} p_i > M^{-d+1}. \tag{7.1}$$

*Then for every $\mathbf{b} = (b_1, \ldots, b_d) \in \mathbb{R}^d$, $b_i \neq 0$ for all $i = 1, \ldots, d$ the sum $E_{\mathbf{b}}^{\text{sum}} = \sum_{i=1}^{d} b_i E^i$ contains an interval almost surely, conditioned on all $E^i$ being nonempty.*

We explain the proof of this theorem in the special case when $d = 3$ and $\mathbf{b} = (1, 1, 1)$. To verify that a certain $a \in E^{\text{sum}}$ we need to prove that the $n$ approximation of the product intersects $S_a$, that is $(E^1 \times E^2 \times E^3)_n \cap S_a \neq \emptyset$ for every $n$. It follows from the dimension formula and (7.1) that we have $M^{n(1+\tau)}$ retained level $n$ cubes for some $\tau > 0$. By the pigeon hole principle for at least one $k = 0, \ldots, 3M^n$ the plane $S_{kM^{-n}}$ intersects at least $M^{n\tau}$ retained level $n$ cubes. For such a $k$ we write $a = kM^{-n}$. So, $\#\{\mathcal{E}_n \cap S_a\} \geq M^{n\tau}$.

Fix an $0 \leq m \leq M$. How many level $n+1$ retained cubes intersect $S_{a+mM^{-(n+1)}}$? If the way $E^1 \times E^2 \times E^3$ develops in every level $n$ cube was independent then we could get that the answer by the large deviation argument: exponentially many except for an event with a super exponentially small probability.

We remind that the cubes are dependent if they have the same $x_1, x_2$ or $x_3$ coordinate. Figure 7 shows the geometric position of (some of: we consider only

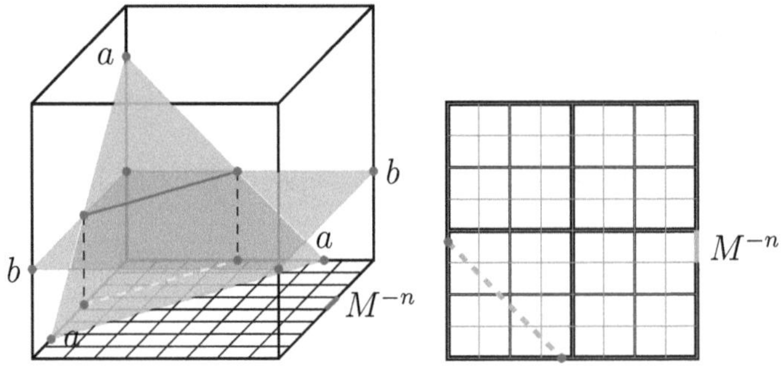


**Fig. 7** The cubes intersecting the *red line* are not independent

the cubes with the same $x_3$ coordinate) cubes dependent on one chosen cube: $x_1 + x_2 + x_3 = $ const and $x_3 = $ const imply $x_1 + x_2 = $ const. Potentially there could be exponentially many such cubes. The key step of the proof is that using a theorem analogous to Theorem 20 for $E^1 \times E^2$ instead of $E^h(2, M, p_1 \cdot p_2)$ one can check that on the red dashed line on Fig. 7 there are only constant times $n$ retained squares, consequently the $M^{n\tau}$ level $n$ cubes having non-empty intersection with $S_a$ (the blue plane on Fig. 7) can be divided into const $\cdot\, n$ classes such that the coordinate axes projection of any two cubes in a class are different. The events inside each class are independent, hence we can use the large deviation theory separately for each class. A technical comment: in order to be able to go with this procedure we may have to decrease $p_1, p_2, p_3$ in such a way that for the modified values we have

$$p_1 \cdot p_2 \cdot p_3 > M^{-2} \text{ but } p_i \cdot p_j < M^{-1} \text{ for distinct } i, j \in \{1, 2, 3\}.$$

That is, $E^1 \times E^2 \times E^3$ is a big set in the sense that it has dimension greater than one but its all coordinate plane projections should be small sets having dimension smaller than one—only then the $n$-th approximates of the coordinate plane projections intersect every line in at most const $\cdot\, n$ retained squares. However, the property of almost surely having intervals in the algebraic sum is monotonous with respect to $\{p_i\}$.

Hence among those level $n$ retained cubes that intersect the blue plane $S_a$ there cannot be more than const $\cdot\, n$ on the red line (any coordinate plane parallel line) which imply that the number of cubes dependent on any one cube is polynomial (const $\cdot\, n$). This bound on the dependency matrix lets us control the dependencies.

## References

[Ar12] Arhosalo, I., Järvenpää, E., Järvenpää, M., Rams, M., Shmerkin, P.: Visible parts of fractal percolation. Proc. Edinburgh Math. Soc. (Series 2), **55**(02):311–331 (2012)

[CCD88] Chayes, J.T., Chayes, L., Durrett, R.: Connectivity properties of mandelbrot's percolation process. Probability theory and related fields **77**(3), 307–324 (1988)

[Ch96] Chayes, L.: On the length of the shortest crossing in the super-critical phase of mandelbrot's percolation process. Stochastic processes and their applications **61**(1), 25–43 (1996)

[DG88] Dekking, F.M., Grimmett, G.R.: Superbranching processes and projections of random cantor sets. Probability theory and related fields **78**(3), 335–355 (1988)

[De09] Dekking, M.: Random cantor sets and their projections. Fractal Geom. Stochast. IV, 269–284 (2009)

[DM90] Dekking, M., Meester, R.W.J.: On the structure of mandelbrot's percolation process and other random cantor sets. J. Stat. Phys. **58**(5), 1109–1126 (1990)

[DS08] Dekking, M., Simon, K.: On the size of the algebraic difference of two random cantor sets. Random Struct. Algorithms **32**(2), 205–222 (2008)

[Fa86] Falconer, K.J.: Random fractals. Math. Proc. Cambridge Philos. Soc **100**(3), 559–582 (1986)

[FG92] Falconer, K.J., Grimmett, G.R.: On the geometry of random cantor sets and fractal percolation. J. Theor. Probab. **5**(3), 465–485 (1992)

[FG94] Falconer, K.J., Grimmett, G.R.: Correction: On the geometry of random cantor sets and fractal percolation. J. Theor. Probab. **7**(1), 209–210 (1994)

[Ha81] Hawkes, J.: Trees generated by a simple branching process. J. Lond. Math. Soc. **2**(2), 373–384 (1981)

[KP76] Kahane, J.-P., Peyriere, J.: Sur certaines martingales de benoit mandelbrot. Adv. math. **22**(2), 131–145 (1976)

[Ma74] Mandelbrot, B.B.: Intermittent turbulence in self-similar cascades- divergence of high moments and dimension of the carrier. J. Fluid Mech. **62**(2), 331–358 (1974)

[Ma83] Mandelbrot, B.B: The fractal geometry of nature/revised and enlarged edition. p. 495. WH Freeman and Co., New York (1983)

[Ma54] Marstrand, J.M.: Some fundamental geometrical properties of plane sets of fractional dimensions. Proc. Lond. Math. Soc. **3**(1), 257–302 (1954)

[MW86] Mauldin, R.D., Williams, S.C.: Random recursive constructions: asymptotic geometric and topological properties. Trans. Amer. Math. Soc **295**(1), 325–346 (1986)

[MSS09] Mora, P., Simon, K., Solomyak, B.: The lebesgue measure of the algebraic difference of two random cantor sets. Indagationes Mathematicae **20**(1), 131–149 (2009)

[RS00] Rams, M., Simon, K.: The dimension of projections of fractal percolations. J. Stat. Phys **154**(3), 633–655 (2014)

[RS00b] Rams, M., Simon, K.: Projections of fractal percolations. To appear in Ergodic Theor. Dyn.Syst. (2013). doi:10.1017/etds.2013.45

# Self-affine Sets and the Continuity of Subadditive Pressure

**Pablo Shmerkin**

**Abstract** The affinity dimension is a number associated to an iterated function system of affine maps, which is fundamental in the study of the fractal dimensions of self-affine sets. De-Jun Feng and the author recently solved a folklore open problem, by proving that the affinity dimension is a continuous function of the defining maps. The proof also yields the continuity of a topological pressure arising in the study of random matrix products. I survey the definition, motivation and main properties of the affinity dimension and the associated SVF topological pressure, and give a proof of their continuity in the special case of ambient dimension two.

**Keywords** Topological pressure · Self-affine sets · Affinity dimension · Subadditive thermodynamic formalism

**2010 Mathematics Subject Classification** Primary 37C45 · 37D35 · 37H15

## 1 Introduction

Let $\mathcal{F} = (f_1, \ldots, f_m)$ be a collection of contractive affine maps on some Euclidean space $\mathbb{R}^d$. That is, $f_i(x) = A_i x + t_i$, where $A_i \in \mathbb{R}^{d\times d}$ are linear maps, $t_i \in \mathbb{R}^d$ are translations, and $\|A_i\| < 1$, where $\|\cdot\|$ denotes Euclidean operator norm (although any other operator norm would work equally well). It is well known that there exists a unique nonempty compact set $E = E(\mathcal{F})$ such that

$$E = \bigcup_{i=1}^{m} f_i(E) = \bigcup_{i=1}^{m} A_i E + t_i. \tag{1.1}$$

The author was partially supported by a Leverhulme Early Career Fellowship.

P. Shmerkin (✉)
University of Surrey, Guildford, UK
e-mail: pshmerkin@utdt.edu

P. Shmerkin
Torcuato Di Tella University, Buenos Aires, Argentina



D.-J. Feng and K.-S. Lau (eds.), *Geometry and Analysis of Fractals*,
Springer Proceedings in Mathematics & Statistics 88,
DOI 10.1007/978-3-662-43920-3_12

Such sets are called *self-affine*. The tuple $\mathcal{F}$ is termed as an *iterated function system*, and $E$ is the *attractor* or *invariant set* of $\mathcal{F}$.

An important special case is that in which the maps $f_i$ are all similarities; in this case $E$ is known as a *self-similar set*. It is known that for self-similar sets, Hausdorff, lower and upper box counting dimensions all agree. Moreover, if $s$ is the only real solution to $\sum_{i=1}^{m} r_i^s = 1$, where $r_i$ is the similarity ratio of $f_i$, then $s$ is an upper bound for the Hausdorff dimension of $E$, and equals the Hausdorff dimension of $E$ under a number of "controlled overlapping" conditions, the strongest and simplest being the strong separation condition, which requires that the basic pieces $f_i(E)$ are mutually disjoint. The number $s$ is called the *similarity dimension* of the system $\mathcal{F}$, and is clearly continuous, and indeed real-analytic, as a function of the maps $f_i$ (identified with the Euclidean space of the appropriate dimension).

The situation is dramatically more complex for general self-affine sets. It is well-known that the Hausdorff and box counting dimensions of self-affine sets may differ, and that each of them is a discontinuous function of the defining maps, even under the strong separation condition. Strikingly, it is not even known whether lower and upper box dimensions always coincide for self-affine sets. No general formula for either the Hausdorff or box counting dimension is known or expected to exist, again even in the strongly separated case. However, although the dimension theory of self-affine sets may appear at first sight like a bleak subject, many interesting and deep results have been obtained. Among these, Falconer's seminal paper [Fal92] has been highly influential. There, Falconer introduces a number $s = s(\mathcal{F})$ associated to an affine IFS $\mathcal{F}$, which we will term *affinity dimension* (no standard terminology exists; the term singularity dimension is also often used). As a matter of fact, $s$ depends only on $A = (A_1, \dots, A_m)$, i.e. the linear parts of the affine maps $f_i$, and is independent of the translations.

Falconer proved that the affinity dimension is always an upper bound for the upper box-counting, and therefore the Hausdorff, dimension of $E$, and in some sense, "typically" equals the Hausdorff dimension of $E$; his result is described in more detail in Sect. 2.3 below. The definition of the affinity dimension is rather more involved than the definition of similarity dimension, which it extends, and is postponed to Sect. 2.3.

The question of whether the affinity dimension is continuous as a function of the generating maps has been a folklore open question in the fractal geometry community for well over a decade (I learned it from B. Solomyak around 2000), and was raised explicitly in [FS09]. Recently, together with Feng and Shmerkin [FS13] we proved that the answer is affirmative:

**Theorem 1.1** *The affinity dimension $s$ is a continuous function of the linear maps $(A_1, \dots, A_m)$.*

A related but in some sense simpler result concerns the norms of matrix products. Again let $A = (A_1, \dots, A_m)$ be a finite collection of invertible linear maps on $\mathbb{R}^d$. Given $s \geq 0$, define

$$M(A,s) = \lim_{n\to\infty} \frac{1}{n} \log \left( \sum_{(i_1\cdots i_k)} \|A_{i_k}\cdots A_{i_1}\|^s \right), \tag{1.2}$$

where $\|\cdot\|$ is the standard Euclidean norm (the limit is easily seen to exist from subadditivity). The reader familiar with the thermodynamic formalism may note that this definition resembles the definition of topological pressure of a continuous potential on the full shift on $m$ symbols, except that here we consider norms of matrix products instead of Birkhoff sums; this is an instance of the topological pressure in the setting of the *subadditive* thermodynamic formalism. This will be discussed in Sect. 3.1.

The quantity $M(A,s)$ is rather natural. On one hand, in the thermodynamic setting it is closely linked to the Lyapunov exponent of an IID random matrix product (with respect to ergodic measures under the shift). On the other hand, the "zero temperature limit" as $s \to \infty$ is the joint spectral radius of the matrices $(A_1,\ldots,A_m)$, which is an important quantity in a wide variety of fields. Although the joint spectral radius is well-known to be continuous, it is far from clear from the definition whether $M(A,s)$ is always continuous. Together with Feng [FS13], we have proved that it is:

**Theorem 1.2** *$M(A,s)$ is jointly continuous in $(A,s)$.*

Although Theorems 1.1 and 1.2 are in effect linear algebraic statements, the proofs make heavy use of dynamical systems theory, and in particular the variational principle for sub-additive potentials.

The goal of this survey is twofold. On one hand, it is an overview of the definition and main properties of the affinity dimension and the closely related singular value pressure, and the geometric reasons why it comes up naturally in the study of self-affine sets. On the other hand, it contains a full proof of Theorems 1.1 and 1.2 in the case of ambient dimension $d = 2$ (for $d = 1$, both results are trivial). The two-dimensional case captures many of the main ideas of the general case, while being technically much simpler.

I note that De-Jun Feng [private communication] has observed that a result of Bocker-Neto and Viana [BV10] on continuity of Lyapunov exponents for IID $\mathbb{R}^2$ matrix cocycles can be used to give a short alternative proof of Theorems 1.1 and 1.2 in the case $d = 2$. However, that proof does not generalize to any other dimensions.

## 2 SVF, Topological Pressure, and Affinity Dimension

### 2.1 Definition and Basic Properties of the SVF

Recall that given a linear map $A \in GL_d(\mathbb{R})$, its *singular values* $\alpha_1(A) \geq \cdots \geq \alpha_d(A) > 0$ are the lengths of the semi-axes of the ellipsoid $A(B^d)$, where $B^d$ is the unit ball of $\mathbb{R}^d$. Alternatively, the singular values are the square roots of the eigenvalues of $A^*A$ (where $A^*$ is the adjoint of $A$). In particular, $\alpha_1(A)$ is nothing else than the Euclidean norm of $A$:

$$\alpha_1(A) = \sup\{\|Av\| : \|v\| = 1\},$$

where $\|v\|$ denotes the Euclidean norm of $v \in \mathbb{R}^d$. Likewise,

$$\alpha_d(A) = \inf\{\|Av\| : \|v\| = 1\} = \|A^{-1}\|^{-1}.$$

Also,

$$\det(A) = \det(A^*A)^{1/2} = \prod_{i=1}^{d} \alpha_i(A).$$

Given $s \in [0, d)$, we define the *singular value function* (SVF) $\varphi^s : \mathrm{GL}_d(\mathbb{R}) \to (0, \infty)$ as follows. Let $m = \lfloor s \rfloor$. Then

$$\varphi^s(A) = \alpha_1(A) \cdots \alpha_m(A)\alpha_{m+1}(A)^{s-m}.$$

An alternative way of expressing this is:

$$\varphi^s(A) = \|A\|_m^{m+1-s} \cdot \|A\|_{m+1}^{s-m}, \tag{2.1}$$

where

$$\|A\|_k = \alpha_1(A) \cdots \alpha_k(A).$$

The reason why (2.1) is useful is that $\|A\|_k$ is a sub-multiplicative seminorm ($\|AB\|_k \le \|A\|_k \|B\|_k$). Indeed, $\|A\|_k$ is the operator norm of $A$ when acting on the space of exterior $k$-forms. Alternatively, $\|A\|_k = \sup\{\det(A|_\pi) : \pi \in G(d, k)\}$ where $G(d, k)$ is the Grassmanian of $k$-dimensional linear subspaces of $\mathbb{R}^d$. As an immediate consequence of (2.1), we get the following key property of the SVF:

**Lemma 2.1** (Sub-multiplicativity of the SVF) *$\varphi^s(AB) \le \varphi^s(A)\varphi^s(B)$ for all $A, B \in GL_d(\mathbb{R})$ and $s \in [0, d)$.*

It is also clear that $\varphi^s(A)$ is jointly continuous in $s$ and $A$; it is also jointly real-analytic for non-integer $s$, but in general it is not even differentiable at $s = 1, \ldots, d-1$ for a fixed $A$. We note also that $\varphi^1(A) = \alpha_1(A) = \|A\|$ and $\lim_{s \to d} \varphi^s(A) = \det(A)$. For completeness we define $\varphi^s(A) = \det(A)^{s/d}$ for $s \ge d$, and note that this definition preserves all of the previous properties when $s \ge d$.

## 2.2 SVF Topological Pressure

Let $I = \{1, \ldots, m\}$. We denote by $I^*$ the family of finite words with symbols in $I$, and write $|\mathtt{i}|$ for the length of $\mathtt{i} \in I^*$. The space $X := I^{\mathbb{N}}$ of right-infinite sequences is endowed with the left-shift operator $\sigma$, i.e. $\sigma(i_1 i_2 \cdots) = (i_2 i_3 \cdots)$. Given $\mathtt{i} \in X$, the restriction of $\mathtt{i}$ to its first $k$ coordinates is denoted by $\mathtt{i}|k$. Finally, if $\mathtt{j} \in I^*$, then $[\mathtt{j}] \subset X$ is the family of all infinite sequences which start with $\mathtt{j}$.

Given $A = (A_1, \ldots, A_m) \in (\mathrm{GL}_d(\mathbb{R}))^m$ and $\mathtt{i} = (i_1 \cdots i_n) \in I^*$, we denote $A(\mathtt{i}) = A_{i_n} \cdots A_{i_1}$. The next lemma introduces the main concept of this article.

**Lemma 2.2** *Given* $A \in (GL_d(\mathbb{R}))^m$ *and* $s \geq 0$, *let*

$$S_n(A, s) = \log \sum_{\mathtt{i} \in I^n} \varphi^s(A(\mathtt{i})).$$

*Then the limit*

$$P(A, s) := \lim_{n \to \infty} \frac{S_n(A, s)}{n} \tag{2.2}$$

*exists and equals* $\inf_{n \geq 1} S_n(A, s)/n > -\infty$.

*Proof* Lemma 2.1 implies that the sequence $S_n = S_n(A, s)$ is subadditive, i.e. $S_{n+k} \leq S_n + S_k$. But it is well known that for any subadditive sequence $S_n$, the limit of $S_n/n$ exists and equals $\inf_{n \geq 1} S_n/n$. Finally, since $\varphi^s(A) \geq \det(A)^s$ and we are assuming that the maps $A_i$ are invertible, one can easily check that

$$\lim_{n \to \infty} \frac{S_n}{n} \geq \log \left( \sum_{i \in I} \det(A(\mathtt{i}))^s \right) > -\infty.$$

□

**Definition 2.1** The function $P(A, s)$ defined in (2.2) is called the *SVF topological pressure*.

It is instructive to compare the definitions of $P(A, s)$ and $M(A, s)$ given in (1.2). Both quantities coincide for $0 \leq s \leq 1$; for $s > 1$, the definition of $P(A, s)$ takes into account different singular values of the matrix products involved. Since $\varphi^s(A) \leq \|A\|^s$, one always has $P(A, s) \leq M(A, s)$.

The following lemma summarizes some elementary but important continuity properties of the topological pressure.

**Lemma 2.3** *The following hold:*

(1) *Given* $A = (A_1, \ldots, A_m) \in (GL_d(\mathbb{R}))^m$, *let*

$$\alpha_* = \min_{i \in I}\{\alpha_d(A_i)\}, \quad \alpha^* = \max_{i \in I}\{\alpha_1(A_i)\}.$$

*Then*

$$(\log \alpha_*)\varepsilon \le P(A, s+\varepsilon) - P(A, s) \le (\log \alpha^*)\varepsilon.$$

(2) *For fixed* $A = (A_1, \ldots, A_m) \in (GL_d(\mathbb{R}))^m$*, the function* $s \to P(A, s)$ *is Lipschitz continuous; the Lipschitz constant is uniform in a neighborhood of* $A$*.*
(3) $P(A, s)$ *is upper semicontinuous (as a function of both* $A$ *and* $s$*).*

*Proof* Note that $\varphi^s(B)\alpha_d(B)^\varepsilon \le \varphi^{s+\varepsilon}(B) \le \varphi^s(B)\alpha_1(B)^\varepsilon$ for any $s, \varepsilon > 0$ and $B \in \mathrm{GL}_d(\mathbb{R})$. Also, $\alpha_1(B) = \|B\|$ is sub-multiplicative and $\alpha_d(B) = \|B^{-1}\|^{-1}$ is super-multiplicative. Combining these facts yields

$$n\varepsilon \log \alpha_* + S_n(A, s) \le S_n(A, s+\varepsilon) \le n\varepsilon \log \alpha^* + S_n(A, s),$$

which yields the first claim. The second claim is immediate from the first, and the fact that $\alpha^*, \alpha_*$ are continuous functions of $A$.

Finally, upper semicontinuity follows since $P(A, s) = \inf_{n \ge 1} S_n(A, s)/n$ is an infimum of continuous functions. □

In light of the previous lemma, it seems natural to ask whether $P(A, s)$ is not just upper semicontinuous but in fact continuous. As we will see in the next section, this question is closely linked to Theorem 1.1. Falconer and Sloan [FS09] proved continuity of $P$ at tuples of linear maps satisfying certain assumptions, and raised the general continuity problem. Feng and the author [FS13] recently proved that continuity always holds:

**Theorem 2.1** *The map* $(A, s) \to P(A, s)$ *is continuous on* $(GL_d(\mathbb{R}))^m \times [0, +\infty)$*.*

A proof of this theorem in dimension $d = 2$ will be presented in Sect. 4.

## 2.3 Affinity Dimension and Self-affine Sets

So far, no assumptions have been made on the maps $A_i$, other than invertibility. However, the motivation for the study of the SVF topological pressure came from the theory of self-affine sets, and in this context the maps $A_i$ are strict contractions.

**Lemma 2.4** *If* $A = (A_1, \ldots, A_m) \in (GL_d(\mathbb{R}))^m$ *and* $\|A_i\| < 1$ *for all* $i \in I$*, then* $s \to P(A, s)$ *is a continuous, strictly decreasing function of* $s$ *on* $[0, \infty)$*, and has a unique zero on* $(0, \infty)$*.*

*Proof* That $P(A, s)$ is continuous and strictly decreasing in $s$ follows immediately from Lemma 2.3, since $\alpha^* < 1$ when the maps are strict contractions. By definition $P(A, 0) = \log m > 0$. On the other hand,

$$P(A, s) \le \log\left(\sum_{i \in I} \varphi^s(A_i)\right) \to -\infty \quad \text{as } s \to \infty.$$

Hence $P(A, \cdot)$ has a unique zero. □

**Definition 2.2** Given $A = (A_1, \ldots, A_m) \in (\mathrm{GL}_d(\mathbb{R}))^m$ with $\|A_i\| < 1$ for all $i \in I$, its *affinity dimension* is the unique positive root $s$ of the pressure equation $P(A, s) = 0$.

Note that Theorem 1.1 is in fact an immediate corollary of Theorem 1.2.

In the rest of this section we indicate why this number is relevant in the study of self-affine sets; this material is by now standard. To begin, let us recall the definition of Hausdorff dimension in terms of Hausdorff content: $\dim_H(E) = \inf\{s : \mathcal{H}^s_\infty(E) = 0\}$, where

$$\mathcal{H}^s_\infty(E) = \inf\left\{\sum_i r_i^s : E \subset B(x_i, r_i)\right\}.$$

Now suppose $E = \bigcup_{i\in I} A_i E + t_i$ is the invariant set of the IFS $\{A_i x + t_i\}_{i\in I}$. Since the maps $f_i(x) = A_i x + t_i$ are strict contractions, for all large enough $R$ the closed ball $B_R$ of radius $R$ and center at the origin is mapped into its interior by all the maps $f_i$. Let $E_0 = B_R$ and define inductively $E_{k+1} = \bigcup_{i\in I} f_i(E_k)$. By our choice of $R$, it is easy to see inductively that $E_k$ is a decreasing sequence of compact sets; moreover, if we call $F = \bigcap_{k=0}^\infty E_k$, then one can check that $F = \bigcup_{i\in I} f_i(F)$. Thus $E = F$ by uniqueness.

The above discussion shows that $E$ is covered by $E_k$ for any $k$; moreover, by construction

$$E_k = \bigcup_{i\in I^k} f_{i_1}\cdots f_{i_k}(B_R) =: \bigcup_{i\in I^k} U_{i_1\cdots i_k},$$

where $U_{i_1\cdots i_k}$ is an ellipsoid with semi-axes $R\alpha_1(A_{i_1}\cdots A_{i_k}) \geq \cdots \geq R\alpha_d(A_{i_1}\cdots A_{i_k})$. This shows that there is a natural cover of the self-affine sets by *ellipsoids*. In order to estimate Hausdorff content (and hence Hausdorff dimension) effectively, one needs to find efficient coverings by *balls*. What Falconer observed is that we can cover each ellipsoid efficiently by balls, in a way that depends on the dimension of the Hausdorff content we are trying to estimate. Namely, for each integer $1 \leq m < d$, we can cover an ellipsoid in $\mathbb{R}^d$ with semi-axes $\alpha_1 \geq \cdots \geq \alpha_d$ by a parallelepiped with sides $2\alpha_1 \geq \cdots \geq 2\alpha_d$. In turn, this can be covered by at most

$$(4R\alpha_1/\alpha_m)\cdots(4R\alpha_{m-1}/\alpha_m)(4R)^{d-m+1}$$

cubes of side length $\alpha_m$, each of which is contained in a ball of radius $\sqrt{d}\alpha_m$. It turns out that if we want to estimate $\mathcal{H}^s_\infty(E)$ by covering each of the ellipsoids that make up $E_k$ in this way, independently of each other, the optimal choice of $m$ is $\lfloor s\rfloor$. This particular choice yields (after some straightforward calculations) a bound

$$\mathcal{H}^s_\infty(E) \leq C_{R,d} \sum_{(i_1,\ldots,i_k)\in I^k} \varphi^s(A_{i_1}\cdots A_{i_k}).$$

From here one can deduce that if $P(A,t) < 0$, then $\mathcal{H}^t_\infty(E) = 0$, whence $\dim_H(E) \leq t$. Letting $t \to s$, the affinity dimension, finally shows that $\dim_H(E) \leq s$.

The argument above can be modified to reveal that the affinity dimension is also an upper bound for the upper box counting dimension of $E$. Thus, we can say that the affinity dimension is a *candidate* to the (Hausdorff, or box-counting) dimension of a self-affine set, obtained by using the most natural coverings, and is always an upper bound for both the box-counting and Hausdorff dimension. In general, these natural coverings may be far from optimal. For example, many of the ellipsoids making up $E_k$ may overlap substantially or be aligned in such a way that it is far more efficient to cover them together rather than separately. Also, most of the cubes that we employed to cover each ellipsoid might not intersect $E$ at all. And indeed, it may happen that the Hausdorff dimension, and/or the box-counting dimension are strictly smaller than the affinity dimension; this is the case for many kinds of self-affine carpets, see e.g. [Ba07] and references therein. However, it is perhaps surprising that, as discovered by Falconer [Fal92], *typically* the Hausdorff and box-counting dimensions of self-affine sets do coincide with the affinity dimension, in the following precise way:

**Theorem 2.2** *Let $A = (A_1, \ldots, A_m)$, with the $A_i$ invertible linear maps on $\mathbb{R}^d$. Assume further that $\|A_i\| < 1/2$ for all $i \in I$. Given $t_1, \ldots, t_m \in \mathbb{R}^d$, denote by $E(t_1, \ldots, t_m)$ the self-affine set corresponding to the IFS $\{A_i x + t_i\}_{i \in I}$.*

*Then for Lebesgue-almost all $(t_1, \ldots, t_m) \in \mathbb{R}^{md}$, the Hasudorff dimension of $E(t_1, \ldots, t_m)$ equals the affinity dimension of $A$.*

We remark that Falconer proved the second part under the assumption $\|A_i\| < 1/3$. Solomyak [Sol98] later pointed out a modification in the proof that allows to replace $1/3$ by $1/2$. By an observation of Edgar [Edg92], the bound $1/2$ is optimal. Since Falconer's pioneering work, many advances have been obtained in this direction. A natural question is whether one can give *explicit* conditions under which the Hausdorff and/or box-counting dimensions equal the affinity dimension; this was achieved in [Fal92, HL95, KS09]. In a different direction, Falconer and Miao [FM08] provided a bound on the dimension of exceptional parameters $(t_1, \ldots, t_m)$. For other recent directions in the study of the dimension of self-affine sets, see the survey [Fal13].

# 3 Further Background

## *3.1 Subadditive Thermodynamic Formalism*

The topological pressure $P(\varphi)$ of a Hölder continuous potential $\varphi$ is a key component of the thermodynamic formalism, which in turn, as discovered by Bowen, is a formidable tool in the dimension theory of conformal dynamical systems. In the classical setting, the functional $P$ is continuous as a function of $\varphi$ in the appropriate topology.

It is well-known that the dimension theory of *non-conformal* dynamical systems is far more difficult, and the classical thermodynamic formalism is no longer the appropriate tool. Instead, starting with the insights of Barreira [Ba96] and Falconer [Fal88], a *sub-additive* thermodynamic formalism has been developed. Both the thermodynamic formalism itself and its application to the dimension of invariant sets and measures is far more difficult in the non-conformal case. The proofs of Theorems 2.1 and 1.2 depend crucially on this subadditive thermodynamic formalism, and hence we review the main elements in this section.

We limit ourselves to potentials defined on the full shift on $m$ symbols $X = I^{\mathbb{N}}$. Let $\Phi = \{\varphi_n\}_{n=1}^{\infty}$ be a collection of continuous real-valued maps on $X$. We say that $\Phi$ is *subadditive* if

$$\varphi_{k+n}(\mathtt{i}) \leq \varphi_k(\mathtt{i}) + \varphi_n(\sigma^k \mathtt{i}) \quad \text{for all } \mathtt{i} \in X. \tag{3.1}$$

Important examples of subadditive potentials, which will be relevant in the proofs of Theorems 2.1 and 1.2, are

$$\varphi_n(\mathtt{i}) = s \log \|A_{i_n} \cdots A_{i_1}\|, \tag{3.2}$$

$$\varphi_n(\mathtt{i}) = \log \varphi^s(A_{i_n} \cdots A_{i_1}). \tag{3.3}$$

We note that the order of the products is the reverse of the order usually considered in the IFS literature; the reason for this will become apparent later when we apply Oseledets' Theorem. Let $\mathcal{E}$ denote the set of probability measures ergodic and invariant under the shift $\sigma$. The thermodynamic formalism consists of three main pieces: the *entropy* $h_\mu$ of a measure $\mu \in \mathcal{E}$, the *topological pressure* $P(\Phi)$ of a subadditive potential $\Phi$, and the *energy* or *Lyapunov exponent* $E_\mu(\Phi)$ of $\Phi$ with respect to a measure $\mu \in \mathcal{E}$. These are defined as follows:

$$h_\mu = \lim_{n\to\infty} \frac{1}{n} \sum_{\mathtt{i}\in I^n} -\mu[\mathtt{i}] \log \mu[\mathtt{i}],$$

$$P(\Phi) = \lim_{n\to\infty} \frac{1}{n} \log \sum_{\mathtt{i}\in I^n} \sup_{\mathtt{j}\in\mathtt{i}} \varphi_n(\mathtt{j}),$$

$$E_\mu(\Phi) = \lim_{n\to\infty} \frac{1}{n} \log \int \varphi_n \, d\mu.$$

By standard subadditivity arguments, all the limits exist. Moreover, if $\mu \in \mathcal{E}$, then for $\mu$-almost all $\mathtt{i}$,

$$h_\mu = \lim_{n\to\infty} \frac{-\log \mu[\mathtt{i}|n]}{n}, \tag{3.4}$$

$$E_\mu(\Phi) = \lim_{n\to\infty} \frac{\log \varphi_n(\mathtt{i})}{n}. \tag{3.5}$$

The first equality is a particular case of the Shannon-McMillan-Breiman, while the second is a consequence of Kingman's subadditive ergodic theorem.

These quantities are related via the following *variational principle* due to Cao, Feng, and Huang [CF08]:

**Theorem 3.1** ([CF08], Theorem 1.1) *If $\Phi$ is a subadditive potential on $X$, then*

$$P(\Phi) = \sup\left\{h_\mu + E_\mu(\Phi) : \mu \in \mathcal{E}\right\}.$$

Particular cases of the above, under stronger assumptions on the potentials, were previously obtained by many authors, see for example [Kae04, Mum06, Ba10] and references therein.

It follows from the semicontinuity of the entropy with respect to the shift map that the supremum in Theorem 3.1 is in fact a maximum; measures which attain the supremum are known as *equilibrium measures* or *equilibrium states* (for the potential $\Phi$). The variational principle and the existence of equilibrium measures for the potentials given in (3.2) and (3.3) go back to [Kae04]. We remark that, unlike the classical setting, equilibrium measures do not need to be unique in the subadditive setting, not even in the locally constant case. Feng and Käenmäki [FK11] characterize all equilibrium measures for potentials of the form $\varphi_n(\mathtt{i}) = s \log \|A_{i_1} \cdots A_{i_n}\|$.

## 3.2 Oseledets' Multiplicative Ergodic Theorem

We recall a version of the Multiplicative Ergodic Theorem of Oseledets. For simplicity we state it only in dimension $d = 2$; see e.g. [Kre185, Theorem 5.7] for the full version.

**Theorem 3.2** *Let $B_1, \ldots, B_m \in GL_2(\mathbb{R})$, and for $\mathtt{i} \in X$ write*

$$B(\mathtt{i}, n) = B_{\mathtt{i}_n} B_{\mathtt{i}_{n-1}} \cdots B_{\mathtt{i}_1}.$$

*Further, let $\mu$ be a $\sigma$-invariant and ergodic measure on $X$. Then, one of the two following situations occur:*

(A) *(Equal Lyapunov exponents). There exists $\lambda \in \mathbb{R}$ such that for $\mu$-almost all $i$,*

$$\lim_{n\to\infty} \frac{\log |B(\mathtt{i}, n)v|}{n} = \lambda \quad \textit{uniformly for } |v| = 1.$$

(B) *(Distinct Lyapunov exponents). There exist $\lambda_1 > \lambda_2$ and measurable families $\{E_1(\mathtt{i})\}, \{E_2(\mathtt{i})\}$ of one-dimensional subspaces such that for $\mu$-almost all $\mathtt{i}$:*

(a) $\mathbb{R}^2 = E_1(\mathtt{i}) \oplus E_2(\mathtt{i})$.
(b) $B_{i_1} E_j(\mathtt{i}) = E_j(\sigma\mathtt{i})$, $j = 1, 2$.

(c) *For all* $v \in E_j(\mathtt{i}) \setminus \{0\}$, $j = 1, 2$,

$$\lim_{n\to\infty} \frac{\log|B(\mathtt{i}, n)v|}{n} = \lambda_j.$$

## 3.3 The Cone Condition

The proof of Theorems 1.2 and 2.1 rely on finding a subsystem (after iteration) which is better behaved than the original one and captures almost all of its topological pressure. In the case of distinct Lyapunov exponents (with respect to a measure chosen from an application of the variatonal principle), the good behavior of this subsystem will consist in satisfying the (strict) cone condition: all the maps will send some fixed cone into its interior (except for the origin). Recall that a *cone* $K \subset \mathbb{R}^d$ is a closed set such that $K \cap -K = \{0\}$ and $tx \in K$ whenever $t > 0, x \in K$ (here $-K = \{-x : x \in K\}$).

This kind of cone condition is ubiquitous in the study of dynamical systems and associated matrix cocycles. In our situation, its usefulness will be derived from the following lemma.

**Lemma 3.1** *Let* $K', K \subset \mathbb{R}^d$ *be cones such that* $K' \setminus \{0\} \subset$ *interior*$(K)$*. There exists a constant* $c > 0$ *(depending on the cones) such that*

$$\|A\| \geq c\,\frac{\|Aw\|}{\|w\|} \tag{3.6}$$

*for all* $w \in K$ *and all* $A \in \mathbb{R}^{d\times d}$ *such that* $AK \subset K'$.

*In particular, there is* $c' > 0$ *such that if* $A_1, A_2 \in \mathbb{R}^{d\times d}$ *are such that* $A_j K \subset (K' \cup -K')$, $j = 1, 2$, *then*

$$\|A_1 A_2\| \geq c' \|A_1\| \|A_2\|.$$

*Proof* Suppose (3.6) does not hold. Then, for all $n$ we can find a linear map $A_n$ of norm 1 with $A_n(K) \subset K'$, and $w_n \in K'$ also of norm 1, such that $\|A_n w_n\| < 1/n$. By compactness, this implies that there are a linear map $A$ on $V$ of norm 1 (in particular nonzero) such that $A(K) \subset K'$, and a vector $w \in K'$ such that $Aw = 0$. Now pick $u \in K$ such that $Au \neq 0$ and $w - u \in K$; this is possible since $K' \setminus \{0\} \subset \text{interior}(K)$. It follows that $A(w - u) = -Au \in -K'$, whence $Au \in K' \cap -K'$, contradicting that $K'$ is a cone.

For the second claim, we may assume (replacing $A_j$ by $-A_j$ if needed) that $A_j K \subset K'$ for $j = 1, 2$. The claim now follows from the first one, since for fixed $w \in K'$ of norm 1,

$$\|A_1A_2\| \geq c\|A_1(A_2w)\| \geq c^2\|A_1\|\|A_2w\| \geq c^3\|A_1\|\|A_2\|.$$

□

A tuple $A = (A_1, \ldots, A_m)$ is said to satisfy the *cone condition* if there exist cones $K', K$ such that $K' \setminus \{0\}$ is contained in the interior of $K$, and $A_i K \subset (K' \cup -K')$ for all $i \in I$. The relevance of this condition can be seen from the following lemma.

**Lemma 3.2** *If $A$ satisfies the cone condition, then $M$ is continuous on $\mathcal{U} \times [0, +\infty)$ for some neighborhood $\mathcal{U}$ of $A$, and the same holds for $P$ if $d = 2$.*

*Proof* We know from Lemma 2.3 that $P$ is upper semicontinuous and Lipschitz continuous in $s$, with the Lipschitz constant locally uniformly bounded. The same arguments show that the same is true for $M$. Hence the task is to prove the claim with "lower continuous" in place of "continuous", with the value of $s$ fixed.

A trivial but key observation is that the cone condition is robust, in the following sense: if $A = (A_1, \ldots, A_m)$ satisfies the cone condition with cones $K, K'$, then there are a neighborhood $\mathcal{U}$ of $A$ and cones $\widetilde{K}, \widetilde{K}'$ such that any $B \in \mathcal{U}$ satisfies the cone condition with cones $\widetilde{K}, \widetilde{K}'$. In particular, applying Lemma 3.1 we find that there exists a constant $c = c(\mathcal{U}) \in (0, 1)$, such that if $B = (B_1, \ldots, B_m) \in \mathcal{U}$, then

$$\|B_{\mathtt{i}} B_{\mathtt{j}}\| \geq c\, \|B_{\mathtt{i}}\| \|B_{\mathtt{j}}\| \quad \text{for all } \mathtt{i}, \mathtt{j} \in I^*. \tag{3.7}$$

Now, for this constant $c$, let

$$\widetilde{S}_n(B, s) = c \sum_{\mathtt{i} \in I^n} \|B_{\mathtt{i}}\|^s,$$

and observe that if $B \in \mathcal{U}$ then, thanks to (3.7), $\widetilde{S}_{n+k}(B, s) \geq \widetilde{S}_n(B, s)\widetilde{S}_k(B, s)$. Therefore, for $B \in \mathcal{U}$,

$$M(B, s) = \lim_{n \to \infty} \frac{1}{n} \log \widetilde{S}_n(B, s) = \sup_n \frac{1}{n} \widetilde{S}_n(B, s).$$

Since a supremum of continuous functions is lower semicontinuous, this yields the claim for $M$.

Suppose now $d = 2$. Since $\alpha_2(B) = \|B^{-1}\|^{-1}$ for $B \in \mathrm{GL}_2(\mathbb{R})$, we have that $\alpha_2(B_1B_2) \geq \alpha_2(B_1)\alpha_2(B_2)$ for any $B_1, B_2 \in \mathrm{GL}_2(\mathbb{R})$. Let $\widetilde{K}, \widetilde{K}', \mathcal{U}, c$ be as before. Then

$$\varphi^s(B_{\mathtt{i}} B_{\mathtt{j}}) \geq c\, \varphi^s(B_{\mathtt{i}})\varphi^s(B_{\mathtt{j}}) \quad \text{for all } \mathtt{i}, \mathtt{j} \in I^*.$$

Thus, arguing as before,

$$P(B, s) = \sup_n \frac{1}{n} \log\left(\sum_{\mathtt{i}\in I^n} c\,\varphi^s(B_{\mathtt{i}})\right),$$

which is lower semicontinuous as a supremum of continuous functions. □

Although we will not use this result directly, the ideas in its proof will arise in our proof of continuity of $M$ and $P$ in dimension $d = 2$.

# 4 Proof of the Continuity of Subadditive Pressure in $\mathbb{R}^2$

## *4.1 General Strategy and the Case of Equal Lyapunov Exponents*

In this section we prove Theorems 1.2 and 2.1 in dimension $d = 2$ (recall that Theorem 1.1 is an immediate corollary of Theorem 2.1). We are going to give the details of the continuity of $P(A, s)$; the proof for $M(A, s)$ is essentially identical. We have already observed that $P$ is upper semicontinuous, hence it is enough to prove it is lower continuous. Moreover, by the second part of Lemma 2.3, it is enough to prove continuity in $A$ for a fixed value of $s$.

Fix $\varepsilon > 0$ for the course of the proof. Consider the potential $\Phi = \{\varphi_n\}$ where $\varphi_n(\mathtt{i}) = \varphi^s(A(\mathtt{i}, n))$ (this is the potential given in (3.3)). Thanks to the variational principle for subadditive potentials (Theorem 3.1), we know that there exists an ergodic measure $\mu$ on $X$, such that

$$h_\mu + E_\mu(\Phi) \geq P(\Phi) - \varepsilon = P(A, s) - \varepsilon. \tag{4.1}$$

(In fact, by the remark after Theorem 3.1, we can take $\varepsilon = 0$ in the above, but we do not need this). The potential $\Phi$ and the measure $\mu$ will remain fixed for the rest of the proof; we underline that they depend on $s$ and $A$.

We apply Oseledets' Theorem (Theorem 3.2) to the the matrices $(A_1, \ldots, A_m)$ and the measure $\mu$. The proof splits depending on whether the resulting Lyapunov exponents are equal or distinct. However, in both cases we will rely on the general scheme given in the next lemma.

**Lemma 4.1** *Suppose there are $n = n(\varepsilon)$, $Y_n \subset I^n$, and a neighborhood $\mathcal{U}$ of $A$ such that the following hold:*

(1) $\log|Y_n| \geq n(h_\mu - \varrho_1(\varepsilon))$,

(2) *If $\mathtt{i}$ is a juxtaposition of $k$ words from $Y_n$, and $B \in \mathcal{U}$, then*

$$\log \varphi^s(B_{\mathtt{i}}) \geq nk(E_\mu(\Phi) - \varrho_2(\varepsilon)).$$

*Then $P(B, s) \geq P(A, s) - \varepsilon - \varrho_1(\varepsilon) - \varrho_2(\varepsilon)$ for all $B \in \mathcal{U}$.*

*Proof* Let $Y_n^k$ denote the family of juxtapositions of $k$ words from $Y_n$. If $B \in \mathcal{U}$, then

$$
\begin{aligned}
P(B,s) &\geq \limsup_{k\to\infty} \frac{1}{nk} \log \sum_{\mathtt{i}\in Y_n^k} \varphi^s(B_{\mathtt{i}}) \\
&\geq \lim_{k\to\infty} \frac{1}{nk}\left( k\log|Y_n| + \min_{\mathtt{i}\in Y_n^k} \log\varphi^s(B_{\mathtt{i}})\right) \\
&\geq h_\mu + E_\mu(\Phi) - \varrho_1(\varepsilon) - \varrho_2(\varepsilon) \\
&\geq P(A,s) - \varepsilon - \varrho_1(\varepsilon) - \varrho_2(\varepsilon),
\end{aligned}
$$

where in the last line we have used (4.1). □

In practice we will take $\varrho_i(\varepsilon)$ to be a multiple of $\varepsilon$, so that in the limit as $\varepsilon \to 0$ we obtain the required lower semicontinuity. Finding a set $Y_n$ such that (4.1) holds is not difficult, and likewise if we also require (4.1) only for $\mathtt{i} \in Y_n$ (rather than $Y_n^k$). The challenge is to make (4.1) stable under compositions of the maps $B_{\mathtt{j}}$, $\mathtt{j} \in Y_n$ as well, and for this we will need geometric and ergodic-theoretic arguments depending on Oseledets' Theorem.

First we deal with the simpler case in which the Lyapunov exponents are equal; the case of different exponents is addressed in the next subsection.

Suppose then that there is a single Lyapunov exponent $\lambda$. It follows easily from (3.5) and Theorem 3.2 that

$$
\lim_{n\to\infty} \frac{1}{n} \int \log\varphi^s(A(\mathtt{i},n))d\mu(\mathtt{i}) = s\lambda. \tag{4.2}
$$

By Theorem 3.2, (3.4) and Egorov's Theorem, we can find a set $Y \subset X$ such that $\mu(Y) \geq 1/2$, and $n_0 \in \mathbb{N}$ such that if $\mathtt{i} \in Y$ and $n \geq n_0$ then

$$
\begin{aligned}
|A(\mathtt{i},n)v| &\geq e^{\lambda n} e^{-\varepsilon n}|v| \quad \text{for all } v \in \mathbb{R}^2 \setminus \{0\}, \\
\mu[\mathtt{i}|n] &\leq e^{-n(h_\mu-\varepsilon)}.
\end{aligned} \tag{4.3}
$$

Fix some $n \geq n_0$ and write $Y_n = \{i|n : \mathtt{i} \in Y\}$. Note that

$$
\frac{1}{2} \leq \sum_{\mathtt{j}\in Y_n} \mu[\mathtt{j}] \leq |Y_n| e^{-n(h_\mu-\varepsilon)},
$$

whence (if $n$ is large enough)

$$
|Y_n| \geq e^{n(h_\mu-2\varepsilon)}.
$$

On the other hand, since $Y_n$ is finite, we can find a neighborhood $\mathcal{U}$ of $A$ such that if $B = (B_1, \ldots, B_m) \in \mathcal{U}$, then

$$
|B_{\mathtt{j}} v| \geq e^{n\lambda} e^{-2\varepsilon n}|v| \quad \text{for all } v \in \mathbb{R}^2 \setminus \{0\}, \mathtt{j} \in Y_n.
$$

Therefore

$$|B_{\mathtt{i}}v| \geq e^{kn\lambda}e^{-2\varepsilon kn}|v| \quad \text{for all } v \in \mathbb{R}^2 \setminus \{0\},\ \mathtt{i} \in Y_n^k,$$

where $Y_n^k \subset I^{kn}$ is the set of all juxtapositions of $k$ words from $Y_n$. In particular,

$$\varphi^s(B_{\mathtt{i}}) \geq e^{kns\lambda}e^{-2\varepsilon kns} \quad \text{for all } \mathtt{i} \in Y_n^k.$$

We have therefore established the hypotheses of Lemma 4.1, with $\varrho_1(\varepsilon) = 2\varepsilon$ and $\varrho_2(\varepsilon) = 2s\varepsilon$. Applying that lemma and letting $\varepsilon \to 0$ establishes lower semicontinuity when the Lyapunov exponents are equal.

## *4.2 The Case of Distinct Lyapunov Exponents*

Suppose now that the Lyapunov exponents are $\lambda_1 > \lambda_2$. We will again construct sets $Y_n$ so that we can apply Lemma 4.1; this is trickier in this case, and the main idea is to use Oseledet's Theorem, Egorov's Theorem and recurrence, to find sets $Y_n$ (for $n$ arbitrarily large) so that the hypotheses of Lemma 4.1 hold when $k = 1$, *and in addition* $\{A_{\mathtt{i}} : \mathtt{i} \in Y_n\}$ *satisfies the cone condition.* The cone condition will allow us to pass to a neighborhood of $A$ first, and to iterates of the $B_{\mathtt{i}}$, $\mathtt{i} \in Y_n$, later.

Recall that for $\mu$-almost all $\mathtt{i}$ there is an Oseledets splitting $\mathbb{R}^2 = E_1(\mathtt{i}) \oplus E_2(\mathtt{i})$. The family of splittings $\mathbb{R}^2 = E_1 \oplus E_2$ has a natural separable metrizable topology; for example, we can take $d(E_1 \oplus E_2, E_1' \oplus E_2') = \max(\angle(E_1, E_1'), \angle(E_2, E_2'))$, where $\angle$ is the angle between two lines. We can then find a fixed splitting $\mathbb{R}^2 = F_1 \oplus F_2$ which is in the support of the push-forward of $\mu$ under the Oseledets splitting or, in other words,

$$\mu(\mathtt{i} : d(E_1(\mathtt{i}) \oplus E_2(\mathtt{i}), F_1 \oplus F_2) < \eta) > 0 \quad \text{for all } \eta > 0.$$

We write $F_i^\gamma = \{E : \angle(E, F_i) < \gamma\}$.

**Lemma 4.2** *There are $R, \eta > 0$ and two cones $K', K \subset \mathbb{R}^2$ with $K' \setminus \{0\} \subset K$, such that the following holds. Suppose that $A \in GL_2(\mathbb{R})$ is such that $Av_j \in F_j^\eta$ for some $v_j \in F_j^\eta$ of unit norm, $j = 1, 2$, and moreover $|Av_1| > R|Av_2|$. Then $AK \subset (K' \cup -K')$.*

*Proof* The lemma is essentially a consequence of compactness. Let $v$ be a unit vector in $F_1$, and let $K, K'$ be any cones such that

$$v \in \text{interior}(K') \setminus \{0\} \subset K' \setminus \{0\} \subset K \subset \mathbb{R}^2 \setminus F_2.$$

Suppose the claim fails with this choice of cones. Then for each $n$ there are $A_n \in \mathrm{GL}_d(2)$ and $v_{n,j} \in F_j^{1/n}$ such that

$$1 = |A_n v_{n,1}| \geq n|A_n v_{n,2}|, \tag{4.4}$$

and $A_n K \not\subset K' \cup -K'$. By passing to a subsequence (and replacing $v_{n,1}$ by $-v_{n,1}$ whenever needed), we may assume that $v_{n,1} \to v$, $v_{n,2} \to w$ and $A_n \to A$ for some $w \in F_2$ of unit norm, and $A \in \mathbb{R}^{2\times 2}$. Moreover, there is $z \in K$ of unit norm such that $Az \notin \text{interior}(K' \cup -K')$. However, (4.4) implies that $Az$ is a non-zero multiple of $v$ for any $z \notin F_2$, which contradicts our choice of cones. This contradiction finishes the proof of the lemma. □

From now on let $R$, $\eta$, $K$, $K'$ be as in the statement of the Lemma. By our choice of $F_1$, $F_2$, we have $\mu(\Delta) > 0$, where

$$\Delta = \{\mathtt{i} \in X : E_j(\mathtt{i}) \in F_j^\eta, \quad j = 1, 2\}.$$

At this point we recall the following quantitative version of Poincaré recurrence due to Khintchine, see [Pet89, Theorem 3.3] for a proof. Although it applies to any set of positive measure in a measure-preserving system, we state only the special case we will require.

**Lemma 4.3** *For every $\delta > 0$, the set $\{n : \mu(\sigma^{-n}\Delta \cap \Delta) > \mu(\Delta)^2 - \delta\}$ is infinite (and it even has bounded gaps).*

In particular, if we set $\kappa := \mu(\Delta)^2/2 > 0$, then the set $S := \{n : \mu(\sigma^{-n}\Delta \cap \Delta) > \kappa\}$ is infinite. On the other hand, by (3.4), (3.5), the last part of Oseledets' Theorem, and Egorov's Theorem, we may find $n_0 \in \mathbb{N}$ and a set $\Sigma \subset X$ with $\mu(\Sigma) > 1 - \kappa/2$, such that if $n \geq 0$ and $\mathtt{i} \in \Sigma$, then:

$$\mu[\mathtt{i}|n] \leq e^{-n(h_\mu - \varepsilon)}, \tag{4.5}$$
$$\log \varphi^s(A(\mathtt{i}, n)) \geq n(E_\mu(\Phi) - \varepsilon), \tag{4.6}$$
$$|A(\mathtt{i}, n)\widehat{E}_1(\mathtt{i})| \geq R\,|A(\mathtt{i}, n)\widehat{E}_2(\mathtt{i})|,$$

where $\widehat{E}_j(\mathtt{i})$ is a unit vector in $E_j(\mathtt{i})$. This is the point where we use that the Lyapunov exponents are different.

Taking stock, we have seen that if $n \geq n_0$, and $\mathtt{i} \in \Delta \cap \sigma^{-n}\Delta \cap \Sigma$ then, by Lemma 4.2, the map $A(\mathtt{i}, n)$ satisfies

$$A(\mathtt{i}, n)K \subset (K' \cup -K').$$

Hence for $n \in S \cap [n_0, \infty)$, we define $Y_n = \{\mathtt{i}|n : \mathtt{i} \in \Delta \cap \sigma^{-n}\Delta \cap \Sigma\}$. We will show that these sets meet the conditions of Lemma 4.1, with suitable functions $\varrho_j(\varepsilon)$. Firstly, note that

$$\begin{aligned}\kappa/2 &\leq \mu(\Delta \cap \sigma^{-n}\Delta \cap \Sigma)\\ &\leq \sum_{\mathtt{j}\in Y_n} \mu[\mathtt{j}]\\ &\leq |Y_n| e^{-n(h_\mu - \varepsilon)}.\end{aligned}$$

Hence $\log|Y_n| \geq n(h_\mu - 2\varepsilon)$, provided $n$ is taken large enough that $e^{-\varepsilon n} < \kappa/2$.

On the other hand, $\{A_{\mathtt{j}} : \mathtt{j} \in Y_n\}$ satisfies the cone condition with cones $K, K'$ (these cones are independent of $n$). Then there are a neighborhood $\mathcal{U}$ of $A$ in $(\mathrm{GL}_d(\mathbb{R}))^m$ and cones $\widetilde{K}, \widetilde{K'}$ such that if $B \in \mathcal{U}$, then $\{B_{j_n} \cdots B_{j_1} : (j_1 \dots j_n) \in Y_n\}$ satisfies the cone condition with cones $\widetilde{K}, \widetilde{K'}$. In particular, by Lemma 3.1, there exists $c > 0$ (depending only on $\mathcal{U}$, and not on $n$) such that

$$\|B_{\mathtt{i}}\| \geq c^{k-1}\left(\min_{\mathtt{j}\in Y_n} \|B_{\mathtt{j}}\|\right)^k \quad \text{for all } \mathtt{i} \in Y_n^k.$$

Arguing as in the proof of Lemma 3.2,

$$\varphi^s(B_{\mathtt{i}}) \geq c^{k-1}\left(\min_{\mathtt{j}\in Y_n} \varphi^s(B_{\mathtt{j}})\right)^k \quad \text{for all } \mathtt{i} \in Y_n^k.$$

By taking $n$ large enough, we may assume that $\log c/n > -\varepsilon$. Furthermore, in light of (4.6) we may find a neighborhood $\mathcal{V} \subset \mathcal{U}$ containing $A$, such that if $B \in \mathcal{V}$ and $\mathtt{j} \in Y_n$, then $\log \varphi^s(B_{\mathtt{j}}) > n(E_\mu(\Phi) - 2\varepsilon)$. We conclude that if $B \in \mathcal{V}$ and $\mathtt{i} \in Y_n^k$, then

$$\log \varphi^s(B_{\mathtt{i}}) > kn(E_\mu(\Phi) - 3\varepsilon).$$

We are now able to apply Lemma 4.1 to conclude that if $B \in \mathcal{V}$, then

$$P(B, s) \geq P(A, s) - 6\varepsilon.$$

As $\varepsilon > 0$ was arbitrary, this yields the desired lower semicontinuity.

## 4.3 Some Remarks on the Higher-Dimensional Case

We finish the paper with some brief remarks on the proof of the continuity of $M$ and $P$ in any dimension.

The proof of the continuity of $M$ in dimension 2 extends fairly easily to arbitrary dimension $d$ (using the general version of Oseledets' Theorem): if all $d$ Lyapunov exponents are equal, then the proof is identical to the two-dimensional case. If not all exponents are equal, let $1 \leq k < d$ be the multiplicity of the largest Lyapunov exponent. Then the argument is very similar, except that one uses cones around $k$-planes.

For $d \geq 3$, one cannot reduce $\varphi^s$ to a quantity involving only matrix norms (of the given maps and their inverses), so it is clear that some new tools are required to prove continuity of $P$ in general dimension. The proof follows the same outline, but it involves working with higher exterior powers of the maps $A_i$, and proving cone conditions for two different exterior powers simultaneously. Altough passing to exterior products is a common trick in the area, this makes the general proof far more technical. The reader is referred to [FS13] for further details, as well as consequences and generalizations of these results.

**Acknowledgments** I am grateful to De-Jun Feng for useful comments on an earlier version of the article.

## References


[Ba07] Barański, K.: Hausdorff dimension of the limit sets of some planar geometric constructions. Adv. Math. **210**(1), 215–245 (2007)

[Ba10] Barreira, L.: Almost additive thermodynamic formalism: some recent developments. Rev. Math. Phys. **22**(10), 1147–1179 (2010)

[Ba96] Barreira, L.M.: A non-additive thermodynamic formalism and applications to dimension theory of hyperbolic dynamical systems. Ergodic Theor. Dyn. Syst. **16**(5), 871–927 (1996)

[BV10] Bocker-Neto, C., Viana, M.: Continuity of lyapunov exponents for random 2d matrices (preprint) (2010). http://arxiv.org/abs/1012.0872

[CF08] Cao, Y.-L., Feng, D.-J., Huang, W.: The thermodynamic formalism for sub-additive potentials. Discrete Contin. Dyn. Syst. **20**(3), 639–657 (2008)

[Edg92] Edgar, G.A.: Fractal dimension of self-affine sets: some examples. Rend. Circ. Mat. Palermo **2**(28), 341–358 (1992) (Measure theory (Oberwolfach, 1990))

[Fal88] Falconer, K.J.: A subadditive thermodynamic formalism for mixing repellers. J. Phys. A **21**(14), L737–L742 (1988)

[Fal13] Falconer, K.: Dimensions of self-affine sets: a survey. In Barral, J., Seuret, S. (eds.) Further Developments in Fractals and Related Fields, pp. 115–134. Springer, Berlin (2013)

[FM08] Falconer, K., Miao, J.: Exceptional sets for self-affine fractals. Math. Proc. Camb. Philos. Soc. **145**(3), 669–684 (2008)

[FS09] Falconer, K., Sloan, A.: Continuity of subadditive pressure for self-affine sets. Real Anal. Exchange **34**(2), 413–427 (2009)

[Fal92] Falconer, K.J.: The Hausdorff dimension of self-affine fractals. Math. Proc. Cambridge Philos. Soc. **103**(2), 339–350 (1988)

[FK11] Feng, D.-J., Käenmäki, A.: Equilibrium states of the pressure function for products of matrices. Discrete Continuous Dyn. Syst. (DCDS-A) **30**(3), 699–708 (2011)

[FS13] Feng, D.-J., Shmerkin, P.: Non-conformal repellers and the continuity of pressure for matrix cocycles. Geom. Funct. Anal. (2014)

[HL95] Hueter, I., Lalley, S.P.: Falconer's formula for the Hausdorff dimension of a self-affine set in $R^2$. Ergodic Theor. Dyn. Syst. **15**(1), 77–97 (1995)

[Kae04] Käenmäki, A.: On natural invariant measures on generalised iterated function systems. Ann. Acad. Sci. Fenn. Math. **29**(2), 419–458 (2004)

[KS09] Käenmäki, A., Shmerkin, P.: Overlapping self-affine sets of Kakeya type. Ergodic Theor. Dyn. Syst. **29**(3), 941–965 (2009)

[Krel85] Krengel, U.: Ergodic theorems, vol. 6 of de Gruyter Studies in Mathematics. Walter de Gruyter & Co., Berlin (1985) (with a supplement by Antoine Brunel)

[Mum06] Mummert, A.: The thermodynamic formalism for almost-additive sequences. Discrete Continuous Dyn. Syst. **16**(2), 435–454 (2006)

[Pet89] Petersen, K.: Ergodic theory, volume 2 of Cambridge Studies in Advanced Mathematics. Cambridge University Press, Cambridge (1989) (corrected reprint of the 1983 original)

[Sol98] Solomyak, B.: Measure and dimension for some fractal families. Math. Proc. Camb. Philos. Soc. **124**(3), 531–546 (1998)

# Stability Properties of Fractal Curvatures

**Martina Zähle**

**Abstract** Lipschitz-Killing curvatures of singular sets are known from geometric measure theory. These are extensions of classical notions from convex and differential geometry. In recent years their fractal versions have been introduced via approximation by parallel sets of small distances. In the present paper stability properties of the corresponding limits under small perturbations of these neighborhoods are studied. The well-known Minkowski content may be considered as marginal case.

## 1 Introduction

Fractal versions of the Lipschitz-Killing curvatures of singular sets known from geometric measure theory [Fed59] have been introduced via approximation by parallel sets of small distances. This construction generalizes that of the Minkowski content considered by many authors, which is a marginal case. A survey on related developments over the last years as well as on some background from classical convex and differential geometry is given in [Zah13]. Moreover, in this paper we outline an approach for self-similar sets by means of an associated dynamical system, which leads to short proofs for the total curvatures, their densities and the curvature measures.

An extension of the average limits of the curvature-direction measures to self-conformal sets is given in [Boh13]. There the tools of thermodynamic formalism are more involved, but the proofs are shorter and the results are more general than in the literature.

An advantage of the approaches via renewal theory, e.g. in [Gat00, Win08, Win11, WZ13, Zah11, Zah13], or gap structures [FK12, KK12, Kom11, LPW11, LPW13,

M. Zähle (✉)
Institute of Mathematics, Friedrich Schiller University Jena, Jena, Germany
e-mail: martina.zaehle@uni-jena.de

 
D.-J. Feng and K.-S. Lau (eds.), *Geometry and Analysis of Fractals*,
Springer Proceedings in Mathematics & Statistics 88,
DOI 10.1007/978-3-662-43920-3_13

PW12], are the more practicable integral representations of the fractal limits for special classes of self-similar or self-conformal sets.

The aim of the present paper is to show that the difference between the approximation by parallel sets and by neighborhoods arising from certain small perturbations of these sets is asymptotically vanishing. To this aim we use suitable Lipschitz mappings between the unit normal bundles of the parallel sets and those of the close neighborhoods describing the perturbations. They have to be chosen in such a manner that the geometry is asymptotically the same. In this approach the representation of the curvature measures by means of associated normal cycles, i.e. by integration of the Lipschitz-Killing curvature forms over the unit normal bundles, appears as a useful tool. We present here only the main ideas. More precise error estimates can be derived.

In Chap. 2, Lemma 2, a general estimate of the difference between the curvature-direction measures of two subsets of $\mathbb{R}^d$ admitting normal cycles which are close to each other is derived. It is assumed that the second normal cycle is the pushforward of the first one under a Lipschitz mapping $f$, and there distance is measured in terms of the distance between the orthogonally transformed approximate differential of $f$ and the identity in $\mathbb{R}^d \times S^{d-1}$.

If we are only interested in Lebesgue measure or surface area, which corresponds to the Minkowski content in applications to fractals, Lemma 1 provides evident tools. The main result is formulated in Theorem 1, where we prove stability of the above mentioned fractal curvatures under asymptotically vanishing perturbations. Here we specify Lemma 2 to this situation. Finally we demonstrate on the example of the Sierpinski gasket how the theorem can be applied.

## 2 Some Background from Geometric Measure Theory

### 2.1 *Normal Cycles and Curvatures*

Throughout the paper we use notions and notations from geometric integration theory in the sense of Federer [Fed69] (see also Morgan [Mor88], Krantz and Parks [KP08]). In particular, we consider multivector fields and differential forms in $\mathbb{R}^d \times \mathbb{R}^d$ in the language of the exterior algebra and the alternating algebra, respectively. The symbol $\langle \eta, \varphi \rangle$ denotes dual pairing of a multivector $\eta$ and a differential form $\varphi$. Currents are continuous linear functionals on spaces of differential forms. In our case they are given by integrating differential $(d-1)$-forms over certain Hausdorff-$(d-1)$-rectifiable subsets of $\mathbb{R}^d \times S^{d-1}$, namely the unit normal bundles of geometric sets in $\mathbb{R}^d$. $\mathscr{H}^{d-1}$ denotes the corresponding Hausdorff measure.

We next recall some results from [RZ01, Zah86], and [RZ05].

For general $X$ with reach $X > 0$ there is an associated rectifiable current called the *unit normal cycle* of $X$ which is given by

$$N_X(\varphi) := \int_{\text{nor}X} \langle a_X(x,n), \varphi(x,n)\rangle \mathscr{H}^{d-1}(d(x,n))$$

for an appropriate unit simple $(d-1)$-vector field $a_X = a_1 \wedge \ldots \wedge a_{d-1}$ associated a.e. with the approximate tangent spaces of the unit normal bundle nor$X$ and for integrable differential $(d-1)$-forms $\varphi$. In these terms for $k \le d-1$ the curvature measure may be represented by

$$C_k(X,B) = N_X \llcorner \mathbf{1}_{B\times\mathbb{R}^d}(\varphi_k) = \int_{\text{nor}X\cap(B\times\mathbb{R}^d)} \langle a_X(x,n), \varphi_k(n)\rangle \mathscr{H}^{d-1}(d(x,n))$$

for any bounded Borel set $B \subset \mathbb{R}^d$, where the *k-th Lipschitz-Killing curvature form* $\varphi_k$ does not depend on the points $x$ and is defined by its action on a simple $(d-1)$-vector $\eta = \eta_1 \wedge \ldots \wedge \eta_{d-1}$ as follows: Let

$$\pi_0(y,z) := y \text{ and } \pi_1(y,z) := z$$

be the *coordinate projections* in $\mathbb{R}^d \times \mathbb{R}^d$, $e_1', \ldots, e_d'$ be the *dual basis* of the standard basis $e_1, \ldots, e_d$ in $\mathbb{R}^d$ and $\mathscr{O}_k$ be the surface area of the $k$-dimensional unit sphere. Then we have

$$\langle \eta, \varphi_k(n)\rangle := \mathscr{O}_{d-k}^{-1} \sum_{\varepsilon_i\in\{0,1\},\, \sum\varepsilon_i=d-1-k} \langle \pi_{\varepsilon_1}\eta_1 \wedge \ldots \wedge \pi_{\varepsilon_{d-1}}\eta_{d-1} \wedge n, e_1' \wedge \ldots \wedge e_d'\rangle. \tag{1}$$

Moreover, in these notations the sign of the above unit simple $(d-1)$-vector field $a_X$ is a.e. determined by

$$\left\langle \Lambda^{d-1}(\pi_0 + \varepsilon\pi_1)a_T(x,n) \wedge n, e_1' \wedge \ldots \wedge e_d' \right\rangle > 0$$

for sufficiently small $\varepsilon$. It can be represented in the form $a_X = a_1 \wedge \ldots \wedge a_{d-1}$ with

$$a_i(x,n) = \left( \frac{1}{\sqrt{1+\varkappa_i(x,n)^2}} b_i(x,n), \frac{\varkappa_i(x,n)}{\sqrt{1+\varkappa_i(x,n)^2}} b_i(x,n) \right), \tag{2}$$

where the $\varkappa_i(x,n) \in (-\infty, \infty]$ are the *generalized principal curvatures* of $X$ at $(x,n)$ (with convention $\frac{\infty}{\infty} = 1$), and their *direction vectors* $b_i(x,n)$ form together with the unit normal vector $n$ a positively oriented orthonormal basis in $\mathbb{R}^d$, i.e., $\langle b_1 \wedge \ldots \wedge b_{d-1} \wedge n, e_1' \wedge \ldots \wedge e_d'\rangle = 1$. (Note that for convex sets all $\varkappa_i(x,n)$ are either 0 or $\infty$.)

From this one obtains the integral representation of the *curvature-direction measures* $\widetilde{C}_k(X,\widetilde{B}) := N_X \llcorner \mathbf{1}_{\widetilde{B}}(\varphi_k)$ on $\mathbb{R}^d \times S^{d-1}$:

$$\widetilde{C}_k(X,A)=\mathcal{O}_{d-k}^{-1}\int\limits_{\text{nor}X}\mathbf{1}_A(x,n)\sum_{i_1<\cdots<i_{d-1-k}}\frac{\varkappa_{i_1}(x,n)\cdots\varkappa_{i_{d-1-k}}(x,n)}{\prod_{i=1}^{d-1}\sqrt{1+\varkappa_i(x,n)^2}}\,\mathcal{H}^{d-1}(d(x,n))$$

for any bounded Borel set $A$ in $\mathbb{R}^d\times S^{d-1}$. We define its *strong variation measure* by

$$\widetilde{C}_k^{svar}(X,A):=\int_{\text{nor}X}\mathbf{1}_A\|\varphi_k\|\,d\mathcal{H}^{d-1},$$

where $\|\varphi(x,n)\|$ denotes here the comass norm of an alternating $(d-1)$-form $\varphi(x,n)$ on the tangent space $\text{Tan}(\text{nor}F_\varepsilon,(x,n))$, defined for a.a. $(x,n)\in\text{nor}X$. Then one infers

$$\widetilde{C}_k^{svar}(X,A):=\mathcal{O}_{d-1-k}^{-1}\int\limits_{\text{nor}X}\mathbf{1}_A\sum_{i_1<\cdots<i_{d-1-k}}\frac{\left|\varkappa_{i_1}\cdots\varkappa_{i_{d-1-k}}\right|}{\prod_{i=1}^{d-1}\sqrt{1+\varkappa_i^2}}\,d\mathcal{H}^{d-1}.$$

$C_k^{svar}(X,\cdot):=\widetilde{C}_k^{svar}(X,(\cdot)\times S^{d-1})$ is called *strong variation measure* of the curvature measure $C_k(X,\cdot)$.

For $d$-dimensional domains $X$ denote the *closure of the complement* of $X$ by $\widetilde{X}$. If the latter has positive reach then the normal cycle of $X$ may be introduced by the pushforward of that of $\widetilde{X}$ under the normal reflection $\rho(x,n):=(x,-n)$:

$$\text{nor}X:=\rho(\text{nor}\widetilde{X})\text{ and }N_X:=\rho_\# N_{\widetilde{X}},$$

which yields

$$C_k(X,\cdot)=(-1)^{d-1-k}C_k(\widetilde{X},\cdot),$$

and

$$C_k^{svar}(X,\cdot)=C_k^{svar}(\widetilde{X},\cdot)$$

for the strong variation measures. Note that for such $X$, $C_{d-1}(X,\cdot)$ agrees with half the surface are measure $\mathcal{H}^{d-1}$ on the boundary of $X$.

Normal cycles and curvatures have been introduced for various other classes of singular sets $X$. (These are $(d-1)$ rectifiable currents without boundaries vanishing on the contact 1-form.) Then the above representations remains valid, but in general, the orienting $(d-1)$-vector field $a_X$ has to be multiplied be an associated integer-valued topological index function $i_X$, i.e.,

$$N_X(\varphi):=\int\limits_{\text{nor}X}\langle i_X(x,n)a_X(x,n),\varphi(x,n)\rangle\,\mathcal{H}^{d-1}(d(x,n))$$

and the strong variation measure of the $k$-th curvature (-direction) measure is given by

$$\int\limits_{\mathrm{nor}X} \mathbf{1}_{(\cdot)}(x,n)|i_X(x,n)|\,||\varphi_k||(x,n)\,\mathscr{H}(d(x,n))\,.$$

The *index function* is determined by

$$i_X(x,n) := \mathbf{1}_X(x)\left(1 - \lim_{\varepsilon\to 0}\lim_{\delta\to 0}\chi\big(X\cap B(x+(\varepsilon+\delta)n,\varepsilon)\big)\right)$$

for the Euler-Poincaré characteristic $\chi$ in the sense of singular homology, where $B(z,r)$ denotes the closed ball with centre $z$ and radius $r$.

In the general case one introduces additionally $C_d(X,\cdot) := \mathscr{L}^d(X\cap(\cdot))$ for the *Lebesgue measure*.

Then the *total values* $C_i(X) := C_i(X,\mathbb{R}^d)$, $i = 0,\dots d$, for certain classes of sets $X$ form a complete system of motion invariant Euclidean valuations, which are continuous with respect to the flat seminorms of the associated normal cycles $N_X$, see [Zah90]. (In the convex setting this is a well-known theorem of Hadwiger.)

Below we will choose for X the parallel sets $F_\varepsilon$ of small distances $\varepsilon$ or different neighborhoods $F^\varepsilon$ of self-similar and other fractals $F$ with normal cycles $N_{F_\varepsilon}$ and $N_{F^\varepsilon}$, respectively. We are seeking for conditions that guarantee the same fractal curvatures arising from the average limits if in the approximation procedure the parallel sets $F_\varepsilon$ are replaced by the $F^\varepsilon$. Here we need a stronger continuity version than that of flat convergence.

## 2.2 Distance Estimates for Curvature Measures

Let us first mention some simple distance estimate for the Hausdorff measures of rectifiable sets which can be applied to the above cases $k = d-1$ and $k = d$. They easily follow from the Area theorem [Fed69, 3.2.22]. Here and in the following $I$ denotes the *identity map* and ap $Df$ the *approximate differential* of a Lipschitz mapping $f$.

**Lemma 1** *If $X$ and $Y$ are $(\mathscr{H}^m, m)$-rectifiable and $\mathscr{H}^m$-measurable subsets of $\mathbb{R}^d$ and there exists a Lipschitz mapping $f$ from $X$ onto $Y$ such that for $\mathscr{H}^m$-a.a. $y\in Y$ the set $f^{-1}(y)$ is a singleton, then we get for any Borel set $B$,*

$$\left|\mathscr{H}^m(X\cap f^{-1}(B)) - \mathscr{H}^m(Y\cap B)\right| \le \int\limits_{X\cap f^{-1}(B)} \sum_{i=1}^{m}\binom{m}{i}\Delta_f^i\, d\mathscr{H}^m\,,$$

*where $\Delta_f(x) = ||I - ap\ Df(x)||$ at a.a. $x\in X$, for the operator norm in the corresponding approximate tangent spaces.*

A similar version for the curvature measures can be formulated as follows. For a rectifiable set $Z\subset\mathbb{R}^d\times\mathbb{R}^d$ in the above sense, a Lipschitz mapping $f : Z \to$

$\mathbb{R}^d \times \mathbb{R}^d$, and any family of orthogonal mappings $O(n) : \mathbb{R}^d \to \mathbb{R}^d$, measurable in $n \in S^{d-1}$, such that

$$\pi_1(f(x,n)) = O(n)n \text{ at a.a.}(x,n) \in \text{nor}X \tag{3}$$

we use the notation

$$\widetilde{D}f(x,n) := \widetilde{O}(n)^{-1} \circ \text{ap}\, Df(x,n) \tag{4}$$

for a.a. $(x,n) \in Z$, where the approximate differential of $f$ is extended to the whole $\mathbb{R}^d \times \mathbb{R}^d$ by the identity on the orthogonal complement of the approximate tangent space of $Z$ at $(x,n)$. The mappings $\widetilde{O}(n) : \mathbb{R}^d \times \mathbb{R}^d \to \mathbb{R}^d \times \mathbb{R}^d$ are given by

$$\widetilde{O}(n)(u,v) := (O(n)u, O(n)v), \quad u, v \in \mathbb{R}^d .$$

**Lemma 2** *If $X$ and $Y$ admit normal cycles $N_X$ and $N_Y$ and there exists a Lipschitz mapping $f : \text{nor}X \to \text{nor}Y$ such that*

$$f_{\#}N_X = N_Y ,$$

*then we get for $k = 0, \ldots, d-1$ and any bounded Borel set $A$ in $\mathbb{R}^d \times S^{d-1}$,*

$$\left|\widetilde{C}_k(X, f^{-1}(A)) - \widetilde{C}_k(Y, A)\right| \leq \sum_{l=0}^{d-1} \int_{f^{-1}(A)} \sum_{m=\max(k-l,1)}^{d-1} c(d,k,l,m)\, \Delta_f^m \, d\widetilde{C}_l^{svar}(X, \cdot) ,$$

*for certain constants $c(d,k,l,m)$ and*

$$\Delta_f(x,n) := ||I - \widetilde{D}f(x,n)||$$

*at a.a. $(x,n) \in \text{nor}X$, in the above notations.*

*Proof* Since $N_Y = f_{\#}N_X$ we obtain (in short notation ap $Df = Df$)

$$\widetilde{C}_k(Y,A) = \int_{\text{nor}X} \mathbf{1}_A(f(x,n)) 1_X(x,n) \langle \Lambda^{d-1} Df(x,n) a_X(x,n), \varphi_k(f(x,n)) \rangle \, \mathcal{H}^{d-1}(d(x,n)) .$$

Moreover,

$$\widetilde{C}_k(X, f^{-1}(A)) = \int_{\text{nor}X} \mathbf{1}_A(f(x,n)) 1_X(x,n) \langle a_X(x,n), \varphi_k(x,n) \rangle \, \mathcal{H}^{d-1}(d(x,n))$$

Below we shall show that for $\mathcal{H}^{d-1}$-a.a. $(x,n) \in \text{nor}X$,

$$\left|\langle a_X(x,n),\varphi_k(x,n)\rangle-\langle \Lambda^{d-1}Df(x,n)a_X(x,n),\varphi_k(f(x,n))\rangle\right|\le\prod_{j=1}^{d-1}\left(1+\varkappa_j(x,n)^2\right)^{-1/2}$$

$$\sum_{i_1<\ldots<i_{d-1-k}}\sum_{m=1}^{d-1}\Delta_f^m\sum_{j_1<\ldots<j_m}\prod_{l=1}^{m}(1+|\varkappa_{j_l}(x,n)|)\prod_{j\in\{i_1,\ldots,i_{d-1-k}\}\setminus\{j_1,\ldots,j_m\}}|\varkappa_j(x,n)|\,, \tag{5}$$

where all indices are positive and do not exceed $d-1$, and the last product over the empty set equals 1. Recalling that

$$|1_X|\prod_{j=1}^{d-1}\left(1+\varkappa_j^2\right)^{-1/2}\sum_{i_1<\ldots<i_{d-1-l}}|\varkappa_{i_1}\cdots\varkappa_{i_{d-1-l}}|$$

equals the density of the strong variation measure $\widetilde{C}_l^{svar}(X,\cdot)$ with respect to Hausdorff measure $\mathscr{H}^{d-1}$ and rearranging the summands in (5) according to the number of $\varkappa_j$ in the corresponding products the assertion follows by integration. (Note that the constants $c(d,k,l,m)$ can be determined by combinatorial arguments.)

In order to prove inequality (5) we rewrite the operators $\widetilde{D}f$ in $\mathbb{R}^d\times\mathbb{R}^d$ in matrix representation as

$$\widetilde{D}f=\begin{pmatrix} I+D_{00} & D_{01}\\ D_{10} & I+D_{11}\end{pmatrix},$$

for linear operators $D_{ij}:\mathbb{R}^d\to\mathbb{R}^d$ and note that $||D_{ij}||\le\Delta_f,\ i,j=0,1$. In this notation we obtain for the basis vectors $a_1,\ldots,a_{d-1}$ in the approximate tangent spaces of nor$X$ chosen according to (2)

$$\pi_0(\widetilde{D}fa_i)=\frac{1}{\sqrt{1+\varkappa_i^2}}(I+D_{00}+\varkappa_jD_{01})b_i$$

$$\pi_1(\widetilde{D}fa_i)=\frac{1}{\sqrt{1+\varkappa_i^2}}(\varkappa_jI+D_{10}+\varkappa_jD_{11})b_i\,.$$

Setting $I_i^0:=I$ and $I_i^1:=\varkappa_iI$, $D_i^0:=D_{00}+\varkappa_iD_{01}$, $D_i^1:=D_{10}+\varkappa_iD_{11}$ we get

$$\pi_{\varepsilon_i}(\widetilde{D}fa_i)=\frac{1}{\sqrt{1+\varkappa_i^2}}(I_i^{\varepsilon_i}+D_i^{\varepsilon_i})b_i\,,\quad \varepsilon_i=0,1\,,$$

and

$$||D_i^{\varepsilon_i}||\le(1+|\varkappa_i|)\,\Delta_f\,,\quad ||I_i^1||=|\varkappa_i|\,. \tag{6}$$

Furthermore, recalling $a_X=a_1\wedge\ldots a_{d-1}$ and the definition of the Lipschitz-Killing curvature form $\varphi_k$ we infer (for the sum running over $\varepsilon_i=0,1$ with $\sum_{i=1}^{d-1}\varepsilon_i=$

$d-1-k$ and the for volume form $\Omega_d := e_1' \wedge \ldots \wedge e_d'$:

$$\begin{aligned}
&\langle \Lambda^{d-1} Df(x,n) a_X(x,n), \varphi_k(f(x,n))\rangle \\
&= \sum \langle \pi_{\varepsilon_1}(Df(x,n)a_1(x,n)) \wedge \ldots \wedge \pi_{\varepsilon_{d-1}}(Df(x,n)a_{d-1}(x,n)) \wedge \pi_1(f(x,n)), \Omega_d\rangle \\
&= \sum \langle \pi_{\varepsilon_1}(\widetilde{D}f(x,n)a_1(x,n)) \wedge \ldots \wedge \pi_{\varepsilon_{d-1}}(\widetilde{D}f(x,n)a_{d-1}(x,n)) \wedge n, \Omega_d\rangle ,
\end{aligned}$$

because of the invariance of $\Omega_d$ under the orientation preserving orthogonal mapping $O(n)$ and the definition of $\widetilde{D}f(x,n)$. Using the above representation of $\pi_{\varepsilon_i}(\widetilde{D}f(x,n)a_i)$ (and omitting the arguments $(x,n)$ for brevity) the last expression may be rewritten as

$$\begin{aligned}
&\prod_{i=1}^{d-1}(1+\varkappa_i^2)^{-1/2} \sum \langle (I_1^{\varepsilon_1} + D_1^{\varepsilon_1}))b_1 \wedge \ldots \wedge (I_{d-1}^{\varepsilon_{d-1}} + D_{d-1}^{\varepsilon_{d-1}})b_{d-1} \wedge n, \Omega_d\rangle \\
&\qquad = \prod_{i=1}^{d-1}(1+\varkappa_i^2)^{-1/2} \sum \langle I_1^{\varepsilon_1} b_1 \wedge \ldots \wedge I_{d-1}^{\varepsilon_{d-1}} b_{d-1} \wedge n, \Omega_d\rangle \\
&\qquad\quad + \prod_{i=1}^{d-1}(1+\varkappa_i^2)^{-1/2} \sum \sum \langle E_1^{\delta_1 \varepsilon_1} b_1 \wedge \ldots \wedge E_{d-1}^{\delta_{d-1}\varepsilon_{d-1}} b_{d-1} \wedge n, \Omega_d\rangle ,
\end{aligned}$$

where the last inner sum runs over $\delta_j = 0, 1$ with $\sum_{j=1}^{d-1} \delta_j \geq 1$, and $E_j^{\delta_j \varepsilon_j} := I_j^{\varepsilon_j}$ if $\delta_j = 0$, $E_j^{\delta_j \varepsilon_j} := D_j^{\varepsilon_j}$ if $\delta_j = 1$.

The first summand of the last expression agrees with $\langle a_X, \varphi_k\rangle$. The absolute value of the second summand coincides with the left hand side of the asserted inequality. In view of the above operator norm estimates (6) it does not exceed

$$\begin{aligned}
&\prod_{i=1}^{d-1}(1+\varkappa_i^2)^{-1/2} \sum_{\varepsilon_i} \sum_{\delta_j} \prod_{i=1}^{d-1} |E_i^{\delta_i \varepsilon_i} b_i| \leq \prod_{i=1}^{d-1}(1+\varkappa_i^2)^{-1/2} \\
&\sum_{i_1<\ldots<i_{d-1-k}} \sum_{\sum \delta_j \geq 1} \prod_{j\in\{i_1,\ldots,i_{d-1-k}\}} ((1-\delta_j)|\varkappa_j| + \delta_j(1+|\varkappa_j|)\Delta_f) \\
&\prod_{j\notin\{i_1,\ldots,i_{d-1-k}\}} (1-\delta_j + \delta_j(1+|\varkappa_j|)\Delta_f) = \prod_{i=1}^{d-1}(1+\varkappa_i^2)^{-1/2} \\
&\sum_{i_1<\ldots<i_{d-1-k}} \sum_{m=1}^{d-1} \Delta_f^m \sum_{j_1<\ldots<j_m} \prod_{l=1}^{m}(1+|\varkappa_{j_l}(x,n)|) \prod_{j\in\{i_1,\ldots,i_{d-1-k}\}\setminus\{j_1,\ldots,j_m\}} |\varkappa_j(x,n)| .
\end{aligned}$$

The last expression equals the right hand side of the asserted inequality (5), thus it is proved.

Note that the formal multiplication by $\prod_{i=1}^{d-1}(1+\varkappa_i^2)^{-1/2}$ everywhere has to be understood in the commutative sense, i.e., the factors with $\varkappa_i = \infty$ should be canceled with the corresponding $|\varkappa_i|$ in the following products. □

## 3 Stability of Fractal Curvatures Under Approximate Perturbations

We now turn back to the problem of approximating fractal curvatures by different neighborhoods. For simplicity we here consider only the total values. Recall that for certain classes of compact fractal sets $F$ the *average Minkowski content* ($k = d, d-1$) and *fractal curvatures* of order $0 \le k \le d-2$ have been introduced by means of the parallel sets $F_\varepsilon$ as

$$C_k^{frac}(F) = \operatorname*{ess\,lim}_{\delta\to 0} \frac{1}{|\ln \delta|}\int_\delta^{\delta_0} \varepsilon^{D-k} C_k(F_\varepsilon)\, \varepsilon^{-1} d\varepsilon\ , \tag{7}$$

where $D$ is the Hausdorff or Minkowski dimension of $F$ and the essential limit is taken with respect to the Lebesgue measure. Under additional assumptions (non-arithmetic cases) the ordinary limits are shown to exist. Then the question arises whether one can choose other approximating sets, say $F^\varepsilon$, in order to obtain the same limits. For example, for practical purposes, it would be useful to replace the parallel sets by close polyhedra. The above estimates can be applied to this and more general situations, where the $F_\varepsilon$ are not necessarily parallel sets:

**Theorem 1** *For Lebesgue a.a.* $\varepsilon < \varepsilon_0$ *let* $F_\varepsilon$ *and* $F^\varepsilon$ *be any compact subsets of* $\mathbb{R}^d$ *admitting normal cycles* $N_{F_\varepsilon}$ *and* $N_{F^\varepsilon}$*, resp., such that there exists a Lipschitz mapping* $f_\varepsilon : \operatorname{nor} F_\varepsilon \to \operatorname{nor} F^\varepsilon$ *satisfying*

$$(f_\varepsilon)_\# N_{F_\varepsilon} = N_{F^\varepsilon} \text{ and } ||I - \widetilde{D} f_\varepsilon|| \le \alpha(\varepsilon)\varepsilon \tag{8}$$

*for some bounded function* $\alpha(\varepsilon)$*, where* $\widetilde{D} f_\varepsilon$ *is defined as in* (4)*. Then we get for* $k = 0, \dots, d-1$,

$$\left| \frac{1}{|\ln\delta|}\int_\delta^{\delta_0} \varepsilon^{D-k} C_k(F_\varepsilon)\, \varepsilon^{-1} d\varepsilon - \frac{1}{|\ln\delta|}\int_\delta^{\delta_0} \varepsilon^{D-k} C_k(F^\varepsilon)\, \varepsilon^{-1} d\varepsilon \right|$$

$$\le \text{const} \sum_{l=0}^{d-1} \frac{1}{|\ln\delta|}\int_\delta^{\delta_0} \alpha(\varepsilon)\varepsilon^{D-l} C_l^{svar}(F_\varepsilon)\, \varepsilon^{-1} d\varepsilon \tag{9}$$

*provided the first two integrals exist for some* $D > 0$. *If*

$$\max_{0\le l\le d-1}\sup_{0<\delta<\delta_0}\frac{1}{|\ln\delta|}\int_\delta^{\delta_0}\varepsilon^{D-l}C_l^{svar}(F_\varepsilon)\,\varepsilon^{-1}d\varepsilon<\infty \text{ and } \operatorname{ess}\lim_{\varepsilon\to o}\alpha(\varepsilon)=0 \quad (10)$$

*then we get the same limits*

$$\operatorname{ess}\lim_{\delta\to 0}\frac{1}{|\ln\delta|}\int_\delta^{\delta_0}\varepsilon^{D-k}C_k(F^\varepsilon)\,\varepsilon^{-1}d\varepsilon=\operatorname{ess}\lim_{\delta\to 0}\frac{1}{|\ln\delta|}\int_\delta^{\delta_0}\varepsilon^{D-k}C_k(F_\varepsilon)\,\varepsilon^{-1}d\varepsilon \quad (11)$$

*provided one of them exists.*

*Analogous statements hold for the non-averaged versions.*

*Proof* We substitute in Lemma 2 the set $X$ by $F_\varepsilon$, the set $Y$ by $F^\varepsilon$ and the mapping $f$ by $f_\varepsilon$. Using the estimate

$$||\Delta_{f_\varepsilon}||\le\varepsilon\alpha(\varepsilon)$$

we then infer for $\varepsilon\le\delta_0$ and varying constant factors independent of $F_\varepsilon$, $F^\varepsilon$,

$$|C_k(F_\varepsilon)-C_k(F^\varepsilon)|\le\text{const}\left(\sum_{l=0}^{k-1}C_l^{svar}(F_\varepsilon)\sum_{m=k-l}^{d-1}\varepsilon^m\alpha(\varepsilon)^m+\sum_{l=k}^{d-1}C_l^{svar}(F_\varepsilon)\sum_{m=1}^{d-1}\varepsilon^m\alpha(\varepsilon)^m\right)$$

$$\le\text{const}\left(\sum_{l=0}^{k-1}C_l^{svar}(F_\varepsilon)\varepsilon^{k-l}\alpha(\varepsilon)^{l-l}\sum_{m=0}^{d-1-(k-l)}\varepsilon^m\alpha(\varepsilon)^m+\sum_{l=k}^{d-1}C_l^{svar}(F_\varepsilon)\varepsilon\alpha(\varepsilon)\sum_{m=0}^{d-2}\varepsilon^m\alpha(\varepsilon)^m\right)$$

$$\le\text{const}\left(\varepsilon^k\alpha(\varepsilon)\left(\sum_{l=0}^{k-1}\varepsilon^{-l}C_l^{svar}(F_\varepsilon)+\varepsilon\sum_{l=k}^{d-1}\varepsilon^{-l}C_l^{svar}(F_\varepsilon)\le\text{const}\,\varepsilon^k\alpha(\varepsilon)\sum_{l=0}^{k-1}\varepsilon^{-l}C_l^{svar}(F_\varepsilon)\right.\right..$$

This implies

$$\left|\frac{1}{|\ln\delta|}\int_\delta^{\delta_0}\varepsilon^{D-k}C_k(F_\varepsilon)\,\varepsilon^{-1}d\varepsilon-\frac{1}{|\ln\delta|}\int_\delta^{\delta_0}\varepsilon^{D-k}C_k(F^\varepsilon)\,\varepsilon^{-1}d\varepsilon\right|$$

$$\le\text{const}\sum_{l=0}^{k-1}\frac{1}{|\ln\delta|}\int_\delta^{\delta_0}\alpha(\varepsilon)\varepsilon^{D-l}C_l^{svar}(F_\varepsilon)\,\varepsilon^{-1}d\varepsilon\,,$$

i.e. (9). The limit equality (11) is a consequence. Finally note that the above estimates also work without averaging over $\varepsilon$. □

*Remark 1* For $F_\varepsilon$ we can take the parallel sets of a *self-similar or self-conformal fractal* with the open set condition and Hausdorff dimension $D$. If the parallel sets are regular in the above mentioned sense and the integrability condition in (10) is satisfied then the average limits

$$\operatorname{ess\,lim}_{\delta \to 0} \frac{1}{|\ln \delta|} \int_{\delta}^{\delta_0} \varepsilon^{D-k} C_k(F_\varepsilon)\, \varepsilon^{-1} d\varepsilon$$

do exist (see [Boh13, BZ13, RZ12, Win08, WZ13]). Note that the integrability condition (10) can be checked for many classical examples, where even existence of the average limits of the strong variation measures can be proved. Then the theorem states that one obtains the same limits if one considers small perturbations $F^\varepsilon$ of $F_\varepsilon$ in the sense of (8).

For example, in the case of a *general Sierpinski gasket* in the plane only the interior triangular holes of the parallel sets determine the limit behaviour. If one investigates small affine perturbations of these triangles such that distances between the corresponding vertices are of order $o(\varepsilon)$, then the mappings $f_\varepsilon : \operatorname{nor} F_\varepsilon \to \operatorname{nor} F^\varepsilon$ can locally be chosen as follows

$$f_\varepsilon(x,n) := \left( \Pi_{\widetilde{F^\varepsilon}}\big(x - \beta(\varepsilon)n\big), -\frac{x - \Pi_{\widetilde{F^\varepsilon}}\big(x - \beta(\varepsilon)n\big)}{|x - \Pi_{\widetilde{F^\varepsilon}}\big(x - \beta(\varepsilon)n\big)|} \right).$$

Here $\Pi_{\widetilde{F^\varepsilon}}$ denotes the metric projection onto $\widetilde{F^\varepsilon} = \overline{(F^\varepsilon)^c}$ and the constant $\beta(\varepsilon)$ depends on the corresponding affine transformation. (The parallel set of distance $\beta(\varepsilon)$ of the triangle of $\widetilde{F}_\varepsilon$ under consideration must contain the corresponding perturbed triangle of $\widetilde{F^\varepsilon}$.) Note that in view of convexity in such a case the estimates are simpler than in general.

*Remark 2* The error estimates in Lemma 2 and Theorem 1 are rough. Our method of proof shows that they can be improved and the corresponding constants can be computed. For the cases $k = d-1, d$, which yield up to some constant the same limit, namely the (average) *Minkowski content* (see [RW13]), Lemma 1 provides better tools than Lemma 2 and the above conditions are not needed. The remaining arguments in this case are similar and we do not expose the details.

## References


[Boh13] T. Bohl: Fractal curvatures and Minkowski content of self-conformal sets. Preprint, arXiv:1211.3421v1, PhD thesis, University of Jena (2013)

[BZ13] Bohl, T., Zähle, M.: Curvature-direction measures of self-similar sets. Geom. Dedicata **167**, 215–231 (2013)

[Fed59] Federer, H.: Curvature measures. Trans. Amer. Math. Soc. **93**, 418–491 (1959)

[Fed69] Federer, H.: Geometric Measure Theory. Springer, Berlin (1969)

[FK12] Freiberg, U., Kombrink, S.: Minkowski content and local Minkowski content for a class of self-conformal sets. Geom. Dedicata **159**, 307–325 (2012)

[Gat00] Gatzouras, D.: Lacunarity of self-similar and stochastically self-similar sets. Trans. Amer. Math. Soc. **352**, 1953–1983 (2000)

[KK12] Kesseböhmer, M., Kombrink, S.: Fractal curvature measures and Minkowski content for self-conformal subsets of the real line. Adv. Math. **230**, 2474–2512 (2012)

[Kom11] S. Kombrink: Fractal Curvature Measures and Minkowski Content for Limit Sets of Conformal Function Systems. PhD thesis, University of Bremen (2011) http://nbn-resolving.de/urn:nbn:de:gbv:46–00102477-19

[KP08] Krantz, S.G., Parks, H.R.: Geometric Integration Theory. Birkhäuser, Boston (2008)

[LPW11] Lapidus, M.L., Pearse, E.P.J., Winter, S.: Pointwise tube formulas for fractal sprays and self-similar tilings with arbitrary generators. Adv. Math. **227**, 1349–1398 (2011)

[LPW13] Lapidus, M.L., Pearse, E.P.J., Winter, S.: Minkowski measurability results for self-similar tilings and fractals with monophase generators. Contemp. Math. **600**, 185–203 (2013)

[Mor88] Morgan, F.: Geometric Measure Theory–A Beginner's Guide. Academic Press, Boston (1988)

[PW12] Pearse, E.P.J., Winter, S.: Geometry of canonical self-similar tilings. Rocky Mt. J. Math. **42**, 1327–1357 (2012)

[RW13] Rataj, J., Winter, S.: Characterization of Minkowski measurability in terms of surface area. J. Math. Anal. Appl. **400**, 120–132 (2013)

[RZ01] Rataj, J., Zähle, M.: Curvatures and currents for unions of sets with positive reach. Ann. Glob. Anal. Geom. **20**, 1–21 (2001)

[RZ05] Rataj, J., Zähle, M.: General normal cycles and Lipschitz manifolds of bounded curvature. Ann. Glob. Anal. Geom. **27**, 135–156 (2005)

[RZ12] Rataj, J., Zähle, M.: Curvature densities of self-similar sets. Indiana Univ. Math. J. **61**, 1425–1449 (2012)

[Win08] Winter, S.: Curvature Measures and Fractals. Diss. Math. **453**, 1–66 (2008)

[Win11] Winter, S.: Curvature bounds for neighborhoods of self-similar sets. Comment. Math. Univ. Carolin. **52**, 205–226 (2011)

[WZ13] Winter, S., Zähle, M.: Fractal curvature measures of self-similar sets. Adv. Geom. **13**, 229–244 (2013)

[Zah86] Zähle, M.: Integral and current representation of Federer's curvature measures. Arch. Math. **46**, 557–567 (1986)

[Zah90] Zähle, M.: Approximation and characterization of generalized Lipschitz-Killing curvatures. Ann. Global Anal. Geom. **8**, 249–260 (1990)

[Zah11] Zähle, M.: Lipschitz-Killing curvatures of self-similar random fractals. Trans. Amer. Math. Soc. **363**, 2663–2684 (2011)

[Zah13] Zähle, M.: Curvature measures of fractal sets. Contemp. Math. **600**, 381–399 (2013)

# Participants

- Shigeki Akiyama, Tsukuba University
- Patricia Alonso Ruiz, University of Siegen
- Peter Arzt, University of Siegen
- Jung-Chao Ban, National Dong Hwa University
- Christoph Bandt, Arndt University
- Michael Barnsley, Australian National University
- Julien Barral, Université Paris 13
- Yann Bugeaud, Université de Strasbourg
- Yongluo Cao, Suzhou University
- Joe Po-Chou Chen, Cornell University
- Zhen-Qing Chen, University of Washington
- Thierry Coulhon, The Australian National University
- Eva Curry, Acadia University
- Xinrong Dai, Sun Yat-Sen University
- Guotai Deng, Central China Normal University
- Qirong Deng, Fujian Normal University
- Xinhan Dong, Hunan Normal University
- Yangge Du, The Chinese University of Hong Kong
- Kenneth J. Falconer, University of St. Andrews
- Ai-Hua Fan, Université de Picardie Jules Verne
- Chun Fang, University of Helsinki
- Ábel Farkas, University of St Andrews
- De-Jun Feng, The Chinese University of Hong Kong
- Axel Françoise, Université Paris-Sud
- Jonathan Fraser, University of St Andrews
- Uta Freiberg, University of Siegen
- Xiaoye Fu, The Chinese University of Hong Kong
- Jean-Pierre Gabardo, McMaster University
- Peter Grabner, Technische Universität Graz
- Alexander Grigor'yan, Universität Bielefeld
- Deguang Han, University of Central Florida



D.-J. Feng and K.-S. Lau (eds.), *Geometry and Analysis of Fractals*,
Springer Proceedings in Mathematics & Statistics 88,
DOI 10.1007/978-3-662-43920-3

- Douglas Hardin, Vanderbilt University
- Xinggang He, Central China Normal University
- Joachim Heinze, Springer-Verlag
- Masanori Hino, Kyoto University
- Ching Wei Ho, The Chinese University of Hong Kong
- Mike Hochman, The Hebrew University
- Jiaxin Hu, Tsinghua University
- Wen Huang, University of Science and Technology of China
- Satoshi Ishiwata, Yamagata University
- Yu Ito, Kyoto University
- Stéphane Jaffard, Université Paris-Est-Créteil Val-de-Marne
- Esa Järvenpää, University of Oulu
- Maarit Järvenpää, University of Oulu
- Xiong Jin, University of St Andrews
- Timo Jolivet, Université Paris Diderot
- Thomas Jordan, University of Bristol
- Palle E. T. Jorgensen, University of Iowa
- Antti Käenmäki, University of Jyväskylä
- Naotaka Kajino, Universität Bielefeld
- Charlene Kalle, Leiden University
- Tamás Keleti, Eötvös Loránd University
- Tom Kempton, Utrecht University
- Jun Kigami, Kyoto University
- Henna Koivusalo, University of Oulu
- Shilei Kong, The Chinese University of Hong Kong
- Takashi Kumagai, Kyoto University
- Chun Kit Lai, McMaster University
- Wai Kit Lam, The Chinese University of Hong Kong
- Ka-Sing Lau, The Chinese University of Hong Kong
- Gregory F. Lawler, University of Chicago
- François Ledrappier, University of Notre Dame
- Woon Yin Paul Lee, The Chinese University of Hong Kong
- Chi Wai Leung, The Chinese University of Hong Kong
- King Shun Leung, The Hong Kong Institute of Education
- Bing Li, South China University of Technology & University of Oulu
- Wenxia Li, East China Normal University
- Xiangyu Liang, Université Paris-Sud
- Lingmin Liao, Université Paris-Est-Créteil Val-de-Marne
- Yong Lin, People's University of China
- Jingcheng Liu, Hunan Normal University
- Yuan Liu, Chinese Academy of Sciences
- Qinghui Liu, Beijing Institute of Technology
- Quansheng Liu, Université de Bretagne-Sud
- Chiu Hong Lo, The Chinese University of Hong Kong
- Fan Lü, Huazhong University of Science and Technology

- Jun Luo, Zhongshan University
- Jun Jason Luo, Shatou University
- Jihua Ma, Wuhan University
- Peter Massopust, Technische Universität München & Helmholtz Zentrum München
- Pertti Mattila, University of Helsinki
- John Ling-Lei Meng, Springer-Verlag
- Tuen Wai Ng, University of Hong Kong
- Sze-Man Ngai, Georgia Southern University
- Jacques Peyrière, Université Paris-Sud
- Mark Pollicott, University of Warwick
- Hua Qiu, Nanjing University
- Yanhui Qu, Tsinghua University
- Hui Rao, Central China Normal University
- John A. Rock, Cal Poly Pomona
- Luke Rogers, University of Connecticut
- Huojun Ruan, Zhejiang University
- Tuomas Sahlsten, University of Helsinki
- Jörg Schmeling, Lund University
- Stéphane Seuret, Université Paris-Est-Créteil Val-de-Marne
- Narn-Rueih Shieh, National Taiwan University
- Pablo Shmerkin, University of Surrey
- Lin Shu, Peking University
- Károly Simon, Technical University of Budapest
- Örjan Stenflo, Uppsala University
- Weiyi Su, Nanjing University
- Qiyu Sun, University of Central Florida
- Yuhua Sun, Universität Bielefeld
- Ville Suomala, University of Oulu
- Hanako Takahashi, Kyoto University
- Bo Tan, Huazhong University of Science and Technology
- Donglei Tang, Nanjing Audit University
- Tai Man Tang, Xiangtan University
- András Telcs, University of Pannonia
- Alexander Teplyaev, University of Connecticut
- Jonathan Tsai, The University of Hong Kong
- Kam Piu Tsang, The Chinese University of Hong Kong
- Baris Ugurcan, Cornell University
- Baowei Wang, Huazhong University of Science and Technology
- Xiangyang Wang, Sun Yat-Sen University
- Yang Wang, Michigan State University
- Zhi-Ying Wen, Tsinghua University
- Shengyou Wen, Hubei University
- Ting Kam Wong, University of Washington
- Dongsheng Wu, University of Alabama in Huntsville

- Haihua Wu, Hunan Normal University
- Jun Wu, Huazhong University of Science and Technology
- Min Wu, South China University of Technology
- Chang Xiao, The Chinese University of Hong Kong
- Yimin Xiao, Michigan State University
- Jiansheng Xie, Fudan University
- Ying Xiong, South China University of Technology
- Jian Xu, Huazhong University of Science and Technology
- S. C. P. Yam, The Chinese University of Hong Kong
- Dawei Yang, Jilin University
- Xiangdong Ye, University of Science and Technology of China
- Yuanling Ye, South China Normal University
- Zuguo Yu, Xiangtan University
- Martina Zähle, University of Jena
- Rüdiger Zeller, Universität Greifswald
- Guohua Zhang, Fudan University
- Pengfei Zhang, The Chinese University of Hong Kong
- Xinghua Zheng, Hong Kong University of Science and Technology
- Ding-Xuan Zhou, City University of Hong Kong
- Zuoling Zhou, Zhongshan University

MIX
Papier aus verantwortungsvollen Quellen
Paper from responsible sources
FSC® C105338

If you have any concerns about our products,
you can contact us on
**ProductSafety@springernature.com**

In case Publisher is established outside the EU,
the EU authorized representative is:
**Springer Nature Customer Service Center GmbH**
**Europaplatz 3, 69115 Heidelberg, Germany**

Printed by Libri Plureos GmbH
in Hamburg, Germany